RESOURCES, ENVIRONMENT AND ENGINEERING II

PROCEEDINGS OF THE 2ND TECHNICAL CONGRESS ON RESOURCES, ENVIRONMENT AND ENGINEERING (CREE 2015), HONG KONG, 25–26 SEPTEMBER 2015

Resources, Environment and Engineering II

Editor

Liquan Xie
Department of Civil Engineering, Tongji University, Shanghai, China

CRC Press is an imprint of the
Taylor & Francis Group, an **informa** business

A BALKEMA BOOK

CRC Press/Balkema is an imprint of the Taylor & Francis Group, an informa business

© 2016 Taylor & Francis Group, London, UK

Typeset by V Publishing Solutions Pvt Ltd., Chennai, India

All rights reserved. No part of this publication or the information contained herein may be reproduced, stored in a retrieval system, or transmitted in any form or by any means, electronic, mechanical, by photocopying, recording or otherwise, without written prior permission from the publisher.

Although all care is taken to ensure integrity and the quality of this publication and the information herein, no responsibility is assumed by the publishers nor the author for any damage to the property or persons as a result of operation or use of this publication and/or the information contained herein.

Published by: CRC Press/Balkema
              P.O. Box 11320, 2301 EH Leiden, The Netherlands
              e-mail: Pub.NL@taylorandfrancis.com
              www.crcpress.com – www.taylorandfrancis.com

ISBN: 978-1-138-02894-4 (Hbk)
ISBN: 978-1-315-64722-7 (eBook PDF)

Table of contents

Preface	xi
The danger of the cohesive bonding released by microseisms: Experimental evidence for monogranular cohesive materials V. Pasquino, E. Ricciardi & D. Cancellara	1
CFD research of hydrodynamic parameters of Artificial Ventricles for pulsating LVAD V. Morozov & A. Zhdanov	9
Analysis of development of sight distance assessment at intersection R. Matuszkova, M. Radimsky, T. Apeltauer & O. Budik	15
Desired and actual bid participation behaviour of Japanese construction companies E. Morimoto & K. Arai	21
The skid resistance evaluation in accordance with the Wehner/Schulze method J. Daskova, P. Nekulova & J. Kudrna	27
Site selection optimization of the cavern in the mountain of jointed rock masses with different slope degrees C. Li, W. Zhu, B.X. Li & D. Zhang	35
Compute development length for BFRP bars based on the bond-slip constitutive model T. Liu, B. Jia, C. Zhang, X. Liu & N. Hou	41
Study on spatial prediction of landslide hazard risk based on GIS H.X. Gao & K.L. Yin	47
The relation between tension, flange thickness and prying force for T-type bolt joints Z.-X. Hou, G.-H. Huang, C. Gong & Y. Zheng	53
Calculation for the distortional effect of curved composite beams based on the elastic foundation beam analogy method Y.L. Zhang, Y.L. Liu, Z.M. Hou & Y.S. Li	59
Dynamic response analysis and fatigue life evaluation of heavy haul railway steel truss bridges Y.S. Li, Y.F. Xiao, Y.F. Diao & Y.L. Zhang	69
Design actuator pump systems Left Ventricle Assist Devices V. Morozov, A. Zhdanov, L. Belyaev & I. Volkova	77
Resistance of asphalt binders to formation of frost cracks P. Coufalik, O. Dasek, P. Hyzl, D. Stehlik, J. Kudrna, I. Krcmova & P. Sperka	83
Experimental studies on pore water pressure changes at structure-soil interface under dynamic soil cutting G.X. Liu, X. Liang, X. Wang, G.J. Hong, L.Q. Xie & P.P. Zhang	91

Numerical analysis of flow effects on water interface over a submarine pipeline *Y.H. Zhu & L.Q. Xie*	99
Research on importance sampling technique in Monte Carlo method for structural reliability *Y.-F. Fang*	105
Optimal sizing of urban drainage systems using heuristic optimization *J.J. Yu, X.S. Qin & R. Min*	111
On the damping of wave propagation over porous bottom *Y.L. Ni, Y. Shen & J.F. Yu*	117
Research on settlement deformation aging characteristics of the concrete faced rockfill dam built on deep overburden layer *L. Jiang, X.G. Wang, X.P. Wei, J. Jian, W. Zhai & H.J. Zhang*	123
The application research of coal gangue in Yiqi Highway subgrade construction *Y. Liu*	129
Turbulent flow in open channel with different Froude numbers *S. Sarkar*	135
Seismic analysis for a new-type hybrid structure of Steel Reinforced Concrete frame-four corner tubes-bidirectional steel truss under rare earthquake *Q. Chen & W. Zhang*	141
The forecasting application of Beijing Urban Waterlogging risk warning in 2012–2014 *Z.C. Yin & N.J. Li*	149
Bearing capacity calculation of U-shaped steel damper via assumed stress-strain curve *H.K. Du, M. Han & R.Q. Zhang*	157
The application research of geogrids in road broadening engineering *Q.B. Tian*	163
Study on behaviors of space frame structures under fire *L. Qiu, S. Xue & Y. Zheng*	169
Emulational analysis of contact open-close considering transverse joints with trapeziform key hydraulic grooves *Y.-z. Yan, L.-j. Yu, X.-c. Wen & G.-b. Wei*	177
Evaluation of comprehensive treatment measures and stability property of Mayanpo of Xiangjiaba Hydropower Station *L. Jiang, X.G. Wang, J.J. Zhang, J.H. Sun & H.J. Zhang*	185
Constant-ductility Residual Displacement Ratio response spectrum *X. Hu, H. He & W. Jiang*	191
Numerical simulation of flow across a transversely oscillating wavy cylinder at low reynolds number *L. Zou, M. Wang, C. Xiong & H. Lu*	197
Bond behavior of Near Surface Mounted Carbon Fiber Reinforced Polymer rod under temperature cycles *H. Lee, Y. Song, C. Park, W. Jung & W. Chung*	203
Study on the wave maker stroke and the energy dissipation of SPH numerical wave flumes *Y. Xu, Y. Pan & Y.P. Chen*	209
Experimental study on the hydraulic performance of baffle-drop shafts *Z.G. Wang, D. Zhang, H.W. Zhang & R. Zhang*	217

Residual bearing capacity of stud connectors for steel-concrete composite beam bridges subjected to fatigue damage *X.-l. Rong, C.-f. Song & P. Zhao*	223
Kinematics study of 6-DOF spatial mechanism *E. Novikova, V. Morozov, A. Zhdanov & I. Volkova*	229
Inhibiting the spontaneous combustion of sulfide ores by bacteria desulfurization *C.M. Ai, A.X. Wu, H.J. Wang & J.D. Wang*	235
Analysis of superposition of wave load and wind load on offshore wind turbine based on load simulation *B.W. Jiang, M.J. Zhao, P. Liu, P. Jin & Z.Y. Tang*	243
Zr modified disordered mesoporous materials for CO_2 adsorption *W. Wu, Z. Tang & X. Zhang*	249
The bending of triangular plate under a concentrated bending moment *Y.J. Chen & Z. Lei*	255
Nonlinear mechanics analysis of a thin plate under different loads in a magnetic field *Y.H. Bian*	261
Numerical simulation on flow characteristic of Newton anti-icing fluids *J.X. Shan, S.H. Song, L. Feng & X. Peng*	267
The empirical study of hierarchical linear model of partitioned water consumption in the Yangtze River *Y.H. Zhu, P. Gao & X.F. Wu*	273
Bioactive properties of peptides from silkworm pupa protein *Z.Y. Zuo, F. Yu, H. Huang, L.L. Li & X.R. Qin*	279
Influences of coating process on the micro-structures of reconstituted tobacco sheets *Y. Hou, Q.H. Sun, Y. Li & S.Q. Hu*	285
A new evaluation method of heavy metal contamination in soil of urban area *Y.M. Yang & Z.J. Ye*	291
Au and Ag adsorption from a low-concentration sulfuric acid residue leaching solution *C.J. Zhou, G. Zhao, S. Wang & H. Zhong*	297
Swimming exercise and Caloric Restriction alter the serum cholesterol of rats *M. Yu & Y. Liu*	305
Study on dynamic responses of tunnels through fracture zones *X. Yan & H. Xiao*	311
Study on characteristics of lipoxygenase and antioxidant enzyme changes during processing of dry-cured sausage *L. Li*	317
Research of mine environment restoration effects and policies in Beijing since 2004 *N. Sun, G.X. Yan, L. Tang & W.Q. Liu*	323
Application of mathematical model for suspended sediment transport in mud-dumping ground of Sino-Myanmar crude oil pipeline and wharf engineering *N. Zhang, L.C. Sun & Q.X. Pang*	331
Research on hydraulic model application in water shortage dispatch plan implementation *X.-y. Zhang, Y.-h. Chen & Y. Xu*	337

Wind stability analysis and design plan contrast of the main bridge of Pinghai Bridge *P.J. Liu*	343
Numerical study on the street flooding in Huinan, Pudong District during the various short-duration rainstorm events *J. Huang*	351
Spatial distribution and pollution characteristics of Ammonia-N, Nitrate-N and Nitrite-N in groundwater of Dongshan Island *H.Y. Wu, S.F. Fu, X.Q. Cai & K.Z. Zhuang*	357
Rapid quantify HSPs mRNA in *Sebastiscus marmoratus* exposed to crude oil by LAMP *R. Chen, Z.P. Mo & Z.Z. Li*	363
Analysis and evaluation of nutrition composition of *Clinacanthus nutans* *Q. Yu, Z.-h. Duan, W.-w. Duan, F.-f. Shang & G.-x. Yang*	369
Extraction by microwave-ultrasonic assisted enzymatic hydrolysis and functional properties of Insoluble Dietary Fiber from soy sauce residue *W. Li & T. Wang*	375
Determination of Cd, Pb and Cu in chia seeds by microwave digestion-HR-CS GFAAS *Y. Liu, S.L. Chen, Y. Li, Y.X. Zhu & S.Y. Jiang*	383
Determination of vitamin C in intact *Actinidia Arguta* fruits using Vis/NIR spectroscopy *G. Xin, B. Zhang, S.Q. Li, J.J. Mu, C.J. Liu & X.J. Meng*	389
Analysis and control of bacteria flora found in the dish-marinated bitter melon and cucumber *Q.H. Yu, Y.S. Jiang & X. Liu*	397
Efficient asymmetric synthesis of (*S*)-3-phenyllactic acid by using whole cells of recombinant *E.coli* *Y. Zhu, Z.Q. Huang, L. Wang, B. Qi & Y. Wang*	405
Immunohistochemistry applications in breast carcinoma: The Hypoxia Inducible Factor 1α, Estrogen Receptor, Progesterone Receptor and HER2 expression before and after neoadjuvant chemotherapy *Z.F. Zhang, Y.N. Liang, C.H. Zhang, G.P. Wang, S.C. Hou & G.Y. Pan*	411
Effect of ^{60}Co-γ irradiation on quality and physiology of blueberry storage *C. Wang, X.T. Li & X.J. Meng*	417
Determination of eight metal elements in polygonatum tea by microwave digestion-AAS *S.L. Chen, C. Li, A.H. Chen, Y. Shao, J.W. Li, S.B. Liu & X.W. Zheng*	423
Characterization of proteins in Soy Sauce Residue and its hydrolysis by enzymes *J.L. Zhang, J.Y. Zhao, Y. Zhao, C.H. Yang & W.P. Wang*	429
Study on corrosion behavior of stainless steel 316 in low oxygen concentration supercritical water *J.Q. Yang, S.Z. Wang, Y.H. Li, T. Zhang, L.S. Wang & M. Wang*	437
300 MW steam turbine transformation technology using high-temperature circulating water for heating *X.T. Wang, X.D. Wang & Y. Han*	441
Optimization of fracturing penetration ratio and fracturing time in Fang 48 fault block *P. Guo, Z.W. Zhang, L.J. Huo, B. Jiang & Y.Z. Lei*	449

Supercritical Water Oxidation of acrylic acid wastewater and sludge 459
L.S. Wang, S.Z. Wang, Y.H. Li, T. Zhang, J.Q. Yang & M. Wang

Study of relationship of shaft seal steam leakage flow with Variable Steam
Temperature Experiment parameters 465
X.D. Wang, Y.Z. Hao, W. Zhen & J.L. Qu

Analysis on biomass briquette status quo and development problems 475
P.-l. Zuo, B.-j. Han, T. Yue, N. Yang, C.-l. Wang, X.-x. Zhang, Y.-h. Ding & S.-f. Qi

Adsorption of uranium from aqueous solution by bamboo charcoal 481
X.Y. Xiong, D.Y. Chen, J.W. Zhao & Z. Li

Research progress of phytoremediation for heavy metal wastewater 487
M.L. Ji, J. Yan & H.X. Li

Ecosystem health assessment of Dongyang River basin 493
T. Wu, Y. Zhang, G.-j. Jiang, X.-f. Xie & H.-j. Bian

Treatment of sludge and wastewater mixture by Supercritical Water Oxidation 499
T. Zhang, S.Z. Wang, Z.Q. Zhang, J.Q. Yang & M. Wang

Author index 505

Preface

It is strategically important to protect and improve the environment for human survival and the coordinated relationship between human beings and nature for the 21st century. Technical Congress on Resources, Environment and Engineering (CREE) aims to showcase the exciting and challenging developments occurring in the area of resources, environmental protection and associated engineering practice, and serve as a major forum for researchers, engineers and manufacturers to share recent advances, discuss problems, and identify practical challenges associated with the engineering applications.

The second Technical Congress on Resources, Environment and Engineering (CREE 2015) was held in Hong Kong on 25 and 26 September 2015. The 5th International Conference on Applied Mechanics and Civil Engineering (AMCE 2015) is an important track of CREE 2015 and focuses on the frontier research of applied mechanics and civil engineering.

This book represents the congress and publishes 76 papers. Each of the papers has been peer reviewed by recognized specialists and revised prior to acceptance for publication. This book reviews recent technological advances in the fields of resources and environment. The papers mainly focus on civil engineering, water science, hydraulic engineering, energy engineering, environmental engineering, applied mechanics, food and biochemical engineering, computing, and system engineering.

We would like to express our deep gratitude to all our authors and reviewers for their excellent work, and Léon Bijnsdorp, Lukas Goosen, and other editors from Taylor & Francis Group for their wonderful work.

Editor

The danger of the cohesive bonding released by microseisms: Experimental evidence for monogranular cohesive materials

V. Pasquino
Department of Hydraulic Engineering, University of Naples Federico II, Naples, Italy

E. Ricciardi & D. Cancellara
Department of Structures for Engineering and Architecture, University of Naples Federico II, Naples, Italy

ABSTRACT: The authors mean to provide experimental evidence of the danger that the cohesive bonding may be released by microseisms, and to quantize the amplitude/frequency danger range for monogranular agglomerates. Such danger may be of particular interest in all those cases where the foundations of buildings are exposed to prolonged uniform vibrations (traffic, reciprocating machinery, etc...). The mathematical model used in this work is the trispherical one, whose theoretical bases are analyzed in previously works in which a well-know model good for cohesionless monogranular materials is considered: it is provided with inter-central elasto-plastic braces simulating, in harmony with the physical behavior, the cohesive brace.

1 INTRODUCTION

A cohesive soil constituted by elements whose dimensions aren't very variable is considered; idealizing this soil in a set of spherical elements, joined by elasto-plastic braces, a very acceptable Mohr's curve is drawn. The model permits to obtain the disgregation's bond between the frequency and the amplitude of a sinusoidal given motion, i.e. the conditions under which the cohesion disappears.

The soil is idealized as a set of spherical elements having the same radius R and the same unit weight, touching one another in horizontal strata. The center C of the generical sphere (Fig. 1), and the centers A and B of the spheres on which it rests, lie in a vertical plane. Considering the set of spheres contained between two parallel vertical planes, whose distance is 2 R, the cartesian orthogonal axes (x, y, z) are fixed so that x is vertical and downward directed, and the centers of the spheres contained in the plane (y, z).

The static range can be examined fixing the spheres A and B, and charging the sphere C by a radial vertical load N and by a radial horizontal load T parallel to z axis. The system is plane, and the C point is obliged on two circular lines, whose radius in R, and centers are B if $\varphi > 0$, A if $\psi > 0$; we have a one freedom's degree system, but the line on which C is obliged isn't regular in $\varphi = \psi = 0$. Let N be fixed and T increasing from zero in R+; if the surface are smooth, the sphere C doesn't move until:

$$-T\ R\cos 30° + N\ R\ sen30° \geq 0 \qquad (1)$$

That is until $T \leq N\ tg30°$. If $T = N\ tg30°$, the position $\varphi = 0$ is a position of equilibrium, bit this is unstable; if a little displacement in induced, the sphere C steps over the sphere B and all be spheres of the same stratum. According that we can write $\sigma = N/4R^2$, $\tau = T/4R^2$, the

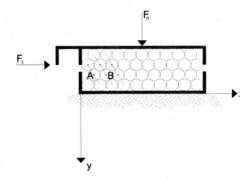

Figure 1. Geometrical model.

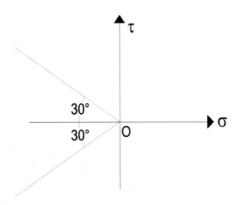

Figure 2. Coulomb's bilateral.

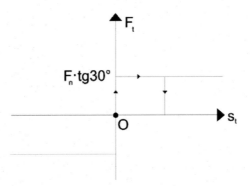

Figure 3. Rigid-plastic behavior.

Coulomb's bilateral is reported in Figure 2 and the diagram τ, γ, for a fixed value of σ, shows a classic rigid-plastic behavior (Fig. 3).

2 PROBLEM FORMULATION

The considered material is made up of steel spheres, whose radius is R. Let the material be put into a box, so that the center C of one sphere, together with the centers A

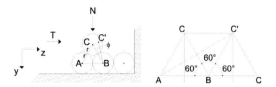

Figure 4. Geometrical model.

and *B* of the spheres it lies on, define a plan that is parallel to one face of the box itself (Fig. 4).

The constraint curve of *C* point is not regular in *C*; as a consequence, the scheme can be simplified and becomes that one represented in Figure 1a, where *C* must be on straight line *z*.

Under the hypothesis of small displacements, one obtains:

$$\varepsilon = \frac{AC' - 2r}{2r} = \frac{\sqrt{7 + 3tg^2(30 - \varphi) - 4\sqrt{3}tg(30 - \varphi)}}{2} - 1 \quad (2)$$

The components (along *z*) S_t of *S* and W_t of *W* are:

$$S_t = -r\frac{\pi}{180°}\frac{4}{3}; \quad W_t = \frac{\pi}{180°}\frac{4}{3}r\varphi \quad (3)$$

If the table undergoes the oscillation:

$$W_p = \eta \sin \omega_p t \quad (4)$$

The displacement *W* of *C* point is:

$$W = W_r + W_p = \frac{\pi}{180°}\frac{4}{\sqrt{3}}r\varphi + \eta \sin \omega_p t \quad (5)$$

Let us choose, as Lagrangian coordinate, the angle CBC' if clockwise ($\varphi > 0$) while the angle CAC' if counterclockwise ($\varphi < 0$); the value of φ and its sign define the position of the upper sphere with respect to two lower ones.

The motion equation is

$$S_t = ma \quad (6)$$

By combining equation (3) and (6) one obtains:

$$\ddot{W} + \omega^2 W = \omega^2 \eta \sin \omega_p t \quad (7)$$

where $\omega^2 = \dfrac{K}{4m}$; $m = \dfrac{4}{3}\pi r^3 \dfrac{\gamma}{g}$.

A particular solution of equation (7) is:

$$W^* = \eta \frac{\omega^2}{\omega^2 - \omega_p^2} \sin \omega_p t \quad (8)$$

And its general solution is

$$W = A\sin \omega t + B\cos \omega t + \eta \frac{\omega^2}{\omega^2 - \omega_p^2} \sin \omega_p t \quad (9)$$

By considering the initial conditions $W = \dot{W} = 0$ and W_r is reported in (10)

$$W_r = W - W_p = \eta \frac{\omega_p^2}{\omega^2 - \omega_p^2}\left(\frac{\omega}{\omega_p}\sin\omega t + \sin\omega_p t\right) \qquad (10)$$

The safety condition is

$$W \leq W_l = \frac{\pi}{180}\frac{4}{3}r\varphi_l \qquad (11)$$

This condition—that is valid for a flattened constraint model—under the hypothesis of small displacements becomes:

$$S_l = -\frac{\pi}{180°}\sin 60° \cdot r \cdot K \cdot \varphi$$

$$W_r = \frac{\pi}{180°}\sin 60° \cdot 2\varphi$$

Equation (11) becomes:

$$\eta \leq \eta_l = \frac{\omega - \omega_p}{\omega_p}\frac{\sqrt{3}\pi}{180°}r\varphi_l \qquad (12)$$

Furthermore:

$$\varphi_l = 30° - \varphi$$
$$\varepsilon_l = 2\cos(60° - \varphi_l/2) - 1$$
$$K = \frac{2rc}{\varepsilon_l}\frac{\sin(60° + \varphi_l)}{\sin(60° - \varphi_l/2)}$$

3 EXPERIMENTAL SETUP

The experimental device used in our tests consists of a shaking table, set up in the laboratory of the Department of Structures for Engineering and Architecture of Federico II University of Naples. It is made up of a steel plate (dimensions cm $40 \times 50 \times 8$, see Fig. 5) whose weight is 9.5 kg; its bottom surface has been conveniently smoothed and lies upon three bearings, arranged as an equilateral triangle, in order to allow horizontal displacements along any direction (see Fig. 6). On the upper surface, some holes have been made to fix the samples (see Fig. 7).

The plate has been put on the center of a frame that can absorb the impressed dynamic impulses by means of a spring system, and can rotate around the static equilibrium position. The unidirectional impulses are impressed by direct contact between a spherical roller bearing—rotating around a vertical axis fixed on the plate—and an eccentric device, conveniently shaped and applied on the top of a speed adjustable driveshaft (1 HP power). The rounds per minute are read on an electric counter and applied on the driveshaft that is governed, in turn, by a speed regulator.

In order to impress more impulses during a whole round, the eccentric device has been shaped as a four-leaved clover, so that when the eccentric has performed one quarter of a round, the plate has performed one whole oscillation.

Figure 5. Experimental setup.

Figure 6. Experimental setup.

Figure 7. Experimental setup.

Figure 8. Experimental setup.

If $r(\Theta)$ is the eccentric device radius, the equation that expresses the variation of r is:

$$r(\Theta) = R - \left[\frac{\Delta}{2} - \frac{\Delta}{2}\cos 4\Theta\right] \qquad (13)$$

where R is the maximum radius of the eccentric device; Δ is the displacement amplitude.

The frequency range has been 0÷100 Hz, while the maximum acceleration has been 10 g. The oscillation amplitude is determined by the used eccentric device (5 mm). All the experimental devices have been put on a table, conveniently stiffened by means of bracing systems, in order to represent the soil that is the reference value to calculate the displacements (see Fig. 8).

4 EXPERIMENTAL RESULTS

The tests have been performed on 15 monogranular samples shaped as truncated cones, made up of leaded spheres (Figs. 9 and 10). In order to prove our theoretical model in relatively short time, the used percentages of binder have been low. Furthermore, by means of uniaxial tensile and compressive testing, the Mohr-Coulomb yield curve has been drawn to obtain the cohesion c and the angle of internal friction ϕ.

Figure 9. Experimental setup.   Figure 10. Experimental setup.

Figure 11. Experimental setup.

Table 1. Experimental results $\gamma = 63.510$ N/m³, $r = 0.0035$ m, $\phi = 24°$, $c = 0.01$ MPa.

Samples	Cohesion (MPa)	Frequency (Hz)	Compression (daN)	Traction (daN)	Breaking time
1	0.010	10	16	7	4′ 20″
2	0.010	10	16	7	13′ 05″
3	0.010	10	16	7	3′ 00″

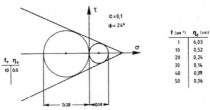

Table 2. Experimental results $\gamma = 63.510$ N/m³, $r = 0.0035$ m, $\phi = 25°$, $c = 0.036$ MPa.

Samples	Cohesion (MPa)	Frequency (Hz)	Compression (daN)	Traction (daN)	Breaking time
1	0.036	18.5	−56	24	10′
2	0.036	18.5	−56	24	19′
3	0.036	18.5	−56	24	17′
4	0.036	18.5	−56	24	18′

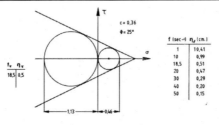

Table 3. Experimental results $\gamma = 63.510$ N/m³, $r = 0.0035$ m, $\phi = 20°$, $c = 0.026$ MPa.

Samples	Cohesion (MPa)	Frequency (Hz)	Compression (daN)	Traction (daN)	Breaking time
1	0.026	20	−36	18	1′ 24″
2	0.026	20	−36	18	0′ 28″
3	0.026	20	−36	18	2′ 35″
4	0.026	20	−36	18	12′ 40″

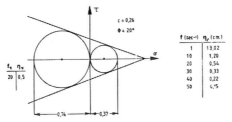

Table 4. Experimental results $\gamma = 63.510$ N/m³, $r = 0.0035$ m, $\phi = 22°$, $c = 0.024$ MPa.

Samples	Cohesion (MPa)	Frequency (Hz)	Compression (daN)	Traction (daN)	Breaking time
1	0.024	20	−35	15	60′

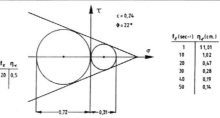

Table 5. Experimental results $\gamma = 63.510$ N/m³, $r = 0.0035$ m, $\phi = 23°$, $c = 0.012$ MPa.

Samples	Cohesion (MPa)	Frequency (Hz)	Compression (daN)	Traction (daN)	Breaking time
1	0.012	12	−18	8	14′
2	0.012	12	−18	8	5′ 30″
3	0.012	12	−18	8	4′ 35″

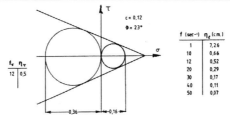

5 CONCLUSIONS

The proposed model provides Mohr's curves which are very close to the real ones. In particular, all these curves have an asymptote at Coulomb's line, passing through origin and inclined at 30°, in accordance with recent results (Adriani 1980, Hess and Stoll 1966, Leonards 1962,

Wai 1975). For a complete variational formulation, see e.g. De Angelis 2000, 2007, 2012. For dynamic nonlinear considerations see among others Cancellara and De Angelis 2012.

The obtained results confirm the validity of the purposed theoretical model and encourage going on with the research activity. Indeed, by impressing multi-directional impulses at tunable frequency and amplitude, the authors mean to study the disaggregation bonding of the mortars, whose characteristics are very interesting, especially in relation to old brick buildings subjected to the vibrations generated by the increasing vehicular traffic and underground lines.

REFERENCES

Adriani L., Franciosi V., Pasquino M. 1980. *Un modello fisico per lo studio degli agglomerati monogranulari coesivi* (*in italian*). Proceedings of the 5th AIMETA Congress, Palermo.

Cancellara D., De Angelis F. 2012. *Dynamic nonlinear analysis of an hybrid base isolation system with viscous dampers and friction sliders in parallel*, Applied Mechanics and Materials, Vol. 234, pp. 96–101.

Cancellara D., De Angelis F. 2012. *Hybrid base isolation system with friction sliders and viscous dampers in parallel: comparative dynamic nonlinear analysis with traditional fixed base structure*, Advanced Materials Research, Vols. 594–597, pp. 1771–1782.

Cancellara D., De Angelis F. 2012. *Seismic analysis and comparison of different base isolation systems for a multi-storey RC building with irregularities in plan*, Advanced Materials Research, Vol. 594–597, pp. 1788–1799.

De Angelis F. 2000. *An internal variable variational formulation of viscoplasticity*, Computer Methods in Applied Mechanics and Engineering, Vol. 190, Nos. 1–2, 35–54.

De Angelis F. 2007. *A variationally consistent formulation of nonlocal plasticity*, Int. Journal for Multiscale Computational Engineering, Vol. 5 (2), pp. 105–116, New York.

De Angelis F. 2012. *On the structural response of elasto/viscoplastic materials subject to time-dependent loadings*, Structural Durability & Health Monitoring, Vol. 8, No. 4, pp. 341–358.

Hess M.S., Stoll R.D. 1966. *Interparticle Sliding in granular materials. Columbia University*, Burm. Lab. Soil Mech. Res. Rep. n. 1.

Leonards G.A. (ed.) 1962. *Foundation Engineering*, Mc Graw Hill, Book Company New York.

Wai-Fah Chen 1975. *Limit Analysis and Soil Plasticity*. Elsevier Publishing Company, Amsterdam. p.27.

CFD research of hydrodynamic parameters of Artificial Ventricles for pulsating LVAD

V. Morozov & A. Zhdanov
Vladimir State University Named Alexander and Nikolay Stoletovs, Russia

ABSTRACT: The results of the studies of hemodynamic characteristics of the Artificial Ventricles of the heart (AV) implantable circulatory support systems and artificial heart pulsating type. A general approach to research individual housing through Computer-Aided Design (CAD) algorithm for finding the index of hemolysis obtained practical results

1 INTRODUCTION

Increased hemolysis and thrombus formation are the main problems impeding the use of implantable Left Ventricle Assist Device (LVAD) in clinical practice [1], [4]. The LVAD include an implantable unit consisting of a pump with control system and power supplies and an Artificial Ventricle (AV) with valves and connecting pipes [5–10]. Blood in such system is contacted with the inner surfaces of AV, valves, and connecting pipes. The degree of hemolysis and thrombus formation are determined by three main factors: individual housing hemocompatibility of artificial materials, valves and connecting pipes; the physiological characteristics of a person; hemodynamic characteristics of individual housing construction, valves and connecting pipes. The first factor is investigated in detail in [1] and is determined by the material, manufacturing technology, and a number of other reasons. The second problem is one of the most important in human physiology [2]. Effect of the third factor can be minimized at the design stage of new constructions in the private housing construction through CAD. In this case we take into account the geometrical design parameters and its hemodynamic properties.

Development and research of individual housing construction is parallel with the creation of the implantable pump, since the characteristics of individual housing largely determine the performance and functional characteristics of the LVAD. This article is an approach to the determination of hemodynamic characteristics of individual AV-housing of pulsating – LVAD.

2 MATERIALS AND METHODS

The main hemodynamic characteristics of AV are speed and direction of flow within the private AV-housing construction; pressure distribution; Normalized Index of Hemolysis (NIH). The most accurate are experimental methods for determining these indicators which require sophisticated laboratory stands and experimental samples of finished individual housing [3], [4]. At the design stage, such an analysis is difficult, so the most convenient method is to analyze the hemodynamic performance through CAD-systems. The inputs to the research are the parameters of blood flow in individual housing: the stroke volume of 85 ml; hematocrit = 45%; flow rate of 5.5 l/min.

The study of hemodynamic characteristics of AV was conducted on the generalized mathematical model of AV with a movable membrane which can be represented as follows. The initial solid model is the internal volume of AV considering the actual geometry of the main body, valves, and the transition radius (Fig. 1). This model was carried out in the form of AV-geometry.

For individual housing of this type requires individual housing main body, inlet and outlet valves, and pipes and is usually a flat movable wall membrane. This volume is divided (Fig. 2) into Finite Elements (FE) subject to the limitations described in Table 1 (ANSYS v.11).

For the research has been accepted the following parameters: viscosity of $6 \cdot 10^{-3}$ Pa/s; density $1,06 \cdot 10^{-3}$ Pa/sec; type of turbulent flow; Model turbulence standard; type of finite element FLUID 142.

Compared to the previous models [4–6] number of refinements and adjustments were made: the mobility condition adopted membrane which has a systole phase velocity of 10.5 mm/sec and a phase of diastole −21 mm/sec; such a condition requires the solution of the geometric and hydrodynamic problem at each step; initial pressure in nozzles and the bulk is taken as zero, in this case, the system itself determines the fluid pressure generated at any point in time of the cardiac cycle; under study it is assumed that the heart rate of the pump 1 Hz (60 beats per minute) for 0.5 seconds occurs systole, and then for the remaining 0.5 seconds—diastole (the ratio of the systole/diastole $\beta = 1:1$); to find the hemodynamic characteristics for individual housing with a movable membrane used ALE-analysis (Arbitrary Lagrangian Eulerian) and transient algorithm for solving.

In the process of solving the problem at each time step, the FE- models and meshes were reconstructed. ALE-analysis used a displacement boundary conditions described in the commands D, DL, and DA to rebuild the FE-model. This movement defines the beginning of each time step relative to the previous step. The proposed algorithm rebuild FE-mesh works best with elements of correct form in which the aspect ratio is 1. Incorrect elongated elements may lead to an error in the process of rebuilding the model. Therefore, building a generalized model focuses on sampling (partition) on a volume of finite elements and boundary conditions.

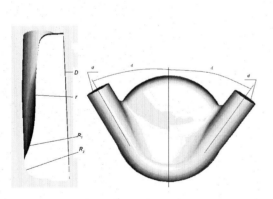

Figure 1. CAD-geometry of individual housing with variable parameters.

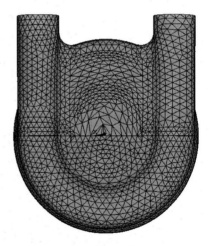

Figure 2. FE-model of AV (number of finite elements—71027, number of units—13599).

Table 1. The boundary conditions for the study of hemodynamic characteristics of AV.

Geometry elements of AV	Velocity (mm/sec)	Moving (mm)	Pressure (Pa)
Inlet pipe	326	0	–
Outlet pipe	326	0	–
Membrane	10,5	21	0
Rest surfaces	0	0	–

3 RESULTS

In the CFD-analysis (ANSYS v.11) were obtained field distribution of the velocity vectors, pressure indicators ENKE, ENDS, EVIS the volume of individual housing construction. Examples of derived distributions are shown in Figures 3 and 4. The analysis of distributions are obtained in stagnation zones (minimum speed and pressure), which can lead to blood clots. In addition the band detected with high turbulence.

The most difficult task is to establish the relationship between hemolysis arising in individual housing, and numerical indicators CFD-calculations. It is known that hemolysis is

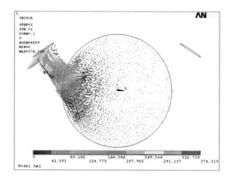 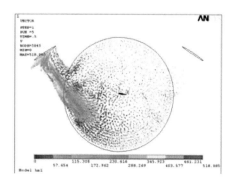

a) 0.1 sec from the beginning of the cycle (systole) b) 0.5 sec from the beginning of the cycle (systole)

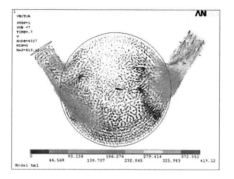 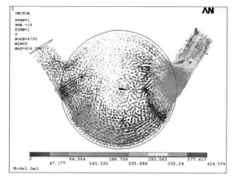

c) 0.7 sec from the beginning of the cycle (diastole) d) 1,0 sec from the beginning of the cycle (diastole)

Figure 3. Results of CFD-analysis (distribution of the magnitude and direction of blood flow rates).

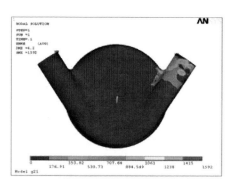 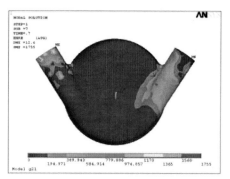

a) 0.1 sec from the beginning of the cycle (systole) b) 0.7 sec from the beginning of the cycle (diastole)

Figure 4. Results of CFD-analysis (distribution index ENKE- turbulent kinetic energy).

dependent on the Reynolds shear stresses, which can be calculated from the values of auxiliary coefficients (turbulent kinetic energy ENKE, the dissipation rate of turbulent kinetic energy ENDS, the effective viscosity EVIS) [2]. The algorithm for determining hemolysis index NIH is shown in Figure 5.

In algorithm made that Reynolds stresses τ and kinetic energy of turbulence associated expression:

$$\tau \cong \rho \left(\overline{V'_x V'_x} + \overline{V'_y V'_y} + \overline{V'_z V'_z} \right) = 2\rho k.$$

Damage to the blood stress-is a function of the amplitude of the shear stresses τ acting on the particles of blood and exposure time t_{exp} for the field of shear stresses. Velocity red blood globules L_{RBC} can be calculated from the experimental expressions:

$$L_{RBC} = \frac{\Delta Hb}{Hb}(\%) = 3{,}62 \cdot 10^{-5} \cdot t_{exp}^{0{,}735} \cdot \tau^{2{,}416}$$

where t_{exp} — exposure time.

The amount of free hemoglobin can be found from the formula:

$$fHb(Time) = V \cdot Ht \cdot \left\{ 1 - \left(1 - \frac{L_{RBC(f)}}{100} \right)^{\frac{FR \cdot t}{V}} \left(1 - \frac{L_{RBC(e)}}{100} \right)^{\frac{FR \cdot t}{V}} \right\},$$

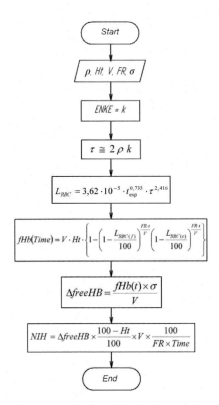

Figure 5. The algorithm for determining hemolysis index NIH.

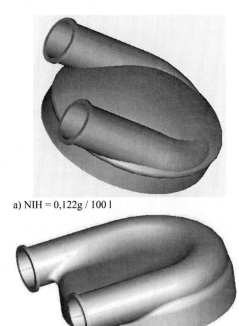

a) NIH = 0,122 g / 100 l

b) NIH = 0,101 g / 100 l

Figure 6. Solid models of existing and future individual AV-housing construction.

Table 2. Specifications for individual AV-housing construction.

AV-parameters	1st variant of AV	2nd variant of AV	Yamagata LVAD
Stroke volume, $V_{уд}$, mL	85	80	75
Residual volume, $V_{ост}$, mL	57	45	50
Angle of the pipes, degrees	14	0	0
Outside diameter AV, D, mm	92	92	90
Thickness AV, Hz, mm	35	32	30
NIH	0,122 g/100 l	0,101 g/100 l	0,098 g/100 l

where V—total blood volume; Ht—hematocrit (%); FR—flow rate (l/min); t—time interval (min); L_{RBC-f}, L_{RBC-e}—rate of red blood cells during the filling phase (*f*) and the ejection phase (*e*), respectively.

We can find the NIH, using the following expression:

$$NIH = \Delta freeHB \times \frac{100 - Ht}{100} \times V \times \frac{100}{FR \times Time}$$

where $\Delta free\ HB = fHb(t) \times \sigma/V$, σ—the density of hemoglobin in red blood cells. In Figure 6 shows a model of the existing and future individual housing construction. They are no areas of stagnation, flow velocity, pressure, meet foreign counterparts. Specifications of AV are presented in Table 2.

REFERENCES

Shumakov, V.I., Tolpekin, V.E., Shumakov, D.V. 2003. Artificial heart and circulatory support (monograph). *Janus-K*, 376.

Sevastyanov, V.I. 1999. Biocompatibility. *IC VNIIGS*, 367.

Zhdanov, A.V. 2007. The study hemodynamic of AV of LVAD and TAH systems pulsating type in CAD/CAM/CAE-systems. *New Technology*, 1.2, 17–19.

Morozov, V.V., Zhdanov A.V. 2007. Synthesis of artificial heart ventricles with specified hemodynamic characteristics (Monograph). *VlSU*, 192.

Okamoto, K., Hashimoto, T., Mitamura, Y. 2003. Design of miniature implantable LVAD using CAD/CAM technology. *Journal of Artificial Organs*, 6, 162–167.

Okamoto, K., Fukuoka, S.-I. 2001. FEM and CAD/CAM technology applied for the implantable LVAD. *Journal of Congestive Heart Failure and Circulatory Support*. 1, 391–398.

Zhdanov, A.V. 2014. Design aspects of implantable mechatronic units for systems of auxiliary blood circulation and TAH. *WIT Transactions on Engineering Sciences (Mechanical and Electrical Engineering)* 96, 295–301.

Morozov, V.V. 2015. Mechatronic unit for pulsative systems of LVAD and TAH. *Advanced mechanical and materials*, 96, 951–957.

Klaus, S., Korfer, S., Mottaghy, K., Reul, H., Glasmacher, B. 2002. In vitro blood damage by high shear flow: Human versus porcine blood. *Int J Artif Organs*, 25:306, 312.

2003. Blood Compatible Design of a Pulsatile Blood Pump Using CFD and CAD and Manufacturing Technology. *Artificial Organs*, 27, 6l–67.

Analysis of development of sight distance assessment at intersection

R. Matuszkova, M. Radimsky & T. Apeltauer
Faculty of Civil Engineering, Brno University of Technology, Brno, Czech Republic

O. Budik
HBH Projekt Spol. s r.o., Brno, Czech Republic

ABSTRACT: Sight distance is one of the key factors of safe traffic movement at intersections. One of the most common reasons of accidents at intersections is insufficient sight distance. This paper maps the development of sight distance assessment at intersections from the first mention in Czech technical standards till today. It is targeting changes in sight triangles definition also as changes in sight obstacle definition or speed and longitude slope consideration. Comparison of standards shows that safety requirements for sight distances at intersections are lowering in time.

1 INTRODUCTION

Road safety is one of the most discussed themes last years. Even though more accidents happen between intersections, intersections are considered as places with the highest density of traffic accidents (when covered length of road is taken in consideration).

Analysis of traffic accidents shows, that 18 497 traffic accidents happens at intersection, which is 23% of all traffic accidents in Czech Republic. From 681 fatalities, 109 (16%) happened at intersections. The most often cause of accident, 40%–60% cause of all accidents at intersections, is failure of yield the right of way.

It is necessary to take into account that essence of cause isn't always a human factor aspect, when driver doesn't follow regulations. Not yielding the right of way may be result of other related factors. One of the key factors is insufficient sight distance at intersections, which would allow the driver to predict and handle traffic situation soon enough to avoid any conflicts.

This paper analyzes the development of sight distance assessment at intersections in Czech Republic. Reaches the first technical standard of road design (1956) and covers progression till today. This paper has been worked out under the project No. LO1408 "AdMaS UP—Advanced Materials, Structures and Technologies", supported by Ministry of Education, Youth and Sports under the National Sustainability Program I".

2 CHANGES OF SIGHT DISTANCE AT INTERSECTIONS DURING YEARS

General definition says that sight triangles are areas that have to be provided with unobstructed stopping sight distance. Vertex of triangle is placed into geometric intersection of lanes' axes, where vehicles which may collide are moving. During changes of standards were made changes of sight triangle side lengths assessment, sight obstacle height assessment or also influence of right of way. Following chapters are going to describe these changes for each related standard which was published in Czech Republic.

2.1 Standard road design (1956)

The first standard defining sight triangles are areas, where no obstacle can be higher than 0.7 m above carriageway. Sight triangle side lengths are defined according to table for 5 different design speeds. Lengths are based on stopping sight distance, however without consideration of longitude slope. Standard doesn't mention different solutions for different rights of way at intersection (Parez, B. et al. 1956).

2.2 Standard intersections at roads and motorways (1964)

First standard dealing with road intersections defined sight triangles more precisely. Height of obstacle was changed from 0.7 m to 0.9 m. Also three basic principles were defined to consider right of way at intersections.

- Roads with same priority (rule of default priority to the right)
- Road, where it is necessary to stop the vehicle before entering to the intersection
- Roads, where one has higher priority than another.

For first option sides of sight triangle are defined as stopping sight distance according to design speed and longitude slope. Second option says, that sight triangle side on main road (road with higher priority) is defined as distance which vehicle covers by driving 0.9V (design speed) for 10 seconds. Length of sight triangle side on side road (road with lower priority) is not defined precisely (Parez, B. et al. 1962; Dopravoprojekt et al. 1963).

Calculation for design speed set as 80 km/h on the main road is as follows:

$$x = 0.9 * V_1 = 72 \text{ km/h} \to 72/3.6 = 20 \text{ m/s} \to t*v = 10*20 = 200 \text{ m}$$

Last option is more complicated. It sets as obligatory that it has to be considered what is longer. Distance covered by vehicle on main road with 0.9V for time necessary for vehicle on side road to stop or stopping sight distance for vehicle on main road.

Calculation of necessary time to stop vehicle on side road:

$$D_z = 1.5 \frac{0.9V}{3.6} + \frac{(0.9V)^2}{2g*3.6^2*(f \pm 0.01s)} + b$$

First part of equation is responsible for distance covered by running even speed for 1.5 s and second part of equation covers distance traveled by vehicle during slowing down.

$$D_z - 1.5 \frac{0.9V}{3.6} = \frac{(0.9V)^2}{2g*3.6^2*(f \pm 0.01s)} + b$$

Equation describing evenly slowed motion:

$$l = \frac{t*v}{2}$$

where l = distance of evenly slowed vehicle (m); t = time necessary to stop the vehicle after start of breaking (s); v = initial speed when breaking is started (m/s).

Equation to solve is:

$$D_z - 1.5 \frac{0.9V}{3.6} = \frac{t*v}{2}$$

After expression t and deal with units of speed we have necessary equations. For total time it is necessary to take into consideration also the react time of driver, generally accepted as 1.5 second. Final equation for time is:

$$t = 1.5 + \frac{8D_z - 3V}{V}$$

Assessment of sight distance is described on following example, where main road has design speed 80 km/h and side road 60 km/h and longitude slope on both roads is set as none.

$$V_1 = 80 \text{ km/h} \rightarrow D_{z1} = 100 \text{ m}$$
$$V_2 = 60 \text{ km/h} \rightarrow D_{z2} = 60 \text{ m}$$

Time calculation of vehicle stopping on side road:

$$t = 1.5 + \frac{8D_{z2} - 3V_2}{V_2} = 1.5 + \frac{8*60 - 3*60}{60} = 1.5 + 5.0 = 6.5 \text{s}$$

Calculation of distance covered by vehicle on the main road with 90% of design speed:

$$x_1 = 0.9 * V_1 = 72 \text{ km/h} \rightarrow 72/3.6 = 20 \text{ m/s} \rightarrow t*v = 6.5*20 = 130 \text{ m}$$

Calculation proves that distance covered by vehicle on the main road during the time necessary to stop the vehicle on the side rotation (130 m) is higher than stopping side distance (100 m). Sight distance necessary for main road is 130 m and for the side road 60 m. This example shows that different design speed makes important impact on calculation.

2.3 Standard road intersection design (1981)

Methodology of sight triangles calculation was slightly changed. First and second option principles stayed and height of sight obstacle also stayed same. However option, when one road is major road and another side road was changed, instead of 0.9V (design speed), 0.75V is used for calculation, but only for side roads. Speed on main road is considered as design speed without reduction. Change will be shown on the very same example from last chapter to be able to compare these standards (Parez, B. et al. 1962; Ustav silnicniho hospodarstvi et al. 1980; Pucek, K. et al. 1974).

$$V_1 = 80 \text{ km/h} \rightarrow D_{z1} = 100 \text{ m}$$
$$V_2 = 60 \text{ km/h} \rightarrow 60*0.75 = 45 \text{ km/h} \rightarrow D_{z2} = 35 \text{ m}$$

Calculation of time necessary for stopping the vehicle on the side road:

$$t = 1.5 + \frac{8D_{z2} - 3V_2}{V_2} = 1.5 + \frac{8*35 - 3*45}{45} = 1.5 + 3.2 = 4.7 \text{s}$$

Calculation of distance covered by vehicle on the main road with full design speed consideration:

$$x_1 = 80 \text{ km/h} \rightarrow 80/3.6 = 22.22 \text{ m/s} \rightarrow t*v = 4.7*22.22 = 105 \text{ m}$$

In this particular example is sight triangle side length on main road 105 m, which is 25 m shorter than according to the calculation from previous standard. In addition to that, sight triangle length on side road is significantly shortened from 60 m to 35 m.

2.4 Standard road intersection design (1995)

This standard didn't bring any new approach considering sight distance assessment (Kucera, V. et al. 1985; Sachlova, Z. et al. 1994; Turcin, V. et al. 1986).

2.5 Standard Road intersection design (2007)

Currently, active standard CSN 73 6102 (including all updates) completely changed sight distance assessment at intersections. Standard defined different groups of vehicles, which differs in height of view point from the vehicle and also differs in acceleration which enters the calculations. Height of view point from vehicle is a major factor when deciding what is, and what is not, sight obstacle. Reason is, that new definition of sight obstacle says, that view obstacle is everything what is higher than 0.25 m under the view beam from view point. This definition means that when something is obstacle for group of vehicles with view point at 1 m high, for another three groups with view point at 2 m it isn't. Turing left from side road (in relation to vehicle coming to intersection from right side) and turning right from side road (in relation to vehicle coming to intersection from left side) are considered as the worst traffic movement related to sight distance. Sight triangles are according to this standard design to permitted speed without any reduction (Sachlova, Z. et al. 2000; Bazant, M. et al. 2004; Muller, M. et al. 2007; Prokes, S. et al. 2006).

Following previous standards, three types of intersection arrangement were kept:

- Intersection with default rule of priority to the right
- Intersection with priority road (main road) and side road, where vehicles have to stop their wheels on side road before entering the main road
- Intersection with priority road (main road) and side road with "Yield" sign (vehicles have to yield right of way, but don't have to stop their movement before entering the main road).

Sight triangle side length at intersections with priority to the right is calculated similar to the intersection with main road and side road with "Yield" sign. The only difference is that sight triangle is necessary to create just to the right side, as view to other vehicles isn't in this case related to traffic movement at intersection.

At intersections, where vehicles are supposed to stop on road with lower priority (side road), is expected, that vehicles stop 3 m before main road carriageway (edgeline, if it was marked). If there is no vehicle in sight, which may interfere its turning movement, after react time (2.5 s) driver starts his maneuver. Standard expect the worst possible situation, so that there is vehicle on main road, which spot turning vehicle (reaction time 2.5 s) slow down to 75% of permitted speed to let turning vehicle finish its turning and accelerate also to 75% of permitted speed. Side of sight triangle is longer when turning left, reason is, that turning left takes longer time and vehicle follows different turning radius (Muller, M. et al. 2007).

At last, type of intersection arrangement are sight triangle sides' length based on assumption that vehicle coming from side road doesn't have to stop and that is why it is moving with speed which is low enough to perform the most inconvenient movement necessary at the intersection. Side of sight triangle based on side road (road with lower priority) is based on assumption, that vehicle coming into intersection spotted vehicle on main road and has to be able to stop. Side of sight triangle length is smaller at main road, than at previous arrangement, because vehicle coming from side road doesn't start from zero, so crossing time is faster.

Even though all these procedures are enumerated in table part of standard, current sight distance assessment is complicated and confusing for designers and administration due to numerous vehicle groups, road categories, and arrangements.

3 CONCLUSION

Almost every technical standard changed the approach to deal with sight distances at intersections. Different methodology was used for definition of sight obstacle. First standard defined sight obstacle by height 0.7 m, following standards used 0.9 m above the carriageway and today sight obstacle is defined through vehicle sight points, which are placed 1 m or 2 m above the ground and sight obstacle than cannot reach out closer than 0.25 m under sight beam from sight point. This means, that height of sight obstacle for passenger cars is 0.75 m.

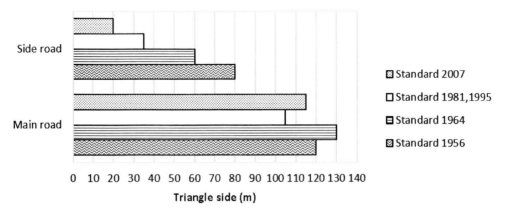

Figure 1. Sight triangle side lengths comparison.

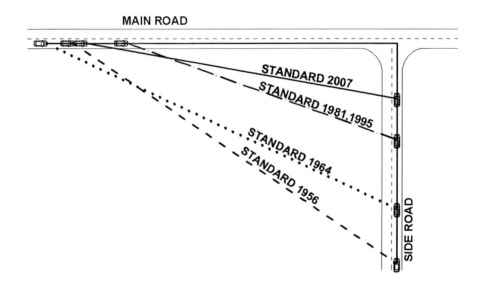

Figure 2. Scheme of sight triangles development for passenger cars and rural areas.

Next, changes were made in sight triangles side lengths determination, at first length was based on stopping sight distance, but with different usage of design speed. Today sight triangles are designed by new procedure, which is targeting to analyze vehicle movement as close as possible while considering the highest permitted speed.

Figures 1 and 2 show changes of technical standards in sight triangles design. Sample of intersection for this comparison was selected, where design (permitted) speed on main road is 80 km/h and design (permitted) speed on side road is 60 km/h. Longitude slope is considered as none and "Yield" sign is placed on the side road. Figures show, that sight triangle's requirements are lowering in time.

REFERENCES

Bazant, M. & Karlicky, P. et al. 2004. *CSN 73 6101 Projektovani silnic a dalnice*. Praha: Cesky normalizacni institut.
Dopravoprojekt et al. 1963. *CSN 73 6102 Krizovatky na silnicich a dalnicich ve volne trati*. Praha: Vydavatelstvi uradu pro normalizaci a mereni.

Kucera, V. & Turcin, V. et al. 1985. *CSN 73 6101 Projektovani silnic a dalnice*. Praha: Vydavatelstvi uradu pro normalizaci a mereni.
Muller, M. & Bartos, L. et al. 2007. *CSN 73 6102 Projektovani krizovatek na silnicnich komunikacich.* Praha: Cesky normalizacni institut.
Parez, B. et al. 1956. *CSN 73 6101 Projektovani silnic.* Praha: Vydavatelstvi uradu pro vynalezy a normalisaci.
Parez, B. et al. 1962. *CSN 73 6101 Projektovani silnic a dalnic*. Praha: Vydavatelstvi uradu pro normalizaci a mereni.
Prokes, S. et al. 2006. *CSN 73 6110 Projektovani mistnich komunikaci.* Praha: Cesky normalizacni institut.
Pucek, K. et al. 1974. *CSN 73 6110 Projektovani mistnich komunikaci.* Praha: Vydavatelstvi uradu pro normalizaci a mereni.
Sachlova, Z. et al. 1994. *CSN 73 6102 Projektovane krizovatek na silnicnich komunikacich.* Praha: Cesky normalizacni institut.
Sachlova, Z. et al. 2000. *CSN 73 6101 Projektovani silnic a dalnice*. Praha: Cesky normalizacni institut.
Turcin, V. et al. 1986. *CSN 73 6110 Projektovani mistnich komunikaci.* Praha: Vydavatelstvi uradu pro normalizaci a mereni.
Ustav silnicniho hospodarstvi et al. 1980. *CSN 73 6102 Projektovani krizovatek na silnicnich komunikacich.* Praha: Vydavatelstvi uradu pro normalizaci a mereni.

Desired and actual bid participation behaviour of Japanese construction companies

Emi Morimoto
The University of Tokushima, Japan

Koki Arai
Shumei University, Japan

ABSTRACT: In the present study, we quantify the desired and actual bidding behaviour of Japanese companies by analysing public procurement data from three regional development bureaus. Our study extends the findings of Iwamatsu et al. (2013), who use a survey questionnaire to determine the desired bidding behaviour of major Japanese construction companies. To compare actual bidding behaviour with desired bidding, we model the probability of participation, which is regressed on the quantified values of the bidding data and other information. The results are then ranked, compared with those of Iwamatsu et al. (2013), and analysed. We focus on the factors on which firms concentrate when determining whether to participate in bidding. Although both Iwamatsu et al. (2013) and our study include widely used high-ranking items, in our analysis, 'company circumstances' are ranked highly at the participation stage. This offers practical justification for including competition circumstances when modelling real-world bidding behaviour.

1 INTRODUCTION

In Japan, bidding systems for government procurement have been undergoing significant changes. For example, designated competitive bidding has been replaced by general competitive bidding, and the method used to select the winning bid has changed from automatic selection of the lowest bidder to a comprehensive assessment system. Under these circumstances, the bidding activities of firms have declined; indeed, firms no longer participate in bidding activities that do not benefit them.

The Japanese construction market saw somewhat of a downward trend during 2002–2011. Before the early 1990s, Japan's economic bubble created extensive domestic demand; thus, many construction companies did not need to develop overseas market strategies. After 1994, some prominent cases of bid rigging came to light and public investment dropped dramatically, as procurement authorities changed their policies to improve input objectivity. However, this policy change may have resulted in decreasing product quality and a suspension of technical progress. The Japanese construction industry struggled with this situation from 2002 to 2011. This is similar to what occurred in the United States (US) and United Kingdom (UK) during the 1980s. The Latham Report (Latham, 1994) and the Egan Report (Egan, 1998), for example, pointed out similar problems.

The most recent examination of bidding behaviour in Japan was conducted by Iwamatsu et al. (2013), who used a questionnaire survey to study the desired bidding behaviour of major Japanese construction companies. They obtained 283 responses on 36 factor keywords in two situations and compared these characteristics with perspectives from Japan, the US, and UK. Similarly, Laryea & Hughes (2008) conducted a review of questionnaire surveys, while a number of empirical studies, such as Ahmad & Minkarah (1988), Shash (1993), and

Mochtar & Arditi (2001), have addressed interview surveys, paying particular attention to two decisions: the decision to participate in the bidding stage.

The present study aims to quantify the desired and actual bidding behaviour of Japanese companies by analysing public procurement bidding data. We use bidding data and other sources to represent each item keyword by an appropriate proxy value, and then we compare this with the questionnaire responses. Thus, these responses are compared effectively with the real intentions of the bidding companies based on regression estimation using actual bidding data. We aim to use this comparative analysis to understand the gap between the desired and actual bidding behaviour displayed by Japanese companies.

This study makes three contributions to construction management. First, a new method of analysis is used. In this research, it is very important for the method of analysis to reveal preferences about bid decisions. Therefore, the analysis includes a discrete logit analysis of the participation decision. Such analysis of participation factors is important in the construction management field. Our method can accommodate bidding in the presence of environmental concerns, and is useful for the comparative analysis of factors that influence bidding.

Second, this research contributes to a new way of thinking. We study the similarities and differences between stated preferences and revealed preferences. Doing so is important for the implementation of institutional reforms, which requires understanding how actual bidding behaviour compares with intended behaviour. Our results indicate the importance of capturing the actual competitive situation in an industry. This study provides insight into the actual impact of bidding behaviour on the global construction industry, allowing the design of a welfare-improving social system.

Third, this study yields unique results. Although we find very little difference between desired and actual bidding, an entity tailors its behaviour to its expectations of its competitors' actual rather than desired behaviour. This finding is similar to that found in the research on US and UK markets. In this regard, competition is a more significant factor than entities are aware of. Our research aims to find ways to use public procurement to improve the construction industry by shedding light on the tools that construction firms use and on their actual preferences. This will aid effective procurement design.

The rest of this article is organized as follows. Section 2 explains the overall Japanese construction industry to understand the background of this study. Section 3 describes the previous study, Iwamatsu, et al. (2013), as a reference point for our comparative study, and illustrates the research design, including quantification strategy. Section 4 deals with the data analysis to estimate regressions. Section 5 demonstrates the results of the estimation, and compares the results with the reference points. Section 6 concludes.

2 JAPANESE CONSTRUCTION INDUSTRY

First, we provide descriptive information about the Japanese construction industry and its circumstances. Since the 1990s, Japanese research on the construction industry and institutional design of public bidding has been conducted in the fields of economics, law, public and civil engineering, and construction engineering.

In particular, Kanemoto (1993) and Kunishima (1995) have had great effects on policy discussions on the institutional design of public bidding. They observed that government procurement operated, in principle, using an openly competitive bidding procedure since Japan became a modern constitutional state in the 1860s. However, this procedure was often of limited use in practice because it allowed anyone to participate and often resulted in the purchase of 'cheap' but 'poor-quality' goods. In practice, the selective bidding procedure played a major role. Contractor selection was left to the discretion of the government agencies that placed the orders. In addition, order-placing decisions were left to the discretion of the agencies, which were, thus, empowered to split an order into smaller units. The number of bidders was restricted through a selection process, and this enabled less efficient, smaller firms to accept orders. This practice gave rise to criticism regarding higher costs, corruption, and illegal cartel agreements among bidders. Since the 1990s, government agencies' ordering discretion has been reduced

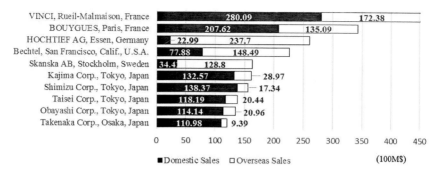

Figure 1. World construction companies' domestic and overseas sales (2009).

through the 'expansion of open competitive bidding' as part of a deregulation and regulatory reform program. Today, the open and competitive bidding procedure is employed for large-scale public works, and the World Trade Organization (WTO) Agreement on Government Procurement applies to most of these works. This procurement situation has established the Japanese construction market, which is among the largest markets but is not the largest market.

Most Japanese construction firms have gained sales not from overseas but from the domestic market (Fig. 1) owing to strong demand for domestic infrastructure until the 2000s, and rather high profit based on coordinated practices in the industry until about 2005 (see Arai & Morimoto, 2014). However, in recent years, major players in the Japanese construction industry have developed their business deployment in overseas market, in particular, the overseas infrastructure market, which is supported by the government of Japan.

This background situation may affect desired and actual bidding behaviour of construction companies in Japan. We undertake a comparative analysis between the two types of behaviour: first, desired bidding behaviour and second, actual bidding behaviour. We describe the former in detail via the previous study of Iwamatsu et al. (2013) and the latter via several regression estimations based on public procurement data.

3 RESEARCH DESIGN

In their comparative study based on surveys of major construction journals and interviews with influential construction managers, Iwamatsu et al. (2013) found that Japan, the US, and UK share certain common bidding characteristics, indicated by relatively high scores for 'type of job' and 'competitiveness in your industry', as well as other features, such as 'labour environment (union, non-union, cooperative)' and 'time of bidding (season)'.

We quantify Iwamatsu et al.'s (2013) keywords and present them in Table 1. Note that this study uses Japanese terms instead of US terms, but the basic concept is common across the three study regions. Comparing the questionnaire results of Iwamatsu et al. (2013) with our quantified values based on bidding data and other information reveals similar characterization of quantified values based on bidding data and other information reveals similar characterization of stated and revealed preferences.

In this analysis, we examine a firm's desired strategy based on the results of the questionnaire survey presented by Iwamatsu et al. (2013). This approach is similar to a stated preference-type method, in which the decision making and conduct of the organization are likely to be more reasonable than those of individuals are because outlier preferences are balanced. However, in addition, we consider three alternative viewpoints. First, no differences exist between the questionnaire survey and the quantified bidding data and other information. Second, stated preferences are upwardly biased because of an embedding effect reflecting altruism. Third, the stated preferences display an additional embedding effect that arises from over-individualism. In considering these viewpoints, our analysis of quantified bidding data and other information is more akin to a revealed preference-type method.

Table 1. Keyword quantification in our study.

	Iwamatsu et al. (2013)	Quantification Method	Remarks
Technical characterization of the project	Type of job	Selecting general civil engineering projects of three regional development bureaus of the Ministry of Land, Infrastructure, Transport and Tourism	
	Location of project	No quantification	
	Degree of difficulty	No quantification	
	Project duration	No quantification	
	Size of job	Dummy of project level (level A=1)	Considering the size of job as the size of the project-level dummy
	Type and amount of equipment required/available	Index of cement price	Representing the cement price index as the equipment required/available
	Designer (A/E)/Design quality	Ratio of the maximum and minimum bid to the value of the predetermined price	Considering the difference of recognition of the project value in the project bidding stage
	Project cash flow	No quantification	
	Rate of return	Gross income on sales	Excluding negative values
	Need for work	No quantification	
Procurement authority and method	Owner	No quantification	
	Type of contract	No quantification	
	Bidding method	Dummy of bidding method (WTO, selecting)	
	Duration	No quantification	
	Time of bidding	No quantification	
	Degree of hazard	No quantification	
	Future perspective of a similar project	Days to next winning bid	Considering days to next winning bid as the future perspective
Circumstances of competition	Number of competitors	Number of participants	Including a declining or invalid number of participants
	Competition	Basic and additional point in comprehensive bidding	Considering the evaluating points
	Strength in the industry	Days since last winning bid	Considering days since last winning bid as the past strength in the industry
	Previous owner	No quantification	
	Overall economy	Nikkei stock average	
	Labor environment	Wage index of construction workers	Considering the wage index of construction workers as the labor environment
	Proportion of work to be	No quantification	
	Reliability of subcontractors	No quantification	
	Situation of the company	Average value of completed general civil engineering works	
Circumstances of the company	Current workload	Days after last bid	
	Uncertainty in the estimate	Gross profit on sales	Considering the total amount of the winning bid benefit as risk-taking in lieu of uncertainty
	Availability of qualified staff	No quantification	
	Type and number of supervisors required/available	Number of first and second supervisors for civil engineering	Number of engineers
	Cost of bidding	No quantification	
	General overheads	General administration cost	
	Capital requirement/availability	No quantification	
	Intuition	No quantification	
	Difference between predetermined total price and internal total price	Difference between predetermined total price and internal total price relative to the value of predetermined price	
	Mathematical model	No quantification	

4 DATA ANALYSIS

This study uses bid data for Levels A and B general public engineering works (>300 million yen) from the Shikoku, Kanto, and Kinki regional development bureaus of the Ministry of Land, Infrastructure, Transport, and Tourism (2002–2011). The study periods are as follows: FY2002–2011 for Shikoku, FY2004-2011 for Kanto, and FY2005-2011 for Kinki. The dependent variables are the probability of participation and the bidding ratio. The probability of participation is measured as the probability of an entity participating in a bid offered by the regional development bureau in a certain period. The bidding ratio is calculated by dividing the bidding price by a predetermined price. Taking the log of both bid numbers allows us to overcome problems posed by price elasticity. The estimation equations for the probability of participation and the bidding ratio are respectively, where indicators i are the individual bidding indices, t is the factor of consideration index, *Prof participation$_i$* is the probability of

$$\log(\textit{Prof participation}_i) = a_1 + \sum b_{1,i,t} \log(\textit{quantified\_data}_{1,i}) + \varepsilon_{1,i} \quad (1)$$

Table 2. Regression results and results comparison with those of Iwamatsu et al. (2013).

Dependent variable method: ordinary least square	Participation n=4586 Coefficient	(standard err.)		Effect Ranking Our Study	Iwamatsu et al.
Size of job	0.587	(0.3438)	*	1	1
Type and no. of equipment	0.182	(0.0395)	***	6	
Designer (A/E)/ Design quality	-0.016	(0.0131)			
Rate of return	0.003	(0.0023)			4
Bidding method (designated dummy)	0.011	(0.0517)			3
Bidding method (WTO dummy)	0.074	(0.0531)			3
Future perspective	0.002	(0.0024)			
Number of competitors	-0.063	(0.0157)	***	8	
Competition	0.029	(0.0535)			7
Overall economy	-0.082	(0.1403)			
Labors environment	-0.190	(0.0364)	***	5	
Situation of the company	0.024	(0.0155)			6
Current workload	-0.008	(0.0012)	***	7	5
Uncertainty the estimate	-0.453	(0.0576)	***	2	
Type of number of supervisory persons	0.346	(0.0170)	***	4	2
General overhead	0.365	(0.0549)	***	3	8
Differences between predetermined and bid	-0.084	(0.1157)			
C	-5.781	(1.9563)	***		
R-square	0.226				
adjusted R-square	0.223				
standard err. Of regressions	0.527				
Akaike Information Criteria	1.559				

Note: The upper values in the cell are the estimated coefficients and the lower values in parentheses are the standard errors. *,**, and *** denote significance at the 10%, 5%, and 1% levels, respectively.

participation, and *quantified_data$_i$* is the factor of consideration. ε is an error term. Ordinary least squares regression analysis is then used to estimate the coefficients a_1, $b_{1,i,t}$. Table 2 presents the results.

Based on these coefficients, we compare the ranking and statistical significance of our results with the ranking of Iwamatsu et al. (2013), as shown in Table 2. According to the comparative analysis presented in Table 2, we examine the items determining a firm's participation in the bidding process.

5 RESULTS

Iwamatsu et al. (2013) and our study have in common some high-ranking items, such as 'size of job' and 'number of supervisors required/available'. Therefore, both analyses are considered to have captured the actual industry situation. In other words, the stated and revealed preferences are similar for these items. Furthermore, both analyses rank highly personnel issues, such as 'type and number of supervisors required/available' and 'labour environment'. Personnel management issues may reduce the probability of participating in the bidding process.

However, there are some differences between the two studies. In our study, the 'uncertainty in the estimate' and 'general overhead' items', which fall into the category of 'company circumstances', are high ranking, whereas they have a relatively low ranking in Iwamatsu et al. (2013). The reason for this difference is that the stated preference approach taken by Iwamatsu et al. (2013) relies on analytical recognition of a person's own ex post behaviour, whereas our study, which uses revealed preferences, tends to capture people's real instincts during bidding.

No self-serving disclosure or non-disclosure tendencies are recognized either in our study or in Iwamatsu et al. (2013). In this regard, the findings of Iwamatsu et al. (2013) are likely to reflect actual business practices.

6 CONCLUSIONS

Our findings add to the literature by accurately describing actual bidding behaviour in Japanese organizations. This is both important and novel in the construction management literature and beneficial for designing bidding institutions. To create a better bidding system, it is necessary to consider actual firm-level behaviour as well as the desires of the companies involved.

The contributions of this study lie in its method of analysis, in the new way of thinking it introduces, and in its findings. While we apply our method to Japanese public procurement, this method can also be applied to bidding in the presence of environmental concerns, or to a comparative analysis of bidding determinants. Such analysis of participation factors is significant for the construction management field. In addition, this study provides insight into the actual impact on the worldwide construction industry, enabling the design of a welfare-enhancing social system.

Our next aim is to have interviews with firm executives and procurement authority officials about pricing strategy in order to verify the inferences of this study and refine the applicable theory. Furthermore, the transition of the desired and actual bidding behaviour would be worth exploring as a future challenge.

ACKNOWLEDGMENTS

We are grateful to Professor Hideo Yamanaka for his advice on data preparation. We gratefully acknowledge support from the Shikoku regional development bureaus of the Ministry of Land, Infrastructure, Transport, and Tourism announcement of Bidding and Contract Data (2002–2005).

This research was partially supported by the Japan Institute of Country-ology and Engineering Grant Number 14013.

REFERENCES

Ahmad, I. & Minkarah, I. 1988. Questionnaire survey on bidding in construction. *Journal of Management in Engineering* 4(3): 229–243.

Arai, K. & Morimoto, E. 2014. A comparative analysis of the desired and actual bidding behaviour of construction companies. *ARCOM Proceedings of the 30th Annual Conference*, Vol.1, 433–442.

Egan, J. 1998. Rethinking construction. *Department of Environment, Transportation and the Region*.

Iwamatsu, J., Morimoto, E., Namerikawa, S., & Endo, K. 2013. A consciousness analysis of a construction company's bidding behaviors. *Journal of JSCE* 69: 62–74 (in Japanese).

Kanemoto, Y. 1993. Design of public procurement. *Government Auditing Review* 7.

Kunishima, M. 1995. Future vision of the system of public engineering works. *Government Auditing Review* 12.

Laryea, S. & Hughes, W. 2008. How contractors price risk in bids: Theory and practice. *Construction Management and Economics* 26(9): 911–924.

Latham, M. 1994. Constructing the team: Joint review of procurement and contractual arrangements in the United Kingdom construction industry: Final Report. HM Stationery Office.

Ministry of Land, Infrastructure, Transport and Tourism. 2005–2011. Announcement of bidding and contract data (EXCEL) http://www.mlit.go.jp/tec/nyuusatu/datalink.html.

Mochtar, K. & Arditi, D. 2001. Pricing strategy in the US construction industry. *Construction Management and Economics* 19(4), 405–415.

Shash, A. 1993. Factors considered in tendering decision by top UK contractors. *Construction Management and Economics* 11(2): 111–118.

The skid resistance evaluation in accordance with the Wehner/Schulze method

J. Daskova, P. Nekulova & J. Kudrna
Faculty of Civil Engineering, Brno University of Technology, Brno, The Czech Republic

ABSTRACT: This paper is concerned with the Wehner/Schulze test method, which allows laboratory measurement of surface skid resistance. Measurements can be performed directly on cores taken from asphalt or concrete roads, or on aggregate test specimens prepared in a laboratory from splits of various fractions. The paper presents the results of measurements of surface skid resistance made on asphalt, concrete and aggregate specimens. Cores were extracted from roads of various age, aggregate types, asphalt mixtures or different type of texturing. Values of friction coefficient measured on cores are compared with longitudinal friction coefficient determined on roads by dynamic measurement system TRT. The results of friction measurements of various types of aggregate samples are compared with the reference method in accordance with the EN 1097-8. Aggregate polishing determined by the Wehner/Schulze device has a higher resolution than the current method given by the EN 1097-8, and based on our results, shows good repeatability and reproducibility.

1 INTRODUCTION

The test has so far been called the "Wehner/Schulze method", and it determines the value of the Laboratory Skid Resistance measurement PWS. The test name changed with the introduction of a new European standard EN 12697-49 "Bituminous mixtures—Test methods for hot mix asphalt—Part 49: Determination of friction after polishing" in January 2014. The new name only refers to measurements performed with asphalt mixtures, but the test is also suitable for concrete surfaces and for determining aggregate polishing. The standard should rather have been classified under tests of pavement surface properties published in the EN 13036 series.

The main aim of the paper is to prove that the Wehner/Schulze method can be used also for concrete roads and for determination of aggregate polishing. Results of measurements of surface skid resistance made by Wehner/Schulze test device were compared with values of longitudinal friction coefficient that were measured by the dynamic measurement system TRT (CEN/TS 15901-4, 2010). TRT device is used to determine skid resistance properties directly on road surface during operation. The results of friction measurements of various types of aggregate samples were compared with the reference method in accordance with the EN 1097-8.

2 TEST DEVICE DESCRIPTION

The test device shown in Figure 1 is composed of two rotating heads, one for polishing and one for measuring the friction at the testing specimen surface. Part of the test device is also a container for mixing the water and quartz powder mixture.

Polishing head is composed of three rubber cones and a device for supplying water and quartz powder mixture. Polishing head is placed on the tested surface with a load of 392 ± 3 N and rotates at a speed of 500 ± 5 RPM during polishing process. The device for water and

Figure 1. Wehner/Schulze test device, polishing head, measuring head.

quartz powder supply comprises a container with a stirrer to ensure mixture homogeneity and a pump with a capacity of 5.0 ± 0.5 l/min, which supplies the water and quartz powder mixture during the polishing process through the center of the polishing head to the pavement surface.

Measuring head is composed of three sliding blocks and a system that measures the moment. The rotating head must be able to reach rotation speed of at least 50 rotations/s, which corresponds to a moving speed of 100 km/h in the measuring track of the sliding blocks and also able to load the test specimen with a static force of 253 ± 3 N. The system must be capable of measuring the moment with an accuracy of ± 1 Nm during breaking. The rotating head is also equipped with a system that measures the rotation speed during breaking with an accuracy of ± 2 rotations/s. Sliding blocks, which are responsible for measuring the friction coefficient, are composed of a circular metal segment covered with rubber of a 5 ± 1 mm thickness.

Clamping system must allow correct fixation of the test specimen with a diameter of at least 225 mm and maximum height of 50 mm and also fixation of a glass-control plate with a textured surface, which functions as a reference surface. Test specimen can be prepared in a laboratory according to EN 12697-33 or it can be core extracted in accordance with EN 12697-27, which must be compatible with the clamping system (EN 12697-49, 2014).

3 TEST METHOD DESCRIPTION

Polishing of the test specimen is performed by rotating cones, which are supplied with a mixture of water and quartz powder. This models accelerated effect of traffic load on the surface of pavement wearing courses. Prior to the actual polishing process, the test specimens must be prepared. Asphalt specimens prepared in a laboratory must be sand blasted by a blast system, so that remaining of any bitumen is removed from the surface. It is also possible to test specimens that haven't had asphalt bitumen remaining removed, and therefore determine the time necessary for it to be removed by traffic. Samples from already used pavements do not have to undergo sand blasting.

After the sample to be polished is fixed in the clamping device, the polishing process starts at a defined speed with the supply of water and quartz powder mixture. This polishing process is divided into cycles (1 cycle = usually 90 000 passes), but depending on the aim of the experiment, also different procedures can be used. After one cycle of the polishing is stopped, the sample is washed with water and the friction coefficient is measured.

Before the friction coefficient of the polished sample is measured it is necessary to verify the correct functioning of the test device by measuring the friction coefficient of the glass-control plate and comparing it with the reference friction value of this plate. If the measured values are more than 10% off from the reference value, the sliding blocks must be replaced.

Once correct functioning of the test device is verified, the friction coefficient of the test sample is measured. First, the measuring head with the sliding blocks starts rotating at a defined speed of 100 ± 5 km/h in its measuring track and then it is pushed on the wet surface of the test specimen with a defined pressure (253 ± 3 N). The speed of the rotating head is gradually decreased until a complete stop. During the breaking phase the test specimen surface is supplied with pure water. During the entire breaking process the moment and speed are measured. After the measuring process of the polished sample is finished, another control measurement is performed with the glass plate. The results of the control measurement must not be deviated from the friction coefficient reference value of the glass plate by more than 10%. This process is repeated after polishing of each of the test samples.

The determined value of moment M is then used to calculate the friction coefficient μ using Eq. 1.

$$\mu = \frac{M}{253 \times 0.09} \tag{1}$$

where M is the measured moment in Nm and μ is the friction coefficient.

The calculated friction coefficient value after polishing (Friction After Polishing, FAP) is an average of two or more separate measurements expressed by the friction coefficient μ_{FAP} calculated using Eq. 2 at a defined speed of 60 km/h. If the difference between the two individual measurements (repeatability) is larger than 0.03, the measurement is considered as invalid and a different test sample is used.

$$\mu_{FAP} = \mu_m - \mu_{km} + \mu_{ref} \tag{2}$$

where μ_{FAP} is a single friction measurement, μ_m is friction coefficient at 60 km/h, μ_{km} is the mean value of the control plate before and after the friction measurement and μ_{ref} is a known friction value of the control plate (EN 12697-49, 2014).

4 RESULTS

The results presented in this work were measured as part of the research project on a Wehner/Schulze test device at the Vienna University of Technology. A newly installed test device in the laboratory of the Faculty of Civil Engineering at the Brno University of Technology is currently in a testing phase with the aim to compare the results measured at the Vienna University of Technology with those measured at this new facility in Brno.

4.1 *Asphalt specimens*

Taking into account the high number of various asphalt mixtures, especially cores from currently used and new wearing courses were chosen. Particular attention was given to ensure various aggregate types are included that reflect the entire scale of skid resistance. Figure 3 shows the progress of the μ_{FAP} values of various types of asphalt wearing courses taken from roads. The figure legend is sorted in a descending order of μ_{FAP} value after the first

Figure 2. Test specimens, from the left: core from the motorway D1 (exposed aggregate), core from asphalt road (BBTM 5), aggregate specimen (gneiss).

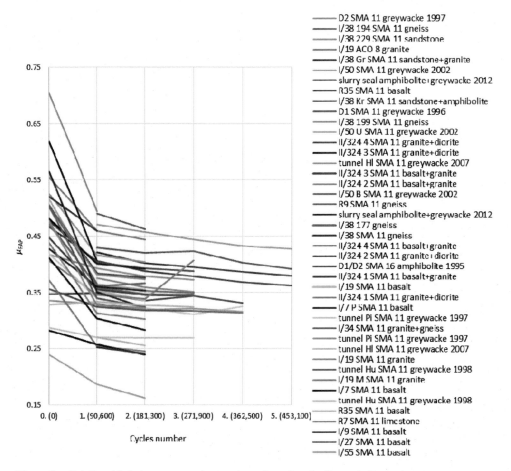

Figure 3. Relationship between μ_{FAP} values and number of cycles for asphalt wearing courses.

polishing cycle. Best results were achieved in case of the slurry seal, which was expected as it is used for restoration of skid resistance.

μ_{FAP} values measured on cores taken from roads were compared with the values of longitudinal friction coefficient F_p determined directly on the road by dynamic measurement system TRT. Figure 4 shows strong correlation between μ_{FAP} and F_p.

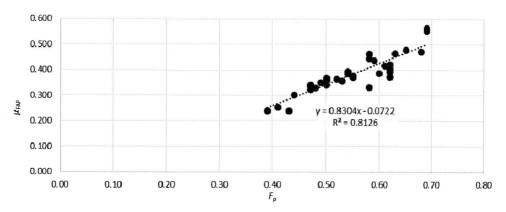

Figure 4. Correlation between μ_{FAP} values and F_p values measured by TRT device.

4.2 *Cores from concrete roads*

In order to measure the friction coefficient after polishing, cores of concrete road surfaces of various age and with various surface texturing technologies were used. The majority of the test samples used was the burlap drag texturing, which is the most used technology at the moment, but rapid decrease of longitudinal friction F_p in time is now being observed. For the comparison purposes we used cores from all currently used technologies for creating surface textures of concrete surfaces, for example exposed aggregates, nylon-tining texturing, transverse tining etc. Also surfaces where high-pressure water blasting was applied to restore skid resistance were tested. The results are summarized in Figure 5, which shows the relationship between the friction coefficient μ_{FAP} value measured by the Wehner/Schulze device and the number of rotation cycles of the polishing head. Some samples were subjected up to 10 cycles, which corresponds to almost one million passes of the polishing head. The figure legend is sorted in a descending order of μ_{FAP} value after the first polishing cycle.

Core samples were taken from both newly laid and long-term used concrete surfaces. The oldest tested sample was taken from a stretch at the D1 motorway (187.6 km) which was laid in 1971. Cores on the pass stretches were taken from the road edges next to the central separating barrier, where it is probable that the surface was not subjected to loads from heavy vehicles. Samples from older stretches are used especially for comparing the friction coefficient measured by the Wehner/Schulze method with the longitudinal friction coefficient F_p measured in long-term by the national reference device TRT.

The results given in Figure 6 shows a good correlation when comparing the results of the Wehner/Schulze test method and the TRT device. Coefficient of determination for concrete specimens has a similar value as the one for asphalt samples.

4.3 *Aggregate specimens*

In order to measure polishing, laboratory samples from 17 different types of aggregates used in asphalt and concrete wearing courses were prepared. Several test samples were prepared for each type. Aggregates came either directly from quarries or they were obtained by dissolving cores from asphalt layers. Test samples were made by manually laying 8/11 aggregate into a form. Aggregate was fixed with epoxide resin and sand 0.2/0.4. Results of the accelerated polishing were compared with data given in the Aggregate catalogue or with data available on webpages of the aggregate manufacturers. The majority of aggregates agreed with the expected results based on PSV. The aggregate is polished by 90 000 passes of the rollers, while being dusted with the quartz powder. Then, the friction is measured and this value expresses the value of polishing μ_{FAP}.

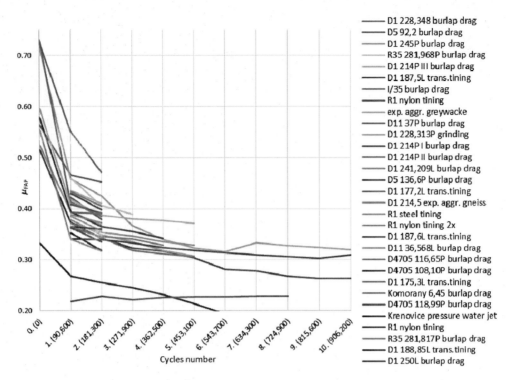

Figure 5. Relationship between μ_{FAP} values and number of polishing cycles for concrete specimens.

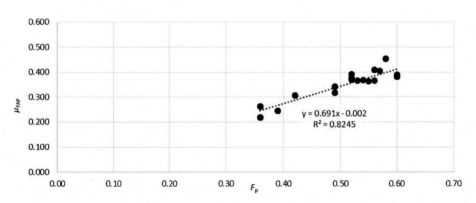

Figure 6. Correlation between µFAP values and Fp values measured by TRT device.

Figure 7 shows that the highest friction values FAP were achieved in case of samples made from sandstone, greywacke and biotic gneiss aggregates, while the lowest were those made from amphibolite, basalt and limestone.

There is a good correlation when comparing the results of accelerated polishing performed by the British method PSV in accordance with EN 1097-8 and the Wehner/Schulze test. The W/S test has a higher resolution. The results given in Figure 8 shows that 10 units of PSV value correspond to 25 units of μ_{FAP}, which means that the aggregates can be more effectively distinguished from one another with respect to their polishing. For reference purposes, the graphs also include correlations from other laboratories (Arampamoorthy H, 2011; Huschek, S., 2007; Dames, J., 1990; Do, M.-T, 2008).

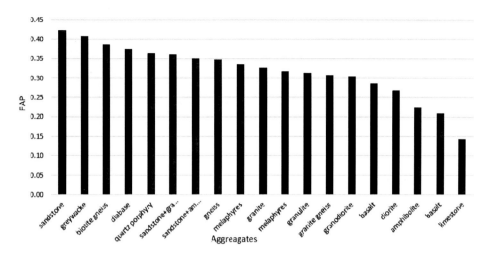

Figure 7. FAP values of aggregate specimens.

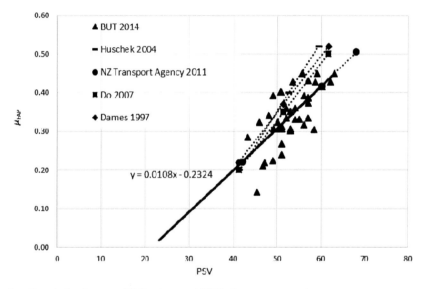

Figure 8. Correlation between FAP values and PSV of aggregate specimens.

5 CONCLUSION

The measured μ_{FAP} values were compared with those from the dynamic test device TRT. Correlation between results was good for asphalt specimens and also for concrete specimens.

Test device described in this paper allows determination of skid resistance of the test specimens prepared in a laboratory or taken from new roads or roads already in operation. It is possible to test not just asphalt, but also concrete surfaces. The advantage of this method is its ability to model long-term effects of traffic load in a short time and therefore allow testing road skid resistance of various aggregate types, fractions, surface textures and bitumen. It is possible to determine the lifetime of the wearing courses and predict the relationship between the values of road surface skid resistance and the intensity of traffic load.

According results presented in this paper the method based on EN 12697-49 allows the determination of aggregate polishing at a higher resolution in comparison with the current method based on EN 1097-8.

ACKNOWLEDGEMENT

This research was supported financially by the Technology Agency of the Czech Republic, project No. TA02030479 "Introduction of laboratory test method according to prEN 12697-49 to determinate the skid resistance and its development depending on traffic loading to decrease traffic accidents and to prolong life-time of pavement wearing courses".

REFERENCES

Arampamoorthy H., Patrick J. (2011) *Potential of the Wehner-Schulze test to predict the on-road friction performance of aggregate*, NZ Transport Agency research report 443, 34 p.

CEN/TS 15901-4 *Road and airfield surface characteristics—Part 4: Procedure for determining the skid resistance of pavements using a device with longitudinal controlled slip (LFCT): Tatra Runway Tester (TRT)*. November 2010.

Dames, J. (1990) *The influence of polishing resistance of sand on skid resistance of asphalt concrete*, Surface Characteristics of Roadways: International research and technologies, ASTM STP 1031, pp. 14–29.

Do, M.-T. et col. (2008) *Laboratory test method for the prediction of the evolution of road skid-resistance with traffic*. SURF 2008. 6th symposium on pavement surface characteristics.

EN 1097-8 *Tests for mechanical and physical properties of aggregates—Part 8: Determination of the polished stone value*. August 2009.

EN 12697-27 *Bituminous mixtures—Test methods for hot mix asphalt—Part 27: Sampling*. February 2001.

EN 12697-33 *Bituminous mixtures—Test method for hot mix asphalt—Part 33: Specimen prepared by roller compactor*.

EN 12697-49 *Bituminous mixtures—Test methods for hot mix asphalt—Part 49: Determination of friction after polishing*. January 2014.

Huschek, S. (2007) *Polishing resistance of mineral aggregates and pavement skid resistance*. Paper presented at the Workshop Measuring and evaluating of road surface skid resistance, Brno University of Technology, ISBN 978-80-214-3429-5, 16th May 2007.

TA02030479 *Introduction of laboratory test method according to prEN 12697-49 to determinate the skid resistance and its development depending on traffic loading to decrease traffic accidents and to prolong life-time of pavement wearing courses*. Provider: Technology Agency of the Czech Republic, Main contractor: Brno University of Technology/Faculty of Civil Engineering, Research years: 2012–2016.

Resources, Environment and Engineering II – Xie (Ed.)
© 2016 Taylor & Francis Group, London, ISBN 978-1-138-02894-4

Site selection optimization of the cavern in the mountain of jointed rock masses with different slope degrees

Chao Li, Weishen Zhu, Bangxiang Li & Dunfu Zhang
Geotechnical and Structural Engineering Research Center, Shandong University, Jinan, P.R. China

ABSTRACT: A study on distribution of characters of initial stress in the condition of gravity effect in the slopes within the alp of which the slope angles poses 30°, 45°, and 60°, respectively, was conducted by numerical analysis. After comparison, it is found that when closing to the region of slope surface, the difference of vertical stress between the values, which are calculated by the direct buried depth and the actual value, is huge. And the difference becomes huger with the slope becoming steeper. Then, a comparison has been made with excavating a cavern which poses 3 different positions separately at different distances from the slope toe in the condition of 45° slope angle and using the equivalent mechanical parameters of jointed rock mass to make numerical analysis to look the rock stability difference of the three schemes. It is found that the closer to slope surface (slope toe), the larger the plastic zone or the damage zone around the caverns become. Stress distribution is also studied and computed in the double slopes for both sides of the valley, and the results with field measured data are compared. It shows it is extremely consistent with each other.

1 INTRODUCTION

China is a mountainous country. Many underground projects of civil engineering and hydropower engineering are often excavated in the alp valley area. Except at some major projects measuring the in-situ stress, most of them should evaluate the rock stability prior to construction, without field stress measurement. At this time, the common method is evaluating or calculating the initial stress of the project area by calculating the direct buried depth (γh) over the cavern generally. Therefore, this article intends to study the distribution of initial stress in the mountain which poses different slop degree in the condition of gravity and with different lateral pressure coefficients of in-situ stress to explain that in some conditions the above common method to calculate σ_y will produce a large error (Zhu,1995).

2 THE DISTRIBUTION OF STRESS FIELD IN DIFFERENT SLOPE DEGREES OF AN ALP

Assuming the model is in two-dimensional and under the plane strain condition, the homogeneous elastic model to analyze the constraining five boundaries of the model contracted in one direction is chosen. Three different slope degrees (30°, 45°, and 60°) are chosen separately to analyze, the parameters as shown in Table 1, in order to observe the value and the distribution characters of the vertical and the maximum stresses which are σ_{yy} and σ_{max}. In the mountain, two parameters are assumed (Yu,1995).

$$N_1 = \frac{\sigma_{max}}{\gamma h} \quad (1)$$

$$N_2 = \frac{\sigma_{yy}}{\gamma h} \quad (2)$$

Table 1. Mechanical parameters of the model.

Density/ [g/cm^3]	Youngs modulus/ [GPa]	Poisson ratio
2610	18	0.21

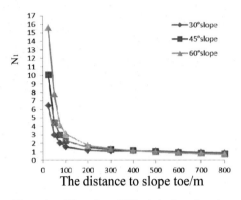

Figure 1. The value of N1 on the 0 m elevation in alps with different slope degrees.

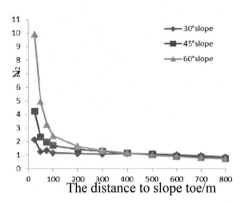

Figure 2. The value of N2 on the 0 m in alps with different slope degrees.

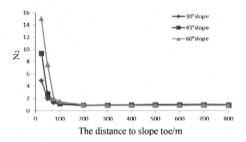

Figure 3. The value of N2 on the 0 m elevation in alps with different slope degrees.

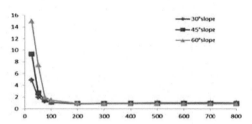

Figure 4. The value of N2 on the 0 m elevation in alps with different slope degrees.

In which: σ_{yy} and σ_{max} are the calculated vertical stress and the calculated maximum stress separately; γh is the value calculated by the direct vertical depth and the weight of rock.

2.1 *The stress field in considering the gravity effect only*

When considering the gravity of mountain only, it is calculated. While the slope poses 30°, 45°, and 60° it is now chosen as the stress at the horizontal elevation of 0 m in the mountain to analyze.

As shown in the Figure 1, 2, in the elevation of 0 m, within the distance of 100 m from the slope toe, the value of N1, N2 change greatly. Near the slope toe, the maximum value of N1 and N2 can be 16 and 10; In the range of 100 m–200 m distance belongs to transition region; Beyond the range of 200 m distance, the value of N1, N2 is always around 1.

2.2 *The case of considering the additional horizontal stress based on the gravity effect*

When adding the additional horizontal tectonic stress by applying the body forces method in the foundation considering gravity effect as an analysis case is computed. Assuming the lateral pressure coefficient k0 is 1.5, choosing the stress distribution on the elevation of 0 m in the mountain to analyze.

Here only the distribution figure of N2 is presented, the distribution regulation of N1 is similar to N2. As shown in the Figure 6, in the elevation of 0 m, with the effect of gravity and lateral stress coefficient of k0 = 1.5, within the distance of 100 m from the slope surface, the value of N1, N2 change greatly; the maximum value of N1, N2 can be 40 and 10; In the range of 100 m–200 m distance belongs to transition region; beyond the range of 200 m distance, the value of N1, N2 is always around 1.

It can be seen from the above, the nearer to the slope toe or the slope surface, due to the stress concentration or the slope surface effect, the larger N becomes. And as shown in the figure, this phenomenon in the slope of 60° is the most obvious. The slope of 45° takes the second place, however, in the slope of 30, the change of N is tempered obviously. It is indicated furthermore that the larger the lateral pressure coefficient is, the more obvious the increase of N1, N2 is due to the slope surface effect.

3 MOUNTAIN GROUND STRESS IN THE SYMMETRICAL DOUBLE SLOPES

Er-tan hydropower station is located in the mountain valley of the Yalong river downstream river. Rock is hard and complete with smaller faults. Late dam site area is mainly for the basalt rock and intrusive in grain of syenite. On both sides of the river slope up to 400–500 m, average slope of 30, and 40. Dam is 245 m high, addressing power plant installed capacity of 300 MW. Dam site area by a large number of field measurement showed that the occurrence of up to 25 to 30 Mpa in-situ stress, the numerical simulation method, deep engineering analysis, and study to lay the foundation for the future.

With FLAC 3D before its handling capacity limitations, adopt pretreatment better modeling of Abaqus, first set up the mountain in Auto CAD 2d floor plan, then put the model into Abaqus, the drawing of the 3d body unit, and then use Abaqus dividing grid meshing, the characteristics of superior finally will be built into FLAC 3D numerical calculation model.

In order to better simulating the field stress status of Er-tan hydropower station, the lateral pressure coefficient is assumed as 2.5 times that of gravity stress for calculation. The density of rock is 2610 kg/m^3, the Young's Modulus is 40 Gpa., Considering the fact, use four gradually varied zones of weathering.

It can be seen from the figure above, the results of numerical simulation and measured results are very closed with each other; the lateral pressure of 2.5 at the time of the simulated results and measured results are most of press close to; this also fully shows that the mountain of stress is not only the weight of the overlying rock mass, but the geological structure stress is also involved.

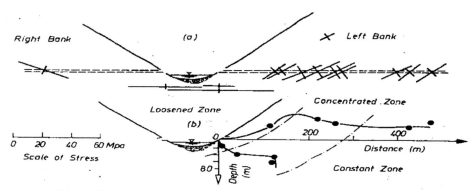

a) Principal stresses in syenite in cross section of the river. b) Three stress zones characterized by measured stresses

Figure 5. Distribution of in-situ stress measured in site of Er-tan project.

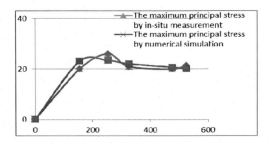

Figure 6. The contrast between measured data and numerical simulation (用该图).

4 THE SITE SELECTION OPTIMIZATION OF THE CAVERN IN THE MOUNTAIN OF JOINTED ROCK MASSES

The jointed rock masses are a very familiar kind of complex rock masses. This section will reference the distribution model of the jointed rock masses to generate REV by stochastic method, then proceeding by the loading test by numerical method to obtain the equivalent mechanical parameters.

4.1 *The parameters of the stochastic distributed jointed rock mass*

By site investigation and statistical analysis in a project, the distribution regularities of the jointed fracture characteristic parameters are obtained. Then programing composition to generate a series of stochastically distributed fracture grid specimens in different dimensions are generated (Guo, 2011). The mechanical parameters of unit body are assigned stochastically according to the gauss normal distribution in the same specimen. The generated specimen is shown as the follow figure. In the simulation, variation coefficient v is defined as the ratio of expected value μ and variance S to describe the discrete degree of normal distribution, that is:

$$v = \frac{S}{\mu} \qquad (3)$$

Then analyzing the generated specimen by numerical method and simulating the compression test in the condition of different confining pressures to obtain the mohr envelope in quadratic form of specimen. The equivalent mechanical parameters in the condition with different confining pressures can be obtained by derivation of the mohr envelope in form:

$$\tau = C + k\sigma$$
$$k = \tan\varphi = \frac{d\tau}{d\sigma} = \frac{a}{2\sqrt{\tau_0 + a\sigma}}, C = \sqrt{\tau_0 + a\sigma} - \frac{a\sigma}{2\sqrt{\tau_0 + a\sigma}} \qquad (4)$$

In which φ is the internal friction angle of rock mass when $\sigma = \sigma_i$, C is the cohesion of rock mass, τ_0 is the shear strength of rock mass, "a" is the fitting parameters.

Programing composition in FISH language which can assign automatically based on where the stress state of the position unit is.

4.2 *The site selection optimization of caverns*

Now the study of relationship between rock masses stability and different cavern position will be conducted. Assuming the cavern is 50 m in height and 20 m in width and there are three different positions of the cavern which is excavated in four steps (as shown in the Fig. 8).

The site scheme I, II, III are 100 m, 150 m, and 200 m far from the slope toe separately.

It can be known from the result, which has been calculated based on the new equivalent mechanical parameters from above, that the regularity of the stress distribution in

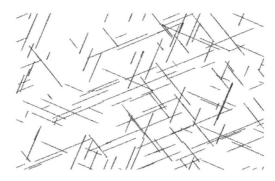

Figure 7. Stochastic fracture networks.

Table 2. The mechanical parameters of jointed rock masses.

Rock classification	Density/ [g/cm³]	Deformation modulus/[GPa]	Poisson ratio	Shear strength φ	Shear strength c/[MPa]	Tensile strength/ [MPa]
II	2610	3.17	0.2	Assigned automatically		0.5

Figure 8. The distribution of plastic zone in scheme I, II, III.

different slope degree mountains as shown in section 1 is basically same as which in intact rock. The only difference is the detail value. In this section an analysis of cavern stability in slope of 45° using the equivalent mechanical parameters of jointed rock masses are conducted (Wang, 2010). The units of model are assigned according to the strength parameter of jointed rock masses (2). The cohesion and the internal friction angle can be obtained:

$$C = \sqrt{2.4265\sigma + 0.6063} - \frac{2.4265\sigma}{2\sqrt{2.4265\sigma + 0.6063}} \quad (5)$$

$$\tan\phi = \frac{2.4265}{2\sqrt{2.4265\sigma + 0.6063}} \quad (6)$$

The other mechanical parameters has been shown in Table 2.

When the effect of mountain gravity is considered only, the plastic zones around caverns in different cavern positions are shown in Figure 8.

It can be seen in figures follow, the plastic zones in schemes I, II, III are 5179 m³, 1559 m³, and 1450 m³, respectively. And it can be known that the closer to slope toe of the cavern, the larger the plastic zone becomes and the more disadvantageous to the stability of jointed rock masses. This result indicates that the normal calculation method, which regards the weight of direct buried depth above cavern as the initial vertical stress, is incorrect within a certain distance to slope surface (slope toe).

When the lateral stress coefficient is of k0 = 1.5, the plastic zone in jointed rock masses becomes much larger. As the same, the plastic zone in scheme I which is the nearest to slope toe is the largest.

5 CONCLUSION AND DISCUSSION

1. On the same elevation in mountain, from the slope surface to the inner mountain, when the slope is steeper, the stress within a distance of 100 m–200 m from slope surface is much larger than expected. Then the maximum stress experiences a process that the value of the max stress decreases in the beginning and then rises. And with the horizontal buried depth increasing, the stress approaches the value of σy = γh gradually. And this regularity becomes more obvious with the slope becoming steeper. So when confirming the value of σy, the slope degree and the mountain height should be taken into consideration, and it is incorrect which calculates the vertical stress according to the vertical buried depth directly.
2. Within the slope surface, when the cavern is excavated on the same horizontal elevation but different distance from the slope surface, the rock masses stability will be very different. In the same working condition, within the distance of 100 m–200 m, the closer the cavern is to slope toe, the larger the plastic zone or the damage zone becomes and the more worse to the stability of jointed rock masses.
3. On the hillside slope within the same elevation, but from different horizontal distance or slope toe, stability of surrounding rock will be very different. Under the same working conditions, in the range of 100 m–200 m from the slope, excavating cavern will cause massive plastic zone and fractured zone.
4. Valley areas are mostly double sloped. To illustrate the importance of the lateral pressure, double slope with high later stress is used to conduct numerical simulation and take the computed results compared with measured stress in site. It is shown that the numerical simulation results and measured results are very similar. This method provides the guiding significance for the future's research in slope.

ACKNOWLEDGMENTS

The work is supported by the National Natural Science Foundation of China 51279095. W.S. Zhu is the corresponding author.

REFERENCES

Guo, Yunhua, Zhu, WS, Yu, DJ, Li, XP. Study of a high slope stability considering the stochastic distribution of rock joint set [C]//International Conference on Civil Engineering and Transportation (ICCET 2011). Switzerland: trans Tech Publications Ltd, 2011.

Wang Li-ge, Zhu Wei-shen, Zhou Kui, et al. Influences on in-situ stress distribution and surrounding rock mass stability of underground cavern groups under different slope inclinations[C]//The 5th International Symposium on In-situ Rock Stress. Beijing: Rock Stress and Earthquakes, 2010: 607–610.

Yu Dajun, Zhu Weishen. Initial geo-stress field characters in mountain with different slopes and its influence on underground projects [J]. Rock and Soil Mechanics, 2011, 32(Suppl.): 609–613.

Zhu Weishen, He Manchao The stability of surrounding rock and the dynamic construction mechanics for rackmass in complicated conditions [M]. Beijing: Science Press Ltd. 1995 (in Chinese).

Compute development length for BFRP bars based on the bond-slip constitutive model

Ting Liu, Bin Jia, Chuntao Zhang, Xiao Liu & Ning Hou
College of Civil Engineering and Architecture, Southwest University of Science and Technology, Mianyang, China

ABSTRACT: It is the basis that enough development length for Bond of Fiber Reinforced Polymer (BFRP) bars to concrete for guaranteeing the normal work of the structure. This paper analyzes the bond behavior of differential element of concrete with BFRP bars in pulled-out test, bases on the actual bond-slip constitutive law, accounts for the modified BPE model and establishes the theoretical formulas of the relative slipping difference between BFRP bars and concrete, the tensile stress of BFRP bars and the bond stress of BFRP bars. According to those theoretical formulas obtain the computational formula of the minimum development length of BFRP bars in the serviceability limit state; analyze and obtain the minimum development length of BFRP bars through using the datum in pulled-out test, compared with development length in the references show they are in good agreement.

1 INTRODUCTION

BFRP bar is a new type of composite material, which is composed of multiple strands of basalt fiber and resin matrix material by the pultrusion and necessary surface treatment. BFRP bar has many unique advantages such as good mechanical property, the corrosion resistance and the fatigue resistance, which can improve the durability of concrete structure instead of steel bar. Compared with the traditional steel bar, BFRP bar is a linear elastic material, the stress-strain curve has no yield stage, and the development properties of BFRP bar to concrete are different from the traditional steel bar as the different mechanical and geometrical properties. An adequate bond behavior between BFRP bar and concrete is the basis of collaborative work, and is the key to instead of traditional steel bar used in reinforced concrete structure. Development length of BFRP bar to concrete will directly depend on the bond behavior between the two. Therefore, the bond property and the evaluation of the development length for BFRP bar based on the actual bond-slip constitutive law have important engineering significance. Usually, the basic development length of BFRP bar is defined as the shortest length when they have reached the ultimate tensile strength, but do not occur in development failure.

Relevant scholars have been carried out a series of researches about the development length of BFRP bar in concrete. The current research methods on development length are dominated by the half-theory and the half-experiment, and establish the theoretical formulas of the developmental length of FRP bar based on the bond performance between FRP bar and concrete. In summary, there are two methods: (1) based on the suitable bond-slip relationship between FRP bar and concrete and the experimental method. S.S. Faza et al. Based on the theory formula of steel bar and used the pulled-out experiment and the cantilever beam experiment to obtain the theory formula of the development length of FRP bar. M.R. Eh-saniet carried out 48 GFRP concrete block beam experiments and 12 GFRP concrete block center pulled-out experiments, and fitted the formula of the development length of GFRP bar. (2) Establish the bond-slip curve model taking into account the existing formulas of development length. E. Cosenza et al. proposed to evaluate the development length of FRP bar based on the modified BPE Model. From the existing research situations we can know the research objects rarely involve the BFRP bar, and the limitation on the development length of BFRP bar is not mentioned by the relevant specification. So establish the theoretical formula based on the actual bond-slip constitutive law is no doubt that BFRP bar are popularized and applied in the engineering and complete specifications.

2 THE MODIFIED BPE BOND-SLIP CONSTITUTIVE MODEL

Cosenza et al. modified the original BPE model to get the revised BPE model, and obtained the typical bond-slip ($\tau-s$) constitutive model between FRP bar and concrete—shown in Figure 1—is considered here in. Usually, the calculation for most of the structures only takes into account for the using stage, which is to simply take $\tau-s$ model rise to:

$$\tau/\tau_1 = (s/s_1)^\alpha \qquad s \leq s_1 \tag{1}$$

where τ_1 is the maximum bond strength, s_1, the slip at peak bond strength, s, the slip and τ is the bond stress.

3 THE DIFFERENTIAL EVALUATION OF BOND-SLIP ($\tau-s$)

Mechanical model of the isolated differential element on the block of FRP bar to concrete, shown in Figure 2.

Analyze the deformation of BFRP bar and concrete, the slip-strain constitutive relationship is governed by the following equilibrium equation:

The difference between BFRP bar slip and concrete slip, is given by:

$$s(x) = s_f(x) - s_c(x) \tag{2}$$

where x is the slip of the element, $s_f(x)$ and $s_c(x)$ are the slip of BFRP bar and concrete, respectively.

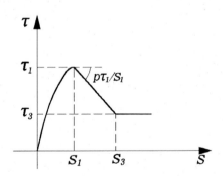

Figure 1. The modified BPE model.

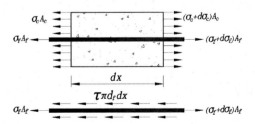

Figure 2. The stress on differential element.

From Eq. (2), the following differential equation is obtained:

$$\frac{ds}{dx} = \frac{ds_f}{dx} - \frac{ds_c}{dx} = \varepsilon_f(x) - \varepsilon_c(x) \qquad (3)$$

where $\varepsilon_f(x)$ is the strain of BFRP bar and $\varepsilon_c(x)$ is the strain of concrete.

By using such laws $\sigma_f = E_f \varepsilon_f$ and $\sigma_c = E_c \varepsilon_c$ to integrate Eq. (3) with the following differential equation:

$$\frac{d^2s}{dx^2} = \frac{d\sigma_f}{E_f dx} - \frac{d\sigma_c}{E_c dx} \qquad (4)$$

where E_f and E_c are the elastic modulus BFRP bar and concrete, respectively, σ_f and σ_c are the tensile stress of BFRP bar and concrete, respectively.

Analyze the stress of the differential element and establish the equilibrium differential equation on the stress of BFRP bar and concrete, is given by:

$$A_c d\sigma_c + A_f d\sigma_f = 0 \qquad (5)$$

And the equilibrium differential equation on the stress and bond of BFRP bar, is given by:

$$\frac{\pi d_f^2}{4} d\sigma_f = \pi d_f \tau dx \qquad (6)$$

where A_f is the area of BFRP bar and A_c is the area of concrete, τ, the bond stress and d_f is the diameter of BFRP bar.

By using Eqs. (5) and (6) to integrate Eq. (4) with the following bond-slip constitutive differential equation is obtained:

$$\frac{d^2s}{dx^2} - \Psi \tau(x) = 0 \qquad (7)$$

where Ψ is a constant and setting $\Psi = \dfrac{4}{E_f d_f}\left(1 + \dfrac{E_f A_f}{E_c A_c}\right)$.

4 THE SOLUTION OF BOND-SLIP DIFFERENTIAL EQUATION BASED ON THE MODIFIED BPE MODEL

Studying on the development length of BFRP reinforced concrete structure in the using stage, only takes the rising step of $\tau - s$ model into account. Therefore, from Eqs. (1) and (7), the following differential equation is obtained:

$$\frac{d^2s}{dx^2} - \frac{\Psi \tau_1}{s_1^\alpha} s^\alpha = 0 \qquad (8)$$

By integrating Eq. (8) with the boundary conditions of $s(0) = 0$, $\left(\dfrac{ds}{dx}\right)_{x=0} = 0$, thus leading to the following expression of $s(x)$:

$$s(x) = \left[\frac{\Psi \tau_1}{2 s_1^\alpha} \cdot \frac{(1-\alpha)^2}{(1+\alpha)}\right]^{\frac{1}{1-\alpha}} x^{\frac{2}{1-\alpha}} \qquad (9)$$

By substituting Eq. (9) into Eq. (7), the distribution law of bond stress along the bar is given by:

$$\tau(x) = \frac{1}{\Psi} \left[\frac{\Psi \tau_1}{2s_1^\alpha} \cdot \frac{(1-\alpha)^2}{(1+\alpha)} \right]^{\frac{1}{1-\alpha}} \frac{2(1+\alpha)}{(1-\alpha)^2} x^{\frac{2\alpha}{1-\alpha}} \qquad (10)$$

By substituting Eq. (10) into Eq. (6), and integrating with the boundary condition of $\sigma(x)\big|_{x=0} = 0$, the distribution law of the tensile stress along the bar is given by:

$$\sigma(x) = \frac{1}{\Psi} \frac{8}{d_f(1-\alpha)} \left[\frac{\Psi \tau_1}{2s_1^\alpha} \cdot \frac{(1-\alpha)^2}{(1+\alpha)} \right]^{\frac{1}{1-\alpha}} x^{\frac{1+\alpha}{1-\alpha}} \qquad (11)$$

To sum up, establish the theoretical formulas of the slip along the BFRP bar, the bond stress of between BFRP bar and concrete and the tensile stress of the bar based on the bond-slip constitutive law.

5 THE DEVELOPMENT LENGTH OF BFRP BAR BASED ON THE BOND-SLIP CONSTITUTIVE LAW

The limited development length of BFRP bar represents an upper bound of the development length related to the ascending branch of the bond-slip constitutive law curve, when a value of the slip $s(x)$ equal to s_1, thus the corresponding value of x represents the ultimate development length l_m. According to Eq. (9), the expression of l_m can be easily given by:

$$l_m = \sqrt{\frac{2s_1}{\Psi \tau_1} \frac{(1+\alpha)}{(1-\alpha)^2}} \qquad (12)$$

When a value of the slip $s(x)$ equal to s_1, solving Eq. (11), the ultimate tensile stress of the bar can be provided by:

$$\sigma_m = \sqrt{\frac{32 s_1 \tau_1}{\Psi d_f^2 (1+\alpha)}} \qquad (13)$$

According to Eq. (13), the coefficient of Ψ can be given by:

$$\Psi = \frac{32 s_1 \tau_1}{\sigma_m^2 d_f^2 (1+\alpha)} \qquad (14)$$

Finally, by substituting Eq. (14) into Eq. (12) and using σ_m to represent l_m, the following equation is obtained:

$$l_m = \frac{\sigma_m d_f}{4\tau_1} \frac{1+\alpha}{1-\alpha} \qquad (15)$$

Considering different modification factors based on the basic development length, therefore, the final value of the minimum development length of BFRP bars in concrete can be obtained by:

$$L = \gamma_g \gamma_t \gamma_c l_m \qquad (16)$$

where γ_g is safety factor, the value can equal to 2.5;

γ_t, bar location modification factor, the value can equal to 1.0;

γ_c, concrete cover modification factor, if $c = d_f$, then the value can equal to 1.5; if $c > 2d_f$, then the value can equal to 1.0; if $d_f < c < 2d_f$, then the value can use the linear interpolation method.

6 A NUMERICAL EXAMPLE ANALYSIS

This paper uses bond performance experimental datum of BFRP bar and concrete in the reference to verify analysis. The experimental parameters present below, having the elastic modulus of BFRP bar equal to 40,000 MPa, the ultimate tensile strength of BFRP bar equal to 700 MPa, and the value of the ultimate compressive strength of concrete equal to 41.3 MPa. This paper gets the value of parameter α equal to 0.183 through fit the bond-slip curve, which obtains from the experimental datum of the reference with the ascending branch of modified BPE model curve. The calculation and comparison of development length for BFRP bar, as shown in Table 1 and Table 2, respectively.

From the Table 2 we can see that the values of development length on this paper are in good agreement with the calculated values of the references, and prove that the established theoretical formula of development length for BFRP bar on this paper has the referenced value.

Table 1. Computation of development length for BFRP bar.

Specimen no.	s_1/mm	τ_1/MPa	d_f/mm	σ_m/MPa	L/mm
B10-2.5d-C30	2.13	29.12	10	700	218
B13-5.0d-C30	1.27	20.83	13	700	396
B16-5.0d-C30	1.28	19.36	16	628	470

Table 2. Comparison of development length for BFRP bar.

Specimen no.	Reference	Formula	L/mm	The difference compares with the paper /%
B10-2.5d-C30	The paper	$L = \gamma_g \gamma_t \gamma_c l_m$	218	0
	Reference	$l_{db} = 0.024 f_{fu} A_b / \sqrt{f_c'}$	231	5.6
	Reference	$l_{db} = 0.028 f_{yf} A_b / \sqrt{f_c'}$	216	0.9
	Reference	$l_{db} = \gamma_g \gamma_t \gamma_c K_1 f_{yu} A_f / \sqrt{f_c'}$	218	0
	Reference	$l_a = f_{fd} d / 13 f_t$	269	18.9
B13-5.0d-C30	The paper	$L = \gamma_g \gamma_t \gamma_c l_m$	396	0
	Reference	$l_{db} = 0.024 f_{fu} A_b / \sqrt{f_c'}$	391	1.3
	Reference	$l_{db} = 0.028 f_{yf} A_b / \sqrt{f_c'}$	364	8.1
	Reference	$l_{db} = \gamma_g \gamma_t \gamma_c K_1 f_{yu} A_f / \sqrt{f_c'}$	366	7.6
	Reference	$l_a = f_{fd} d / 13 f_t$	350	11.6
B16-5.0d-C30	The paper	$L = \gamma_g \gamma_t \gamma_c l_m$	470	0
	Reference	$l_{db} = 0.024 f_{fu} A_b / \sqrt{f_c'}$	530	11.3
	Reference	$l_{db} = 0.028 f_{yf} A_b / \sqrt{f_c'}$	495	5.1
	Reference	$l_{db} = \gamma_g \gamma_t \gamma_c K_1 f_{yu} A_f / \sqrt{f_c'}$	497	5.4
	Reference	$l_a = f_{fd} d / 13 f_t$	386	17.9

7 CONCLUSION

This paper establishes the theoretical formula of development length for BFRP bar based on the actual bond-slip constitutive law and presents numerical example to verify and analyze, allows to draw the following conclusions:

1. This paper analyzes the bond behavior of differential element of concrete with BFRP bar in pulled-out test, bases on the actual bond-slip constitutive law, accounts for the modified BPE model and establishes the theoretical formulas of the relative slipping difference between BFRP bar and concrete, the tensile stress of BFRP bar and the bond stress of BFRP bar.
2. Based on the theoretical formulas of the slip and the tensile stress of BFRP bar, and considering different modification factors, this paper establishes the minimum development length of BFRP bar in the serviceability limit state. In addition, by the comparison and analysis of the experimental datum to prove the established theoretical formula of development length for BFRP bar is observed this paper with reliability.

ACKNOWLEDGMENTS

This work was financially supported by the Science & Technology Research Program of Mianyang Science & Technology Bureau (14G-03-8) and the International Science & Technology Cooperation Program of China (2014HH0062).

REFERENCES

Baena M., Torres L. and Turon A., et al. 2009. *Experimental study of bond behavior between concrete and FRP bars using a pull-out test* (Composites Part B:Engineering) 40(8): 784–797.
Conseza E., Manfredi G. and Realfonzo R. 1997. *Behaviour and modeling of bond of FRP rebars to concrete* (Composites for Construction) 1(2): 40–51.
Conseza E, Manfredi G and Realfonzo R. 2002. *Development length of FRP straight rebars* (Composites Part B: Engineering) 33(7): 493–504.
Ehsaniet M.R., Saadatmanesh H. and Tao S. 1996. *Design recommendation for bond of GFRP rebars to concrete* (Journal of Structural Engineering) 122(3): 247–257.
Liang L.L. 2011. *FRP reinforced concrete beam experimental research and numerical simulation* (China: Xi'an University of Architecture and Technology).
Ministry of Construction of the People's Republic of China. 2010. *Fiber reinforced composite construction engineering technical specifications* (GB50608-2010).
Ou Y., Zhen Y. and Li G.Q. 2009. *Cohesive zone model based analytical solutions for adhensively bonded pipe joints under torsional loading* (International Journal of Solids and Structures) 46(5): 1205–1217.
Wu F. 2009. *The experimental research on bond behavior between BFRP rebars and the concrete* (China: Dalian University of Technology).
Xue W.C. 1997. *The experimental research on the new type of FRP reinforcement in concrete structures* (China: Hehai University Postdoctoral Research Report).
Xue W.C. and Kang Q.L. 1999. *The application of FRP bars in concrete structures* (China: Industrial Architecture) 29(2): 19–21.
Xu F., Wu Z.M. and Zhang J.J, et al. 2010. *Experimental study on the bond behavior of reinforcing bars embedded in concrete subjected to lateral pressure* (China: Journal of Materials in Civil Engineering) 24(1): 125–133.
Zhang H.X. and Zhu F.S. 2000. *Bonding mechanism and calculating method for embedded length of fiber reinforced polymer rebars in concrete* (Journal of Hydraulic Engineering).
Zhang H.X. and Zhu F.S. 2006. *Study on the anchorage length of FRP bars with bond-slip constitutive relationship* (Journal of Shenyang Architectural University: Natural Science Edition).
Zhou Z.P., Li F. and Zhao Q.L. 2011. *Research progress of FRP tendon bonding anchors* (China: Industrial Architecture): 663–666.
Zhu H. and Qian Y. 2006. *The mechanical properties of FRP bars in engineering structures* (China: Journal of Construction Science and Engineering) 23(2): 26–31.

Study on spatial prediction of landslide hazard risk based on GIS

H.X. Gao
Zhejiang Ocean University, Zhoushan, Zhejiang, China

K.L. Yin
China University of Geoscience (Wuhan), Wuhan, Hubei, China

ABSTRACT: The risk evaluation of landslide hazard is a system that contains the assessment of landslide risk and vulnerability assessment of hazard bearing body. The authors have made spatial prediction and division of the regional landslide risk. The division and evaluation of vulnerability of regional hazard bearing body have also been accomplished. According to the basic requirements for risk assessment, the study area is divided into five zones—the extra high risk zone, the high risk zone, the medium risk zone, the low risk zone, and the extra low risk zone. At last, acceptable risk standards and management measures that are suitable for the study area are proposed based on economic development in the study area and disaster prevention efforts of government.

1 INTRODUCTION

Huge economic losses and casualties are caused by the geologic hazards throughout the world every year. With the further growth of social wealth, the absolute risk of natural disaster will generally increase. It shall be increased by 30%–50% in general areas and 80%–100% in serious areas. That is, economic losses which the disasters give rise to may be doubled. The losses caused by urban landslide disasters will be even more significant. Badong County is a typical example which is seriously threatened by the landslide hazard during the migration project and socioeconomic development.

In the presence of serious landslide hazards, traditional methods and means to study and control single landslide have appeared to be inadequate and helpless. Therefore, another effective way of comprehensive prevention and treatment of landslide hazard in Badong County is to conduct risk zoning of landslide hazard and identify dangerous zone or high risk zone of landslide hazard to strengthen monitoring and forecasting the landslide hazard area with serious damage.

In addition, existing research results are obtained to superimpose risk division map of landslide over vulnerability division map of hazard bearing body to conduct risk zoning. Highly dangerous units or highly vulnerable units are viewed as high risk area, namely, high danger or high vulnerability means high risk. It results in varying degrees of exaggeration of prediction unit. The main reason of this is neglecting that landslide risk is an entity of its natural attributes and social attributes. We know that when a landslide occurs in an uninhabited valley, generally it can hardly cause disaster even if it is very dangerous; on the contrary, landslide hazard risk is greater for a city full of population and close-packed buildings. Therefore, the standard for high or low risk cannot be identified by using traditional risk division map.

2 THEORY OF RISK ZONING SYSTEM

The risk evaluation system of landslide hazard includes four parts: the assessment for landslide hazard, the vulnerability assessment, the damage assessment, and the landslide

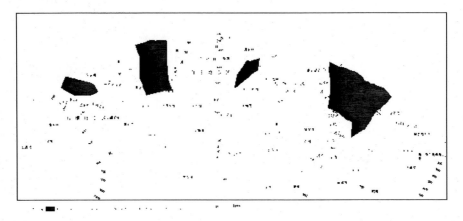

Figure 1. Spatial prediction division map of landslide hazard in Xincheng District of Badong County (reservoir water level elevation of 83 m).

disaster management. Among these, the assessment for landslide hazard, which is the input of system, is primarily the disaster formative environment and disastrous factor, namely natural attributes evaluation of landslide disaster. The vulnerability analysis of landslide hazard, which is the conversion of system, is mainly the hazard bearing body, that is, the social attributes of landslide hazard. The risk assessment of landslide hazard, which is the output of system, is the combination of its natural attributes and social attributes, namely expected loss of disaster. The disaster risk management based on established risk standards is the application of system and ultimate goal of risk assessment, as shown in Figure 1. Therefore, the risk of landslide hazard can be expressed as the product of risk, vulnerability, and value of hazard bearing body, as shown in equation:

$$R = \sum\sum\sum E_i \times H_j \times V_K \qquad (1)$$

R—risk of landslide hazard, E—value of hazard bearing body, H—risk of landside mass, V—vulnerability of hazard bearing body.

In this paper, information content model and calculation method are established on the basis of existing functions of GIS. Secondary development to implement "an integrated management information system includes information management module of landslide, risk evaluation module of landslide, evaluation module of vulnerability and evaluation module of risk prediction" is achieved on GIS platform to output spatial prediction zoning of landslide hazard, economic risk zoning, and other prediction maps.

3 STUDY ON SPATIAL PREDICTION OF LANDSLIDE HAZARD

3.1 Risk zoning method and mathematical model

Landslide hazard (Y) is affected by a variety of factors xi with distinct degrees and properties. In different geological environments, there will always be a "best combination of factors" for landslide disasters. Therefore, "best combination of factors" rather than single factor should be studied comprehensively for the prediction of regional landslide hazards. Based on the standpoint of information prediction, whether landslide hazard occurs or not is connected with the quantity and quality of the information which is obtained in the process of forecast as well as measured by information content

$$I(y, x_1 x_2 \ldots x_n) = \log_2 \frac{P(y \mid x_1 x_2 \ldots x_n)}{P(y)} \qquad (2)$$

It can be written as:

$$I(y, x_1 x_2 \ldots x_n) = I(y, x_1) + I_{x_1}(y, x_2) + \ldots + I_{x_1 x_2 \ldots x_{n-1}}(y, x_n) \qquad (3)$$

where $I(y, x_1 x_2 \ldots x_n)$ is the information content (bit) of landslide provided by specific combination of factors $x_1 x_2 \ldots x_n$, $P(y|x_1 x_2 \ldots x_n)$ is the probability of landslide under the combination condition of factors $x_1 x_2 \ldots x_n$, $P(y)$ is the probability of landslide and $I_{x_1}(y, x_2)$ is the information content (bit) of landslide provided by factor x_2 when factor x_1 exists.

Equation (3) shows that the information content of landslide provided by the combination of factors $x_1 x_2 \ldots x_n$ is equal to the summation of the information content provided by factor x_1 and the information content of landslide provided by factor x_2 with determined x_1, by parity of reasoning, and the information content of landslide provided by factor x_n when $x_1 x_2 \ldots x_{n-1}$ is determined. $P(y|x_1 x_2 \ldots x_n)$ and $P(y)$ can be represented by statistical probability. The information content provided by the combination of various factors to predict landslide can be positive or negative. And $I(y, x_1 x_2 \ldots x_n) > 0$ if $P(y|x_1 x_2 \ldots x_n) > P(y)$, conversely, $I(y, x_1 x_2 \ldots x_n) < 0$. $I(y, x_1 x_2 \ldots x_n) > 0$ indicates that the combination of factors $x_1 x_2 \ldots x_n$ is beneficial to predict landslide, and the opposite situation indicates that the combination of these factors is not conducive for landslide.

Spatial prediction of regional landslide hazards is based on the study of regional grid unit division. According to specific geological and topographical conditions of different areas, corresponding shapes and sizes of grid are adopted. Statistics and analysis of information is carried out with the distribution map of regional landslide hazards. It is assumed that a certain region is divided into N units. There are N_0 units where landslide has already occurred. In M units with the same combination of factors $x_1 x_2 \cdots x_n$, M_0 units have landslide disasters. On the principle that statistical probability represents prior probability, according to equation (3), the information content of landslide provided by factors $x_1 x_2 \ldots x_n$ in the area is:

$$I(y, x_1 x_2 \ldots x_n) = \log_2 \frac{M_0/M}{N_0/N} \qquad (4)$$

If area ratio is used to calculate the information content, then equation (4) can be expressed as:

$$I(y, x_1 x_2 \ldots x_n) = \log_2 \frac{S_0/S}{A_0/A} \qquad (5)$$

A: total area of the units within the region; A_0: area sum of the units where landslide disaster has occurred; S: total area of the units with the same combination of factors $x_1 x_2 \ldots x_n$; S_0: area sum of the units with the same combination of factors $x_1 x_2 \cdots x_n$ where landslide disaster has occurred.

Under normal circumstances, there are many factors that act on landslide disasters, so corresponding combinations of factors are also particularly abundant. Sample size is often limited, then simplified information content model with single factor is used to calculate step by step. Combined with comprehensive analysis, corresponding information content model is rewritten as:

$$I = \sum_{i=1}^{n} I_i \sum_{i=1}^{n} \log_2 \frac{S_0^i/S^i}{A_0/A} \qquad (6)$$

I: predictive value of information content of certain unit in prediction area; S^i: total area of factor x_i; S_0^i: area sum of the units of factor x_i where landslide disaster has occurred.

3.2 *Analysis and evaluation of landslide hazard zonation case of Xincheng District of Badong County*

1. Prediction process and model establishment
   Based on the principle to select evaluation indicators as well as the general features of internal control factors and external trigger factors of landslide hazard in Xincheng District of Badong County, evaluation indicator system is established using analytic hierarchy process to meet the requirements of regional landslide hazard zonation with the scale of 1:10000. Based on spatial analysis function of GIS, historical distribution map of landslide hazards is regarded as the base map and is superimposed over distribution maps of various factors with the identical scale, respectively. Then information content of single-factor is calculated and prediction model is established using information content analysis module. The information content of single-factor represents influence degree of the factor on landslide hazard or weight factor. In general, the larger is this indicator, the greater role of the factor on landslide hazard is it.
2. Grade classification of landslide hazard and prediction division map compilation
   Using information content prediction module, comprehensive indicator of multi-factor (total information content) is obtained by overlying indicator maps of various factors and is divided to determine corresponding risk grade of landslide hazard. Prediction unit is divided into five risk grades (extra low danger, low danger, medium danger, high danger, and extra high danger). The same grade is marked with the same color. Then spatial prediction and evaluation zoning map of risk is generated (Fig. 1).

From Figure 1, it can be seen that extra high dangerous zone is mainly distributed in Huangtupo, Baitupo, Yuntuo, and Xirangpo District. It centers on leading edge of ancient landslide in Huangtupo, Huaping landslide, Zhaoshuling landslide, and Xiasha River-bath brick factory. The area makes up approximately 9% of the study area. Highly dangerous zone is chiefly located in south-east of Huangtupo ancient landslide, Tongjialiangzi, Zhafangping, Tanjiaping, Hongshibao, leading edge of Yuntuo, and Tongjiaping as well as sparsely distributed in other places. Its area is about 4.39 km², accounting for 20% of the entire study area. The distribution is basically consistent with the areas where reservoir bank is reformed after impounding water and potential landslide hazard zones that have been found.

4 STUDY ON VULNERABILITY OF LANDSLIDE HAZARD BEARING BODY

1. Evaluation principle of vulnerability
   Landslide hazard has both natural attributes and social attributes, so the study of landslide hazard includes not only space, time, intensity, and other risks but also the objects threatened by landslide disaster (human life, property, environment, resources, and society, etc.). The vulnerability of landslide hazard is the degree of damage of the objects threatened by landslide disaster with certain intensity. Now it is usually expressed by loss rate with the interval of [0, 1]. 0 means no loss and 1 indicates total loss.
2. Evaluation of economic vulnerability
   To assess vulnerability of an evaluation unit or an area, we should not only know vulnerability subentry of various hazard bearing bodies but also calculate the comprehensive vulnerability of this unit or this area to carry out regional vulnerability zoning through comprehensive vulnerability. Calculation model of comprehensive vulnerability can be determined by equation (7).

$$V_e = \frac{1}{S} \sum_{j \to 1}^{m} \sum_{i \to 1}^{n} P_{ij} \cdot V_{ij} \cdot S_{ij} \qquad (7)$$

V_e—comprehensive vulnerability of a unit, S—area of a unit; V_{ij}—degree of damage of No. *i* hazard bearing body under No. *j* hazard, namely vulnerability subentry, S_{ij}—area

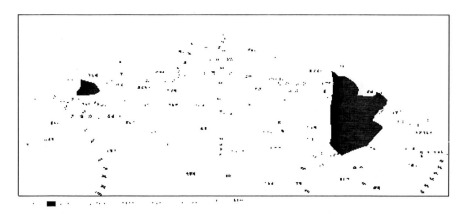

Figure 2. Division map of economic vulnerability of landslide hazard bearing body in Xincheng District of Badong County.

of No. i hazard bearing body which suffers from No. j hazard, P_{ij}—probability of damage of No. i hazard bearing body under No. j hazard.

Substituting relevant data into equation (5), the results are shown in Figure 2. The area with extra high vulnerability or high vulnerability of hazard bearing body in Xincheng District of Badong County is mostly located in Huangtupo District, east of Baitupo District, Yuntuo District, and Xirangpo District. Its area is about 2.92 km², accounting for 13.12% of the study area. This area is political, cultural, economic, and trade center with dense population, building and wealth. The type of hazard bearing body is complicated. The vulnerability of landslide hazard with same intensity is much greater than that of other areas.

5 RISK EVALUATION AND RISK STANDARD OF LANDSLIDE HAZARD

1. To determine the probability of landslide
   In the evaluation of landslide hazard, risk zoning is the relative risk of landslide hazard based on information content. To meet the needs of risk evaluation, absolute quantitative risk factors must also be determined. These absolute quantitative risk factors have different significance for monomer or regional landslide hazard. Failure probability function of landslide with different information content can be determined using information content of spatial risk of landslide hazard.

   1. Extra low danger $P = 0.007I + 0.1$ $I \leq -7$
   2. Low danger $P = 0.056I + 0.44$ $-7 < I \leq -2.5$
   3. Medium danger $P = 0.06I + 0.45$ $-2.5 < I \leq 2.5$
   4. High danger $P = 0.15I + 0.225$ $2.5 < I \leq 4.5$
   5. Extra high danger $P = 0.025I + 0.788$ $4.5 < I$

2. Evaluation of expected loss of regional economy
   Evaluation of expected loss of economy is to assess buildings, indoor property, roads, arable land, and land resources. The authors transform all property and resources threatened by landslide in the districts into economic indicators that can be added or subtracted directly. The economic values of slope units are also found through division by area.

Based on MAPGIS system and calculation model, economic risk prediction can be divided into five levels by superimposing failure probability distribution map of landslide hazard over economic vulnerability distribution map and density distribution map of economic value. The units with same risk level have same color. Distribution map of expected loss of

Table 1. Economic risk distribution table of landslide in Xincheng District of Badong County.

Grade classification of disaster	Extra low risk zone	Low risk zone	Medium risk zone	High risk zone	Extra high risk zone
Area of prediction unit	8.93	2.32	2.33	5.22	3.54
Area ratio (%)	39.97	10.38	10.43	23.37	15.85

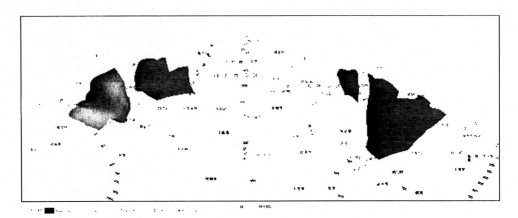

Figure 3. Economic risk distribution map of landslide in Xincheng District of Badong County.

economy of landslide hazard (Fig. 3) and statistical table (Table 1) in Xincheng District of Badong County are compiled. It can be seen in the Figure 3.

6 CONCLUSIONS

1. Using unit information content and grade classification standard for landslide probability, probability function of landslide with different range of information content is determined. So risk division map of regional landslide hazard is compiled to solve the problem that the probability of landslide disaster is too absolute or simplex;
2. By superimposing risk division map of landslide disaster over comprehensive vulnerability division map of hazard bearing body and density distribution map of economic value, combining with grade classification standard of economic risk, economic risk division map of landslide hazard in the study area is obtained. At the same time, by reference to risk standards of landslide hazard of the United States, Australia, and Japan, acceptable economic risk standards that are suitable for the study area are proposed.

REFERENCES

Li, L., Wang R.Z., Sheng, J.B., etc. 2006. *Risk Evaluation and Management of the Dam*. Beijing: Water Power Press.
Xie, Q.M. 2004. Study on the Risk Evaluation and Treatment Decision-making Methods of Landslide Hazard. Wuhan: PH. D. Dissertation of Wuhan University of Technology.
Yin, K.L. 2004. *Landslide Hazard Prediction and Evaluation*. WuHan: China University of Geoscience Press.
Zhang, Y.C. 1998. Comprehensive risk prediction and counter measures for disasters in China. *Journal of geological hazards and environment preservation*. 9(1):1–5.

The relation between tension, flange thickness and prying force for T-type bolt joints

Zhao-Xin Hou
College of Civil Engineering, Tongji University, Shanghai, China
Central Research Institute of Building and Construction, MCC Group, Co. Ltd., Beijing, China

Guo-Hong Huang, Chao Gong & Yun Zheng
Central Research Institute of Building and Construction, MCC Group, Co. Ltd., Beijing, China

ABSTRACT: According to the formula and parameter analysis of high strength bolts in T-type tension joint, this paper reveals the relationships between the tension loading and flange thickness or prying force. The formula for prying force and flange thickness before and after the inflection point is deduced and it is proposed that the relative strength of bolt and flange plate determines the size of the prying force. The formula for the inflection point for the critical tension loading caused by the changing prying force is also derived in the paper. Besides, the effects of bolt diameter, flange plate strength and bolt grade on prying force and thickness is also analyzed. This paper provides a reference for the design of high strength bolts in tension joints.

1 INTRODUCTION

In the current specification "Technical specification for high strength bolt connections of steel structures" JGJ82-2011, the prying force calculation model for T-type tension joint is shown in Figure 1. In the calculation model, flange plate and bolt reach the limits state at the same time, when prying force occurs in the edge of the flange plate, plastic deformation occurs in the flange plate and the plastic hinge occurs near the edge of the web (Yan Wang et al. 2008 & Dongyun Jia 2005). For high strength bolts, the ultimate limit state is bolts pulled out; serviceability limit state means flange plate is separated (Jie Zheng 2006 & Rui Bai et al. 2013).

In JGJ82-2011, the formula of flange thickness t_e and prying force Q can be deduced according to the computational model in Figure 1:

$$t_e = \sqrt{\frac{4e_2 N_t}{\Psi b f}} \qquad (1)$$

$$Q = N_t^b \left[\delta \alpha \rho \left(\frac{t}{t_e}\right)^2 \right] \qquad (2)$$

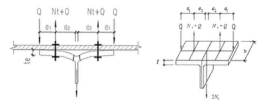

Figure 1. The force calculation model diagram for T-type tension joint.

Table 1. t_e and Q calculation list for T-type tension joint with 8.8 M20 bolt and Q235 flange plate.

N_t (kN)	0.1P	0.2P	0.3P	0.4P	0.5P	0.6P	0.7P
N_t (kN)	12.5	25	37.5	50	62.5	75	87.5
t_e (mm)	7.989	11.299	13.838	15.979	17.865	19.570	24.733
Q (kN)	4.000	8.000	12.000	16.000	20.000	24.000	12.500

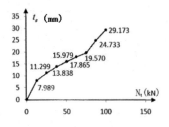
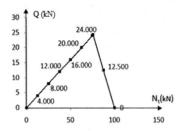

(a) Relationship of tension load and flange thickness (b) Relationship of tension load and prying force

Figure 2. The diagram for tension load and prying force or the thickness of M20 G8.8s Q235 bolt.

In the formula, t_e is the minimum thickness of the flange plate of T-type Joints, without considering the impact of the prying force; N_t is an axial tension of high strength bolt; N_t^b is tension bearing capacity for pulled high strength bolts; t is the thickness of the flange. plate of T-type tension joint; $\delta = 1 - (d_o/b)$ is the cross-section coefficient of flange plate; $\rho = e_2/e_1$ is a coefficient; Q is the prying force; $\alpha = \frac{1}{\delta}\left[\frac{N_t}{N_t^b}\left(\frac{t_e}{t}\right)^2 - 1\right] \geq 0$ is a coefficient greater than or equal to zero, $\beta = \frac{1}{\rho}\left(\frac{N_t^b}{N_t} - 1\right)$.

When $e_1 = 1.25e_2$, t_e and Q of T-type tension joint with M20 8.8 bolts and Q235 flange plate are calculated for N_t from 0.1P to 0.7P according to the above Eqs. (1) and (2) and given in Table 1 below.

According to Table 1 calculation results, the relationship diagrams of tension load, flange plate thickness and prying force are drawn in Figure 2. There are two sections of different curvature line segments in the relationship of tension load and flange plate thickness. The slope before inflection point is less than the slope after the inflection point. There are also two sections of different curvature line segments in the relationship of tension load and prying force. Before the inflection point is proportional relationship and after the inflection point is inverse relationship.

2 DERIVATION OF FLANGE PLATE THICKNESS AND PRYING FORCE CHANGES BEFORE AND AFTER THE INFLECTION POINT

As seen from Figure 2, with the increase of the tensile load, the corresponding thickness of the flange plate is larger accordingly. Thickness with the load change is in accordance with the different curvature before and after the inflection point. The corresponding prying force value is at a watershed before and after the inflection point, upward phase before the inflection point and plummeted stage after the inflection point. Therefore, these two sections of function line need further derivation.

Before the inflection point, $\alpha' = 1.0$, It is known that $\psi = 1 + \delta\alpha' = 1 + \delta$, $\delta = 1 - \frac{d_o}{b} = 1 - \frac{d_o}{3d_o} = \frac{2}{3}$, $t_e = \sqrt{\frac{4e_2 N_t}{\psi bf}}$, $Q = N_t^b \left[\delta\alpha\rho\left(\frac{t_e}{t_{ec}}\right)^2\right] = \rho \cdot N_t - \frac{bf}{4e_1} \cdot t_e^2$.

$$t = \sqrt{\frac{4e_2 N_t}{\psi bf}} = \sqrt{\frac{4e_2 N_t}{(1+\delta)bf}} \qquad (3)$$

$$Q = \rho \cdot N_t - \frac{bf}{4e_1} \cdot \frac{4e_2 N_t}{\psi bf} = \left(\rho - \frac{\rho}{\psi}\right) \cdot N_t = \rho \cdot \left(1 - \frac{1}{\psi}\right) \cdot N_t = \rho \cdot \frac{\delta}{1+\delta} \cdot N_t \qquad (4)$$

After the inflection point, $0 < \alpha' < 1.0$, It is known that: $\alpha' = \frac{1}{\delta}\left(\frac{\beta}{1-\beta}\right)$, $\psi = 1 + \delta\alpha' = 1 + \frac{\beta}{1-\beta} = \frac{1}{1-\beta}$.

$$t_e = \sqrt{\frac{4e_2 N_t}{\psi bf}} = \sqrt{\frac{4e_2 N_t}{\frac{1}{1-\beta}bf}} = \sqrt{\frac{4e_2 N_t(1-\beta)}{bf}} \qquad (5)$$

$$Q = \rho \cdot \left(1 - \frac{1}{\psi}\right) \cdot N_t = \rho \cdot [1 - (1-\beta)] \cdot N_t = \beta\rho N_t \qquad (6)$$

According to the above derivation of Eqs. (3), (4), (5) and (6), corresponding relationship of flange thickness and loading, prying force and loading can be expressed as follow.

3 DOUBLE EFFECTS OF FLANGE PLATE AND BOLTS ON THE PRYING FORCE Q

The analysis shows that the four corners of relationship exist between the bolt, flange plate, prying force and tensile load. Any of these factors are mutually corresponding relationship to the other three. When the tensile load is determined, the size of prying force is determined by the relative strength of bolt and flange plate. Generally for fixed bolt diameter, in this case, when the rigidity of the flange plate is less than the bolt stiffness, the prying force straightly increases with increasing tensile load. When the flange plate stiffness is equivalent with the stiffness of the bolt, the prying force is the maximum value and inflection point occurs. When the flange plate stiffness is larger than bolt stiffness, the prying force presents the trend of plummeting.

According to Eqs. (1), (2) and Figures 2 and 3, the analysis shows that, t_e is turning points and Q is peak value at the inflection point.

It is known that: $\alpha' = \frac{1}{\delta}\left(\frac{\beta}{1-\beta}\right) = 1.0$, $\delta = \left(\frac{\beta}{1-\beta}\right)$, $\beta = \frac{\delta}{1+\delta}$, $\beta = \frac{1}{\rho}\left(\frac{N_t^b}{N_t'} - 1\right)$, And then, deduced the value of the load N_t' at the inflection point.

$$N_t' = \frac{N_t^b}{\left(1 + \frac{\rho\delta}{1+\delta}\right)} \qquad (7)$$

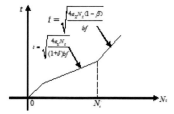
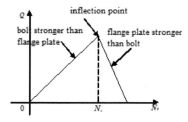

(a) Corresponding relationship diagram for flange thickness and loading

(b) Corresponding relationship diagram for prying force and loading

Figure 3. Relationship diagram for flange thickness and prying force with loading.

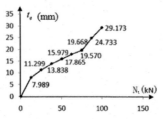 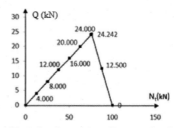

(a) Inflection point diagram for loading and flange thickness

(b) Inflection point diagram for loading and flange thickness

Figure 4. Inflection point diagram for tension load and prying force or the thickness of M20 G8.8s Q235 bolt.

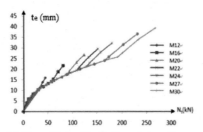 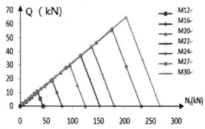

(a) The effect of bolt diameter on flange thickness

(b) The effect of bolt diameter on prying force

Figure 5. The effect of bolt diameter on flange thickness and prying force.

Example: When $e_1 = 1.25e_2$, calculate T-type tension joint of M20 8.8 bolt with Q235 flange. The load value at inflection point is $N'_t = N_t^b / \left(1 + \dfrac{\rho\delta}{1+\delta}\right) = \dfrac{5N_t^b}{2\rho+5} = 75.758\ (kN)$. According to the calculation results shown in Table 1, $t_e = 19.668$ mm and $Q = 24.242$ kN. Based on the calculation result of Table 1 and the point of inflection corresponding to the t_e and Q values, it is possible to draw the inflection point relationship diagram of following tensile load, flange plate thickness and prying force as shown in Figure 4.

4 EFFECT OF BOLT DIAMETER

According to Table 1, the flange thickness with changing tensile load values are listed in Figure 5 for the case of Q345 flange plate, 10.9 bolt diameter, with M12, M16, M20, M22, M24, M27, M30. It can be seen that bolt diameter affect the change of inflection point, but the trend is almost unchanged before and after the inflection point.

5 EFFECT OF FLANGE PLATE MATERIAL

Similarly, flange plate thickness with changing tension loads are shown in Figure 6 for the case of 10.9 M22 bolt with different flange plate materials of Q235, Q345, Q390, Q420. As can be seen, the flange plate material almost does not affect the position of the inflection point, and trends before and after the inflection point is very close. In prying force diagram, each line segment is coincident.

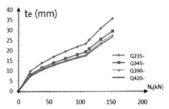 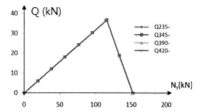

(a) The effect of flange strength on flange thickness   (b) The effect of flange strength on prying force

Figure 6. The effect of flange strength to flange thickness and prying force.

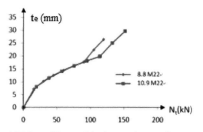 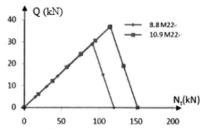

(a) The effect of bolt grade on flange thickness   (b) The effect of bolt grade on prying force

Figure 7. The effect of bolt grade on flange thickness and prying force.

6 EFFECT OF BOLTS GRADE

Similarly, in Q345 flange plate M22 bolts conditions, Figure 7 shows the relationship of prying force, flange thickness and tensile loads of T-type tension joints when the bolts were 8.8 and 10.9. As can be seen, the bolt strength only affect the position of the inflection point, trends before and after the inflection point are almost the same.

7 MAIN CONCLUSIONS

i. In T-type tension joints, the relative strength of bolt and flange plate determines the size of the prying force. When the rigidity of the flange plate is less than the stiffness of the bolt, the prying force straightly increases with the tensile load. When the rigidity of the flange plate is equivalent with the stiffness of the bolt, the prying force is the maximum value and inflection point occurs. When the rigidity of the flange plate is larger than the stiffness of the bolt, the prying force presents the trend of plummeting.
ii. The diameter of the bolt affects the thickness of the flange. Larger the diameter of the bolt, greater the thickness variation of the inflection point and the prying force. But the trend around the inflection point is almost constant.
iii. The flange plate material almost does not affect the position of the inflection point, trends before and after the inflection point is also very close. The flange plate material has some impact on thickness, but did not affect the prying force.
iv. The bolt grade only affect the position of the inflection point for thickness variation and prying force change. The trends before and after the inflection point are not affected.

ACKNOWLEDGEMENT

The work reported in this paper was sponsored by "the National Natural Science Foundation of China (Grant No. 51408620)" and "the Major Science and Technology Project during the

Third Five-Year Plan Period of China MCC (Grant No. 0012013010)". The support is gratefully acknowledged.

REFERENCES

Bai Rui, Hao Jiping, Tian Limin, Huang Yuqi, Cui Yangyang. Modified calculation method for prying action. Building Structure. 2013, 43(9):88–91. (in Chinese).
Jia Dongyun, Calculation of the pry action in steel structure connection. Journal of Hefei University of Technology. 2005, 28(5):514–517. (in Chinese).
Technical Specification for High-Strength Bolt Connection of Steel Structure. JGJ82-2011, China Building Industry Press, 2011. (in Chinese).
Wang Jingye, Zhang Haijun, Liu Wenwu. Comparison of the Resistance of individual Bolts between Codes of Steel Structure of China and Europe. Steel Structure. 2013, 2(28):50~58. (in Chinese).
Wang Yan, Hou Zhaoxin, Zheng Jie. Design Method for High Strength Bolt Connection Considering Prying Action, Journal of Progress in steel building structures, 2007, 9(2):27–33. (in Chinese).
Wang Yan, Zheng Jie, Hou Zhaoxin. Design and finite element study on the prying force of high strength bolt in extended end-plate connections. Building Structure. 2009, 39(5):68–75. (in Chinese).
Wang Yan, Zheng Jie. Study on the prying force of high strength bolt in extended end-plate connection. Industrial Construction. 2008, 38(9):99–103. (in Chinese).
Wu Zhaoqi, Zhang Sumei. Influence of prying force on tension carrying capacity of bolted connections. Journal of earthquake engineering and engineering vibration. 2008, 28(6):220–225. (in Chinese).
Zhao Wei, Jin Xiaoqun, Tong Genshu. Finite Element Analysis of Bolt Force in Extended End-plate Connections. Steel Structure. 2006, S1:112~117. (in Chinese).
Zheng Jie, Wang Yan. Present situation of design on high strength bolt t-shaped tensile connection considering prying action. Steel Construction. 2006, 21(87):15–18. (in Chinese).

Calculation for the distortional effect of curved composite beams based on the elastic foundation beam analogy method

Y.L. Zhang & Y.L. Liu
School of Civil Engineering, Shijiazhuang Tiedao University, Shijiazhuang, China

Z.M. Hou
Department of Civil Engineering, Tsinghua University, Beijing, China

Y.S. Li
School of Civil Engineering, Shijiazhuang Tiedao University, Shijiazhuang, China

ABSTRACT: In order to analyze the distortion effect, the curved steel-concrete composite beam was simulated to an equivalent straight one by Method/radius method. Based on the energy-variational principle, and considering the material difference between the concrete slab and steel girder, the distortional governing differential equation of the composite beam was deduced. A curved composite beam in Harbin-Dalian passenger railway was selected as calculation example. Based on the elastic foundation beam analogy method, the distortional angle and bimoment were obtained by the initial parameter method and the finite element method respectively. The results indicate that, based on the elastic foundation beam analogy method, and simulating the curved composite beam to an equivalent straight one by Method/radius method, both the initial parameter method and the finite element method can deduce the satisfied distortion effect of the curved composed beam.

1 INTRODUCTION

Due to the axial curvature of the curved steel-concrete composite beam, combined effects of bending, shear and torsion will occur even if the curved beams are only subjected to vertical external loads, and apparent distortional effect will produce in the thin-walled composite box section.

For the box beam with diagrams, the analysis methods about the distortional effect include analytic methods, numerical methods and elastic foundation beam analogy method (Guo J.Q., et al., 1982). The generalized coordinate method is a widely-used analytic method, which has clear concept, but can not reflect the actual shape and dimension of the beam section (Xu X., et al., 2013; Robert K.D., et al., 2012). The numerical methods include finite element method, finite strip method, and so on, which can give more accurate results, but will expense rather long time (Zhang L., 2013; Zhang W.X., 2009). Compared with the two former methods, the elastic foundation beam analogy method is easier, and can help designers understand the structural performance, so it is used the most widely (Hu Z.Z., 1987; Yang B.W., et al., 2011).

In present, most of the references about the distortional effect analysis of the box beam based on the elastic foundation beam analogy method focus on the straight and single material box beam, little research has hitherto concerned the distortional effect of the curved composite beam. In this paper, the curved steel-concrete composite beam was simulated to an equivalent straight one by method/radius method. Based on the energy-variational principle, and considering the material difference between the concrete slab and steel girder, the distortional governing differential equation of composite beam was deduced. According to the elastic foundation beam analogy method, the distortional angle and bimoment of a curved composite beam were obtained by the initial parameter method and the finite element method respectively.

2 MODEL SIMPLIFICATION BASED ON THE M/R METHOD

The aim of this paper is to research distortional effect of the simply-supported curved composite beam under the mid-span concentrated load. The plan layout and cross-section charts are shown in Figure 1. The anti-torque supports were set at two ends of the simply-supported curved composite beam. Several diaphragms with upper flanges were set along the beam axis.

In 1970, Tung and Fountain proposed an approximate method to analyze the torsional behaviour of curved girders, which is very simple and efficient and named after M/R method. The general steps are given as follows:

1. To straighten the curved beam along the axis with the same boundary conditions being remained, and to calculate the bending moment (M) of the straight beam under vertical loads;
2. To divide the moment (M) by the radius (r), and the value of M/r being obtained;
3. Taking the arc length l between the anti-torque supports as the calculation span, the value of M/r as the external distributed torque load of the straight beam, and supposing the boundary condition is simply-supported, and then the internal torsional moment being obtained, which is just the torsional moment of the curved beam.

According to the M/r method, the curved beam under mid-span concentrated load in Figure 1 is transformed to an equivalent straight one, the external distributed torque load should be applied on the equivalent straight beam can be written as

$$m(z) = \frac{M(z)}{r} = \begin{cases} \dfrac{Q}{2r}\left(\dfrac{l}{2}+z\right) & \left(-\dfrac{l}{2} \le z < 0\right) \\ \dfrac{Q}{2r}\left(\dfrac{l}{2}-z\right) & \left(0 \le z \le \dfrac{l}{2}\right) \end{cases} \quad (1)$$

where z is the coordinate along the beam axis, l is the span.

3 DISTORTIONAL GOVERNING DIFFERENTIAL EQUATION OF THE COMPOSITE BEAMS

According to Vlasov's theory, it is assumed that the composite section has only one deformation freedom degree. The distortional angle γ_1 at the corner point 1 in Figure 1 is taken as the basic unknown quantity, the distortional angle at other points can be expressed by γ_1. In Figure 1 the upper flange is concrete slab, the web and lower flange belong to steel girder.

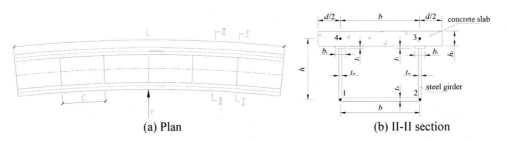

(a) Plan (b) II-II section

Figure 1. Curved composite beam.

3.1 Decomposing of the total torque loads

If the curved composite beam is simulated to an equivalent straight one by using the M/r method, the external torque load can be further transformed to a couple of horizontal antisymmetric loads, see Figure 2(a), then, the torsional loads and distortional loads can be deduced by using load decomposing method, see Figure 2(b) and 2(c). Where the upper flange of the box beam is concrete slab. $P = M/(hr)$; $P_1' = -P_3' = -P_1 = P_3 = M/(2br)$, b is the lower flange width of the composite beam; $P_4' = -P_2' = -P_2 = P_4 = M/(2hr)$, h is the space between the center of the upper and lower flange.

3.2 Distortional strain energy of the composite box beam

The distortional strain energy of the composite box beam includes three parts: the distortional strain energy U_1 produced by the transverse deformation of the cross-section frame, the strain energy U_2 produced by the distortional warp deformation of the cross-section and the potential energy of loads U_3.

3.2.1 The distortional strain energy U_1 produced by the transverse deformation of the cross-section frame

In the following derivation, the difference of the elastic module between the concrete slab and the steel girder is considered, but the slip deformation between them is not taken in account. Take a segment of the composite beam with the length of $dz = 1$ along the beam axis to analyze, Assume that the distortional angle at the corner point 1 as $\gamma_1(z)$, so that the horizontal displacement of the upper flange is $\gamma_1 h$. Because the composite box beam is symmetric, but the external action is antisymmetric, the transverse bending moment in frame produced by the horizontal displacement are antisymmetric, see Figure 3.

According to the symmetric principle, half of the frame is taken to be analyzed (see Fig. 3). Assume that the external horizontal load at the point 'A' is $P = 1$, the support reaction X_1 can be obtained by the force method equation $\delta_{11}X_1 + \Delta_{1P} = 0$, that is

$$X_1 = \frac{-\Delta_{1P}}{\delta_{11}} = \frac{\dfrac{bh^2}{4nI_w} + \dfrac{b^2h}{12nI_b}}{\dfrac{b^3}{24I_c} + \dfrac{b^2h}{4nI_w} + \dfrac{b^3}{24nI_b}} = 2X \quad (2)$$

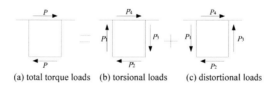

(a) total torque loads (b) torsional loads (c) distortional loads

Figure 2. Decomposition of the total torque loads.

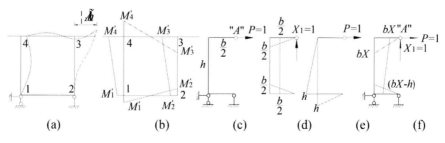

Figure 3. The transverse frame and transverse bending moment of the composite box beam.

If the half-frame is subjected to $X_1 = 2X$ and P at the same time, the bending moment diagram is as shown in Figure 3(f), the horizontal displacement δ_h at the point A can be obtained by self-multiplying of the internal force, that is

$$\delta_h = \frac{1}{6E_c I_c}\left\{\frac{b^3}{I_c}X^2 + \frac{(bX-h)^2}{nI_b}b + \frac{2h}{nI_w}[b^2 X^2 + (bX-h)^2 + bX(bX-h)]\right\} \quad (3)$$

In Equations 2 and 3, E_c is the elastic modular of concrete; E_s is the elastic modular of steel; n is the elastic modular ratio of the concrete to the steel; I_c is the inertia moment of the concrete upper flange in unit length, $I_c = 1 \times h_c^3/12$, h_c is the thickness of the concrete slab; I_w is the inertia moment of the steel girder web in unit length, $I_w = 1 \times t_w^3/12$, t_w is the thickness of the steel girder web; I_b is the inertia moment of the steel lower flange in unit length, $I_b = 1 \times t_b^3/12$, t_b is the thickness of the steel lower flange.

If the horizontal displacement at the point A is δ_h under the unit load $P = 1$, then the external load will be $P_4 = \gamma_1 h/\delta_h$ when the horizontal displacement is $\gamma_1 h$, so the transverse bending moment of the frame can be expressed as follows (see Fig. 3(f)):

$$M_4' = -M_3' = -bX\gamma_1 h/\delta_h = K_1 \gamma_1 \quad (4)$$

$$M_1' = -M_2' = -(bX-h)\gamma_1 h/\delta_h = K_2 \gamma_1 \quad (5)$$

where: $K_1 = -bXh/\delta_h$, $K_2 = -(bX-h)h/\delta_h$.

The distortional strain energy U_1 produced by the transverse deformation of the cross-section frame can be expressed as:

$$U_1 = \int_0^l \int_s \frac{M'^2}{2EI} ds\, dz = \int_0^l \frac{1}{6}\left[K_1^2 \frac{b}{E_c I_c} + K_2^2 \frac{b}{nE_c I_b} + \frac{2h}{nE_c I_b}(K_1^2 + K_2^2 + K_1 K_2)\right]\gamma_1^2 dz = K_3 \int_0^l \gamma_1^2 dz \quad (6)$$

where: K_3 is half of the distortional frame stiffness of the composite box beam, that is

$$K_3 = \frac{1}{6E_c}\left[K_1^2 \frac{b}{I_c} + K_2^2 \frac{b}{nI_b} + \frac{2h}{nI_b}(K_1^2 + K_2^2 + K_1 K_2)\right] \quad (7)$$

3.2.2 The strain energy U_2 produced by the distortional warp deformation of the cross-section

When the distortional deformation is produced in the composite box beam, the warp will be happen in the plane of the flanges and web, and the warp strain energy will be produced. The in-plane warp strain and bending moment of the flanges and web are shown in Figure 4.

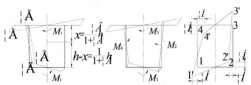

(a) Distortional warp strain (b) Distortional warp bending moment (c) Displacement produced by the distortional deformation

Figure 4. Distortional effect of the composite box beam.

According to Figure 4(a), let $\beta = \varepsilon_4/\varepsilon_1$. Because the sum of the warp bending moment in the cross-section should be zero, i.e. $M_1 + M_2 + M_3 + M_4 = 0$, β can be expressed as:

$$\beta = \frac{\varepsilon_4}{\varepsilon_1} = \frac{E_s b^2 t_b + 3E_s b h t_w}{\dfrac{E_c (b+d)^3 h_c}{b} + 3E_s b h t_w} \tag{8}$$

where, d is the total length of the two cantilever flanges; ε_4 is the warp strain at the point 4; ε_1 is the warp strain at the point 1.

The warp bending moment can be obtained by the distortional warp stress. Assume that the in-plane displacement of the box beam's four slabs is v_1, v_2, v_3 and v_4 respectively (see Fig. 4(c)), which can be regarded as the in-plane deflection of the slabs, the correlation between v_i ($i = 1, 2, 3, 4$) and M_i ($i = 1, 2, 3, 4$) can be expressed as follows:

$$v_4'' = v_2'' = \frac{-M_4}{E_s I_w'} = -\frac{\varepsilon_1}{h}(1+\beta), \quad v_1'' = \frac{-M_1}{E_s I_b'} = \frac{-2\sigma_1}{E_s b} = -\frac{2\varepsilon_1}{b}, \quad v_3'' = \frac{-M_3}{E_c I_c'} = -\frac{2\beta\varepsilon_1}{b} \tag{9}$$

where, I_c' is the in-plane inertia moment of the concrete flange, $I_c' = (b+d)^3 h_c/12$; I_w' is the in-plane inertia moment of the steel girder web, $I_w' = h^3 t_w/12$; I_b' is the in-plane inertia moment of the steel lower flange, $I_b' = b^3 t_b/12$.

According to Figure 4(c), replacing the distortional angle γ_1 by the slabs' displacement along the x and y directions, then γ_1 can be expressed as,

$$\gamma_1 = \frac{dx_4 - dx_1}{h} + \frac{dy_2 - dy_1}{b} = \frac{v_1 + v_3}{h} + \frac{v_4 + v_2}{b} \tag{10}$$

Taking differential transformation to Eq. (10) and substituting Eq. (9) into it, it can be obtained that,

$$\gamma_1'' = -\frac{4\varepsilon_1(1+\beta)}{bh} \tag{11}$$

Multiply the normal strain of the point 1 ε_1 by the elastic modular of steel, then,

$$\sigma_1 = E_s \varepsilon_1 = -E_s \frac{bh}{4(1+\beta)} \gamma_1'' = S\gamma_1'' \tag{12}$$

The strain energy U_2 produced by the distortional warp deformation of the cross-section is,

$$U_2 = \int_0^l \int_F \frac{\sigma^2(z,s)}{2E} dF dz = \Omega \int_0^l (\gamma_1'')^2 dz \tag{13}$$

where, Ω is half of the distortional warp inertia moment of the composite box beam, that is

$$\Omega = \frac{S^2}{6n^2 E_c} \beta^2 h_c (b+d)\left(1+\frac{d}{b}\right)^2 + \frac{S^2}{6nE_c}[bt_b + 2ht_w(\beta^2 - \beta + 1)] \tag{14}$$

3.2.3 *The potential energy of loads U_3*
The potential energy of loads U_3 is,

$$U_3 = -\int_0^l P_4(z)\gamma_1 h dz = -\int_0^l \frac{M(z)}{2r} \gamma_1 dz \tag{15}$$

3.2.4 *Distortional governing differential equation of the composite box beam with constant cross section*

If the strain energy produced by the shear deformation is not taken into account, the total potential energy of the composite box beam under the distortional loads can be obtained as,

$$U = U_1 + U_2 + U_3 = \int_0^l K_3 \gamma_1^2 dz + \int_0^l \Omega(\gamma_1'')^2 dz - \int_0^l \frac{M(z)}{2r} \gamma_1 dz \qquad (16)$$

According to the Euler-Lagrange equation, the distortional governing differential equation of the composite box beam with constant section can be derived as follows:

$$2\Omega\gamma_1'''' + 2K_3\gamma_1 = \frac{M(z)}{2r} \qquad (17)$$

where, Ω can be obtained by Equation 14, K_3 can be obtained by Equation 7.

The distortional bimoment can be defined as

$$B_A = -2\Omega\gamma_1'' \qquad (18)$$

4 SOLVE THE DISTORTIONAL EFFECT OF THE CURVED COMPOSITE BOX BEAM BY USING THE ELASTIC FOUNDATION BEAM ANALOGY METHOD

The distortional governing differential equation of the composite box beam obtained by Equation 17 is similar to the deflection governing differential equation of the elastic foundation beam, where the bending stiffness of the elastic foundation beam EI_b corresponds to the distortional warp inertia moment of the composite box beam 2Ω; the subgrade modulus of the elastic foundation beam K corresponds to the distortional frame stiffness of the composite box beam $2K_3$; the uniform load q applied on the elastic foundation beam corresponds to the distortional horizontal load $M/2r$ applied on the composite box beam; the deflection of the elastic foundation beam y corresponds to the distortional angle of the composite box beam γ_1; the bending moment of the elastic foundation beam M corresponds to the distortional bimoment of the composite box beam B_A.

4.1 *Introduction of the calculation example*

A simply-supported curved composite beam in Harbin-Dalian passenger railway in China is taken as the example to illustrate the application of the analytical and numerical elastic foundation beam method. As shown in Figure 1, the composite beam has single box section, with a calculation span $l = 24$ m and a radius of curve $r = 120$ m. The width of the upper flange is $b_t = 600$ mm and the thickness is $t_t = 30$ mm. The height of the web is $h_w = 2340$ mm and the thickness is $t_w = 16$ mm. The width of the lower flange is $b_b = 3000$ mm and the thickness is $t_b = 30$ mm. The width of the deck is $b_c = 6500$ mm and the height is $h_c = 400$ mm. The steel girder was fabricated from 345 MPa steel, and the bridge deck utilised C50 concrete. Two diaphragms are installed at both end of the composite beam. A concentrated load of $Q = 50$ kN is applied at the middle of the span.

Based on the M/r method, the curved beam is stretched to an equivalent straight beam. According to the Equation 1 and Figure 2, the horizontal distortional load part is a triangle distributed load, which corresponds to the distributed load q applied on the elastic foundation beam, as shown in Figure 5.

4.2 *Initial parameter method*

It is difficult to solve the mid-span distortional angel and bimoment under the triangle distortional load defined in Figure 5 directly. The easier method is to calculate the influence line

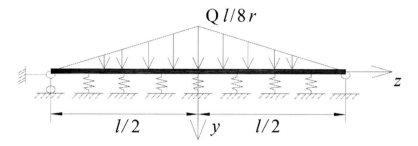

Figure 5. Horizontal distortional load applied on the equivalent straight beam.

of the mid-span distortional angel and bimoment firstly, and then integrate them with the triangle distortional load to obtain the distortional angel and bimoment at the middle of the span.

Dividing both sides of the Equation 17 by 2Ω gives,

$$\gamma'''' + 4\lambda^4 \gamma = \frac{M(z)}{2\Omega \cdot 2r} \tag{19}$$

where:

$$\lambda = \sqrt[4]{\frac{2K_3}{8\Omega}} \tag{20}$$

The Equation 19 can be solved by initial parameter method. Considering the symmetric condition, take the middle span as the longitudinal coordinate origin. The initial parameters are: the mid-span distortional angle γ_0 (i.e. the mid-span deflection y_0 of the elastic foundation beam); the mid-span distortional bimoment B_0 (i.e. the mid-span bending moment M_0 of the elastic foundation beam); the first derivative of the mid-span distortional angle $\gamma'_0 = 0$ (i.e. the mid-span section angle $y'_0 = 0$ of the elastic foundation beam); the first derivative of the mid-span distortional bimoment $B'_0 = -0.5$ (i.e. the mid-span shear $Q_0 = -0.5$ of the elastic foundation beam).

Substituting these initial parameters into Equation 19 gives

$$\left.\begin{array}{l}\gamma(z) = \gamma_0 \cos\lambda z\,\text{ch}\,\lambda z - \dfrac{1}{4\Omega\lambda^2} B_0 \sin\lambda z\,\text{sh}\,\lambda z - \dfrac{B'_0}{8\Omega\lambda^3}(\sin\lambda z\,\text{ch}\,\lambda z - \cos\lambda z\,\text{sh}\,\lambda z) \\[2mm] B(z) = 4\Omega\lambda^2\gamma_0 \sin\lambda z\,\text{sh}\,\lambda z + B_0 \cos\lambda z\,\text{ch}\,\lambda z + \dfrac{B'_0}{2\lambda}(\sin\lambda z\,\text{ch}\,\lambda z + \cos\lambda z\,\text{sh}\,\lambda z)\end{array}\right\} \tag{21}$$

Substituting the geometric parameters to Equations 14 and 7 gives $2\Omega = 3.811 \times 10^7$ kN·m^4, $2K_3 = 6.080 \times 10^2$ kN·m^4, and $\lambda = 0.0447$ m^{-1}. According to the boundary conditions of Equation 21 ($\gamma_{l/2} = 0$ (i.e. $y_{l/2} = 0$), $B_{l/2} = 0$ (i.e. $M_{l/2} = 0$)), γ_0 and B_0 can be obtained as

$$\gamma_0 = 7.173 \times 10^{-6}\,\text{rad} \qquad B_0 = 5.749\,\text{kN}\cdot\text{m}^2$$

Substituting γ_0 and B_0 into Equation 21 gives

$$\left.\begin{array}{l}\gamma(z) = 7.173 \times 10^{-6}\cos\lambda z\,\text{ch}\,\lambda z - \dfrac{1}{4\Omega\lambda^2}5.749\sin\lambda z\,\text{sh}\,\lambda z - \dfrac{Q_0}{8\Omega\lambda^3}(\sin\lambda z\,\text{ch}\,\lambda z - \cos\lambda z\,\text{sh}\,\lambda z) \\[2mm] B(z) = 4\Omega\lambda^2 7.173 \times 10^{-6}\sin\lambda z\,\text{sh}\,\lambda z + 5.749\cos\lambda z\,\text{ch}\,\lambda z + \dfrac{Q_0}{2\lambda}(\sin\lambda z\,\text{ch}\,\lambda z + \cos\lambda z\,\text{sh}\,\lambda z)\end{array}\right\}$$

$$\tag{22}$$

According to the reciprocal theorem, the expression defined in Equation 22 is just the influence line of the mid-spam distortional angle and bimoment of the composite beam (see Figures 6 and 7 in section 3.3.3). Integrating the influence line function with the triangle distortional load, the distortional angel and bimoment at the middle of the span can be obtained as

$$\gamma_A = 2\int_0^{\frac{l}{2}} \frac{Q}{4r}\left(\frac{l}{2}-z\right)\gamma(z)dz = 8.614 \times 10^{-5}\,\text{rad}, \quad B_A = 2\int_0^{\frac{l}{2}} \frac{Q}{4r}\left(\frac{l}{2}-z\right)B(z)dz = 56.947\,\text{kN}\cdot\text{m}^2$$

4.3 Finite element method

The finite element model of the elastic foundation beam was build with Midas/civil software to solve the mid-spam distortional angle and bimoment of the composite beam.

4.3.1 Defining the material and section dimensions of the elastic foundation beam

If the finite element model is build with the elastic foundation beam method, the material and section dimensions should be defined firstly to give the value of EI_b. Due to the elastic modular difference between the steel and concrete in the composite beam, the equivalent elastic modular was assumed as the concrete's elastic modular. Let the product of the EI_b of the elastic foundation beam equal to the corresponding parameter 2Ω of the composite box beam, the equivalent bending inertia moment can be obtained as $I_{be} = 2\Omega/E_c$.

For bending effect analysis of the simple-supported box beam with constant section, the actual section can be replaced by an arbitrary rectangle section i.e. *similar beam* with the bending stiffness unchanged (Cheng X.Y., 2009). Assuming the width of the similar beam as $b_c = 1$ m, the elastic modular as same as the actual elastic foundation beam, and letting the

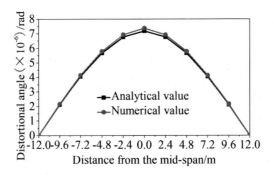

Figure 6. The influence line of the mid-span distortional angel.

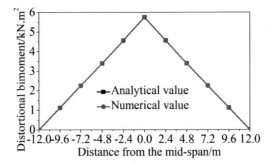

Figure 7. The influence line of the mid-span distortional bimoment.

bending inertia moment of the similar beam equal to $I_{be} = b_e h_e^3/12$, the equivalent height of the similar beam can be obtained as,

$$h_e = \sqrt[3]{12I_{be}/b_e} = \sqrt[3]{12I_{be}} \tag{23}$$

4.3.2 *Simulating of the subgrade modulus of the elastic foundation beam K*

The distortional frame stiffness of the composite box beam $2K_3$ in the distortional governing differential Equation 17 corresponds to the subgrade modulus of the elastic foundation beam K, so the distortional frame stiffness $2K_3$ can be simulated to the distributed elastic bearings which only be pressed along the vertical direction in the elastic foundation beam. The two end diaphragms can be replaced with hinged bearings (Cheng X.Y., 2009). In the finite element model, the distributed elastic bearings were simplified to the node elastic bearings, the concentrated subgrade modulus K_{ij} at each node can be calculated with Equation 24.

$$K_{ij} = \frac{\Delta S_{i-1} + \Delta S_i}{2} \times 2K_3 \tag{24}$$

where, ΔS_{i-1} and ΔS_i is the length of the two adjoining elements respectively.

4.3.3 *Influence line of the mid-span distortional angle and bimoment*

Assume that the elastic modular of the similar beam is $E_c = 3.5 \times 10^4$ MPa, the equivalent inertia moment is $I_{be} = 2\Omega/E_c = 1.089$ m^6, the equivalent width is $b_e = 1$ m, so the equivalent height can be obtained as $h_e = \sqrt[3]{12I_{be}} = 2.355$ m. There are totally 120 elements in the FE model, the length of each element is 0.2 m, so the concentrated subgrade modulus at each node can be calculated as $K_{ij} = 0.2 \times 608 = 121.6$ kN/m.

Letting the elastic foundation beam be subjected to the mid-span unit concentrate load (i.e. unit distortional load on the composite beam), the deflection and bending moment of the elastic foundation beam along the axial direction are shown in Figures 6 and 7, which correspond to the distortional angle and bimoment of the actual composite box beam. According to the reciprocal theorem, the diagrams in Figures 6 and 7 are just the influence line diagrams of the mid-span distortional angel and bimoment respectively.

From Figures 6 and 7 we can see, the analytical results obtained from the initial parameter method are satisfied well with the numerical results obtained from the finite element method, which indicate that the FE model is correct.

4.3.4 *The mid-span distortional angel and bimoment under the triangle distributed distortional load*

Replacing the mid-span concentrated load on the FE model defined in section 3.3.3 by the triangle distributed load which was shown in Figure 5, the mid-span deflection and bending moment of the elastic foundation beam, i.e. the distortional angel and bimoment of the composite box beam can be calculated as $\gamma_A = 8.790 \times 10^{-6}$ rad and $B_A = 56.886$ kN·m^2, which is 2.04% larger and 0.01% smaller than the corresponding analytical value obtained by the initial parameter method. The numerical results are satisfied well with the analytical results.

5 CONCLUSIONS

Based on the M/r method, the curved composite beam can be transformed to an equivalent straight one, the torque load produced by the axial curvature can be obtained and be decomposed to the distortional load. Applying the distortional load on the equivalent straight beam, and according to the elastic foundation beam analogy method, the analytical results obtained from the initial parameter method are satisfied well with the numerical results obtained from the finite element method, which indicates that both the initial parameter method and the finite element method can deduce the satisfied distortion effect of the curved composed beam.

ACKNOWLEDGEMENTS

This study is sponsored by the Natural Science Foundation (51108281), the Natural Science Foundation of Hebei Province (E2014210038), the Key Topics of Science and Technology Research Project for Colleges and Universities of Hebei Province (ZD2014025), and the China Postdoctoral Science Foundation (2014M560088).

REFERENCES

Cheng X.Y., 2009. Rereading distortion calculation of the trapezium-shape single cell box beam. *Highway Engineering*, 2009, 34(3): 27–32.
Guo J.Q., Zhang Z.M., Zhou R.G. 1982. Distortional stress calculation of the box beam bridges. *Highway*, 1982, (4): 11–17.
Hu Z.Z. 1987. Research on two kinds of analogy method of beam on elastic foundation for distortion of single-cell box girder bridge. *Journal of China North-East Forestry University*, 15(1): 80–89. (in Chinese).
Robert K.D., Timothy P.J. 2012. Closed-form shear flow solution for box-girder bridges under torsion. *Engineering Structures*, 34: 383–390.
Tung, D.H., Fountain R.S. 1970. Approximate torsional analysis of curved box girders by the M/R method. *Eng. J.*, 7(3): 65–74.
Xu X., Qiang S.Z. 2013. Research on distortion analysis theory of thin-walled box girder. *Engineering Mechanics*, 30(11): 192–201.
Xu X., Ye H.W., Qiang S.Z. 2013. Distortion analysis of thin-walled box girder taking account of shear deformation. *Chinese Journal of Computational Mechanics*, 30(6): 860–866.
Yang B.W., Li Y.L., Wan S., Zhang J.D. 2011. Stress analysis of box girders with corrugate steel webs under distorsion. *Journal of Southeast University(Natural Science Edition)*, 41(5): 1065–1069.
Zhang L. 2013. Influences of diaphragm plate and geometric characteristics on distortion effect of steel box girder. *Journal of railway engineering society*, (8): 68–73.
Zhang W.X., Pang S., Huang J.F., Zhang W.C. 2009. Experimental investigation on distortion effect of widely-flanged box girders. *Journal of Northeastern University (Natural Science)*, 30(7): 1047–1050.

Dynamic response analysis and fatigue life evaluation of heavy haul railway steel truss bridges

Y.S. Li, Y.F. Xiao, Y.F. Diao & Y.L. Zhang
School of Civil Engineering, Shijiazhuang Tiedao University, Shijiazhuang, China

ABSTRACT: Running heavy haul trains on the existing railway line can satisfy the rapid development of China's railway transport, but it is bound to have an adverse effect on the safety and the fatigue life of existing railway bridges. The finite element model of a four-span continuous steel truss bridge was built with ANSYS software, the dynamic responses were analyzed under ordinary train loads and heavy haul train loads, respectively, and the stress time history of some typical members were obtained. Then, the rain flow method was used to count the stress amplitude number, and the cumulative damage degree was deduced by linear cumulative damage theory. On the base of the cumulative damage degree, the fatigue lives of some typical members were evaluated at last. The analysis results show that, the vertical deformation, transverse amplitude, and vertical acceleration of the steel truss all increase with the train's axial load and speed. In general, the fatigue lives under C96 heavy haul train loads of the typical members in the steel truss bridge in this paper will be about 80% shorter than that of under the ordinary train loads.

1 INTRODUCTION

Under the long period running of the train load, the steel truss bridge will be damaged due to the cumulative fatigue effect, especially under the long marshalling heavy haul train load, the load frequency and axial weight will all have a great improve, so that the fatigue damage will be more serious.

Some researches about dynamic and fatigue analysis of the steel truss bridge have been reported. Xia H. et al. used the vehicle-bridge coupled dynamic analysis to strengthen an old steel bridge; Sun Y. et al. analyzed the dynamic behavior of an existing railway, and proposed a method to strengthen the transverse stiffness; Kaliyaperumal et al. conducted an advanced dynamic finite element analysis of a skew steel railway bridge; Guo W.W. et al. finished the dynamic response analysis of a railway new type steel-concrete composite truss bridge; Peng X.Q. et al. researched the dynamic response properties of the bridge-subgrade transition section in an existing heavy railway line; Zhao Z.W. et al. and Mohammad J. et al. evaluated the fatigue life of bridge based on the fatigue-reliability method and field data, especially. Most of the references above are about the dynamic properties under the ordinary train load, but little research has hitherto concerned the fatigue life evaluation of the steel truss bridge under the heavy haul load.

In this paper, a four-span continuous steel truss bridge—ChangDong yellow river bridge—was selected as the sample, the finite element model was build with ANSYS software, the dynamic responses of the steel truss bridge were analyzed under ordinary trains and heavy haul trains, respectively, and the stress time history of some typical members were obtained. Then, the rain flow method was used to count the stress amplitude number, and the cumulative damage degree was deduced by linear cumulative damage theory. At last, the fatigue lives of some typical members were evaluated.

2 DYNAMIC ANALYSIS OF STEEL TRUSS BRIDGE UNDER TRAIN LOAD

2.1 Finite element model

ChangDong yellow river bridge is a four-span continuous steel truss bridge, its vertical view is shown in Figure 1. The finite element model of the main truss was build with ANSYS software. The truss members were simulated with beam44 elements, among the elements are all rigid connections. The truss member sections are all I-shape. The movable hinge supports were set at the nodes of $E0$, $E18$, $E18'$, and $E0'$, with the translational freedom degree along Y and Z directions restrained. The fixed hinge supports was set at the node of $E36'$, with the translational freedom degree along X, Y, and Z directions restrained.

2.2 Vehicle model

In the dynamic analysis of bridge, the vehicle is often simulated to the movable loads, movable masses, movable harmonic loads, or movable masses on spring. Among these methods, the last method is the most rational, but also the most complex, the two former methods are easier (Liu K. et al., 2009), and have no apparent difference compared with the last method, so the first method, i.e., simulating the vehicle to the movable loads, is selected in this paper.

With the train wheel loads simulated as the concentrated node loads, two train vehicle models, one is the ordinary train with the axle load of 21t (C64K, for example), the other is the heavy haul train with the axle load of 30t (C96, for example), were built. The trains of C64K and C96 types all have four axes (eight wheels), which were simulated as eight concentrated loads to act on the truss bridge FE model, respectively. The vehicle parameters are shown in Table 1. The length between truck pivot centers of each train type is shown in Figure 2, and the vehicle-bridge model is shown in Figure 3.

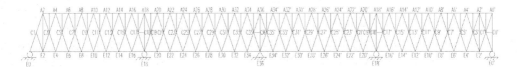

Figure 1. Main truss vertical view of the four-span continuous steel truss bridge.

Table 1. Vehicle parameters of C64K and C96.

Vehicle type	Carrying capacity/t	Self weight/t	Axle load/t	Vehicle length/mm	Length between truck pivot centers/mm	Fixed wheelbase/mm
C64K	61	23	21	13 430	8 700	1 750
C96	96	24	30	13 600	9 800	1 860

Figure 2. Length between truck pivot centers of C64K and C96.

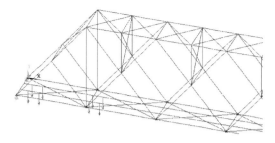

Figure 3. Vehicle-bridge model.

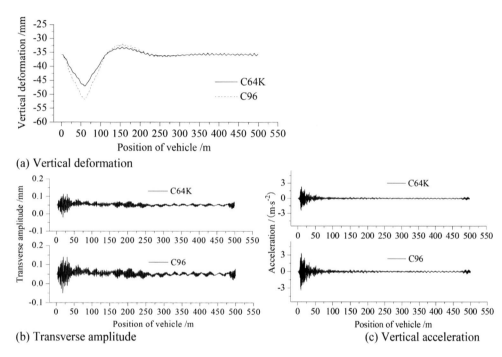

(a) Vertical deformation

(b) Transverse amplitude (c) Vertical acceleration

Figure 4. Dynamic response of the lower chord at the end midspan under two different train loads.

2.3 *Analysis results of dynamic response*

2.3.1 *Comparison of dynamic response under different vehicle loads*

The dynamic responses were calculated, respectively, when the C64K loads and the C96 loads run through the truss bridge with the speed of 100 km/h, the time-history curves of the vertical deformation, transverse amplitude, and vertical acceleration at the lower chord of the end midspan are shown in Figure 4.

From Figure 4 we can see:

1. The maximum vertical deformation is −47.016 mm under C64K load and −51.846 mm under C96 load, the latter is 10.3% more than the former, which indicates that the vertical deformation of the steel truss bridge increases with the axle load. The maximum vertical deformation appears when the train arrives at the middle of the end span;
2. The maximum transverse amplitude is 0.113 mm under C64K load and 0.140 mm under C96 load, the latter is 23.9% more than the former, which indicates that the transverse amplitude of the steel truss bridge increases with the axle load. The maximum transverse

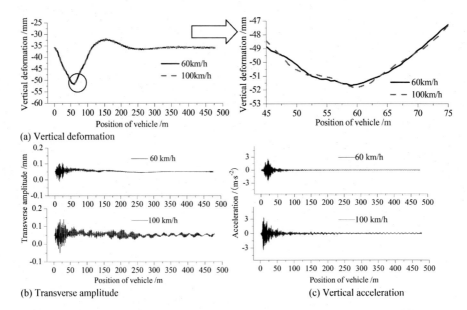

Figure 5. Dynamic response of the lower chord at the end midspan under the C96 load with different speeds.

amplitude appears when the train arrives at the middle of the end span. What needs to be explained is that, the loads used in this paper are only a series of concentrate loads, which cannot induce the wheel-rail interaction, so the calculated transverse amplitude can only reflect the natural transverse vibration mode of the truss itself, but not forced vibration;

3. The maximum vertical acceleration is 2.329 m/s² under C64K load and 3.328 m/s² under C96 load, the latter is 42.9% more than the former, which indicates that the transverse amplitude of the steel truss bridge increases with the axle load. The maximum vertical acceleration appears when the train arrives at the 1/4th length of the end span.

2.3.2 Comparison of dynamic response under different vehicle speed

Under the C96 load, the dynamic responses were calculated with the speed of 60 km/h and 100 km/h, respectively, see Figure 5.

From Figure 5 we can see:

The maximum vertical deformation, maximum transverse amplitude, and maximum vertical acceleration are −51.68 mm, 0.105 mm, and 2.590 m/s², respectively, when the C96 load runs through the bridge with the speed of 60 km/h, and −51.85 mm, 0.140 mm, and 3.328 m/s² with the speed of 100 km/h; the increased amplitudes are 0.33%, 33%, and 30%, respectively, which indicate that the vertical deformation, transverse amplitude, and vertical acceleration of the steel truss bridge all increase with the vehicle speed, and the effects on the transverse amplitude and vertical acceleration are more apparent.

3 FATIGUE LIFE EVALUATION OF SOME TYPICAL MEMBERS

3.1 Miner–Palmgren linear cumulative damage theory

Miner–Palmgren linear cumulative damage theory is the most popular method to evaluate the structure fatigue life (Ni K. 1999), which defines the ratio of the stress cycle number to the fatigue life under the nominal stress as the fatigue damage degree D:

$$\sum_{i=1}^{n} \frac{n_i}{N_i} = D \quad (1)$$

where, n_i and N_i is the cycle number and the fatigue life under the stress amplitude of the *i*th level, respectively.

When $D = 1$, the fatigue damage happens.

3.2 *Time-history analysis of stress for the typical members*

When the Miner–Palmgren linear cumulative damage theory is used to evaluate the structure fatigue life, it is needed to obtain the stress amplitude of each level and its corresponding cycle number. According to the static calculated results, the lower chords of E_8E_{10} and $E_{26}E_{28}$, the diagonals of $E_{14}C_{15}$, $E_2C_{1'}$, and $E_{16}C_{17}$, whose stress condition are more unfavorable, are selected as the typical members.

Taking the lower chords of E_8E_{10} as the sample, let the ordinary train formation consisting of 100 C64K carriages run through the truss with the speed of 100 km/h, the stress time-history curve of the lower chords of E_8E_{10} is calculated and shown in Figure 6. Let the heavy train formation consisting of 100 C96 carriages run through the truss with the speed of 100 km/h, the stress time-history curve of the lower chords of E_8E_{10} is shown in Figure 7.

From Figure 6 and Figure 7 we can see that, the stress time-history curve of the lower chords of E_8E_{10} has the same change law under the C64K and the C96 train formation, but the maximum stress of the latter is 20.9% more than the former.

3.3 *Fatigue life evaluation for the typical members*

Due to some defects such as the stress concentration, the actual fatigue life of bridge usually is shorter than the theoretically evaluated life. Considering the stress concentration coefficient (1.424) at the bolt hole in the steel truss bridge, the nominal stress spectrum of the lower chords of E_8E_{10} is dealt with rain-flow counting method when the C64K and the C96 train formation run through the bridge with the speed of 100 km/h, respectively, see Table 2.

According to the Eq. (1), the fatigue damage degree of the lower chords of E_8E_{10} under the C64K train formation is $D_{C64K} = 1.294 \times 10^{-7}$, while under the C96 train formation is $D_{C96} = 7.079 \times 10^{-7}$.

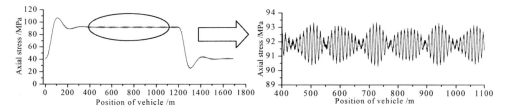

Figure 6. The stress time-history curve of the lower chords of E8E10 under C64K train formation.

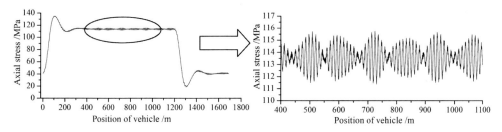

Figure 7. The stress time-history curve of the lower chords of E8E10 under C96 train formation.

Table 2. The nominal stress spectrum of the lower chords of E_8E_{10} under different train formation.

C64K train formation				C96 train formation			
Stress amplitude range	Stress amplitude number n	Mean stress amplitude	Cycle number N	Stress amplitude range	Stress amplitude number n	Mean stress amplitude	Cycle number N
[0.0, 0.5]	1 989	0.25	4.879×10^{22}	[0.0, 0.5]	1 966	0.25	4.879×10^{22}
[0.5, 1.0]	59	0.75	4.406×10^{19}	[0.5, 1.0]	50	0.75	4.406×10^{19}
[1.0, 1.5]	60	1.25	1.691×10^{18}	[1.0, 1.5]	44	1.25	1.691×10^{18}
[1.5, 2.0]	50	1.75	1.975×10^{17}	[1.5, 2.0]	39	1.75	1.975×10^{17}
[2.0, 2.5]	37	2.25	3.972×10^{16}	[2.0, 2.5]	38	2.25	3.972×10^{16}
[2.5, 3.0]	26	2.75	1.104×10^{16}	[2.5, 3.0]	29	2.75	1.104×10^{16}
[3.0, 3.5]	1	3.25	3.800×10^{15}	[3.0, 3.5]	26	3.25	3.800×10^{15}
[4.5, 5.0]	1	4.75	3.372×10^{14}	[3.5, 4.0]	23	3.75	1.525×10^{15}
[5.0, 5.5]	1	5.25	1.780×10^{14}	[4.0, 4.5]	9	4.25	6.858×10^{14}
[18.0, 18.5]	1	18.25	6.263×10^{10}	[6.5, 7.0]	1	6.75	3.580×10^{13}
[19.5, 20.0]	1	19.75	3.783×10^{10}	[7.5, 8.0]	1	7.75	1.482×10^{13}
[65.5, 66.0]	1	65.75	1.754×10^{7}	[26.0, 26.5]	1	26.25	6.155×10^{9}
[68.0, 68.5]	1	68.25	1.382×10^{7}	[27.5, 28.0]	1	27.75	4.317×10^{9}
				[94.0, 94.5]	1	94.25	3.035×10^{6}
				[97.0, 97.5]	1	97.25	2.645×10^{6}

Each train formation running on the ChangDong yellow river bridge is composed of about 60–100 carriages, and every day and night there are about 60–80 train formations, so in the fatigue life evaluation analysis, suppose each train formation is composed of 80 carriages, there are 70 train formations, i.e., 5600 carriages running through the bridge every day and night, so the one-year fatigue damage degree D^{ly}_{C64K} of the lower chords of E_8E_{10} under the C64K train formation is

$$D^{ly}_{C64K} = 1.294 \times 10^{-7} \times 365 \times 5\,600/100 = 2.645 \times 10^{-3}$$

The one-year fatigue damage degree D^{ly}_{C96} of the lower chords of E_8E_{10} under the C96 train formation is

$$D^{ly}_{C96} = 7.079 \times 10^{-7} \times 365 \times 5\,600/100 = 1.447 \times 10^{-2}$$

If $D = 1$, the fatigue damage will happen, the fatigue life N of the lower chords of E_8E_{10} can be evaluated as:
Under the C64K train formation:

$$N_{C64K} = 1/D^{ly}_{C64K} = 378 y$$

Under the C96 train formation:

$$N_{C96} = 1/D^{ly}_{C96} = 69 y$$

From the results we can see that, fatigue life N_{C96} of the lower chords of E_8E_{10} is 81.7% shorter than N_{C64K}, which indicates that the heavy haul train load will shorten the fatigue life of the existing steel truss bridge apparently.

The fatigue lives of all the typical members are shown in Table 3.

From Table 3 we can see that, the most unfavorable member is the diagonals of $E_2'C_1$, for which fatigue life is only 25 years under the C96 train formation. In general, the fatigue lives of all the typical members under the C96 heavy haul train formation are about 80% shorter than that of under the C64K ordinary train formation.

Table 3. The fatigue lives of all the typical members/year.

Typical member	Fatigue life under C64K train formation	Fatigue life under C64K train formation	Decrease amplitude/%
Lower chords of E_8E_{10}	378	69	81.7
Lower chords of $E_{26}E_{28}$	709	106	85.0
Diagonals of $E_{14}C_{15}$	135	26	80.7
Diagonals of $E_2'C_1'$	123	25	79.7
Diagonals of $E_{16}C_{17}$	152	33	78.3

4 CONCLUSIONS

The maximum vertical deformation, maximum transverse amplitude, and maximum vertical acceleration of the steel truss bridge all increase with the axle load. The maximum vertical deformation and the maximum transverse amplitude all appear when the train arrives at the middle of the end span, but the maximum vertical acceleration appears when the train arrives at the 1/4th length of the end span.

The maximum vertical deformation, maximum transverse amplitude, and maximum vertical acceleration of the steel truss bridge all increase with the vehicle speed, and the effects on the transverse amplitude and vertical acceleration are more apparent.

In general, the fatigue lives of all the typical members under the C96 heavy train formation are about 80% shorter than that of under the C64K ordinary train formation; the heavy train load will shorten the fatigue life of the existing steel truss bridge apparently.

ACKNOWLEDGEMENTS

This study is sponsored by the scientific research development program (2013G010-A) of China Railway Corporation.

REFERENCES

Guo W.W., Xia H., Li H.L., Zhang T. 2012. Dynamic analysis of a new type of railway steel-concrete composite trussed bridge under running trains excitation. *Journal of vibration and shock* 31(4): 128–133.

Kaliyaperumal G., Imam B., Righiniotis T. 2011. Advanced dynamic finite element analysis of a skew steel railway bridge. *Engineering Structures* 33(1): 181–190.

Liu K., Reynders E., DeRoeck G., Lombaert G. 2009. Experimental and numerical analysis of a composite bridge for high-speed trains. *Journal of Sound and Vibration*, 320 (1–2): 201–220.

Mohammad J., Guralnick S., Polepeddi R. 1998. Bridge fatigue life estimation from field data. *Practice Periodical on Structural Design and Construction* 23: 128–133.

Ni K. 1999. Advances in stochastic theory of fatigue damage accumulation. *Advances in mechanics*, 29(1): 43–65.

Peng X.Q., Shi J. 2011. Study on the dynamic responses and exchange-filling strengthening measures in embankment-bridge transition sections of existing heavy railway. *Journal of railway science and engineering* 8(4): 7–13.

Sun Y, Gu P. 2007. Study of dynamic characteristics and strengthening lateral stiffness of Existing railway steel truss bridge. *Journal of shijiazhuang railway institute* 20(1): 10–13.

Xia, H., Rocek G.D., Zhang, H.R., Zhang N. 2001. Dynamic analysis of train-bridge system and its application in steel girder reinforcement. *Computers & Structures* 79(21–22): 1851–1860.

Zhao Z.W., Halder A., Breenjr. L. 1994. Fatigue-reliability evaluation of steel bridge. *Journal of Structural Engineering* 120: 1608–1623.

Design actuator pump systems Left Ventricle Assist Devices

V. Morozov, A. Zhdanov, L. Belyaev & I. Volkova
Vladimir State University Named Alexander and Nikolay Stoletovs, Russia

ABSTRACT: The article discusses the design of actuators pump systems of left ventricle assist device. Presents the results of fusion power for the drive systems, of left ventricle assist devices.

1 INTRODUCTION

The implantable Left Ventricle Assist Device (LVAD) used a pulsating type pump modules—Mechatronic Drives (MD) comprising a multi-pole DC motor valve, working in reverse mode, and an Actuating Mechanism (AM) convert rotary motion into linear (Morozov, 2006). Currently, the most common and reliable design of these modules are AM Roller Screw Mechanism (RSM), and as the DC motor used torque multi-pole motors. Examples of such structures can be: Baylor LVAD, TAH, USA (RSM); Yamagata LVAD, TAH, Japan (RSM); Japan Hokkaido LVAD (BSM); Swiss LVAD, Switzerland (RSM), and others. (Morozov, 2005) The design of such systems is only possible with the combined influence of the DC motor and AM. To accurately determine the parameters of AM and DC motor it is necessary to limit the synthesis according to the criteria of dynamic capabilities, low Power Consumption (PC).

2 MATERIALS AND METHODS

Determination of the allowable range Kinematic Transfer Function (KTF) AM them to perform according to the formula:

$$S_1 \leq S_X \leq S_2$$

where $S_{1,2} = \frac{1}{2}\frac{\eta M_\Pi}{F_H}\left(1 \pm \sqrt{1 - \frac{P_H}{P_0}}\right)$.

There M_Π—the starting point of DC motor; F_H—force load acting on the output member; η—efficiency of AM; P_H—load capacity $P_H = F_H v_H/\eta$; P_0—DC motor power at the rated voltage $P_0 = M_\Pi \omega_{XX}/4$; ω_{XX}—angular speed of idling DC motor; moving speed $v_H(t)$ in the ejection phase has the form $v_H(t) = v_{sys}\left[1 - \exp(-t/T_\gamma)\right]$; v_{sys}—amplitudnoe value of systolic velocity, T_γ—electromechanical time constant DC motor T_γ.

Optimal KTF for a minimum PC is given by:

$$S_P = \sqrt{p_\Pi v_H T_y \frac{\eta M_\Pi}{F_H}\frac{U}{U_0}},$$

where p_Π—number of pole pairs; T_y—time constant of the control system; U—voltage on the output of the control system; $U_0 = 12$ V—voltage at the input of the DC motor.

Motor selection with the necessary capacity should provide the required law of displacement load range KTF $S_1 \leq S_X \leq S_2$, because it is not always technically possible to observe the exact value of the KTF, as a single value it may be less than optimal. For MD LVAD valve needs a contactless phase DC motor permanent magnet with a design that is optimized for the time (Morozov, 2009). For MD with RSM needs hollow rotor motor. Because of the machines best meets DC motor flanged EC 45 Flat (Maxon motor, USA), the DC motor with a hollow rotor 563-06B (Kollmorgen Magnedyne, USA) and on the basis of domestic DC motor DBM50 (Table 1).

Table 2 summarizes the results of the power synthesis for MD LVAD as described in (Morozov, 2006) on the basis of Table engines 1. As basic, operation mode is selected with the characteristics of: heart rate $f = 120$ min^{-1}; the ratio systole/diastole $\beta = 1:3$; systolic pressure $p_{sys} = 120$ mm Hg; discharge quantity $V = 80$ ml; diaphragm diameter $d = 90$ mm.

Engines have approximately equal coefficients of losses, i.e. usefulness of the criterion of heat can only be tested experimentally. The disadvantage of engine EC 45 Flat is a great inertia that requires a large expenditure of energy, but it is compensated by the high engine power. Version with motor 563-06B is the most optimal (because on the one hand provides the largest number of modes, and on the other—quite economical in terms of energy).

Table 1. Specifications DC motor MD LVAD.

Name engine	The number of poles	Power, W	Starting torque, Nm	Idle speed, min^{-1}	The moment of inertia of the rotor, kg·mm^2	Starting current, A	Voltage	Thermal resistance, °C/W	Length mm	External diameter, mm	Inner diameter, mm	Weight, g
DBM50	14	11,78	0,3	1500	0,45	2	12	1,4	30	50	29,6	180
563-06B	14	26,5	0,53	1910	14,00	7,9	11,56	3,5	20,3	45,7	25,0	130
EC 45 flat	16	29,95	0,26	4400	9,25	10,2	12	2,4	16,3	45,0	–	88

Table 2. Results of the power of synthesis.

Parameter	Value		
Type of engine	563-06B	DBM50	EC 45 flat
	Load settings		
Maximum speed, mm/sec t	113,6	80,27	139,88
Load, N	101,493	101,493	101,493
Power (considering efficiency),	16,47	11,64	20,28
The required voltage, V	9,11	11,93	9,87
	The parameters of the mechanism and the working point		
Range KTF, mm^{-1}	4,42–18,55	5,78–7,22	2,43–8,83
Optimal KTF, mm^{-1}	4,42	5,78	3,25
Voltage, V	11,56	12	10,5
The frequency at the operating point, mm^{-1}	1542	833	2582
The moment at the operating point, Nm	0,102	0,133	0,075
Power loss at the operating point, W	8,01	7,24	11,23
Heating temperature, °C	28,0	25,35	31,0

As AM MD LVAD proposes two schemes: (1) RSM recessed into the hollow rotor motor, and (2) a planetary screw mechanism with rack and pinion on the basis of the motor flanged.

In the first scheme RSM obtained using simple planetary gear consisting of one central wheel and satellite set in cage. The role of the central wheel carries a threaded screw with a screw thread, and the role of satellites—rollers with annular ring nails, which are mounted in the supports of the carrier. Upon rotation of the driving member, which may be a screw, and the carrier, the driven member will be forward movement in their guides. If constructively provided axially adjusting the position of the rollers in the assembly, the number of rollers can be limited to only the neighborhood condition, and the screw may have any number of entries cuts. The need to comply with any relationship between the number of entries with the number of screw rollers in this case is unnecessary. An important advantage with the ring transfer rollers is the presence of only one engagement in each roller instead of four, which drastically reduces the number of passive connections in transmission. From design to RSM ring rollers, it follows that each turn of the driving member corresponds to the axial movement of the driven member at the stroke of screw thread screw. RSM efficiency is defined as:

$$\eta = \frac{\mathrm{tg}\,\gamma}{\mathrm{tg}(\gamma + \rho)},$$

where ρ—angle of friction engaging screws and clips, γ—angle of deviation of the normal at the point of application of force from the rotational axis of the carrier.

Figure 1 shows the design of the MD system LVAD-based RSM and RSM options for different engines.

In the second scheme, as a downshift "spin-spin" in the MD, used Planetary Reduction (PR), made under the scheme A_{ha}^b (Fig. 2). PR is the main chevron pinion shaft with external teeth, three satellites, the central wheel with the right and left cuts. The satellites are installed in cage and rotate about their axes, making planetary motion around a central shaft gear. When fixed wheel movement can be transmitted from the pinion shaft to the carrier, and vice versa. Efficiency PR with the rack looks like:

$$\eta = \eta_{\text{Пл}}\eta_{\text{p}},$$

where η_{pl}—the efficiency of the PR; η_r—the efficiency of the rail.

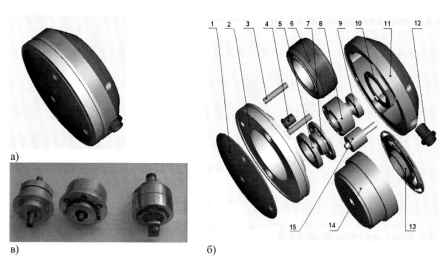

Figure 1. Structure MD system LVAD based RSM: a) CAD-model assembled; b) CAD-model is disassembled; c) options RSM: 1—pusher; 2—body of the MD; 3—rail; 4—bush; 5, 9—Bearings (schematic); 6—DC motor hollow rotor; 7—housing cover RSM, 8—body of the RSM; 10—ring gauges; 11—a cover of MD; 12—output socket, 13—cover; 14—body; 15—RSM.

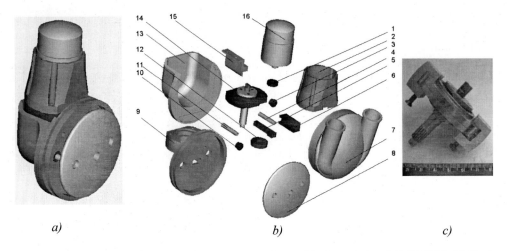

Figure 2. The design of the MD system LVAD-based PR with the rack (Pro/ENGINEER WF4) a) CAD-model assembled; b) CAD-model is disassembled; c) PR: 1,12—bearings; 2—motor housing; 3,10—anti-friction bushings; 4,11—guides; 5—rack; 6—anti-friction strap; 7—artificial heart ventricle; 8—membrane; 9—prefabricated housing; 13—to the outer cover; 14—PR; 15—clamp; 16—DC motor.

The procedure for the design of AM systems MD LVAD can be represented by the following steps:

1. Preliminary stage, which includes the creation of them in well-known design techniques, development of mathematical models of the mechanism, the creation of a database of standard components and assemblies, series engines;
2. The conceptual design phase, which is a structural synthesis of MD and AM;
3. The stage of parametric synthesis, which includes the power synthesis MD, calculation parameters AM, detailed design and assembly, geometrical analysis of solid models;
4. Analysis of the structure, consisting of finite element analysis design and analysis of dynamic AM characteristics of the MD;
5. Synthesis and optimization criteria can include optimizing design of reliability, durability, wear, smoothness, and minimize the energy consumed al., Synthesis mechanical adjusting devices synthesis algorithms MD digital control system;
6. Production engineering is the development of manufacturing technology and assembly AM and MD, as well as preparation of design documentation;
7. Fabrication Step.

3 RESULTS

The proposed new functional arrangement converting the rotational movement of the shaft in the movement of the diaphragm DC motor heart pump allows you to use the advantages of both PR screw units and rack and pinion. Using a helical gear transmission with large angles of inclination of the teeth provide noise reduction by reducing the gap in the engagement and qualitative changes kinematics transmission elements, thereby increasing the efficiency of MI, which increases the load carrying capability and reduces the size of transmission. Quiet and small relative size of the MD can use them in devices and mechanisms operating in a deficit of weight and volume, i.e., as part of an implantable system LVAD. Table 3 shows the comparative characteristics of the MD systems based on different LVAD AM.

Research carried out in the framework of a grant from the President of the Russian Federation for the state support of young Russian scientists MK-5860.2015.8.

Table 3. Specifications MD-based systems LVAD AM.

Parameter	Baylor LVAD, Baylor Medical Center, US (RSM)	Yamagata LVAD, Yamagata, Japan (RSM)	Vlsu (RSM)	Vlsu (PR with rail)
Diameter, mm	97	90	92	92
Thickness, mm	70	56	56	50
Volume of the artificial ventricle, ml	250	256	250	250
Weight, g	620	380	500	450
Productivity, l/min	7,5	7,5	7,4	7,2
Efficiency	45	20	42	56

REFERENCES

Morozov, V.V. 2005. Roller. The kinematic characteristics: monograph. *Vladimir Univ Vladim.gos.un Press*, Vladimir: 78.

Morozov, V.V., Zhdanov, A.V., Novikov, E.A. & Kosterin A.B. 2006. The implantable system VC-based mechatronic modules: Monograph. *Vladimir State University Publishing House*, Vladimir: 134.

Morozov, V.V., Zhdanov, A.V. & Belyaev L.V. 2009. Development of mechatronic module implantable artificial heart system and laboratory bench tests. *New technologies*, 5 (98). - 58–61.

Resistance of asphalt binders to formation of frost cracks

Pavel Coufalik, Ondrej Dasek, Petr Hyzl, Dusan Stehlik, Jan Kudrna,
Iva Krcmova & Pavel Sperka
Faculty of Civil Engineering, Brno University of Technology, Veveri, Czech Republic

ABSTRACT: Monitoring low-temperature properties of asphalt binders is fundamental for determining resistance of asphalt pavements to the formation of frost cracks. This paper compares the properties of paving bitumen, two bitumens modified with various amounts of Crumb Rubber (CRmB) and three bitumens with variable amount of SBS. Also a simulation of short-term and long-term ageing was performed in order to determine the resistance of these bitumens to the ageing process. The individual bitumens, including the effect of ageing were compared using empirical tests, in particular, the needle penetration test and the softening point and also the relaxation test performed using Dynamic Shear Rheometer (DSR). The results of this test for the individual bitumens were subsequently correlated with the results from Bending Beam Rheometer (BBR). A strong correlation was found between the flexural creep stiffness and the maximum shear stress in DSR, between the m value and the area of the relaxation curve, and between the m value and the residual shear stress.

1 INTRODUCTION

Asphalt mixtures are one of the most frequently used materials in pavement design. There are over 5 million km of roads in Europe, out of which more than 90% are made of bitumen. Approximately 275 million tons of asphalt mixtures were produced in Europe during 2012 (EAPA, 2012). In the United States there are 4.2 million km of roads, and asphalt mixtures make up 93% of these. In the U.S., approximately 500 million tons of asphalt mixtures are produced every year, costing over 30 billion dollars (National Asphalt Pavement Association, 2012). It is therefore essential to pay close attention to their quality and life-time, where the key determinants of these are the actual properties of these asphalt mixtures.

Bituminous binder is a material of organic origin made of various hydrocarbon compounds that under certain conditions can undergo degradation processes as a result of heat, oxygen in the air, ultraviolet electromagnetic radiation, or combination of these. Bituminous binders are commonly exposed to these factors during production and laying of the asphalt mixture and during its life time. The effect of all these factors together can overall be referred to as binder ageing. The process of binder ageing causes chemical, as well as physical changes in the binder properties, which makes the bituminous binder stiffer and more brittle, therefore more vulnerable to the formation of cracks in the pavement, especially at lower temperatures.

Laboratory methods that simulate change in properties are used in order to determine the changes in bituminous binder properties. Such methods can be divided into two categories: short-term and long-term binder ageing. Short-term binder ageing corresponds to the time of production and the laying of the binder. The most common method used for short-term binder ageing simulation is the Rolling Thin Film Oven Test (RTFOT) (EN 12607-1, 2007). Long-term binder ageing then simulates change in properties of bituminous binders during its life time and the most commonly used method is the Pressure Ageing Vessel (PAV) (EN 14769, 2006). An alternative method to PAV is the modified RTFOT, where the exposure period is prolonged three times (3xRTFOT). This method of long-term binder ageing simulation is used for example in Austria, where the maximum acceptable increase in softening point during the 3xRTFOT is 15 °C (Hospodka, 2013; Spiegl, 2009; RVS 08.97.05, 2010). Detailed information regarding the correlation between PAV ageing method and modified RTFOT can be found for example in paper (Muller and Jenkins, 2011).

Change in properties of bituminous binders during ageing can have major effect on its resistance against crack formation. This paper compares properties of selected paving bitumen, two Crumb Rubber Modified Bitumens (CRmBs) and finally three bitumens with variable amounts of Styrene Butadiene Styrene (SBS).

A device referred to as Dynamic Shear Rheometer (DSR) can be used to determine the relaxation properties of bituminous binders that are related to low-temperature behavior. This test determines the maximum stress when a shear strain is applied and the subsequent rate of relaxation of shear strain can be an indicator of the ability of the material to reduce the strain and also an indicator of quality of the modification system. The ability of a bituminous binder or asphalt mixture to quickly absorb strain is an important factor, which determines the ability of such mixture to resist crack formation. (Narayan et al., 2012; Coufalik et al., 2014; Dasek et al., 2014).

When a shear strain is applied to an ideally elastic material, the shear stress relaxation does not occur and the shear strain remains constant. In case of an ideally viscous behavior, upon application of shear strain there is almost immediate disappearance of inner stress. Application of shear strain on a viscoelastic material results in a delayed relaxation of shear stress, which is either partially, or after some time, full. The relaxation curve has exponential shape and the degree of shear stress relaxation depends on the viscous component of the material. (Mezger, 2011).

The aim of this paper was to compare properties of the individual binders and determine their resistance to ageing, which is a factor affecting their resistance against crack formation. Another aim was to compare the results of flexural creep stiffness and the m-value (tangent of the flexural creep stiffness curve at t = 60 s) determined using the Bending Beam Rheometer (BBR) with the results from the relaxation test using DSR including determination of the effect of ageing using the RTFOT and 3xRTFOT methods for selected binders. These results can be used to estimate, whether a binder is resistant against crack formation at low temperatures.

2 METHODS

Simulation of short-term binder ageing was performed using the RTFOT method in accordance with the EN 12607-1 (2007) standard (75 minutes at 163 °C, 8 × 35 g of bitumen). Long-term ageing was modeled by the modified 3xRTFOT method (225 minutes at 163 °C, 8 × 35 g of bitumen).

In order to compare the properties of the bituminous binders before and after ageing, the binders were subjected to empirical tests, in particular to the needle penetration test in accordance with EN 1426 (2007) and the softening point was tested by the ring and ball method in accordance with the EN 1427 standard (2007). Low-temperature characteristics of paving bitumen 50/70 and CRmBs (crumb rubber ratio of 11% and 17%) were determined using BBR in accordance with EN 14771 (2012) at temperatures of −10 °C, −16 °C, and −22 °C. Based on the SHRP program (Strategic Highway Research Program), temperatures where the flexural creep stiffness S_m, at time t = 60 s was approaching 300 Mpa, and temperature at which the sample shows m-value of 0.3 were determined. Based on the findings from the SHRP research, cracks form in bitumen surface when it is strained for a period of 2 seconds by the lowest estimated temperature and when the flexural creep stiffness exceeds 300 MPa. During laboratory testing it is assumed that the same flexural creep stiffness is determined for t = 60 s if the temperature is increased by 10 °C (Harrigan, 1994; Valentin, 2003). Each result was determined using at least two test samples so that reproducibility of measurement was in accordance with the EN 14771 standard (2012).

All binders were subsequently subjected to a relaxation test using DSR. During the determination of the rate of shear stress relaxation in DSR, a method was chosen in which the bituminous binder sample is first subjected to shear strain at a particular temperature and with particular shear rate and the shear strain lies in the linear viscoelastic region. After the initial increase of shear stress, which corresponds to the shear strain and shear rate, the shear strain is maintained for the entire duration of the test at a constant value and the decrease of shear stress in time is monitored. The result of the test is a shear stress relaxation curve, which shows the relationship between shear stress decrease and time. Residual stress is a ratio of shear stress after a particular relaxation period and maximum shear stress expressed in %.

Shear stress relaxation in DSR was determined at temperatures −10 °C and −16 °C using parallel plate-plate geometry with a diameter of 4 mm and 2 mm gap size. Relaxation test was performed in controlled shear strain mode. The initial shear strain was set to a value of 0.5% of the sample thickness and load time set to 10 s. Relaxation period was set to 15 minutes so that it was possible to calculate the area under the curve at this time. Each result was determined using two test samples. A diagram of the test procedure is shown in Figure 1.

3 MATERIALS

A paving bitumen 50/70 was chosen for the laboratory tests. Two Crumb Rubber Modified Bitumens (CRmBs) were produced in the laboratory. 11% and 17% of crumb rubber with a particle size 0/1 mm was added to paving bitumen. CRmBs were prepared at a temperature of 175 °C for 60 minutes. Properties of the selected bitumens were subsequently compared to paving bitumen 70/100 with 1, 3 and 5% addition of SBS.

All bitumens were aged in a laboratory by the RTFOT or 3xRTFOT method. Table 1 lists selected properties of the individual binders with respect to laboratory ageing.

4 RESULTS AND DISCUSSION

First part of this section presents the results of bituminous binder empirical tests and the effect of ageing by RTFOT or its modified version 3xRTFOT on bituminous binder properties. The next part then describes the results of tests performed using BBR and DSR.

4.1 Empirical tests of bituminous binders

Highest penetration values and lowest softening point values were observed for the paving bitumen. Increasing the degree of modification by crumb rubber and also SBS led to decrease in penetration and softening point also increased. Table 1 shows the effect of ageing of bituminous binders on the needle penetration values and the softening point of the individual binders. As the extent of ageing increased, the needle penetration value decreased while the softening point value increased. Only in the case of binders with 3 and 5% SBS the ageing by 3xRTFOT method caused slight decrease in softening point in comparison to the RTFOT method.

By comparing the penetration values and softening point values of binders aged by the 3xRTFOT and non-aged binders it is possible to determine resistance of the individual

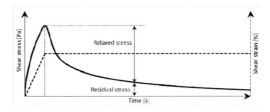

Figure 1. Course of the shear stress relaxation test.

Table 1. Summary of bituminous binder properties.

Binder	Ageing	Empirical tests Needle pen. 25 °C [0,1 mm]	Softening point [°C]	BBR S_{60s}/m_{60s} [°C]	DSR Relative residual stress t = 60 s −10/−16 °C [%]	Maximum shear stress −10/−16 °C [MPa]	Area under relaxation curve −10/−16 °C [%]
50/70	–	64	46.7	−16.8/−19.3	18.6/31.0	0.31/0.80	6.93/13.84
	RTFOT	41	51.1	−15.8/−17.2	25.1/36.9	0.47/0.94	10.58/18.47
	3xRTFOT	27	55.6	−15.1/−15.2	30.4/43.2	0.47/1.03	13.72/23.61
CRmB 11%	–	39	56.1	−21.7/−19.7	27.2/36.2	0.16/0.40	12.75/18.42
	RTFOT	27	64.0	−19.9/−15.3	28.1/37.1	0.25/0.44	12.85/19.47
	3xRTFOT	20	67.1	−20.1/−14.8	31.4/39.2	0.31/0.62	15.28/21.49
CRmB 17%	–	31	63.5	−22.7/−18.4	29.6/34.7	0.15/0.31	14.21/18.81
	RTFOT	18	71.6	−24.1/−17.3	28.7/37.4	0.16/0.36	13.54/18.89
	3xRTFOT	19	74.6	−23.0/−16.3	29.0/37.3	0.18/0.42	13.71/19.75
1% SBS	–	56	48.3	–	30.0/44.7	0.45/0.99	13.91/24.43
	RTFOT	33	54.5	–	25.0/39.9	0.37/0.79	10.55/20.00
	3xRTFOT	16	64.1	–	35.9/47.5	0.45/1.08	18.17/27.00
3% SBS	–	50	51.8	–	29.2/43.5	0.38/0.88	13.66/24.16
	RTFOT	30	68.2	–	32.1/44.6	0.37/0.78	15.18/24.96
	3xRTFOT	18	67.1	–	38.8/49.6	0.52/0.99	20.09/28.81
5% SBS	–	29	67.4	–	32.5/47.0	0.37/0.81	15.35/26.01
	RTFOT	22	79.6	–	34.3/45.6	0.36/0.70	16.78/25.01
	3xRTFOT	15	79.4	–	36.5/48.0	0.40/0.78	18.80/28.38

binders to ageing. Needle penetration value of paving bitumen 50/70 decreases to 42.2% of its original value. Addition of crumb rubber to the bitumen causes decrease in resistance to ageing. For CRmBs with 11% or 17% crumb rubber, the needle penetration value decreases to 51.3 or 61.3% of the original value respectively. 70/100 binder with 1%, 3% or 5% SBS causes decrease in needle penetration to 28.6%, 36.0% or 51.7% of the original value, respectively. This shows that higher ratio of SBS leads to increased resistance to ageing.

Same comparison using the softening point shows that the value of softening point of the paving bitumen 50/70 increases after ageing using 3xRTFOT by 19.1% in comparison to non-aged binder. Softening point of a binder with crumb rubber added is higher than in case of paving bitumen. For CRmB with 11% crumb rubber, the value of softening point after performing 3xRTFOT ageing increased by 19.6% in comparison to non-aged binder. Softening point of CRmB with 17% crumb rubber increased by 17.5%. In case of binders with SBS added, higher ratio of this modifier causes lower sensitivity towards potential change in softening point after ageing process by 3xRTFOT. Addition of 1% SBS causes increase in softening point by 32.7%, addition of 3% SBS leads to a softening point 29.5% higher and addition of 5% SBS causes increase in softening point by 17.8%.

4.2 Results of low-temperature behavior determined using BBR

Determination of stiffness modulus upon bending using BBR was performed for paving bitumen 50/70 and for both CRmBs. Also the resistance to ageing of bituminous binders was analyzed for the individual binders using the RTFOT and 3xRTFOT methods.

Ageing by RTFOT or 3xRTFOT causes gradual worsening of the lower critical temperature, which increases as the ageing process progresses. An exception is the CRmB with 17% crumb rubber, where there was a slight decrease in the lower critical temperature value after ageing.

Determination of critical temperature was performed by interpolation of flexural creep stiffness values and the m-value, which were determined at temperatures −10 °C, −16 °C and −22 °C. The lower critical temperature of bituminous binders was determined for flexural creep stiffness after reaching S_{60s} = 300 MPa and for m-value after reaching m_{60s} = 0.3. The results of the lower critical temperature values are given in Table 1. These results show that increasing the amount of crumb rubber in a binder leads to desirable decrease in lower critical temperature for both the flexural creep stiffness (S_{60s}) and the m-value. An exception was the lower critical temperature of CRmB with 17% crumb rubber determined from the m-value, which was slightly higher than the lower critical temperature of the remaining binders. This could be explained by the non-homogeneity in properties of binders containing high ratio of crumb rubber.

4.3 Results of shear stress relaxation determined using DSR

Selected values of shear stress relaxation determined using DSR are given in Table 1. The values correspond to residual stress at relaxation time t = 60 s given as a percentage of maximum achieved shear stress value, the maximum value of shear stress during stressing of the specimen and finally percentage ratio of area under the relaxation curve for relaxation time less than 15 minutes. The area under the curve is given in percentages, where 100% would correspond to an ideally elastic material (stress would not disappear).

Lowering the test temperature from −10 °C to −16 °C caused 34% increase in relative residual stress at time 60 s in case of the paving bitumen 50/70. For the CRmBs, the level of increase was only 22%. This is due to the positive effect of crumb rubber in the binder, which decreases temperature sensitivity. Increasing ratio of SBS in a bituminous binder leads to smaller change in relative residual stress. For the binder 70/100 with 5% SBS there was an increase in stress by 27%. Increasing the amount of crumb rubber causes increase in relative residual stress after relaxation period of 60 s at temperatures −10 °C, as well as −16 °C. This is also associated with increased area under the relaxation curve. However, the initial maximum shear stress decreases for both temperatures as the ratio of crumb rubber increases, even though higher ratio of crumb rubber causes large increase in viscosity and complex shear moduli at high temperatures (Dasek, 2014).

During ageing of the binders compared it was possible to see increase in relative residual stress after 60 s relaxation and also increase in the area under the relaxation curve at both test temperatures. Only in case of the CRmB with 17% crumb rubber, the increase in relative residual stress and area under relaxation curve was not very obvious, which again proves the high resistance of this binder against ageing. Maximum shear stress of all binders increased as the duration of ageing increased in case of all binders and at both temperatures tested.

When comparing the time of decrease of shear stress to 25% of its maximum value (relaxation rate) it can be seen that at a temperature of −10 °C it is the paving bitumen 50/70 that relaxes fastest (in 38 s). Shear stress of CRmB with 11% and 17% crumb rubber decreases to 25% of its original value in 72 s or 87 s, respectively. Simulation of binder ageing by 3xRTFOT causes the shear stress of paving bitumen 50/70 to relax to 25% of its original value in 90 s. CRMBs with 11% and 17% show a decrease to such value in 101 s or 82 s, respectively. This fact proves the higher resistance of paving bitumen to binder ageing. When comparing the results at a temperature of −16 °C, the relaxation of shear stress in case of paving bitumen 50/70 decreases to 25% of its original value after ageing by 3xRTFOT method in 251 s, in case of CRmB with 11% crumb rubber in 196 s and for CRmB with 17% crumb rubber in 164 s. It is therefore obvious that CRmB with 17% crumb rubber is able to relax the applied stress at −16 °C fastest. One can therefore assume that modifications with higher ratios of crumb rubber increase the resistance of bituminous binders against frost crack formation in comparison to paving bitumen and low-viscous CRmB with 11% crumb rubber.

If the effect of variable ratio of SBS in paving bitumen 70/100 is evaluated in a similar manner, it can be seen that relaxation of shear stress in case of these binders is slowest.

For example, in case of bituminous binders with 1% to 5% SBS aged by the 3xRTFOT method at −16 °C, the shear stress decreases to 25% of the maximum measured shear stress in 359 s to 430 s. It is therefore obvious that at low temperatures, the CRmBs are able to eliminate the applied stress fastest.

4.4 *Correlation between the results of low temperature behavior measured using BBR and relaxation of shear stress measured using DSR*

The following chapter describes correlation between the results from BBR and DSR. Figure 2 shows the strong correlation coefficient between flexural creep stiffness and maximum shear stress ($R^2 = 0.96$ at temperature −10 °C and $R^2 = 0.98$ at temperature −16 °C). Figure 2 also proves that as the flexural creep stiffness S_{60s} in BBR increases, also the maximum shear stress during relaxation test in DSR increases in case of both test temperatures. The correlation is quite strong and test geometry as well as the way of loading had negligible effect on correlation results of the test. Area under the curve during relaxation test was indirectly related to the m-value at both test temperatures—as the m-value determined using BBR increases, the area under relaxation curve determined during test of shear stress relaxation using DSR decreases (Fig. 3). In case of the temperature −10 °C the correlation coefficient $R^2 = 0.81$, for temperature −16 °C the determined correlation coefficient

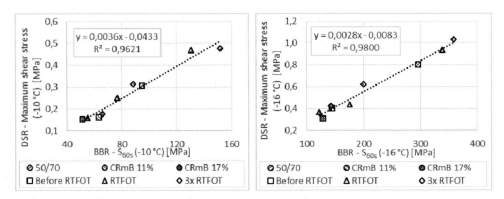

Figure 2. Relationship between maximum shear stress in DSR and the value of S_{60s} in BBR at temperatures −10 and −16 °C.

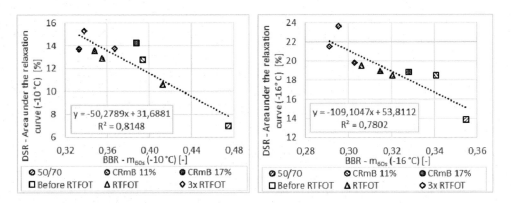

Figure 3. Relationship between area under the relaxation curve determined using DSR and the value of m_{60s} determined using BBR at temperatures −10 and −16 °C.

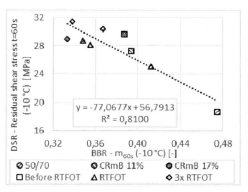 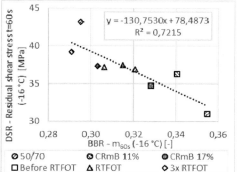

Figure 4. Relationship between residual shear stress measured in DSR and m_{60s} value measured using BBR at temperatures −10 °C and −16 °C.

$R^2 = 0.78$. Figure 4 shows that increase of the m-value is associated with decrease in relative shear stress determined using DSR at relaxation time 60 s and in case of both test temperatures (−10 °C and −16 °C). The correlation coefficient at −10 °C $R^2 = 0.81$, for −16 °C is the $R^2 = 0.72$.

5 CONCLUSION

This paper analyzes the properties of paving bitumen 50/70, two crumb rubber modified bitumens with 11% and 17% crumb rubber and three 70/100 bitumens, which were modified with various amounts of SBS (1%, 3% and 5%). The effect of short-term ageing (RTFOT) and long-term ageing (3xRTFOT) on changes in properties (needle penetration, softening point, BBR, shear stress relaxation using DSR) is described for all the individual binders.

Empirical tests show that increasing the extent of modification by both crumb rubber and SBS leads to decreased penetration and increase in softening point, which proves their higher resistance against deformations. Binder ageing also increases the softening point, which has a positive effect on the increase in resistance against deformations, but can also cause decrease in resistance against frost crack formation at low temperatures.

Using a BBR device a comparison was performed between the low temperature properties of paving bitumen 50/70 and two crumb rubber modified bitumens with 11% and 17% crumb rubber. The results of these measurements show that ageing by the RTFOT and 3xRTFOT method causes gradual increase in lower critical temperature, which is undesirable with regards to the risk of frost crack formation. An exception is the CRmB with 17% crumb rubber, where after ageing by RTFOT there was a slight decrease in lower critical temperature, which could be due to high resistance of CRmB with higher crumb rubber content against ageing. The actual increase in crumb rubber content in CRmBs causes decrease in lower critical temperature for flexural creep stiffness (S_{60s}) as well as the m-value.

The main aim of this paper was determination of shear stress relaxation in DSR. For bituminous binders it is desirable that at low temperatures, the induced stress disappears as quickly as possible. The results show that decreasing the temperature and ageing of binders slow down shear stress relaxation. Slowest (i.e. worst) shear stress relaxation was observed in case of the SBS modified bitumens. In contrast, addition of crumb rubber led to a lower value of shear stress at the same extent of applied shear strain and shear stress relaxed quicker than in case of paving bitumen 50/70 and SBS-modified paving bitumen 70/100.

Comparison of the results determined using BBR and DSR shows a strong correlation between the initial maximum shear stress determined during the shear stress relaxation test and the flexural creep stiffness was measured with BBR at temperatures −10 °C and −16 °C.

A relationship was also observed between the relative area under relaxation curve and the m-value determined at t = 60 s using BBR at same test temperatures. It was also found that as the m-value increases, the relative residual stress determined using DSR at relaxation time 60 s decreases in case of both test temperatures (−10 °C and −16 °C).

ACKNOWLEDGMENT

The paper was created with the support of the FAST-J-15-2901 project "Reduction of crack formation in asphalt pavements based on the used asphalt binder" and the project TA03030381 "New Test Methods for Asphalt Binders and Mixtures Allowing the Extension of Asphalt Pavements Performance".

REFERENCES

Coufalik, P., Dasek, O., Kachtik, J., Kudrna, J. & Stoklasek, S. 2014. The Stress Relaxation of Modified Bitumens. IN: *Proceeding of The 13th Annual International Conference on Asphalt, Pavement Engineering and Infrastructure*. Liverpool: Liverpool Centre for Materials Technology, pp. 14.
Dasek, O. 2014. *Uplatneni pryzoveho granulatu v asfaltovych pojivech a hutnenych asfaltovych smesich*. Disertační prace. Brno: Vysoke uceni technicke v Brne.
Dasek, O., Hyzl, P., Varaus, M., Coufalik, P., Spacek, P. & Hegr, Z. 2014. Usage of advanced functions of Dynamic Shear Rheometer for the selection of a suitable binder for asphalt mixtures. IN: *Asphalt Pavements*. Netherlands: CRC Press/Balkema, Taylor & Francis Group, pp. 10.
Harrigan, E., Leahy, R. & Youtcheff, J. 1994. The SUPERPAVE Mix Design System Manual of Specification, Test Methods, and Practices. IN: *SHRP-A-379* [online]. Washington. Available from: http://onlinepubs.trb.org/onlinepubs/shrp/SHRP-A-379.pdf [Accessed 2015-05-11].
Hospodka, M. 2013. *Alterungsmechanismen von Bitumen und Simulation der Alterung im Labor*. Masterarbeit. Wien: Universität für Bodenkultur Wien.
Mezger, T. 2011. *The rheology handbook: for users of rotational and oscillatory rheometers*. 3rd rev. ed. Hanover, Germany: Vincentz Network.
Muller, J. & Jenkins, K. 2011. The Use of an Extended Rolling Thin Film Ageing Method as an Alternative to the Pressirised Ageing Vessel Method in the Determination of Bitumen Durability. *10th Conference on Asphalt Pavements for Southern Africa*, no. 1, pp. 14.
Narayan, S., Krishnan, J., Deshpande, A. & Rajagopal, K. 2012. Nonlinear viscoelastic response of asphalt binders: An experimental study of the relaxation of torque and normal force in torsion. IN: *Mechanics Research Communications* [online]. pp. 9. Available from: http://1url.cz/355V [Accessed 2015-06-22].
Spiegl, M. 2009. The road to performance-related bitumen specification. IN: *12th Colloquium on Asphalt and Bitumen*. Kranjska Gora: ZAS, pp. 10.
Valentin, J. 2003. *Uzitne vlastnosti a reologie asfaltovych pojiv a smesi: charakteristiky, nove zkusebni metody, vyvojove trendy*. Praha: Ceske vysoke uceni technicke v Praze.
2004. EN 13398 *Bitumen and bituminous binders—Determination of the elastic recovery of modified bitumen*. Praha: Český normalizační institut.
2006. EN 14769 *Bitumen and bituminous binders. Accelerated long-term ageing conditioning by a Pressure Ageing Vessel (PAV)*. Praha: Český normalizační institut.
2007. EN 12607-1 *Bitumen and bituminous binders. Determination of the resistance to hardening under influence of heat and air. RTFOT method*. Praha: Český normalizační institut.
2007. EN 1427 *Bitumen and bituminous binders. Determination of the softening point. Ring and Ball method*. Praha: Český normalizační institut.
2010. RVS 08.97.05 *Anforderungen an Asphaltmischgut*. Wien.
2012. Asphalt. *EAPA* [online]. Available from: http://www.eapa.org/asphalt.php [Accessed 2015-05-12].
2012. EN 14771 *Bitumen and bituminous binders—Determination of the flexural creep stiffness—Bending Beam Rheometer (BBR)*. Praha: Český normalizační institut.
2015. Engineering Overview. *National Asphalt Pavement Association* [online]. Available from: http://1url.cz/bXB0 [Accessed 2015-05-11].

Experimental studies on pore water pressure changes at structure-soil interface under dynamic soil cutting

Gongxun Liu
Key Laboratory of Dredging Technology of CCCC, CCCC National Engineer Research Center of Dredging Technology and Equipment Co. Ltd., Shanghai, China

Xin Liang & Xiwei Wang
Department of Hydraulic Engineering, Tongji University, Shanghai, China

Guojun Hong
Key Laboratory of Dredging Technology of CCCC, CCCC National Engineer Research Center of Dredging Technology and Equipment Co. Ltd., Shanghai, China

Liquan Xie & Peipei Zhang
Department of Hydraulic Engineering, Tongji University, Shanghai, China

ABSTRACT: In the trailing suction dredging, the mechanical ploughing method is mainly used to loosen the hard sediment on the seabed. Cutter teeth are the structures which are used to loosen material from the seabed. The interaction between seabed soil and structure is an important issue in geotechnical engineering as well as coastal engineering. It is necessary to study the pore water pressure changes on the tooth-soil interface when subjected to mechanical ploughing, so as to study the required drag force exerted on the cutter teeth. The experiments were conducted in this paper to study the variation trend of pore water pressure in the soil while the cutter tooth moving forward at a specified speed. The experimental results show that the pore water pressure was decreased to be negative, even close to be a complete vacuum, while the seabed soil was subjected to a complex load of squeezing, raising and crushing.

1 INTRODUCTION

Structure-soil interaction is of great importance in the improvement of soil cutting tools and has attracted a lot of researchers to conduct deep studies. Tool-soil interaction is also important in dredging engineering since the compacted soil needs to be loosened before taking away from the seabed by suction force. Most existing soil loosening tools in dredging engineering were derived from tillage tools (see Fig. 1). The equation for soil-ploughing resistance was first derived in 1965 (Reece, 1965). Two-dimensional cutting theory (Van Os, 1997) to calculate the soil-cutting resistance of cutter teeth (Hettiaratchi et al, 1974 Godwin, 1977, Miedema, 2004), and the ploughing thickness was done (Van Os, 1987). Many experiments were done with saturated soil (He and Vlasblom, 1998) and unsaturated soil (Aluko et al, 2004). Some numerical simulations were done to study the characters of soil (Jiang, 2003, Usama et al, 1999).

The mechanical behavior of the seabed soil is complex to predict, different from the soil in agricultural engineering, so the dredging efficiency is difficult to improve. For fine sandy soil, the response of the soil includes the dilatancy effects. It is difficult to study and investigate quick cutting-induced pore water pressure changes in the seabed while considering the dilatancy effect, large deformation and fluid-solid coupling etc. Up till now, few experiments have been conducted to study pore water pressure which alters the stress state of the seabed

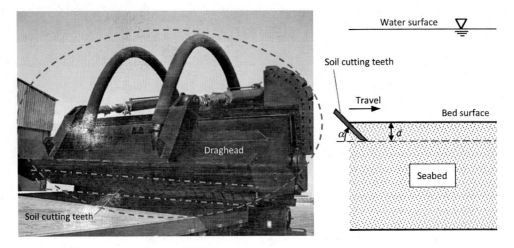

Figure 1. Draghead installed with soil cutting teeth and the working method.

soil, and accordingly changes the value of cutting resistance. It is important to conduct experimental studies for finding the soil cutting mechanism in dredging engineering.

The soil cutting forces and soil loosening efficiency have not been well documented, and they are critical for manure incorporations. The objectives of this study were: (1) to find the changing trend of the pore water pressure during the continuous soil cutting, and (2) to investigate effects of the tool working depths and speeds on cutting forces and soil surface disturbance. In this paper, a set of experiments were designed to study the developing characteristics of pore water pressure in the soil on the contact surface of cutter teeth and soil.

2 MATERIALS AND METHODS

The laboratory experiments in this study were carried out at the National Engineer Research Center of Dredging Technology and Equipment, Shanghai in China.

2.1 Equipment and testing facilities

The dredging test system (Fig. 3) includes moving trolley, control and monitoring system, control rig for cutting tool's characteristics, a soil bin. The dimensions of the soil bin were 12 m length, 0.8 m width, and 0.5 m depth.

The testing apparatus setup is shown in Figure 2. The main pieces of equipment used were the soil tank, a large cutter trolley, and a cutter tooth. The cutter trolley has three parts, a ladder, a trolley, and the controlling device. The trolley controlling device is used to control the cutter trolley moving forward and back on the guide rails. And the ladder is used to install the cutter tooth, and we could change the position and the rake angle of the cutter tooth by stretching and drawing hydraulic cylinders.

Figure 3 shows the schematic view of testing. The cutter tooth moves from left to right at the velocity of V_c with a rake angle of a. The horizontal motion trail of the cutter tooth end is called the cutting line. The soil bed in the soil tank was 0.5 m deep (h_2 in Fig. 2), 1.0 m wide, and 13.0 m long. And the water surface is 5 cm (h_1) higher than the soil surface. The thickness of soil above the cutting line is called soil cutting depth, h_1. The cutter tooth used in the experiment is approximately 16.0 cm long, 11.2 cm wide, and 3.0 cm of thickness. Two gauges #A and #B were installed into the holes of the cutter tooth in order to figure out the pore water pressure changing process, as shown in Figure 2. And two observers were also installed in the middle of the cutter tooth to observe the movement of soil grains by taking a video. These gauges were installed under the water surface to make sure that there was no air in the holes.

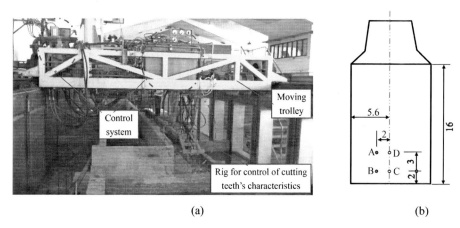

Figure 2. Dredging test system (cm).

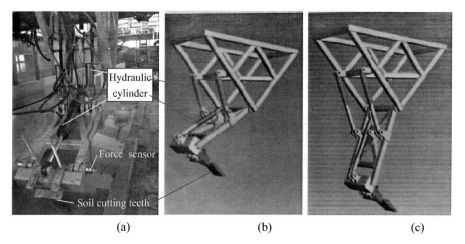

Figure 3. Adjusting mechanism for cutter tooth characteristics.

The data of pore water pressure were collected by SDS-810C dynamic strain detector. And the velocity is measured by a laser sensor monitoring which is installed on the trolley.

The soil-cutting resistance of the cutter tooth is closely related to the effective stress of soil. According to the principle of effective stress (as show in Eq. (1)), as the total stress σ remains unchanged, the pore water pressure u becomes smaller, and the effective stress σ' becomes larger. Therefore, the study of pore water pressure changing process can help to figure out the soil-cutting resistance of the cutter tooth.

$$\sigma = \sigma' + u \qquad (1)$$

2.2 Soil preparation

The same soil was used in these tests. In the Table 1, d_{50} is the grain size at which 50% of the soil is finer, ρ_d is the dry density of the soil. The coefficient e in the table is the void ratio. The soil used in the experiments was poorly graded sand.

The soil bed was paved according to the density of soil as guideposts. And water came from the bottom of the soil bed to make soil saturated. And the soil was made saturated for at least 48 hours before tests began.

Table 1. Physical properties of the soil.

Dry density ρ_d, g/cm³	Median particle diameter d_{50}, mm	Particle specific weight Gs	Void ratio e
1.29	0.078	2.69	1.02

Table 2. Test conditions for all the tests.

Test number	Velocity V_c, m/s	Soil cutting depth h_i, cm	Rake angle a, °
1	0.2	5	75
2	0.4	5	75
3	0.6	5	75
4	0.8	5	75
5	0.2	3	75
6	0.2	4	75
7	0.2	6	75
8	0.2	5	30
9	0.2	5	45
10	0.2	5	60

2.3 *Experimental design and measurements*

The procedure used for the tests is summarized as follows: (1) Place the soil in the tank; (2) Compact the soil with a large vibrator; (3) Fill up the tank gradually and make sure that the water surface is much higher than the soil surface; (4) Install the gauges into the cutter tooth; (5) Wait until the soil is totally saturated; (6) Plough the soil and collect the data; and (7) Empty the tank and prepare for the next test.

As the soil-cutting resistance is affected by the velocity of the cutter tooth, the thickness of cut and the rake angle of the cutter tooth (Miedema, 1984) (Eq. (2)), 10 tests were undertaken, as shown in Table 2.

$$F_{ci} \propto \frac{\rho_w \cdot g \cdot v_c \cdot h_i^2 \cdot b \cdot e}{k_m} \qquad (2)$$

In which F_{ci} = soil-cutting resistance; ρ_w = the specific weight of the water; g = acceleration due to gravity; b = width of the cutter tooth; and k_m = coefficient of permeability.

3 RESULTS AND DISCUSSIONS

During the continuous moving of the cutter tooth, the soil ahead of the tooth will be spilt away from the bed, and be lifted until the soil body has been destroyed. The soil which is spilt away will crack in the middle and be separated in two parts (as shown in Fig. 5). The mechanism of the negative pore water pressure can be represented as shown in the previous study (Qin, S., 2013). When the soil at the cutter tip is split away from the bed, we can see the shear zone above the soil cutting line. It is obvious that the arrangement of sediments in shear zone will change to a looser arrangement after shearing deformation. The phenomenon shows that the porosity increases at shear zone after the sediments moving freely. Along with the phenomenon the apparent pore volume increases obviously, which leads to expansion of soil volume, termed dilatancy effect. Water cannot go into the pore volume and pore water pressure is decreased to be negative, termed effect of water absorption (Zhang, 1999). Mechanical changes in pore volume will cause changes in pore pressure.

3.1 Changes of pore water pressure responding to different travel speeds

Figure 4 shows the monitored results of pore water pressure changes in Test 1. The pore water pressure decreased greatly at the points of A and B. The maximum change for A is 36.85 kPa while the change for B is 43.57 kPa. Point B is closer to the cutting blade tip than point A. Also we can see that the pressure for A is much steadier than that for B, which means that the pore water pressure around the cutting blade tip changes frequently with much higher range. The pore pressure changes for point A falls behind that for B, as point A is located higher than point B.

The soil failure pattern is cracking in the tests, shown as in Figure 5. The soil which was spilt away by the cutter tooth cracked in the middle and was separated into two piles of broken soil.

Figure 6 shows the peak pressure value of test 1, 2, 3, and 4. The absolute #A pressure values were all larger than these of #B. When the velocity changed from 0.2 m/s to 0.6 m/s, the absolute peak pore water pressure values increased linearly. But when the velocity changed from 0.6 m/s to 0.8 m/s, the absolute peak values became smaller on the contrary. As the

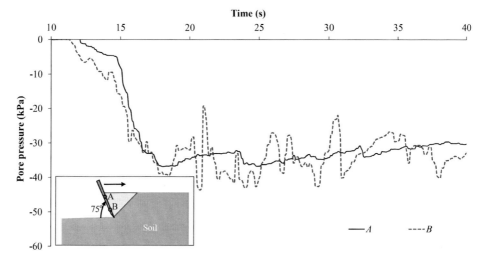

Figure 4. Changes of pore water pressure under dynamic soil cutting (Test 1).

Figure 5. Dynamic loosening process of the saturated soil.

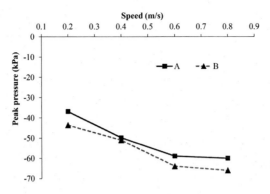

Figure 6. The peak pressure value (Test 1, Test 2, Test 3 and Test 4).

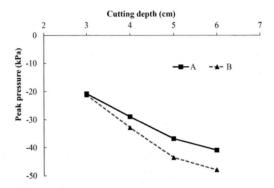

Figure 7. The peak pressure value (Test 1, Test 5, Test 6 and Test 7).

velocity of the cutter tooth increased, the dilatancy effect became obvious and the effect of water absorption was intensified, unless the velocity was higher than the limiting value. But when the velocity was higher than the limiting value, the soil which was spilt away was susceptible to break and the absolute values of pore water pressure dropped.

3.2 *Changes of pore water pressure responding to different soil cutting depth*

Figure 7 shows the peak pressure value of test 1, 5, 6, and 7. The absolute #A pressure values were all larger than these of #B. When the ploughing changed from 3 cm to 6 cm, the absolute peak pore water pressure values increased approximately linearly. But unlike the velocity group, the differences between #A pressure peak value and #B pressure peak value grew as the soil cutting depth increased. The difference was only 0.44 kPa while the soil cutting depth was 3 cm. And the differences for 4 cm, 5 cm, and 6 cm were 3.96 kPa, 6.72 kPa, and 7.02 kPa, respectively.

As the soil cutting depth increases, the effect of water absorption is aggravated. Therefore, not only the negative pore water pressure but also the difference between #A pressure and #B pressure grow as the soil cutting depth increases.

3.3 *Changes of pore water pressure responding to different rake angles*

Figure 8 shows the peak pressure value of test 1, 8, 9, and 10. The absolute peak pore water pressure values increased approximately linearly as the rake angle grew. And the #A pressure increased faster than #B. An interesting phenomenon was found. The absolute #A pressure value was smaller than #B, as shown in Figure 7.

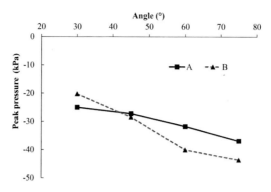

Figure 8. The peak pressure value (Test 1, Test 8, Test 9 and Test 10).

The pore water pressure time series of Test 8 is shown in Figure 8. The #B pore water pressure was susceptible to dissipate. The soil which was spilt by the cutter tooth could easily break from the middle and slide down from both sides of the cutter tooth. And water could go into the ditch through the break and the below pore volume could get water readily. Therefore, the #B pore water pressure was susceptible to dissipate.

4 CONCLUSIONS

A large scale experiment of soil ploughing by a cutter tooth was conducted to study the pore water pressure changes, and some important results from the test are listed here.

1. The plough movement will cause the sediments in shear zone to rearrange to a looser condition. The phenomenon shows that the porosity increases at shear zone after the sediments moving freely. Along with the phenomenon the apparent pore volume increases obviously, which leads to expansion of soil volume, termed dilatancy effect. Water cannot go into the pore volume and pore water pressure is decreased to be negative, termed effect of water absorption.
2. The soil which is spilt away will crack in the middle and be separated in two parts and the pore water pressure in the soil will decrease to be negative when the cutter tooth plowing around.
3. The pore pressure ~ time curve can be divided into three stages: (1) the pore water began to decrease to negative and then (2) remained stable for a short time and then (3) the negative pore water pressure began to dissipate.
4. First the below negative pore water pressure (#B) is larger than the above one (#A) because it is harder for the deeper soil to get water upside. And then, the soil begins to break from the middle and go into the ditch. The below negative pore water pressure is obviously dissipated while the above one dissipates slowly.
5. The negative pore water pressure values are influenced by velocity of the cutter trolley, the soil cutting depth, and the rake angle of the cutter tooth.

ACKNOWLEDGMENTS

Funding for this research was provided by Key Lab of Dredging Technology CCGRP and the National Natural Science Foundation of China (NSFC 11172213 and NSFC 51479137).

REFERENCES

Aluko, H.W. Chandler. Characterisation and Modelling of Brittle Fracture in Two-dimensional Soil Cutting, Biosystems Engineering, 2004, Vol.88(3):369~381.

Godwin, R.J, Spoor G. Soil Failure with Narrow Tines. Agric Engineering Research, 1977, 22(4):213~228.

Hettiaratchi, D.R.P. and Reece, A.R. The calculation of passive soil resistance. Geotechnique, 1974, 24(3):45~67.

Jiang, M.J. An efficient technique for generating homogeneous specimens for DEM studies. Computers and Geotechnics, 2003, 30:579~597.

Jisong He & W.J. Vlasblom. Modelling of saturated sand cutting with large rake angle. 15th world dredging congress, Las Vegas, Nevada, USA, June 1998.

Miedema, S.A. Mathematical Modelling of the Cutting of Densely Compacted Sand Under Water. Dredging & Port Construction, July 1985:22~26.

Miedema, S.A. The cutting mechanisms of water saturated sand at small and large cutting angels. Proceedings of the International Conference on Coastal Infrastructure Development Challenges in the 21st Century. Hong Kong, 2004: 40–40.

Qin, S., Xie, L., Hong, G. and Wang, J. Numerical Studies of Cutting Water Saturated Sand by Discrete Element Method. Applied Mechanics and Materials, 2013, 256–259, 306–310.

Reece A.R. The fundamental equation of earth-moving mechanics. Symp. on Earth-Moving Machinery, Proc. Inst. Mech. Eng: 1965, 16~22.

Usama EI Shamy, Mourad Zegllal. Coupled Continuum-Discrete Model for Saturated Granular Soils. Journal of Engineering Mechanics, 2005, 131:413~426.

Van Os A.G., Leussen W. Basic research on cutting forces in saturated sand. Journal of Geotechnical Engineering, 1987, 113(12):151–159.

Van Os A.G. Behavior of soil when excavated underwater. International Course Modern Dredging, Hague, The Netherlands, 1997:135–141.

Zhang, J.M., Tokimatsu, K. and Taya, Y. Effect of water absorption in shear of post-liquefaction. Chinese Journal of Geotechnical Engineering, 1999, 21(4):398–404. (in Chinese).

Numerical analysis of flow effects on water interface over a submarine pipeline

Yehui Zhu & Liquan Xie
College of Civil Engineering, Tongji University, Shanghai, China

ABSTRACT: Numerical simulations are conducted on unidirectional flow over a submerged pipeline. Results indicated that remarkable changes of water interface profile show up over the pipeline due to the choking effects. The flow experiences a drawdown across the pipeline and the drawdown curve is asymmetry in that the rise of water level is well smaller than the drop one. The variation of water interface becomes smaller as pipeline embedment grows larger or flow depth climbs up. The variation of water interface is less than 0.5% when the pipeline is half-buried or the flow depth is large. The variation of water interface becomes much more significant when the flow velocity increases. The nadir point on the curve moves downstream with the increment of flow velocity.

1 INTRODUCTION

Submarine pipelines are virtual parts of oil and gas transportation systems in the sea. Nevertheless, they are always facing with severe working conditions. Scours may show up and extend beneath the pipeline due to pressure difference upstream and downstream the pipeline. When the scour hole is long enough, it will lead to pipeline span and sagging, which is an important attribution to pipeline failures. Installation of the pipeline changes the flow structure significantly, causing choking of the flow. The water level rises up on the upstream side of the pipeline and experiences a drawdown across the pipeline, creating water level changes and thus static pressure difference on two sides of the pipeline, which is an important contribution to the onset of scour under the pipelines.

Previous studies on pressure difference upstream and downstream show the pipeline to be affluent. Mao (1988) described the three vortex systems proximate to the pipeline and linked the onset of scour with the pressure difference between upstream and downstream of the pipeline. Chiew (1990) stated that the hydraulic gradient induced by the pressure drop on two sides of the pipeline was the predominant cause of piping, which caused the onset of scour. Zang et al. (2009) examined the effects of flow parameters on pressure drop coefficient over the pipeline in current and wave conditions, respectively, with a numerical model. Zang et al. (2010) studied the onset of scour experimentally and numerically, and investigation on pressure coefficient difference over a pipeline under combined wave and current conditions, which was included. However, the documentations about the water interface are limited. Chiew (1991) conducted a series of experiments in shallow flow conditions, finding that low flow depths caused weir-flow conditions at the cylinder and the drawdown increased with the decrease of flow depth. Zhang et al. (2013) proposed a numerical model on the basis of Finite Volume Method (FVM). The results showed that a hydraulic jump took place when the relative flow depth was 2.08 and pipeline embedment was zero. For larger water depths, there was no such phenomenon.

In this study, the numerical investigations on unidirectional flow over a submerged pipeline are carried out based on VOF method. The effects of flow parameters including pipeline embedment, flow depth, and flow velocity on water interface profiles are examined.

2 MODEL DESCRIPTION

2.1 *Governing equations*

FLOW-3D, a piece of general-purpose Computational Fluid Dynamics (CFD) software, is utilized for the simulation on flow adjacent to the submerged pipeline in steady currents. Finite difference and finite volume methods are used to discretize the governing equations. The Volume of Fluid (VOF) method, first introduced by Hirt & Nichols (1981) is utilized to trace the free surface of fluid. Geometric regions in the simulations are modeled with Fractional Area/Volume Obstacle Representation (FAVOR) Method (Hirt & Sicilian, 1985).

In this study, water is considered as an incompressible viscous fluid and the Cartesian right-handed coordinate system is adopted. The governing equations are two-dimensional Reynolds-Averaged Navier Stocks (RANS) equations, which are expressed as (Flow Science Inc, 2012):

Mass continuity equation.

$$\frac{\partial}{\partial x}(uA_x) + \frac{\partial}{\partial y}(vA_y) + \frac{\partial}{\partial z}(wA_z) = 0 \qquad (1)$$

where u, v, w are velocity components in x, y, z directions in Cartesian coordinate system; A_x, A_y, and A_z are fractional area open to flow in x, y, and z directions.

Momentum equations

$$\begin{cases} \dfrac{\partial u}{\partial t} + \dfrac{1}{V_F}\left(uA_x\dfrac{\partial u}{\partial x} + vA_y\dfrac{\partial u}{\partial y} + wA_z\dfrac{\partial u}{\partial z}\right) = -\dfrac{1}{\rho}\dfrac{\partial p}{\partial x} + G_x + f_x \\ \dfrac{\partial v}{\partial t} + \dfrac{1}{V_F}\left(uA_x\dfrac{\partial v}{\partial x} + vA_y\dfrac{\partial v}{\partial y} + wA_z\dfrac{\partial v}{\partial z}\right) = -\dfrac{1}{\rho}\dfrac{\partial p}{\partial y} + G_y + f_y \\ \dfrac{\partial w}{\partial t} + \dfrac{1}{V_F}\left(uA_x\dfrac{\partial w}{\partial x} + vA_y\dfrac{\partial w}{\partial y} + wA_z\dfrac{\partial w}{\partial z}\right) = -\dfrac{1}{\rho}\dfrac{\partial p}{\partial z} + G_z + f_z \end{cases} \qquad (2)$$

where V_F is the fractional volume open to flow; ρ is the fluid density; G_x, G_y, G_z are the body accelerations; f_x, f_y, f_z are viscous accelerations.

Renormalized-Group turbulence (RNG) model, which is revised from standard k-ε turbulent model, is chosen for closure of RANS equations for the k-ε model has that been shown to provide reasonable simulation results in channel flows (Rodi, 1980) and the RNG model enjoys more accuracy in low intensity turbulence flows. The transport equations are:

$$\frac{\partial k_T}{\partial t} + \frac{1}{V_F}\left(uA_x\frac{\partial k_T}{\partial x} + vA_y\frac{\partial k_T}{\partial y} + wA_z\frac{\partial k_T}{\partial z}\right) = P_T + \text{Diff}_{k_T} - \varepsilon_T \qquad (3)$$

$$\frac{\partial \varepsilon_T}{\partial t} + \frac{1}{V_F}\left(uA_x\frac{\partial \varepsilon_T}{\partial x} + vA_y\frac{\partial \varepsilon_T}{\partial y} + wA_z\frac{\partial \varepsilon_T}{\partial z}\right) = \frac{\text{CDIS1}\cdot\varepsilon_T}{k_T}P_T + \text{Diff}_\varepsilon - \text{CDIS2}\frac{\varepsilon_T^2}{k_T} \qquad (4)$$

where k_T is the turbulent kinetic energy; ε_T is the rate of turbulent energy dissipation; P_T is the turbulent kinetic energy production; Diff_{kT} is the diffusion term; $\text{Diff}_{\varepsilon T}$ is the diffusion of dissipation; CDIS1 is a dimensionless parameter, the default of which is 1.44; CDIS2 is another dimensionless parameter, which is computed from the turbulent kinetic energy and turbulent production terms.

2.2 *Computational domain and boundary conditions*

A two-dimensional numerical flume with a length of 9.0 m and a height of 1.0 m (see Fig. 1) is established as the computational domain. A Cartesian right-handed coordinate system is built with the origin set at the midpoint and on the bottom of the flume and the coordinate

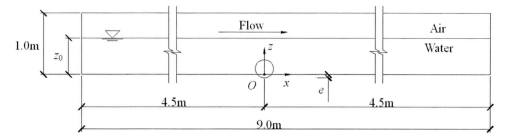

Figure 1. Layout of the numerical flume.

Figure 2. Computational mesh near the pipeline.

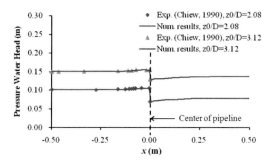

Figure 3. Pressure water head distribution on the surface of sand bed.

axes in the following directions: x = streamwise, z = bed-normal (see Fig. 1). The pipeline, with a diameter of 110 mm in all cases, is located at $x = 0$. The horizontal distances from the pipeline to the entrance and the exit of the flume are both over $40D$. The position of the pipeline in z direction varies in different cases.

The boundary conditions are set as follows: the entrance of the flume (Xmin) is set to be specified velocity boundary, and the fluid elevation and velocity are specified; the exit of the flume (Xmax) is considered as specified velocity boundary, and the fluid elevation is specified; the bottom of the flume (Zmin) is considered as wall boundary conditions; the top of the flume (Zmax) is set to be symmetry boundary condition. Structured rectangular grids are used to mesh the computational domain. Denser grids are used in the area which is in proximity to the pipeline (Fig. 2). The total mesh grids add up to 130,000. The initial time step is 0.001 s; the minimum time step is 1×10^{-7} s; and the total simulation time is 300 s.

2.3 Model validation

Numerical model is validated with the experiment results reported by Chiew (1990) in this study. Figure 3 shows the pressure water head distribution along the surface of sand bed. As is depicted in Figure 3, the pressure water head experienced a great drop when crossing the pipeline and almost keeps constant in other places. The drop of pressure water head is smaller in the case with larger flow depths. The numerical results of pressure water head agree well with the experimental data.

3 RESULTS AND ANALYSIS

Due to the choking of flow and the weir flow over the pipeline, changes take place in the water interface adjacent to the pipeline. Figure 4 shows the flow conditions in the flume when $e/D = 0.045$, $z_0/D = 2.73$, and $V_0 = 0.6$ m/s, where the red block represents water and the blue block represents air. In this figure, the change of water profile over the pipe is remarkable. The water level rises up on the upstream side of the pipeline and experiences a drop on the downsteam side. The sharp rise and drop of the water level is an important source of the great pressure difference on two sides of the pipeline, which will lead to the seepage flow in the sediment under the pipeline.

3.1 Effects of pipeline embedment on water interface profile

The effect of embedment-to-diameter ratio e/D on water profile is considered. Figure 5 shows the water profiles for various embedment-to-diameter ratios. In calculation, the embedment of pipeline varies from $0.045D$ to $0.5D$. The diameter of the pipeline D is 110 mm; the undistributed relative flow depth z_0/D is 3.64 and the undistributed depth averaged velocity V_0 is 0.4 m/s. The normalized water depth z/z_0 is chosen as the vertical axis. As is shown in Figure 5, the shape of water profile is quite regular under different pipeline embedment. The water profile starts to drop about (2~4) D upstream to the pipeline and reaches an extreme value at a point somewhere downstream to the pipeline. After that, the water interface gradually climbs up and eventually gets to the undistributed water level.

It is indicated that the variation of water interface becomes smaller as e/D grows larger. When the pipeline is half-buried (i.e. $e/D = 0.5$), the variation of water level is less than 0.4% of the undistributed flow depth. This phenomenon is not far from understanding. As the pipeline embedment grows larger, the projected area of pipeline on the normal plane of streamwise direction gets smaller; the choking effect of pipeline decreases and the pipeline is less effective to the main current. As a result, the variation of water level is less violent.

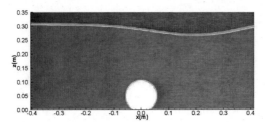

Figure 4. Water interface profile ($e/D = 0.045$, $z_0/D = 2.73$, $V_0 = 0.6$ m/s).

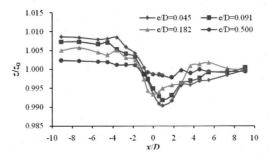

Figure 5. Water interface profile for different e/D.

3.2 Effects of flow depth on water interface profile

The effect of undistributed relative flow depth z_0/D on water profile is considered. Figure 6 shows the water profiles for various undistributed relative flow depths. In calculation, the undistributed flow depth varies from $2.73D$ to $7.27D$. The embedment-to-diameter ratio e/D is 0.045 and the undistributed depth averaged velocity V_0 is 0.4 m/s. The normalized water depth z/z_0 is chosen to be the vertical axis. It is indicated that the variation of water interface becomes smaller as z_0/D climbs up. When $z_0/D = 2.73$, the variation of water interface is over 6% of the undistributed flow depth, and when $z_0/D = 7.27$, the variation is less than 0.3%. In addition, the drop rate of water interface variation decreases with the increase of flow depth.

As is shown in Figure 6, the shape of water profile winds in the same pattern as is mentioned above. Further observation of the figure shows that the curve is asymmetry, for margin of rise and fall on two sides of the pipeline is not equal. For example, in Figure 6, when $z_0/D = 2.73$, the rise of water level on the upstream side of pipeline is about 2% of the undistributed flow depth, but the drop on the downstream side is more than 4%. The rise of water level is well smaller than the drop.

3.3 Effects of flow velocity on water interface profile

The effect of undistributed depth averaged velocity V_0 on water profile is considered. Figure 7 shows the water profiles for various undistributed depth averaged velocities. In calculation, V_0 varies from 0.2 m/s to 0.6 m/s. The embedment-to-diameter ratio e/D is 0.045 and the undistributed relative flow depth z_0/D is 2.73. The normalized water depth z/z_0 is chosen for the vertical axis. It is indicated that the variation of water interface becomes much greater when the V_0 increases. When V_0 reaches 0.6 m/s, the variation of the water interface overtakes 10% of the undistributed flow depth.

In addition, the shape of the curve generally changes with the increase of V_0. For example, the nadir point on the curve moves downstream with the increment of V_0. When V_0 increase from 0.2 m/s to 0.3 m/s and 0.4 m/s, the nadir point moves from $x/D = 0.545$ to $x/D = 0.909$

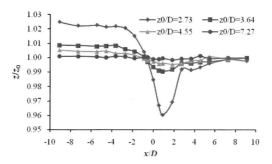

Figure 6. Water interface profile for different z_0/D.

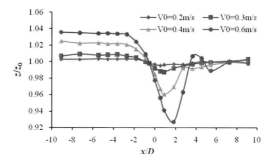

Figure 7. Water interface profile for different V_0.

and 1.818, respectively. In addition, when $V_0 = 0.6$ m/s, the water interface does not return to the undistributed water level immediately downstream to the drawdown curve as it does in other cases. On the contrary, it winds up and down before going back to the undistributed water level. This may be associated with the phenomenon of hydraulic jump.

4 CONCLUSIONS

The following conclusions can be drawn on the basis of numerical results:

1. Remarkable changes of water interface profile shows up over the partially buried submerged pipeline due to the choking effect. Water level experiences backup on the upstream side and drawdown at the downstream side. The margin of rise and fall on two sides of the pipeline are not equal, for the rise of water level is well smaller than the drop.
2. The variation of water interface becomes smaller as pipeline embedment grows larger. The variation of water interface is neglectable when the pipeline is half-buried.
3. The variation of water interface becomes smaller as flow depth climbs up. The variation of water interface is less than 0.3% when the relative flow depth is 7.27. The drop rate of water interface variation decreases with the increase of flow depth.
4. The variation of water interface becomes much more significant when the flow velocity increases. The nadir point on the curve moves downstream with the increment of flow velocity. The water interface winds up and down before it goes back to the undistributed water level when the flow velocity is 0.6 m/s, which can be associated with the phenomenon of hydraulic jump.

ACKNOWLEDGMENTS

Funding for this research was provided by the National Natural Science Foundation of China (NSFC 11172213 and NSFC 51479137).

REFERENCES

Chiew, Y. 1990. Mechanics of Local Scour Around Submarine Pipelines. *Journal of Hydraulic Engineering,* 116, 515–529.
Chiew, Y. 1991. Flow Around Horizontal Circular Cylinder in Shallow Flows. *Journal of Waterway, Port, Coastal, and Ocean Engineering,* 117, 120–135.
Flow Science, Inc. 2012. FLOW-3D Documentation: Release 10.1.0.
Hirt, C.W. & Nichols, B.D. 1981. Volume of fluid (VOF) method for the dynamics of free boundaries. *Journal of Computational Physics,* 39:1, 201–225.
Hirt, C.W. & Sicilian, J.M. 1985. A porosity technique for the definition of obstacles in rectangular cell meshes. *International Conference on Numerical Ship Hydrodynamics.*
Mao, Y. Seabed scour under pipelines. Proceedings of 7th International Symposium on Offshore Mechanics and Arctic Engineering (OMAE), Houston, 1988. 33–38.
Rodi, W. 1980. Turbulence models and their application in hydraulics: A state of the art review. *International assocation of hydraulic research.*
Yakhot, V. & Orszag, S.A. 1986. Renormalization group analysis of turbulence. I. Basic theory. *Journal of Scientific Computing,* 1, 3–51.
Yakhot, V. & Smith, L.M. 1992. The renormalization group, the e-expansion and derivation of turbulence models. *Journal of Scientific Computing,* 7, 35–61.
Zang, Z., Cheng, L. & Zhao, M. Onset of Scour Below Pipeline Under Combined Waves and Current. ASME 2010 29th International Conference on Ocean, Offshore and Arctic Engineering, 2010. 483–488.
Zang, Z., Cheng, L., Zhao, M., Liang, D. & Teng, B. 2009. A numerical model for onset of scour below offshore pipelines. *Coastal Engineering,* 56, 458–466.
Zhang, Z.Y., Shi, B., Guo, Y.K. & Yang, L.P. 2013. Numerical investigation on critical length of impermeable plate below underwater pipeline under steady current. *Science China Technological Sciences,* 56, 1232–1240.

Research on importance sampling technique in Monte Carlo method for structural reliability

Yan-Feng Fang
School of Naval Architecture and Civil Engineering, Zhejiang Ocean University, Zhoushan, China

ABSTRACT: In this paper, analysis on importance sampling in Monte Carlo method for structural reliability is presented, and variance of simulation result under Monte Carlo method is deduced. It's acceptable to take the design point as the center of sampling if the variables follow normal distribution and the type of sampling function and the variances of variables can be the same as the distribution function. For correlated variables, the sampling center should be moved to the point where the line at an angle of 45 degree to the coordinate axis joins when the correlation coefficient is high.

1 INTRODUCTION

To some extent, the calculation under First-Order Second-Moment method is approximate if variables follow no-normal distribution or the limited state equation is nonlinear [Guo Fan Zhao, 2000], so the simulation method (Monte Carlo method) is becoming increasingly popular for it is far more accurate and has been accepted in analysis and design for structural reliability as an important part [Guo Fan Zhao, 1996]. In this paper, the principle on improving accuracy on importance sampling technique is thoroughly analyzed and a novel conception is presented to resolve the problem of locating center of sampling function for both correlated and uncorrelated variables. Formula to control deviation is deduced.

2 PRINCIPLE FOR IMPORTANCE SAMPLING IN MONTE CARLO METHOD

Monte Carlo method is emphasized increasingly for the ability to solve difficult reliability problems with higher accuracy for both certainty problem and uncertainty problem [Jie Li, 2001]. It is convenient in solving complicated mathematical problem for complexity of limit state surface can be avoided here and is usually used to evaluate the calculation accuracy of other methods [Guo Fan Zhao, 1996]. But acceptable result can't be obtained until sampling number is large enough when the failure probability is very small, which will consume a great deal of time and even is almost impossible to achieve on some occasions. Fortunately, the problem can be improved with the development of computer [Ayyub, B. M., 1984]. The basic concept of Monte Carlo method (also called simulation) is relatively straightforward but the procedure can become computationally intensive [Andrzej S. Nowak Kevin R. Collins, 2005].

In Monte Carlo method, problem space is needed to be constructed and a proper statistics value $v(x)$ will be chosen for the convenience of simulation, where $x_i (i=1,2,\ldots,n)$ is variable in the space. We can associate the failure probability with mathematical expectation of $v(x)$ in the following equation

$$p_f = \int v(x)dF(x) = E(v) \qquad (1)$$

As a kind of unbiased estimation, sample mean is safe to be viewed as evaluation of failure probability based on large number of laws [Christian P. Robert, 2009] and the relationship can be expressed as

$$\hat{p}_f = \frac{1}{N}\sum_{1}^{N} v(x_i) = \bar{v} \qquad (2)$$

In general according to Monte Carlo method, failure probability can be expressed by

$$p_f = \int_{D_f} f(x)dx = \int I(g(x))f(x)dx \qquad (3)$$

here, $I[g(x)]$ is the concrete form of $v(x)$, $g(x)$ denotes the limit state, and $I[\bullet]$ is a sign function. When $g(x) \leq 0$, $I[g(x)] = 1$; otherwise, $I[g(x)] = 0$. D_f stands for failure dome where $g(x) \leq 0$ and $f(x)$ is the probability density function. Then the simulation accuracy can be expressed in form of deviation as

$$D(\hat{p}_f) = D(\bar{v}) = \frac{D(v)}{n} = \frac{E(v^2) - E^2(v)}{n} = \frac{p_f(1-p_f)}{n} \qquad (4)$$

Apparently, the simulation result is unstable when the sampling number isn't large enough. If the sampling number is limited, proper sampling techniques such as importance, condition, and duality sampling techniques should be adopted to improve the simulation accuracy. Importance sampling is more efficient than other techniques [Maurice Lemaire, 2009].

For importance sampling method, the simulation efficiency is raised a great deal because more sampling points will fall into failure domain after the sampling center is changed. But it must be clearly understood that the effect of efficiency improvement can't only be evaluated by the number of the points which fall into failure domain for there are other factors too which we will consider. Sampling principle can be deduced as follows

$$p_f = \int_{D_f} f(x)dx = \int I(g(x))f(x)dx = \int I(g(u))\frac{f(u)}{h(u)}h(u)du \qquad (5)$$

here, $h(u)$ is the sampling density function for vector u, a statistics value v is constructed that

$$v(u) = I(g(u))\frac{f(u)}{h(u)} \qquad (6)$$

If the sampling center locates at design point p^* and the limit function is linear, then we get $E(I(g(u))) = p_o = 0.5$. p_o is called median probability which can be found easily. Suppose that $w(u) = \frac{f(u)}{h(u)}$, then the sampling mean can be expressed as

$$E(\bar{v}(u)) = E(I(g(u))w(u)) = p_o E(w(u)) = p_f \qquad (7)$$

The responding deviation can be expressed as

$$D(\bar{v}(u)) = \frac{D(v(u))}{n} = \frac{E(v^2(u)) - E^2(v(u))}{n} = \frac{p_o E(w^2(u)) - p_o^2 E^2(w(u))}{n} \qquad (8)$$

Combining equation (7) and (8) we get

$$D(\bar{v}(u)) = \frac{\frac{E(w^2(u))}{E(w(u))}p_f - p_f^2}{n} \qquad (9)$$

Compared with equation (4), the deviation will be smaller through importance sampling method if $w(u) < 1$, which demands that a proper sampling center should be found. After the sampling center is fixed, from equation (7) we get $E(w(u)) = \frac{p_f}{p_o}$. For smaller deviation, $\frac{E(w^2(u))}{E(w(u))}$ can be deduced as

$$\frac{E(w^2(u))}{E(w(u))} = \frac{nE(w^2(u))}{nE(w(u))} \approx \frac{\sum_{i=1}^{n} w^2((u_i))}{n\frac{p_f}{p_o}} \quad (10)$$

Based on Cauchy-inequality, $\frac{E(w^2(u))}{E(w(u))}$ will come to minimum $\frac{p_f}{p_o}$ when $w(u) \equiv \frac{p_o}{p_f}$ and efficiency will come to maximum at this time. The deviation can be expressed as

$$D(\bar{v}(u)) = \frac{\frac{E(w^2(u))}{E(w(u))} p_f - p_f^2}{n} \geq \frac{\left(\frac{1}{p_o} - 1\right) p_f^2}{n} \quad (11)$$

In order to improve efficiency, we should raise median probability p_o which lead to more points in failure domain. At the same time, we should struggle to keep distribution of $w(u)$ closer to uniform. In order words, two factors weigh heavily in effect of efficiency improvement which will affect each other.

3 SAMPLING FUNCTION AND CENTER

The effect that importance sampling technique improves simulation accuracy depends on the construction of sampling function $h(u)$ and the sampling center. In theory, the accuracy will come to maximum when $p_o = 1$ and $h(u) = \frac{f(u)}{p_f}$ but such a function $h(u)$ can't be found for p_f is unknown. In application, it can be accepted when the deviation of the variables is small enough. Sampling function $h(u)$ can be the same in form as the probability density function $f(u)$ of the variables with a changed center in one case. Design point p^* will be safe to act as the sampling center as shown in Figure 1.

Here $f(u)$ is a standard normal function with uncorrelated variables u_1 and u_2, $h(u)$ is the sampling function which is the same in form as $f(u)$ except that the center is changed from O to p^*. If two cutting planes are passed through the sampling function $h(u)$ along u_1'

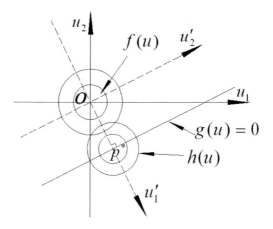

Figure 1. The location of sampling center.

coordinate axis and limit line $g(x)=0$, two sections will be obtained as shown in Figure 2 and Figure 3.

For the purpose of keeping $w(u)$ as closer as possible to uniform, the idea that the outline of sampling function $h(u)$ is moved to dotted line in Figure 2 and Figure 5 by changing deviations σ of variables is natural. As proved in simulation, improved accuracy of simulation can't be obtained if the deviations σ change for the case $w(u)=(f(u)/h(u))>1$ will appear when the sampling points are distant from the sampling center which has great effect on the stability of $w(u)$. From the Figure 2 and Figure 3, it's clear that improved stability can't be obtained if the sampling center leave design point p^*. In other words, the simulation accuracy will be lower in the case.

About correlated variables, the reliability problem can be solved by JC method after they are converted to uncorrelated ones [Breitung K, 1989]. The method of setting up an affine coordinates system can also be used to calculate reliability index through iterative formula [Wong R, 1989]. For higher accuracy and clear concept, simulation method is ideal. The sampling points are distributed in Figure 4 and Figure 5 where correlation coefficient $\rho=\pm 0.99$ and sampling number $n=1000$. All the points concentrate along the lines which incline at angles of $\pm 45°$ to the coordinate axis. When importance sampling technique is talked about, the sampling function $h(u)$ can be the same in form as the density function $f(u)$ with same correlation coefficient ρ and different centers, but the simulation result will be dangerous if design point p^* acts as the sampling center for the value of density function $f(u)$ is very small here and larger reliability index β will be obtained. In order to keep $w(u)$ stable, p_+^* and p_-^* get greater suitability as sampling centers where the coordinate axis and the lines that incline at angles of $\pm 45°$ to the coordinate axis join.

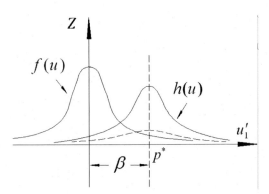

Figure 2. Section along u_1' coordinate axis.

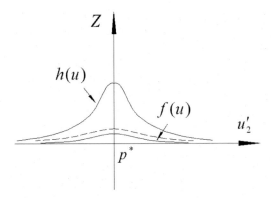

Figure 3. Section along limit line $g(x)=0$.

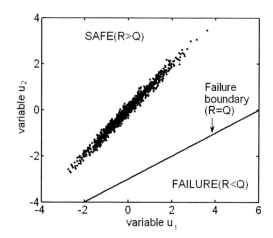

Figure 4. Result of simulation ($n = 1000$, $\rho = 0.99$)).

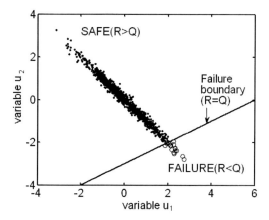

Figure 5. Result of simulation ($n = 1000$, $\rho = -0.99$).

4 CONCLUSION

The effect of accuracy improving is satisfactory when importance sampling technique is adopted in Monte Carlo method on the premise that proper sampling function is constructed. In importance sampling technique, it's acceptable to take the design point as the center of sampling if the variables follow normal distribution. The type of sampling function and the variances of variables can be the same as the distribution function. Apparent improvement for simulation accuracy can't be obtained by changing the sampling center or the variance of variables of sampling function. For correlated variables, the sampling center should be moved to the points where the lines which incline at an angle of 45 degree to the coordinate axis and the limit state equation join when the correlation coefficient is high. The simulation result will be dangerous if wrong sampling centers are accepted.

ACKNOWLEDGMENTS

This work was partially supported by the Zhejiang Education Department, Grant No. Y201018955.

REFERENCES

1. Guo Fan Zhao, Wei Liang Jin, Jin Xin Gong. Theory of Structural Reliability, in Chinese [M]. Beijing: China Architecture & Building Press, 2000.
2. Guo Fan Zhao. Theory of Engineering Structural Reliability, in Chinese [M]. Dalian: Dalian University of Technology Press, 1996.
3. Jie Li. Probability and statistics group of computer center of Chinese Academy of Sciences. Calculation for Probability and Statistics, in Chinese. Beijing: Science press, 2001.
4. Ayyub, B.M. & Haldar, A. Practical structural reliability techniques. Journal of Structural engineering, ASCE, 1984, 110(8).
5. Bjerager, P. On computation methods for structural reliability analysis. Structural Safety, 1990, 9.
6. Engelund, S. & Rackwitz, R. A benchmark study on importance sampling techniques in structural reliability. Structural Safety, 1993, 12.
7. Harbitz, A. An efficient sampling method for probability of failure calculation. Structural Safety, 1986, 3.
8. Schueller, G.I. & Stix, R. A critical appraisal of methods to determine failure probabilities. Structural Safety, 1987, 4.
9. Schueller, G.I. et al. On efficient computational schemes to calculate structural failure probabilities. Stochastic Structural Mechanics (Lecture notes in engineering), 1989, Y.K.L in & G.I. Schueller eds. Springer Verlag.
10. Shinozuka, M. Basic analysis of structural safety. Journal of Structural Engineering, ASCE, 1983, 109.
11. Andrzej S. Nowak Kevin R. Collins. Reliability of Structures. Chongqing University Press, 2005.
12. Christian P. Robert, George Casella. Monte Carlo Statistical Methods. Beijing: The world publisher, 2009, 10.
13. Maurice Lemaire. Structural reliability. John Wiley & Sons, Inc, 2009.3.
14. Breitung K. Asymptotic approximations for probability integrals. Probabilistic Engineering Mechanics, 1989, 4(4).
15. Yun Gui Li, Guo Fan Zhao, Bao He Zhang. Gradual Analysis on Structural Reliability, in Chinese [J]. Journal of Dalian University of Technology, 1994 (4).
16. Yun Gui Li, Guo Fan Zhao. Incremental Analysis for Structural Reliability in generalized Probability Space [J]. Journal of Hydraulic Engineering, 1994(8).
17. Wong R. Asymptotic approximations for integral [M]. New York: Academic Press Incorporation, 1989.

Optimal sizing of urban drainage systems using heuristic optimization

J.J. Yu
School of Civil and Environmental Engineering, Nanyang Technological University, Singapore
Environmental Change Institute, School of Geography and Environment, University of Oxford, Oxford, UK

X.S. Qin
School of Civil and Environmental Engineering, Nanyang Technological University, Singapore

R. Min
DHI-NTU Water and Environment Research Centre and Education Hub, Nanyang Technological University, Singapore

ABSTRACT: Improving the capacity of urban drainage system to accommodate intensive rainfall is one of the effective measures to mitigate the flood risk. In recent years, integration of urban rainfall-runoff simulation into an optimization model for urban drainage planning and design has become an attractive idea but the related studies are still rarely reported. This study proposed a generic heuristic optimization aided framework for urban drainage sizing. An optimization model was formulated to minimize the construction cost whilst limiting flood risk under an acceptable level, where the size (e.g., width and depth) of conduits are decision variables. The genetic algorithm was employed to seek the optimal solution. A typical urban catchment in Singapore was selected for demonstration. The results showed that the proposed method could effectively determine the optimum size for drainage design. Future studies were desired to improve the optimization model more practically.

1 INTRODUCTION

Rapid climate change and urbanization have posed an increasing threat on sustainable urban development due to exacerbated urban flood disasters (Schreider et al. 2000). Improving the capacity of urban drainage system to accommodate intensive rainfall is one of the effective measures to mitigate the flood risk. In such a context, an optimized solution to upgrade existing drainage system is of particular importance to yield a rational tradeoff between drainage construction costs and flood risks. With recent improvement of computational power, optimization techniques have been developed and widely adopted to offer relatively robust solution to such problems (Lund et al. 2002, Voortman & Vrijling 2003, Woodward et al. 2014, Jia et al. 2012, Haghighi & Bakhshipour 2015, Tao et al. 2014, Li et al. 2015). However, most of the previous works focused on optimization of layout and pipe specification of sewage collection networks; few has tackled the optimal sizing of drainage systems and also taken flood risk into consideration. Therefore, this study aims to propose a heuristic optimization framework for urban drainage sizing from engineering design perspective. An optimization model is developed to minimize the urban drainage construction cost, whilst an acceptable level of flooding could be met. The Genetic Algorithm (GA) (Goldberg 1989) is coupled with an urban hydrological model to search the optimum parameters of drainage design. A typical urban catchment in Singapore is used for demonstration.

2 METHODOLOGY

2.1 *Generic framework of heuristic optimization for urban drainage sizing*

Figure 1 illustrates a general framework of heuristic optimization for urban drainage sizing. Firstly, the meteorological, hydrological, topographic, and drainage network data are collected for the study area. They are used to (i) derive the design storm events at various probabilities (i.e., return periods); and (ii) establish and calibrate the urban hydrological model in order to simulate urban rainfall-runoff under various scenarios. In this study, the Storm Water Management Model (SWMM) is employed as the modeling tool, which has the capability of routing runoff and external inflows through the drainage system network of pipes, channels, storage/treatment units, and diversion structures (Gironás et al. 2010). It has been extensively applied into various applications of urban drainage planning, analysis, and design (Jia et al. 2012, Abi Aad et al. 2010). Secondly, according to practical requirements, the calibrated urban hydrological model is driven by a storm event at certain probability (i.e., a 2-years return period in this study) to generate a flooding scenario. The simulation results are then examined to identify the conduits which need to be examined for optimal sizing. It should be noted that it is normally a manual and interactive process and human expertise need to be incorporated to achieve the tradeoff between importance (i.e., select conduits whose overflows have significant social and economic impacts) and efficiency (i.e., computation time to run simulation model). Finally, an optimization model is constructed

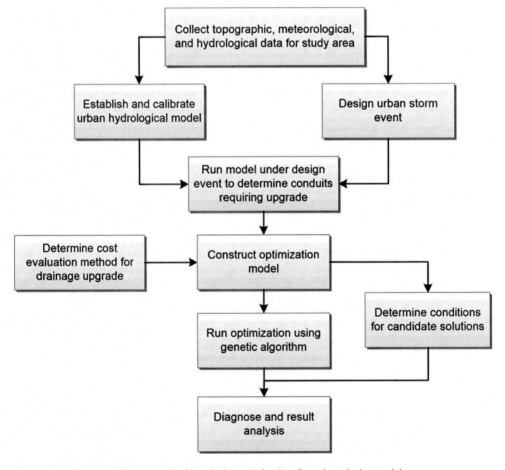

Figure 1. The generic framework of heuristic optimization for urban drainage sizing.

to minimize the construction cost of drainage sizing constrained on an acceptable level of flooding (i.e., total flood volume less than a certain threshold). The GA technique (one of the heuristic approaches) is employed to search the optimal design parameters (i.e., width and depth) of the conduits under consideration. The simulation model will be executed iteratively with the candidate inputs generated by GA through biology-analogical-processes like initiation, evaluation, selection, crossover, and mutation.

2.2 *Formulation of optimization model*

The optimization model for urban drainage sizing problem could be formulated as follows:

$$Min\ c \infty V \approx \sum_{i=0}^{n} \left(w_i^m * d_i^m - w_i^o * d_i^o\right) * l_i \quad (1a)$$

subject to

$$Q = s(w,d) \leq \gamma \quad (1b)$$
$$0 \leq w \leq \delta \quad (1c)$$
$$0 \leq d \leq \varphi \quad (1d)$$

where c is the cost of drainage upgrade work, which should be practically a function of a number of factors including mobilization cost, operating cost, quantity of work required, and unit cost of materials etc. (Woodward et al. 2014). In order to simplify the optimization model for demonstration purpose, we try to formulate the objective function in the form of minimizing total quantity of work (represented by the cut volume, V) instead of the real management cost. It is based on the assumption that the cost c is proportional to V to some extent. The drainage system is considered to consist of open or closed rectangle channel only. Accordingly, the width (**w**) and depth (**d**) of the conduits are the decision variables. They have a dimension of n, representing the widths and depths of the n conduits selected for sizing to reduce the flood risk. It also implies that the locations and lengths (*l*) of the original drainage networks will not be changed. According to Eq. 1a, the cut volume (V) could be calculated approximately as the summation of drainage capacity difference after and before sizing, where i is the index of the conduit, and superscripts m and o denote the candidate widths and depths, and original ones, respectively.

Constraint 1b is used to limit the total flood volume (Q) less than the threshold (γ), which is determined according to the affordable flood damage practically. The total flood volume is calculated based on the simulation result in a way of summating overflows at all concerned junctions, where s denotes the simulation model (i.e., SWMM) having **w** and **d** as inputs. Constraints 1c and 1d are used to limit the candidate size (width and depth) of the conduits into a reasonable range. They should be larger than 0, and less than certain values (δ and φ) to avoid the excessive width and depth affecting the surrounding buildings or infrastructures etc. To use GA for seeking optimal solutions, the constraints 1c and 1d will be checked initially. If successful, the candidate decision variables generated from GA is then used to drive the simulation model to obtain the flood volume. In case any constraint is violated, a penalty will be added to the cost making the value much larger than the possible flood volume for rejection of the candidate variables.

3 RESULT ANALYSIS

As shown in Figure 2a, a typical urban hydrological catchment in Singapore is selected to demonstrate the proposed heuristic optimization framework for urban drainage sizing. This catchment has an area of 0.32 km², including 25 sub catchments, 24 junctions, and 24 conduits. In using SWMM, an urban hydrological model is established and calibrated using three storm events occurred in 2012. An optimal problem is formulated to avoid the flood risk

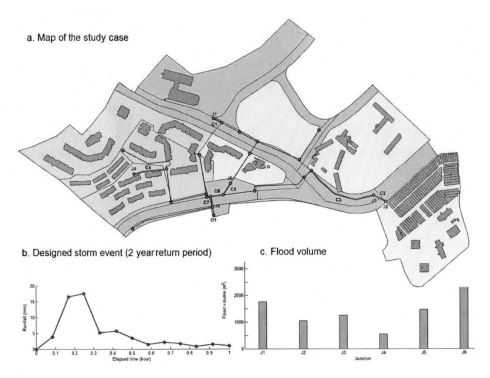

Figure 2. Illustrations of study case: (a) map of a typical urban catchment in Singapore; (b) designed storm event with 2-years return period; and (c) simulated flood volume at selected junctions.

Table 1. Optimized width and depth for nominated conduits under sizing consideration.

Conduits	Type	Original depth (m)	Original width (m)	Optimized depth (m)	Optimized width (m)
C1	Open rectangle	0.68	1.215	1.13	1.247
C2	Closed rectangle	1.2	1.3	1.42	1.302
C3	Closed rectangle	0.9	1.1	1.04	1.104
C4	Closed rectangle	1.2	0.45	1.21	1.662
C5	Open rectangle	1.435	1.5	1.535	1.54
C6	Open rectangle	1.025	2.08	1.029	2.15
C7	Closed rectangle	1.265	2.08	1.265	2.22

under a 2-year-return-period storm event. The corresponding rainfall intensity is around 61 mm/hour, which is adapted from Singapore's IDF curve (PUB 2011). Figure 2b shows the disaggregated rainfall profile according to Huff rainfall distribution type II, which will be used to drive the simulation. The peak of the rainfall occurs in the first 15 minutes at a level of 17 mm. The storm water falling down into the catchment will be routed and discharged to outfall O1 (see Fig. 2a). Under such storm event, the simulated average and peak flow rate at O1 are 0.47 m^3/s and 3.97 m^3/s, respectively. Figure 2c illustrates the overflow at six junctions (named as J1 to J6) with the flood volume ranging from 550 to 3100 m^3. Correspondingly, the associated seven conduits are nominated for sizing, which are marked as red in Figure 2a. In this study, the flood volume threshold (γ) is set as 10 m^3. The maximum width (δ) and depth (φ) are set as 2 and 0.5 meters, respectively, for all the nominated conduits.

Table 1 shows the optimal solution of the conduit size through applying the proposed method. It reveals that all the conduits except for C4 only requires minor construction works, with the increased conduit width and depth less than 0.1 and 0.45 m, respectively. Regarding C4, it needs to be widened approximately 1.21 meters to accommodate excessive storm water to avoid flooding. The optimal solution will lead to a minimum cut volume of 263.4 m^3. Accordingly, with the optimal solution, the average flow rate at outfall O1 increase slightly to 0.5 m^3/s, whereas the peak flow rate remains unchanged. Considering GA is always criticized with its capability of achieving global optimum, it is suggested to run the optimization model several times and then determine the final solution based on result diagnose and comparison.

4 DISCUSSIONS

This study proposed a heuristic optimization framework for urban drainage sizing. The GA is linked with simulation model (i.e., SWMM) to seek the optimal specification (i.e., width and depth) of conduits under sizing consideration. The optimization model tries to minimize the drainage construction cost with the constraint of limiting flooding. It shed some light of using optimization technique to yield relatively rational solution in urban drainage design in a scientific manner, which could thus provide robust evidence to support decision making for flood risk management. It also provides the flexibility to set the predefined parameters (γ, δ, and φ). It thus allows various scenarios analysis based on practical requirements.

However, in applying the proposed method, there are still some concerns. Firstly, the formulated optimization model is simplified for demonstration. In practical applications, urban drainage design is normally a multi-objective problem with multiple criteria, where various impacts of drainage sizing on residents, environments, and infrastructures etc. are expected to be considered. Secondly, inundation is not considered in this study. Linkage with a 2D inundation model could benefit study with awareness of flood depth over pervious areas; whereas how to reduce the corresponding computational cost is still challenging. Finally, uncertainties associated with designed storm events and urban hydrological model may affect the reliability of the solution. A risk-based optimization for urban drainage design is thus desired further exploration.

5 CONCLUSION

A heuristic optimization framework for urban drainage sizing was proposed in this study. The optimization model was formulated to minimize the drainage construction cost constrained on an acceptable level of flood risk. The genetic algorithm was employed to seek the optimal size (i.e., width and depth) of the conduits nominated for sizing consideration, where an urban hydrological model (i.e., SWMM) was driven to simulate the flooding conditions. The results from a case study in a typical urban catchment in Singapore showed that the proposed method could effectively support urban drainage sizing in a scientific manner.

ACKNOWLEDGEMENT

This material is based on research supported by Singapore's Ministry of Education (MOE) AcRF Tier 1 (Ref No. RG188/14; WBS no.: M4011420.030) Project.

REFERENCES

Abi Aad M., Suidan M., Shuster W. 2010. Modeling Techniques of Best Management Practices: Rain Barrels and Rain Gardens Using EPA SWMM-5. *Journal of Hydrologic Engineering* 15(6): 434–443.
Goldberg D.E. 1989. Genetic Algorithms in Search, Optimization and Machine Learning. Addison Wesley Longman, Inc.

Gironás J., Roesnerb L.A., Rossmanc L.A., Davisd J. 2010. A new applications manual for the Storm Water Management Model (SWMM). *Environmental Modelling & Software* 25(6): 813–814.

Haghighi A. & Bakhshipour A.E. 2015. Deterministic Integrated Optimization Model for Sewage Collection Networks Using Tabu Search. *Journal of Water Resources Planning and Management* 114(1), 04014045.

Jia H., Lu Y., Yu S.L., Chen Y. 2012. Planning of LID–BMPs for Urban Runoff Control: The Case of Beijing Olympic Village. *Separation and Purification Technology* 84(9): 112–119.

Li F., Duan H.F., Yan H.X., Tao T. 2015. Multi-Objective Optimal Design of Detention Tanks in the Urban Stormwater Drainage System: Framework Development and Case Study. *Water Resources Management* 29(7): 2125–2137.

Lund J. 2002. Floodplain Planning with Risk-based Optimization. *Journal of Water Resources Planning and Management* 128(3): 202–207.

Public Utility Board (PUB). 2011. Code of Practice–7. Drainage Design and Considerations, Singapore, http://www.pub.gov.sg/general/code/Pages/SurfaceDrainagePart2-7.aspx, updated on 2 Dec, 2011.

Schreider S.Y., Smith D.I., Jakeman A.J. 2000. Climate Change Impacts on Urban Flooding. *Climatic Change* 47: 91–115.

Tao T., Wang J., Xin K., Li S. 2014. Multi-objective Optimal Layout of Distributed Storm-water Detention. *International Journal of Environmental Science and Technology* 11(5): 1473–1480.

Voortman H. & Vrijling J. 2003. Time-dependent Reliability Analysis of Coastal Flood Defense Systems. In Proceedings of European Safety and Reliability Conference. Swets & Zeitlinger, pp.1637–1644.

Woodward M., Gouldby B., Kapelan Z., Hames D. 2014. Multiobjective Optimization for Improved Management of Flood Risk. *Journal of Water Resources Planning and Management* 140(2): 201–215.

Resources, Environment and Engineering II – Xie (Ed.)
© 2016 Taylor & Francis Group, London, ISBN 978-1-138-02894-4

On the damping of wave propagation over porous bottom

Y.L. Ni
Maritime and Civil Engineering School, Zhejiang Ocean University, Zhoushan, China

Y. Shen
Zhoushan Ocean Survey and Design Institute, China

J.F. Yu
Zhoushan Oceanic and Fishery Administration, China

ABSTRACT: Wave damping is a common phenomenon as wave propagation over porous medium. Base on Dean and Dalrymple's theoretical results, the study of wave propagation over rigid, porous bottom is further carried out. First, a simple and efficient method is proposed to solve the dispersion relationship and the solutions are validated by the perturbation method. Then, some two-dimensional numerical simulations are performed and discussed to investigate the damping properties. The results indicate that a more obvious wave damping will occur in a gravel seabed and shallow water than in a coarse/fine sand seabed and deep water.

1 INTRODUCTION

In the classical problem of linear waves in water of uniform depth, the bottom is assumed to be impermeable and the wave are undamped. However, it is usually covered with different types of sediments on seabed, such as gravel, coarse sand and fine sand that can be characterized as porous medium. As wave propagates over porous bottom, wave energy is transmitted into the marine sediments and wave damping will occur owing to the percolation. Therefore, the assumption of impermeability of the bottom is inevitable to bring error more or less, and it is necessary to have a mathematical model to investigate basic damping properties of wave propagation over porous bottom.

Dean and Dalrymple (1991) concentrated on the issue of a progressive wave over rigid, porous bottom and derived a dispersion relationship for porous medium. Based on their theoretical results, further study on the wave damping is carried out in the present paper. We derive the velocity potential and the dispersion relationship for water waves over porous bottom in the following section. In section 3, a simple and efficient method to solve the dispersion equation is presented and the results are verified. The spatial damping rate and wave damping in different porous medium and relative depth is discussed in section 4. Section 5 gives some concluding remarks.

2 THEORETICAL FORMULATION

2.1 *Governing equation for water waves and boundary conditions*

Consider a homogeneous incompressible fluid with irrotational motion travelling over an infinite permeable bottom with a constant depth h. For a two-dimensional problem, the governing equation is derived from the continuity equation and can be written in terms of the complex velocity potential Φ, as

$$\frac{\partial^2 \Phi}{\partial x^2} + \frac{\partial^2 \Phi}{\partial z^2} = 0 \tag{1}$$

where x is the horizontal coordinate, z is the vertical coordinate measured positively upwards with the undisturbed free surface at $z = 0$.

It is reasonable to assume

$$\Phi(x,z,t) = \varphi(x,z)e^{-i\omega t} \tag{2}$$

where $i = sqrt(-1)$, ω is angular frequency and t is the time. The function $\varphi(x, z)$ appearing in Eq. (2), is the normalized potential in the frequency domain, and it can be expressed as

$$\varphi(x,z) = -\frac{ig}{\omega}\left[a\frac{\cosh k(z+h)}{\cosh kh} + b\frac{\sinh k(z+h)}{\sinh kh}\right]e^{ikx} \tag{3}$$

where g is the acceleration due to gravity, a and b are the unknown coefficients, and k is the complex wave number, which will be discussed in more detail later.

The free surface boundary condition, obtained by combining the linear dynamic and kinematic free surface boundary conditions is

$$\frac{\partial \varphi}{\partial z} = \frac{\omega^2}{g}\varphi \quad (z=0) \tag{4}$$

The bottom boundary condition is

$$\frac{\partial \varphi}{\partial z} = u_w \quad (z=-h) \tag{5}$$

where u_w is the velocity outside the porous media, at the interface in $z = -h$.

2.2 Governing equation for the water in porous media

For a fully saturated porous medium, which is assumed to be incompressible, the continuity equation for a two-dimensional problem is

$$\nabla \cdot \overline{u}_s = 0 \tag{6}$$

where $\nabla = (\partial/\partial x, \partial/\partial z)$ is the gradient operator, and $\overline{u}_s$ is the average velocity across a given area of porous media that is related to the pressure gradients in the water by Darcy's law

$$\overline{u}_s = -\frac{K}{\mu}\nabla p_s \tag{7}$$

where K is a constant called the permeability coefficient, which is a characteristic of the porous medium, μ is the dynamic viscosity of the water, and p_s is the pore pressure. By substituting Eq. (7) into Eq. (6), we obtain the governing equation for the water in porous medium

$$\nabla^2 p_s = 0 \tag{8}$$

Hence the pore pressure satisfies the Laplace equation, which is assumed to be

$$p_s = \rho g c e^{k(z+h)} e^{ikx} \tag{9}$$

where ρ is the mass density of the water, and c is the unknown coefficient.

2.3 Matching conditions and solution

At the interface, the continuity of pressure and velocity require that

$$p_w = p_s \quad (z=-h) \tag{10}$$

and

$$\overline{u_w} = \overline{u_s} \quad (z=-h) \tag{11}$$

where p_w is the dynamic pressure outside the porous media, at the interface in $z = -h$. And based on the linearized Bernoulli equation, we have

$$p_w = i\omega\rho\varphi \quad (z=-h) \tag{12}$$

Substituting Eq. (5), (7) and (12) into Eq. (11), yields

$$\frac{\partial \varphi}{\partial z} = -\frac{K}{\mu}\frac{\partial p_s}{\partial z} = -\frac{Kk}{\mu}p_s = -\frac{i\omega\rho Kk}{\mu}\varphi \quad (z=-h) \tag{13}$$

We introduce a new constant defined by

$$T = \frac{K}{\upsilon} \tag{14}$$

where $\upsilon = \mu/\rho$ is the kinematic viscosity of the water. Rewriting the Eq. (13), we obtain

$$\frac{\partial \varphi}{\partial z} = -i\omega k T \varphi \quad (z=-h) \tag{15}$$

To solve the Eq. (1), we assume the wave amplitude is A, satisfying

$$A = a + b \tag{16}$$

And substituting Eq. (3) into Eq. (4) and (15) respectively, we have

$$ak\tanh kh + \frac{bk}{\tanh kh} = \frac{\omega^2}{g}(a+b) \tag{17}$$

and

$$\frac{bk}{\sinh kh} = -i\omega k T \frac{a}{\cosh kh} \tag{18}$$

Solving for a and b give us

$$a = \frac{A}{1 - i\omega T \tanh kh} \tag{19}$$

and

$$b = \frac{-i\omega A T \tanh kh}{1 - i\omega T \tanh kh} \tag{20}$$

Substituting for a and b in Eq. (17), results in

$$\omega^2(1-i\omega T\tanh kh)=gk(\tanh kh-i\omega T) \tag{21}$$

This is the dispersion relationship. And as mentioned above, k is the complex wave number, which may be written as $k=k_r+ik_i$. The real part of k represents the real wave number, related to the wavelength, equaling 2π/wavelength, while the imaginary component determines the spatial damping rate.

3 THE SOLUTION OF THE DISPERSION EQUATION

3.1 *The iteration scheme*

As pointed out by many authors, it is numerically difficult to locate the root of the dispersion equation in the complex plane. A proper initial guess is required to solve Eq. (21) by the iteration scheme. The way to obtain the complex wave number, for a determined ω, T and h, is as follows:

1. Calculate the wave number k_0 for the standard dispersion equation.
2. Let $k_r^0=k_0$ as for the initial value of the real part of complex wave number k. Then set the initial value of the imaginary component $k_i^0=2\omega Tk_r^0/(2k_r^0h+\sinh 2k_r^0h)$. Combining the two initial values, we have

$$k^0=k_r^0+ik_i^0 \tag{22}$$

3. Rewrite Eq. (21) in iterative algorithm

$$k^{n+1}=\frac{\omega^2(1-i\omega T\tanh k^n h)}{g(\tanh k^n h-i\omega T)} \tag{23}$$

where the superscript $n=0, 1, 2\ldots$ denotes the number of iterations.

4. Enter the iteration loop to calculate k^{n+1} using Eq. (23). If $|k^{n+1}-k^n|<\varepsilon$, where ε is some prescribed tolerance, the iteration loop terminates and k^{n+1} is the desired complex wave number. Else, update $k^n=k^{n+1}$ to repeat the procedure.

3.2 *Some results*

In order to verify the present solutions for different permeability coefficients and dimensionless depth, comparison with the perturbation method proposed by Mendez and Losada is made in Table 1, in which $L_0=40$ m, is the incident wave length. As shown in the table, the solutions calculated by the present method and the results by the perturbation method agree very well in different cases, which indicate the validity of the present method to solve the dispersion relationship.

Table 1. Comparison of the solutions by the present method and the perturbation method.

Permeability coefficients	Dimensionless depth	Present method		Perturbation method	
T	h/L_0	k_r	k_i	k_r	k_i
0.000	0.375	0.15707963	0	0.15707963	0
0.005	0.375	0.15707946	$3.199*10^{-5}$	0.15707946	$3.199*10^{-5}$
0.001	0.375	0.15707962	$0.640*10^{-5}$	0.15707962	$0.640*10^{-5}$
0.001	0.500	0.15707963	$0.142*10^{-5}$	0.15707963	$0.142*10^{-5}$
0.001	0.050	0.15707953	$1.655*10^{-4}$	0.15707953	$1.655*10^{-4}$

4 NUMERICAL EXAMPLES AND DISCUSSIONS

In this section, a two-dimensional problem of wave propagation over porous bottom is simulated to investigate the damping properties. We will first discuss some influence factors of the spatial damping rate, and then calculate wave damping for different cases.

4.1 *The spatial damping rate*

As mentioned previously, the image part of complex wave number represents the spatial damping rate, commonly related with the permeability coefficient of porous medium (represented by T defined in Eq. (14)) and relative depth h/L_0. Figure 1 illustrates the spatial damping rate increases with the increase of permeability coefficient. And when T is equal to zero, the spatial damping rate is zero, too, namely the special case of wave on impermeable bottom.

The spatial damping rate versus relative depth is shown in Figure 2. It is noted that the deeper the relative depth is, the smaller the spatial damping rate is.

4.2 *Simulation of wave damping*

The damping of wave propagation over porous bottom with different relative depth and permeability coefficients are illustrated in Figure 3. It can be seen from the figure, the wave

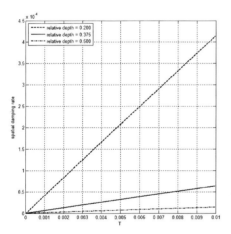
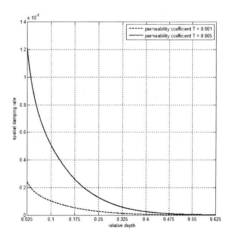

Figure 1. Spatial damping rate versus T. Figure 2. Spatial damping rate versus relative depth.

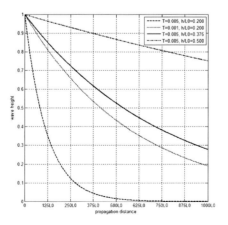

Figure 3. Wave damping over porous bottom.

height has an exponential decay with distance. And the intensity of wave damping increases with an increase in permeability coefficient, but decreases with relative depth increasing. This implies that a more obvious wave damping will occur in a gravel seabed and shallow water than in a coarse/fine sand seabed and deep water.

5 CONCLUSIONS

In this paper, the study of wave propagation over rigid, porous bottom is further developed. In order to obtain the complex wave number for given wave period, water depth and porous medium, a simple and efficient method is proposed to solve the dispersion relationship, and the solutions agree very well with the results by the perturbation method, indicating the validity of the present method. The spatial damping rate, image part of the complex wave number, increases with an increase in permeability coefficient, but decreases with relative depth increasing. That is, a more obvious wave damping will occur in a gravel seabed and shallow water than in a coarse/fine sand seabed and deep water.

REFERENCES

Dalrymple, R.A., & Dean, R.G. 1991. Water wave mechanics for engineers and scientists. *Prentice-Hall*.

Mendez, F.J., & Losada, I.J. 2004. A perturbation method to solve dispersion equations for water waves over dissipative media. *Coastal engineering*, 51(1), 81–89.

Putnam, J.A. 1949. Loss of wave energy due to percolation in a permeable sea bottom. *Eos, Transactions American Geophysical Union*, 30(3), 349–356.

Reid, R.O., & Kajiura, K. 1957. On the damping of gravity waves over a permeable sea bed. *Eos, Transactions American Geophysical Union*, 38(5), 662–666.

Research on settlement deformation aging characteristics of the concrete faced rockfill dam built on deep overburden layer

Long Jiang & Xiaogang Wang
China Institute of Water Resources and Hydropower Research, Beijing, China

Xiaoping Wei & Jianbo Jian
Henan Hekou Village Reservoir Engineering Construction Management Bureau, Henan, China

Wei Zhai
Beijing Jing Fourth Construction and Engineering Co. Ltd., Beijing, China

Huijuan Zhang
Institute of Remote Sensing and Digital Earth Chinese Academy of Sciences, Beijing, China

ABSTRACT: With the rapid development of national economic construction and infrastructure construction, the construction of water conservancy project ushered in the peak, and the engineering geological problems became more and more complicated. In order to solve one of the three major engineering geological problems in building concrete faced rockfill dam, settlement deformation aging characteristics of the concrete face rockfill dam built on deep overburden layer were systematically studied. The results show that: ①settlement deformation of dam foundation and dam body was closely related to the geological conditions, filling height of dam foundation, and time. When filling height reached 20%, 50%, 75%, 100% of the overall height of the dam, the settlement was respectively around 35%, 65%, 85%, and 90% of the maximum settlement; ②settlement deformation of dam foundation and dam body increased with the increasing height of the rockfill dam. When the overburden layer was thicker, the settlement deformation was delayed, but the settlement rate dropped significantly. The largest dam settlement accounts for about 0.72% of the dam height, in accordance with the general settlement deformation law of earth-rock dams; ③the improved horizontal fixed inclinometer observation system is proved effective on monitoring the wide dam foundation settlement, which can well reflect the engineering construction.

1 INTRODUCTION

With the rapid development of national economic construction and infrastructure construction, the construction of water conservancy project ushered in the peak, and the engineering geological problems became more and more complicated. There exist three major engineering geological problem areas, namely, settlement, seepage, and earthquake liquefaction. This paper systematically studied settlement deformation aging characteristics of the concrete face rockfill dam built on a deep overburden layer relying on the reservoir project of Henan Hekou village, which provided technical support for the engineering construction, and provided scientific basis for similar future projects.

Hekou village reservoir, located in the last paragraph exit of Qinhe River canyons of the primary tributaries of the Yellow River, about 9 km away from the Wulongkou hydrometric station downward and belonging to Kejing town of Jiyuan city in Henan province, was a key project for controlling flood and runoff and also an important part of flood control engineering system at the lower reaches of the Yellow River, with project scale of large(2) type and engineering grade of class II, composed of a concrete faced rockfill dam, spillway tunnel of 1#, spillway tunnel of 1#, water diversion power tunnel, spillway, and hydropower station. The foundation of the concrete faced rockfill dam was made up of cohesive soil in a deep overburden layer, upstream of the dam body filled with graded ingredients and downstream with non-graded ingredients, thickness in general 30 m, and the lithology of gravel including boulders and mud, with 4 layers of discontinuous cohesive soil and a number of sand lens.

2 PROJECT MONITORING ARRANGEMENT AND MONITORING METHOD

2.1 Project monitoring design

Considering the deep overburden layer, the dam type and filling properties, a horizontal fixed inclinometer was set in the foundation, and three vibrating string settlement instruments lay out in different elevation of the dam to be used for monitoring settlement deformation of dam foundation and dam body. The monitoring layout diagram is shown in Figure 1.

2.2 Project monitoring method

1. Horizontal fixed inclinometer

According to the working principle of observation system of horizontal fixed inclinometer, starting point A or ending point B can be used as the starting point of calculating accumulated settlement, the calculated result is the relative settlement value relative to point A or point B. As long as the absolute settlement value of point A or point B is measured, the absolute settlement amount of each monitoring point in the system can be estimated. The horizontal fixed inclinometer device and calculation diagram are shown in Figure 2. In Figure 1, 1 is the fixed end, 2 is for an anchorage block, 3 for guide slot protection tube, 4 for connecting rod joint, 5 for the horizontal fixed inclinometer, 6 for the protection pipe joint, 7 for connecting rod, 9 for the transmission rod, 10 for protection tube, 11 for transition end, 12 for a universal joint, and 13, 14 and 15 separately for the connecting rod, the displacement sensor, and protecting box.

2. Vibration string settlement instrument

The working principle of the observation system of vibration string settlement instrument is consistent with the inclinometer.

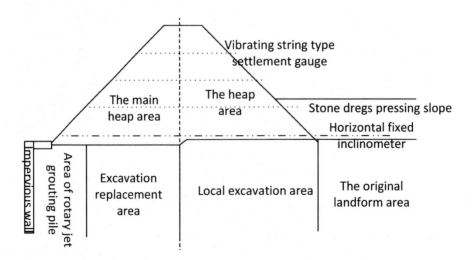

Figure 1. Concrete face rockfill dam of different elevation monitoring arrangement.

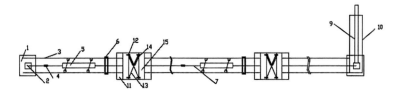

Figure 2. Horizontal fixed inclinometer installation diagram.

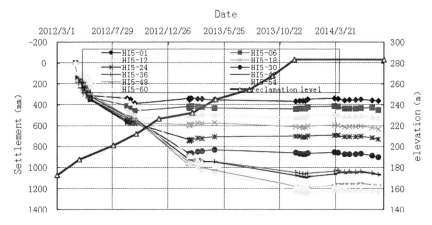

Figure 3. The foundation settlement curve.

3 ANALYSIS OF MONITORING DATA

3.1 *The dam foundation settlement characteristics*

A set of horizontal fixed inclinometers from upstream to downstream was buried in the 173 m elevation of the dam section 0+140. Sixty three horizontal fixed inclinometers were lay out at every 5, 6, and 7 m spacing to monitor the dam foundation settlement of more than 350 m. The monitoring results were shown as Figure 3 and Figure 4.

From Figure 3 and Figure 4 we can see that the dam foundation settlement increases gradually with the filling height and increasing time. When filling to the elevation of 225 m, 240 m, and 286 m, the maximum settlement deformations were 461 mm (D0-182), 651 mm (D0-51), and 789 mm (D0-51) separately; when filling to the elevation of 286 m and after quiescence, the settlement deformation amount fluctuated between −5 mm and 15 mm. Settlement deformation was related to geological conditions and ways of rockfill filling of the dam foundation, in accordance with deformation characteristics of general earth-rock dams.

3.2 *The dam body settlement characteristics*

Three sets of vibrating string settlement instruments were, respectively, buried in the 221.5 m, 241.5 m, and 260.0 m elevation of the dam section 0+140. The monitoring results were shown as Figure 5 and Figure 6.

From Figure 5 and Figure 6 we can see that the dam settlement increases gradually with increasing filling height and time. Settlement deformation of 221.5 m, 241.5 m, and 260.0 m of the dam section 0+140 was, respectively, 502.0 mm, 424.5 mm, and 238.1 mm.

3.3 *The dam settlement characteristics*

Through overall reorganization and comprehensive analysis of settlement deformation data of the fixed level inclinometer and the horizontal fixed inclinometer of the D0+140 dam

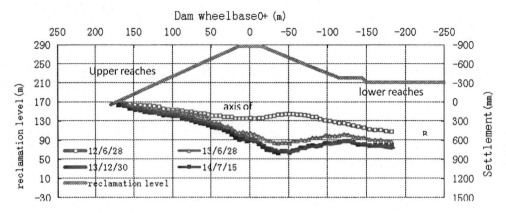

Figure 4. Settlement curves of each measuring point section of dam foundation.

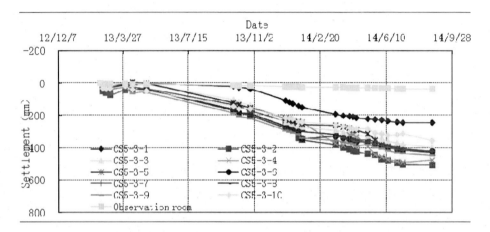

Figure 5. The settlement time curve of dam.

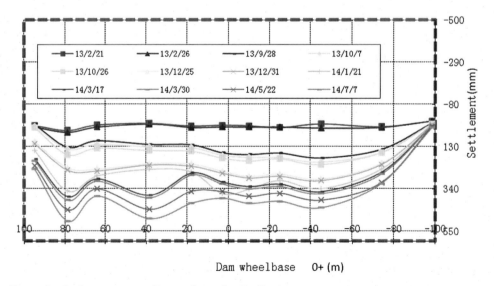

Figure 6. Settlement curves of measuring point distribution.

126

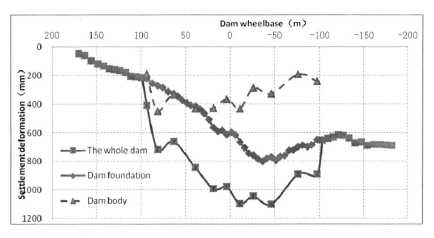

Figure 7. Distribution map of settlement deformation of rockfill dam of Hekou Village Reservoir.

section, the vibrating subsidence overall reorganization and analysis of settlement deformation of the dam settlement monitoring instrument, the distribution curve of deformation was shown in Figure 7.

Figure 7 shows that the maximum settlement of dam foundation was 800 mm (D0-36), and the maximum settlement of dam body was 447 mm (D0+82), and the maximum settlement of the whole dam was 1097 mm (D0-46). Comprehensively considering Hekou village reservoir rock fill dam with maximum dam height of 112 m, maximum thickness of overburden of dam foundation 40 m and the largest settlement of the whole dam accounting for about 0.72% of the dam height at the present stage, the overall settlement amount was in accordance with settlement deformation law of general earth rock dam.

4 CONCLUSIONS

1. Settlement deformation of dam foundation and dam body was closely related to the geological conditions, filling height of dam foundation and time, and increased with the rockfill dam height increase. When the overburden layer was thicker, the settlement deformation was delayed, but the settlement rate dropped significantly.
2. The largest dam settlement accounts for about 0.72% of the dam height, in accordance with the general settlement deformation law of earth-rock dam.
3. The improved horizontal fixed inclinometer observation system is proved effective on monitoring the wide dam foundation settlement, which can well reflect the engineering construction.

REFERENCES

Ai Bin. Deformation of concrete face rock fill dam and dam safety monitoring problems with [J], 1996.6. (in Chinese)

China Institute of Water Resources and Hydropower Research. Hekou Village Reservoir in Henan province impoundment safety appraisal of safety monitoring reports [R]. 2014. (in Chinese).

Shao Yu, Li Haifang, Deng Gang. Characteristics of concrete faced rockfill dam surface [M]. Beijing: China Water Conservancy and Hydropower Press, 2011. (in Chinese).

The Yellow River Engineering Consulting Co., Ltd. Hekou Village Reservoir in Henan province water storage security identification design self-test report [R]. 2014. (in Chinese).

The application research of coal gangue in Yiqi Highway subgrade construction

Y. Liu
Geotechnical and Structural Engineering Research Center, Shandong University, Ji'nan, China
Department of Civil Engineering, Shandong Jiaotong University, Ji'nan, China

ABSTRACT: According to the characteristics of coal gangue itself, in accordance with the case study of Yiqi Highway subgrade coal gangue construction, various performance indicators in highway construction of coal gangue were explored. Combined with test road paving, coal gangue subgrade construction technology and quality control method were confirmed. The results show that: Coal gangue as subgrade filling material is characteristic of small compressibility and good stability. It can be popularized and applied in mining cities as a highway subgrade filling material.

1 INTRODUCTION

Coal gangue is a kind of solid waste in the process of coal exploitation. Accumulation of coal gangue will not only occupy land resources but produce exhaust gases and dust pollution once spontaneous combustion occurs. How to make rational use of coal gangue in mining cities is a big concern for its atmospheric environment.

This research aims at providing theoretical and practical basis for the further rational application of coal gangue in mining cities. Based on Yiqi Highway subgrade construction, an indoor test to coal gangue was conducted, construction technology and quality control of coal gangue subgrade were analyzed, and evaluation of compaction quality was confirmed.

Yiqi Highway spans more than 117 kilometers, is designed with four totally enclosed lanes, intercommunicated overpass and speed limit of 100 kilos per hour. It plays an important role in the communication and transportation in Heilongjiang Province.

City Qitaihe, where Yiqi Highway traverses, is rich in coal resources, piles of burned coal gangue have accumulated for more than ten years. With the purpose of saving cost and protecting environment, K102~K117 subgrade construction made use of coal gangue as filling material.

2 PAVEMENT PERFORMANCE OF RAW MATERIAL

2.1 *Sieve test*

Well graded subgrade filling material is a key to the insurance of subgrade compaction (Shao, 2005). Coal gangue from different mines has different particle sizes, so sieve test is needed before filling. Particle composition of coal gangue samples from different mines are shown in Figure 1.

According to the test result, coal gangue from different mining sites has different graduation. Before coal gangue being taken as filling material, its particle graduation is suggested to meet the value that nonuniformity coefficient is more than or equal to 5 and coefficient

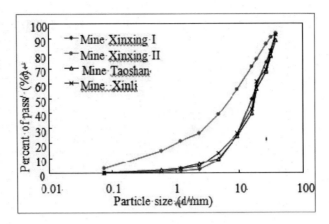

Figure 1. Coal gangue particle size curve.

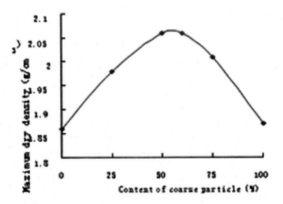

Figure 2. Dry density change curve based on content of coarse particle.

of curvature is between 1~3. The disproportionately graded coal gangue can be cracked to insure both the maximum and minimum particle sizes meet the existing standard.

2.2 *Moisture-density test*

The dry density of coal gangue changes along with content change of coarse and fine particles (Jiang, 2012). A moisture-density test was conducted on coarse particles sizing more than or equal to 5 mm. The test result is shown in Figure 2. Along with the content increase of coarse particles, the dry density of coal gangue increased first and then decreased. When the content reached 60%, the maximum dry density was attained. Based on this, content of coarse particles can be reasonably increased when choosing raw material.

2.3 *Bearing-ratio test*

Highway Subgrade Construction Specifications set bearing-ratio index for highway subgrade filling materials. Coal gangues with different ignition loss were selected in the bearing-ratio test and the result showed that their bearing ratios were obviously greater than the specified maximum value 8%, shown as Table 1. Other research also shows that coal gangue bearing-

Table 1. Coal gangue Bearing-Ratio results (CBR).

Sample section	CBR with different compaction (%)			Ignition loss (%)
	93%	96%	100%	
Mine Taoshan	31.6	46.2	55.6	8.76
Mine Xinli	25.4	37.4	47.0	11.2
Mine Xinxing I	20.9	42.1	58.7	15.6
Mine Xinxing II	36.9	52.0	61.2	1.74

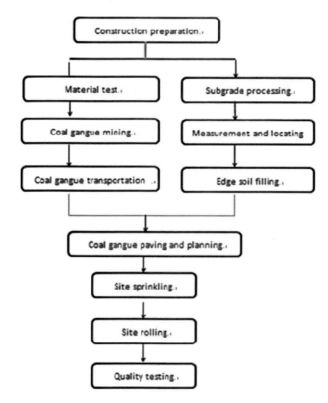

Figure 3. Process drawing for coal gangue subgrade construction.

ratio index should be reasonably increased considering the difference between gangue and other common filling materials (Shi, 2009). Coal gangue with greater ignition loss has a lower bearing ratio, combined with the test result, the bearing ratio of coal gangue for highway subgrade filling is suggested to be above 25%.

3 SITE OPERATION

3.1 *Construction technology*

To meet the performance requirement for coal gangue subgrade, a 100~200 m test section will be selected in the construction site before the extensive paving, which is the basis to determine construction data including machineries, optimum water content, stratified paving thickness, loose paving coefficient, etc. The whole process drawing is shown as Figure 3.

3.1.1 Selection of raw material

According to the existing specifications, coal gangue can't be used as subgrade filling material when it is soft or its ignition loss is above 15%. To assure coal gangue compaction effect and subgrade strength, content of coarse particles sizing 5 mm or above should be controlled to be or more than 60% and the bearing ratio should be or more than 20%.

3.1.2 Coal gangue mining and transporting

Naturally accumulated gangue mountain is characteristic of being unstable and easy to collapse, having low water content and high dust loading. In mining process, certain people should be designated to take charge to avoid injuries and losses. In transportation, it should be covered with canvas. In mining site, effective methods should be adopted to lower its temperature, such as sprinkling or airing. Coal gangue with temperature above 50 °C should not be transported to the construction site.

3.1.3 Filling of surrounding soil

The global stability of subgrade suffers a weakening due to immergence of rainwater or surface water and the weathering of naked coal gangue (Lei, 2011). To avoid this, surrounding soil is designed beside subgrade with the width of 1~2 m and paving thickness of 10%~15%. Surrounding soil and coal gangue should also be compacted synchronously after their stage construction and leveling.

3.1.4 Site sprinkling

Considering coal gangue's character of low water content, site sprinkling is needed after the paving and leveling, to assure the compaction can be conducted with optimum water content. Specific amount of sprinkling needs calculating based on the optimum water content specified in the test section. Site rolling begins when the surface of coal gangue feels a little bit dry, about 20 minutes after the sprinkling.

3.1.5 Site rolling

To achieve the global compaction quality, keywords for compaction should be synchronous, from outside to inside, and static-vibrating-static. To improve the occlusion among particles, one-or-two-times static press is needed, then is the six-to-eight-times vibrating compaction from outside to inside with the vibratory roller, whose wheel-mark should overlap 30 cm. To make the compaction of surrounding soil in place, a second compaction with small roller is needed.

3.2 Construction quality control method

As a kind of subgrade filling material, coal gangue has more coarse particles than fine particles. Therefore, adopting regular sand cone method to test degree of compaction will lead to the situation that compaction cannot meet the requirement or exceeds 100. Instead, other methods can be adopted, such as irrigation method, solid volume ratio method, settlement method, etc.

3.2.1 Irrigation method

The maximum particle size of site filling coal gangue is generally within 100 mm. Before irrigation method, heavy-duty compaction test should be adopted to confirm the maximum dry density. Size of test pit is at least three times as big as the maximum particle size of coal gangue (Qin, 2008). Wet density of coal gangue can be calculated based on the test pit volume tested through irrigation method and the gross mass of coal gangue. Dry density of coal

Table 2. Control parameters of compaction quality of coal gangue subgrade.

Test items	Compaction degree (%)	Solid volume rate (%)	Settling index (unit: mm) Average settlement for the last two times	Standard deviation
Span	≥96	≥85	≤3	≤1
Testing frequency	random inspection or 1 spot/50 m²		5~6 spots/500 m (per layer)	
Site compaction parameters	particle size ≤80 mm, rolling times 6~8, rolling speed 2~4 km/h			

Table 3. Result of deflection test.

Subgrade class	Actual deflection (0.01 mm)	Designed deflection (0.01 mm)	Pass percent (%)
Original subgrade	192.1	232.1	100%
Coal gangue	87.3		100%

gangue can be attained based on the water content of gangue sample through indoor experiment. The degree of compaction can be attained finally.

3.2.2 *Solid volume ratio method*

Solid volume ratio method is often adopted to test compaction quality of rock-fill embankment. As for particle size, coal gangue, a coarse-particle filling material, has similar characteristics to rock-fill embankment. Solid volume ratio is the ratio of solid volume of coal gangue and site-measured test pit volume. The formula is shown as Equation 1. The solid volume ratio of coal gangue tested in Yiqi Highway subgrade construction site is ranging from 85% to 89%, meeting the requirement of compaction.

$$T = \frac{\sum_{i=1}^{n} \frac{m_i}{p_i}}{V} \quad (1)$$

where T = solid volume ratio (%); V = test pit volume (m³); m_i = dry quality of sample (Kg) and p_i = bulk density of coal gangue (kg/m³).

3.2.3 *Settlement method*

With the increasing times of subgrade rolling, the accumulated compaction settlement of stage-paved coal gangue increases gradually and tend toward a stable value that meets the requirement of compaction. By testing the last two compaction settlement ratios, it can be confirmed whether the coal gangue subgrade meets the requirement of compaction.

Based on the site paving of test road and the data analysis, compaction quality control index of coal gangue can be attained. The specific data is shown in Table 2.

3.3 *Subgrade construction quality test*

After the coal gangue subgrade construction of Section K102~K117 of Yiqi Highway, the deflection test was conducted on the finished subgrade, the result, as shown in Table 3, suggests that coal gangue subgrade possess the characteristics of high intensity and good stability.

4 CONCLUSIONS

Accumulation of coal gangue will not only occupy land resources but produce exhaust gases and dust pollution. The recycling use of coal gangue in highway subgrade construction can solve not only the environmental problems but also the earth fetching difficulty in subgrade construction.

REFERENCES

Jiang, Li. Dong, Jianxun. & Zhang, Jinsheng. 2012. Application Research of Coal Gangue on Highway Bridge Culvert Deck Backfilling. *Highway*. 9: 23.
Lei, Jian. & Wang Zhaohui. 2011. Analysis on Tipping Soil Thickness of Coal Gangue Subgrade under Freeway. *Journal of Chang'an University (Natural Science Edition)*. 31(1): 22.
Qin, Shanglin. &Chen Shanxiong. 2008. Research on Filling Test of High Embankment with Over Coarse-grained Soil .*Chinese Journal of Rock Mechanics and Engineering*. 27(10): 2101.
Shao, Lageng. & Fu, Jianyong. 2005. Compaction Quality Control of Expressway Rock Embankment in Mountain Area. *Journal of Highway and Transportation Research and Development*. 22(2).
Shi, Chenglin. & Zhu, Kan. 2009. Research on Classification Technology of Highway Coal Gangue. *Journal of China & Foreign Highway*. 29(4): 401.

Turbulent flow in open channel with different Froude numbers

Sankar Sarkar
Physics and Applied Mathematics Unit, Indian Statistical Institute, Kolkata, West Bengal, India

ABSTRACT: The paper presents the detailed measurements of turbulence spectra, turbulence anisotropy, and turbulent bursting for flows in a rough open channel with respect to different Froude numbers under a shallow water depth. The anisotropy analysis suggests that with increase in the Froude number the flow follows three-dimensional isotropic characteristics at the middle of the flow depth and with increase in vertical distance, lower Froude number flow follows the one-component anisotropy. The quadrant analysis suggests that as the Froude number increases, in the neighborhood of the rough bed the sweep becomes the governing event in turbulent bursting. Also, no significant changes were observed in the power and coherence spectra due to changes in Froude numbers.

1 INTRODUCTION

Turbulence in open channel is a matter of continuing interest to the hydraulic engineers due to the practical application of this in river mechanics, applied hydrogeology, and environmental sciences. With the advancement of the technology, different measuring instruments and techniques are introduced to study the turbulent flow. Particularly in last few decades, researchers have conducted a considerable amount of experimental and theoretical studies to solve many problems of turbulent flow but till date, there is no general theory of turbulence. The problem becomes more critical for the flow with hydraulically rough beds. This is the main reason for that the turbulence gets important to the engineers, mathematicians, physicists, and others communities of scientists. To describe the turbulent flow, in most of the cases researchers have studied the velocity distribution, Reynolds shear stress, turbulence intensity, turbulent kinetic energy, turbulent kinetic energy budget, turbulent bursting, length and time scales, etc. The importance of power spectra at different Froude numbers is not well explored so far and the same stands for the turbulent anisotropy also. According to many researchers, Turbulence anisotropy is one of the important tolls to describe the turbulence in open channel. For a better understanding of the behavior of the turbulent flow characteristics, a knowledge of the anisotropy is required (Lumley and Newman 1977; Shafi and Antonia 1995). It is a useful tool to determine the sensitivity of the turbulence structure to different boundary conditions required by the turbulence models. According to the best knowledge of the author, very limited studies have been done so far to show the anisotropy in open channel with respect to Froude numbers although some important studies include Lumley and Newman (1977) and Smalley et al. (2002). The present study is therefore undertaken to observe the behavior of power and coherence spectra, turbulence anisotropy at different Froude numbers and to see the general behavior of the flow, turbulent bursting was also examined by using the quadrant analysis.

2 EXPERIMENTAL SETUP

Experiments were carried out in a 0.5-m-wide, 0.5-m-deep, and 10-m-long recirculating closed-circuit rectangular flume at the Fluvial Mechanics laboratory of Indian Statistical

Institute Kolkata, India. The transparent side-walls of the flume facilitated the clear observation of the flow. Uniform streamwise bed slopes of 0.35% and 0.075% were made with arbitrarily spreaded uniform crushed stones of a median size d_{50} = 10 mm over the flume bottom in layers. The bed surface fluctuations and flow depth were measured using a Vernier point gauge with a precision of ±0.1 mm. The three-dimensional (3D) instantaneous velocity components (u = the streamwise velocity, v = the lateral velocity and w = the vertical velocity component, respectively) were captured along the centerline of the flume within the fully-developed zone by a 4-beam 50-mm down-looking Vectrino probe. Experiments were carried out with two different velocities and flow depth was 0.125 m above the bed level. For the simplicity of the explanation they are called as Case1 and Case2 in the paper. Details of the experimental parameters are tabulated below. Here, d_{50} = mean diameter of the crushed stone, S = the streamwise bed slope, h = total flow depth, Fr = Froude number ($U/(gh)^{0.5}$, and b = width of the channel. Both the experiments were conducted under a fixed bed condition.

The 3-D velocity measuring instrument, Vectrino is an acoustic Doppler velocimeter manufactured by Nortek that was operated with an acoustic frequency of 10 MHz and a sampling rate of 100 Hz. Its four acoustic beams meet at a distance of 50 mm below the probe head and form a vertical cylindrical of 6 mm diameter and 1–4 mm user adjustable height. As the measuring location was 50 mm below the probe, the probe could not affect the measured data and obviously the measurement was not possible for the 50 mm heights from the top free surface. It is important to mention here that the Vectrino very often captures spiked data. It was therefore important to despike the data for the correct estimation of the turbulent behaviors. However, before starting the main experiments several trail was undertaken which suggested that a duration of sampling 300 s to be adequate to achieve time-independent averaged velocity components at a measuring location. Considering the importance of the near-bed profile and accuracy of the data, at the closer vertical distance from the bed level,

Table 1. Experimental parameters.

Case	d_{90}, m	S%	b, m	h, m	U, m/s	Fr
1	0.01	0.1	0.5	0.125	0.35	0.32
2	0.01	0.35	0.5	0.125	0.80	0.72

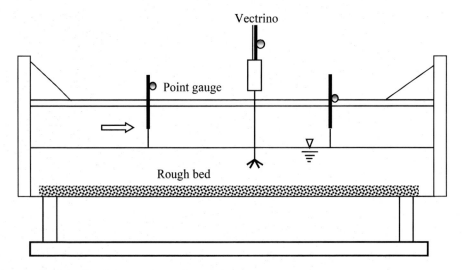

Figure 1. Schematic of the experimental set-up.

the lowest sampling height of the Vectrino measurements was used as 1 mm. The lowest measuring point was 2 mm above the crest level of the bed. The measurement ensured that the sampling volume did not touch the rough bed and also at the same time, the points of measurements were quite close to the bed. For the flow levels above the shear layer, the sampling height used was 4 mm. The Vectrino signal correlation values greater than 70% were used as the minimum value. However, nearest the bed, the range of the Vectrino signal correlation was sometime less than 70% because of the shear layer, and the minimum SNR (signal-to-noise ratio) was maintained as 17 throughout the depth. After capturing the data, the velocity power spectra was noticed, they were processed further with a high-pass filter and the data despiking was done using Acceleration threshold method.

3 RESULTS

The major findings of the works are described below in this section.

3.1 Spectral behavior at different flow conditions

The spectral behavior of the flows at different nondimensional depths was observed. As an example two of them are shown in Figure 2 for lower and higher Froude number flows respectively. The figures here are presented for $z/h = 0.1$, where z = any vertical distance where measurements were taken. Both the cases, the power spectra of these signals exhibited a satisfactory fit with Kolmogorov $-5/3$-law in the inertial subrange of frequency without any discrete spectral after using the despiking algorithm. In this study, the raw captured data were processed by using a high-pass filter with a cut-off frequency of 0.5 Hz. Importantly, no obvious differences in $S_{ii}(f)$ were observed in the flows with different Froude numbers. Thus, it was revealed that the velocity power spectra were not influenced by the Froude numbers. Although they are having differences in the values, their basic trends are same throughout the depth for both the cases.

After plotting the spectral graphs, the analyses were done for the coherence spectra also. For an instance, two coherence spectral graphs are shown in the Figure 3(a) and 3(b). From the Figure 3(a) and 3(b), it is understood that the coherence spectra graphs show the similar characteristics for both the flows. For all the cases, there values remain within unity. This is another proof of the accuracy of the Vectrino data because for the spiked results they sometime show the values more than unity.

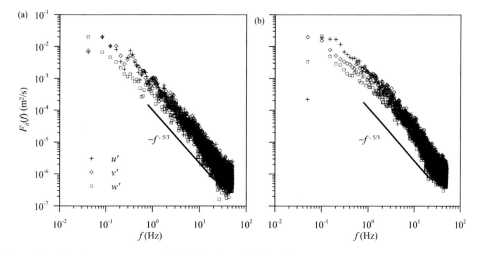

Figure 2. Velocity power spectra for (a) Case1 and (b) Case2.

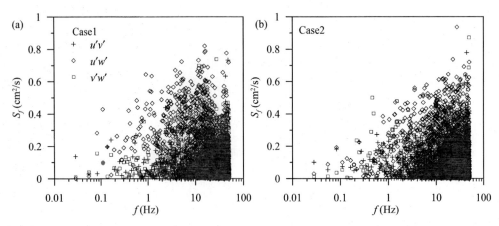

Figure 3. Coherence spectra at $z/h = 0.25$ for (a) Case1 and (b) Case2.

3.2 *Turbulent anisotropy*

The departure from isotropy (known as turbulent anisotropy) for flows with two different Froude numbers are shown in Figures 4–5 by the use of different anisotropic parameters, Anisotropy Invariant Map (AIM) and Anisotropy Invariant function (*F*), respectively.

Figure 4 shows the distribution of b_{ik} for case1 and case2. It is evident from the figure that close to the bed the values of b_{11} are more for higher Froude number than that of the lower Froude number, whereas b_{22} plays the opposite role. Close to the bed b_{13} is more with negative values for flow with lower Froude number than that of higher Froude number. Values of b_{33} are almost same for both the cases. All these values take part in the changes of overall anisotropy.

Plotting the Anisotropic Invariant Map or AIM is an important and easy method to explain the overall anisotropy in a turbulent fluid flow. According to the AIM, the anisotropy of a turbulent flow can be explained using two invariants *II* and *III*. The invariants are defined as: $II = -b_{ik}b_{ik}/2$ and $III = b_{ij}b_{jk}b_{ki}/3$. The invariants can also be transformed in ξ and η and to get the anisotropy in terms of ξ vs. η graph, where $\xi = (III/2)^{1/3}$ and $\eta = (-II/3)^{0.5}$. The ξ vs. η-plot forms a triangle known as modified Lumley triangle as shown in Figure 5(a). In the AIM, the ordinate refers to the degree of anisotropy and the abscissa corresponds to the type of anisotropy in a fluid flow. The left and right boundaries of the modified Lumley triangle correspond to axisymmetric turbulence and are given by $\eta = \pm\xi$, and the triangle is bounded by the upper boundary defined by $\eta = (1/27 + 2\xi^3)^{0.5}$. The bottom vertex $\xi = 0$, $\eta = 0$ represents isotropic turbulence; whereas the right vertex $\xi = 1/3$, $\eta = 1/3$, and left vertex $\xi = -1/6$, $\eta = 1/6$ represent one-component turbulence, and two-component axisymmetric turbulence, respectively.

It is observed from the ξ vs. η-plot that close to the bed, the data plots follow two-dimensional turbulence as they are close to the left curved boundary. With the increase in depth, the data plots shift towards right curved boundary showing the three-dimensional turbulence. For lower Froude number flow, the data follows the one-component anisotropy with increase in depth. On the other hand, for higher Froude number flows, the data plots have a tendency to move towards bottom cusp showing a better three-dimensional isotropy.

The AIM plotted in the ξ-η plane are shown in Figure 5 (a). Another method of determining anisotropy is the use of Invariant function $\mathbf{F} = 1 + 9II + 27III$. The value of $\mathbf{F} = 0$ for two-dimensional turbulence and $\mathbf{F} = 1$ for a 3D isotropic state. Figure 5(b) shows that for both the cases, the anisotropy starts from the two-dimensional state and follows the 3D state as the vertical distance increases. Close to the bed, for higher Froude number a better two-dimensional turbulence isotropy is observed, whereas above $0.1\,h$, a better 3D isotropic state in flow is prevalent for higher Froude number flow.

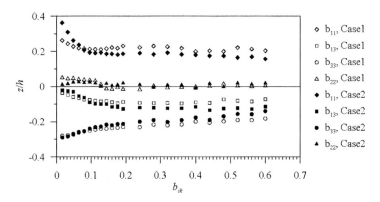

Figure 4. Distribution of b_{ik} for different cases.

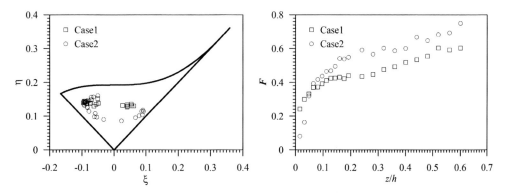

Figure 5. (a) Anisotropy Invariant Map (AIM) and (b) distribution of Invariant Function (F).

3.3 *Turbulent bursting*

Investigations of 1960s and 70s (Kline et al., 1967; Corino and Brodkey, 1969; Grass, 1971) illustrated that the nature of the flow pattern near the wall in a turbulent boundary layer is repetitive and the flow occurs in the form of a quasi-cyclic process, called the bursting process. Turbulent events occurring in the boundary layer can be well defined by the quadrant analysis, introduced by Willmarth and Lu (1972). In the quadrant analysis the local flow behavior is divided into quadrants, depending on the sign of u' and w', where they are the fluctuations of the velocity components in the streamwise and vertical directions. Four quadrants are defined as: Q_1, first-quadrant $(u'w')_1$, where $u' > 0$ and $w' > 0$, an event in which high speed fluid moves toward the center of the fluid flow field; Q_2, second-quadrant $(u'w')_2$, where $u' < 0$ and $w' > 0$, denotes an event in which low-speed fluid moves toward the center of the fluid field, away from the wall (ejection); Q_3, second-quadrant $(u'w')_3$, where $u' < 0$ and $w' < 0$, denotes an event in which low-speed fluid moves toward the wall; and Q_4, second-quadrant $(u'w')_4$, where $u' > 0$ and $w' < 0$, denotes an event in which low-speed fluid moves toward the wall (sweep). A hole-size parameter H, determined by the curve $|u'w'| = H(\overline{u'u'})^{0.5}(\overline{w'w'})^{0.5}$ is used to differentiate the fractional contributions of $-\overline{u'w'}$ at different quadrants and be presented by the Figure 6(a) and 6(b). According to the Figure 6(a) and 6(b), at $z/h \leq 0.1$, sweep is the governing mechanism for flows with higher Froude number, whereas for lower Froude number flows, outward interactions dominate marginally at very near bed. On the other hand, ejection dominates for both the cases above $0.1\ h$.

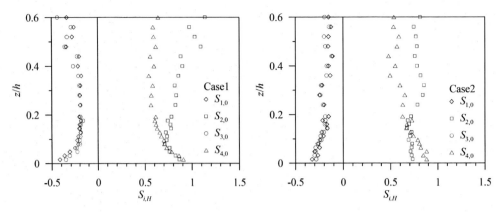

Figure 6. Variations of bursting events with vertical distance for $H = 0$. For (a) lower Froude number and (b) higher Froude number at $z/h = 0.02$ and (c) lower Froude number and (d) higher Froude number at $z/h = 0.5$.

4 CONCLUSIONS

Turbulent spectra, anisotropy and bursting for flows in rough open channel with respect to different Froude numbers under a shallow water depth are examined. The anisotropy analysis suggests that with increase in the Froude number the flow follows three-dimensional isotropic characteristics at the middle of the flow depth; whereas with increase in vertical distance lower Froude number flow shifts toward one-component anisotropy. Considering the velocity and coherence spectra, no significant changes were observed at different Froude numbers although their values change at different layers of the flow. As the Froude number increases, in the vicinity of the rough bed the sweep becomes the governing event in the turbulent bursting.

REFERENCES

Corino, E.R., and Brodkey, R.S. (1969). "A visual investigation of the wall region in turbulent flow." *J. Fluid Mech.*, 1–30.

Grass, A.J. (1971). "Structural features of turbulent flow over smooth and rough boundaries." *J. Fluid Mech.*, 233–255.

Kline, S.J., Reynolds, W.C., Schraub, F.A., and Runstadler, P.W. (1967). "The structure of turbulent boundary layers." *J. Fluid Mech.*, 30, 741–773.

Lumley, J.L., and Newman, G.R. (1977). "The return to isotropy of homogeneous turbulence." *J. Fluid Mech.*, 82 (1), 161–178.

Shafi, H.S., and Antonia, R.A. (1995). "Anisotropy of the Reynolds stresses in turbulent boundary layer on a rough wall." *Exp. Fluids*, 18 (3), 213–215.

Smalley, R.J., Leonardi, S., Antonia, R.A., Djenidi, L., and Orlandi, P. (2002). "Reynolds stress anisotropy of turbulent rough wall layers." *Exp. Fluids*, 33 (1), 31–37.

Willmarth, W.W., and Lu, S.S. (1972). "Structure of the Reynolds stress near the wall." *J. Fluid Mech.*, 55, 65–92.

Seismic analysis for a new-type hybrid structure of Steel Reinforced Concrete frame-four corner tubes-bidirectional steel truss under rare earthquake

Q. Chen & W. Zhang
Hong Kong Hua Yi Design Consultants (Shenzhen) Ltd., China

ABSTRACT: The podium of CNOOC Shenzhen Building is a four-storey large-span structure situated between two super high-rise towers. To fulfill the requirements of the local municipal planning and architecture functions, a new-type structure was proposed for the podium, named as Steel Reinforced Concrete (SRC) frame-four corner tubes-bidirectional steel truss hybrid structural system. Elasto-plastic time-history analysis for the structure under rare earthquake was carried out using the finite element method. The damage of the four corner tubes focuses on coupling beams. Large-span steel trusses keep non-yielding state. The damage on RC beams and steel beams can dissipate some dynamic energy. Therefore the rationality of the structural arrangement is verified.

1 INTRODUCTION

The CNOOC Shenzhen Building consists of two towers and a podium in between, as shown in Figure 1. The podium is located between two towers with 45 floors and 200 m in height, whereas it is an independent structure because of the aseismic joints that are set between them. The plane of the podium structure is approximately rectangular, and the size is 78.1 m by 51.8 m. The podium is a four-storey building with 24.5 m height. A planning road will cross the middle of the podium along the east-west direction in accordance with municipal planning as shown in Figure 2, so the span of the column grid along the south-north direction in the middle of the podium should be 33.6 meters. To meet with the requirement of elevation of the building, the span of the west side along the south-north direction should be 50.4 m. The center of the podium at the third and fourth floors forms a column free space measuring 27.2 m by 32.6 m. In addition, the joints which connect the towers and the podium on the south side and north side should supply large span to fulfill the use function. Therefore, the structural system of the podium should have large span space along the east-west and south-north directions based on the objective conditions described above.

To resolve these problems a new hybrid structure is proposed finally: steel reinforced concrete frame-four corner tubes-bidirectional steel trusses. The three-dimensional model and the structural plan for the second and third floor are shown as Figure 3 to Figure 5. The main feature of the structural system is that the most important vertical components resisting lateral force are grouped on the four corners of the building. The four corner tubes refer to the shear wall tubes providing the space for staircases and elevator shafts on the four corners, shown as the tube 1 to tube 4 at the Figure 4 and Figure 5. There are three trusses set along the south-north direction; the truss 1 to truss 3 is used to connect corner tubes or SRC columns. Four trusses with span of 27.2 m along the ①, ②, ③, and ④ axes are set at the east-west direction, named as truss 4 to truss 7. Steel main beams are used for connecting both steel reinforced concrete columns and trusses, and while for trusses the steel secondary beams are set at equal distance along the span of the steel main beams. Corrugated steel sheets laid on the top of steel beams are cast with concrete to form deck concrete composite slabs.

Figure 1. The construction effect chart.

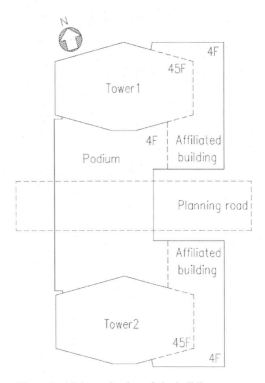

Figure 2. Schematic plan of the building.

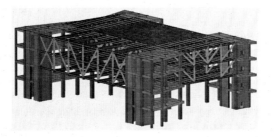

Figure 3. Three-dimensional model.

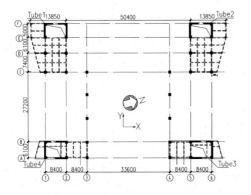

Figure 4. Plan of the second floor.

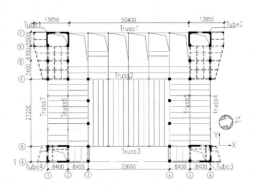

Figure 5. Plane of the third floor.

According to the code for seismic design of buildings of China [GB 50011 2010], the seismic precautionary intensity of the building is seven degree. The elasto-plastic time history analysis using Finite Element Analysis (FEA) method is carried out in this paper to study the seismic behavior of the structure under rare earthquake.

2 DESCRIPTION OF THE ANALYSIS

The Abaqus Software (ABAQUS 2006) is used to carry out the elasto-plastic time history analysis for the hybrid structure under rare earthquake, and the whole structural analysis model is shown in Figure 6. The material behaviors provided by Code for Design of Concrete Structures [GB 50010 2010] and Code for Design of steel structures [GB 50017 2003] are adopted here. Earthquake effects both from horizontal and vertical directions are considered for the large span structure, and the ratio of peak acceleration for three-direction earthquake is X:Y:Z = 1:0.85:0.65 when the earthquake under X-direction is specified as the main horizontal earthquake; while the ratio is X:Y:Z = 0.85:1:0.65 when the earthquake under Y-direction is specified as the main horizontal earthquake. Three earthquake waves are applied to conduct time history analysis represented as T1-X, T1-Y, T2-X, T2-Y, W-X, and W-Y. The peak accelerations for all waves are adjusted to 220 gal according to the code for seismic design [GB 50011 2010].

3 THE TIME HISTORY ANALYSIS RESULTS

3.1 *The whole structural seismic performance*

The regions for maximum displacement of the structure under six earthquake conditions are located in the middle of the rooftop of the large span column free space (large span space), and obvious dents are observed on the rooftop of the large span space. The deformation diagram of the structure when the maximum displacement is reached (at 12.38 s) under T1-Y earthquake condition is shown in Figure 7.

The displacement time history of the node 1070 on the rooftop (see Fig. 6) which has the maximum displacement under T1-Y, T2-Y, and W-Y earthquake conditions is illustrated in Figure 8, where 1070-X, 1070-Y, and 1070-Z are displacement component time history in X, Y, and Z direction, respectively. It is observed that: the maximum absolute value of Z direction displacement is about 70–80 mm after subtracting the initial displacement caused by gravity load, the maximum absolute value of X and Y direction displacement is about 30–40 mm, and the displacement component in Z direction has the maximum weight for the general displacement compared with the other two horizontal displacement components. Therefore vertical earthquake effects cannot be ignored for this large span building. The displacement time history of another node 1298 on the rooftop (see Fig. 6) which is

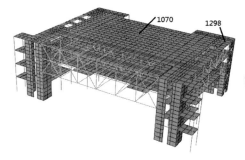

Figure 6. Three-dimensional model of analysis.

Figure 7. Deformation diagram under T1-Y condition.

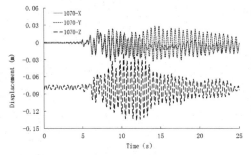 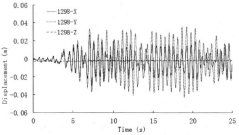

Figure 8. Displacement time history of node 1070 under T1-Y condition.

Figure 9. Displacement time history of point 1298 under T2-Y condition.

Table 1. Maximum inter storey drift of vertical components.

Components	T1-X	T1-Y	T2-X	T2-Y	W-X	W-Y
Shear walls	1/514	1/533	1/444	1/423	1/549	1/578
Floor	3	3	3	3	3	3
Columns	1/432	1/476	1/348	1/358	1/495	1/481
Floor	4	3	4	3	4	4

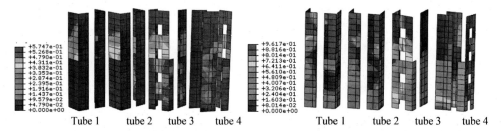

Figure 10. (a). Damage variables in compression. (b). Damage variables in tension of tubes.

not located in the large span space and has relatively large displacement under T2-X and T2-Y earthquake conditions is given in Figure 9. It is illustrated that displacement component in Z direction of node outside the large span space is very small, so this effect can be ignored, meanwhile the maximum displacement component in X and Y direction is about 40–50 mm.

The vertical lateral force resisting components of the building are shear wall tubes and SRC frame columns, the maximum inter storey drifts of the two lateral force resisting components under six earthquake conditions are given in Table 1. The analysis results indicated that the maximum inter storey drift of vertical components has happened under T2 earthquake condition, all of the storey drifts of shear walls are less than those of frame columns, and all the storey drifts of vertical components are less than 1/300 which is far less than the ultimate value of 1/100 required by the code for seismic design of buildings. Therefore, deformation of the building under rare earthquakes can satisfy the requirements of code for seismic design.

3.2 *Seismic performance of components*

The discussion on the seismic performance of components hereinafter is based on the analysis results under T2 wave. As shown in Figure 10, the maximum damage in compression of concrete in tubes whose value is 0.57, is found in coupling beams on the third floor, and the corresponding stress reaches the descending part of stress-strain curve of concrete, and about $0.56 f_c$. The maximum damage in tension of concrete is found in the ends of shear walls on the third and fourth floors. It is shown in Figure 11 that the maximum stress of vertical distribution steels is 337 Mpa closing to yield strength, and the corresponding location is the same with that where damage in tension of concrete is found (the third and fourth floor of tube 2), and is also the connecting parts of tube 2 and truss 1. The result indicates that the connecting parts of the tube and large span have to transfer large internal forces. The maximum stress of horizontal distributing steels appears at the second floor of the tubs which is only 206 MPa, while the stress values on the third and fourth floors are relatively small. So it can be concluded that the first and second floors of tubes exhibit distinct shear deformation while distinct flexural deformation appear on the third and fourth floors.

Figure 12 shows the stress cloud chart of shape steels in the restrained edge members when the maximum value is reached. The maximum stress of shape steels appears at the root of shear wall in tube 3, the value is 243 Mpa which is still an elastic value.

It is indicated from the damage and stress results of tubes that the walls of the tubes on the third and fourth floor suffer more serious damage than the walls on the first and second floor. This result can be explained by following two factors: 1. There are only few slabs on the second floor, the earthquake actions caused by ground motions are focused on the third and fourth floor which is laid with full horizontal components. 2. The vertical components on the third and fourth floor are designed with less stiffness and load-carrying capacity compared with those on the first and second floor. With considerable stiffness, the four tubes on the first and second floor act as supports under rare earthquakes. So damage on the first and second floor is not significant.

The stress cloud chart of shape steels and damage variable in compression of concrete of the maximum values in SRC frame columns are shown in Figure 13. The maximum stress of shape steels appears at the joints, where the bottom chords of truss 2 connect with the frame columns on the either end of the truss, and the stress value is 141 MPa which means that all the shape steels in the frame columns remain elastic during rare earthquakes. The maximum value of damage variable in compression of concrete is 0.147, corresponding to about $0.88 f_c$, and the concrete of column does not reach compressive strength. The concrete suffering relatively serious damage in compression are in the roots of frame columns,

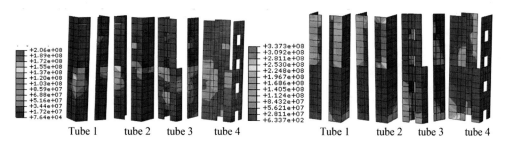

Figure 11. (a). Stress of horizontal distribution steels of tubes (Pa). (b). Stress of vertical distribution steels of tubes (Pa).

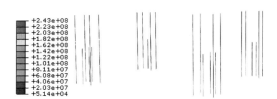

Figure 12. Stress cloud charts of shape steels in the restrained edge members (Pa).

Figure 13. (a). Stress of shape steels in SRC columns (Pa). (b). Damage in compression of SRC columns.

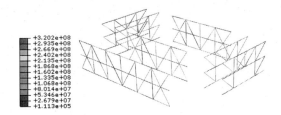

Figure 14. Stress of steel trusses (Pa).

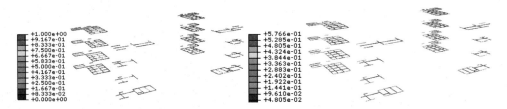

Figure 15. (a). Damage variables in tension of beams. (b). Damage variables in compression of beams.

the joints of the bottom chords of truss 2 and truss 3, and the columns on the either end of these trusses.

According to behavior of the stress of shape steels and damage of concrete that, all the materials of SRC frame columns do not reach material strength and no failures occur in columns during rare earthquakes. It is also revealed that the joints of trusses and frame columns are the key regions for force transfer between them, where relatively high stress is induced.

Figure 14 displays the stress cloud chart of steel trusses when maximum stress is attained under earthquake wave T2. The maximum value is 320 MPa that appeared at the truss 2, which means no yield occurs in trusses. The stresses of the trusses surrounding large span space, i.e., truss 2, truss 3, truss 5, and truss 6, are commonly larger than those of the peripheral trusses, as truss 1, truss 4, and truss 7. So it is indicated that cross-layer steel trusses connecting vertical components act as mega beams to realize the effective force transfer, and have high strength reserve.

The beams discussed here include SRC frame beams, RC beams, slender coupling beams modeled by beam elements and steel beams. The results in Figure 15 indicated that tensile damage is more serious than compressive damage, some concrete in frame beams and coupling beams attain the descending branch of concrete uniaxial compressive constitutive relation and the stress is only half of the compressive strength. Figure 16 and Figure 17 give the stress cloud charts of shape steels in SRC beams and coupling beams and steel beams, respectively. The shape steels in beams remain elastic, for the maximum value is only 311 MPa. However, some steel beams reach plastic states, as the maximum stress of steels is larger than the yielding strength to be 364 MPa. Therefore,

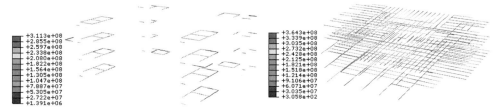

Figure 16. Stress of steels in SRC beams and coupling beams (Pa).

Figure 17. Stress of steel beams (Pa).

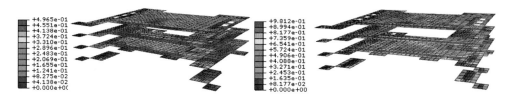

Figure 18. (a). Damage variables in compression of slabs. (b). Damage variables in tension of slabs.

Figure 19. Stress of distributed steels in slabs (Pa).

the concrete of lateral components suffers serious damage and obviously yield occurs in steel beams.

Figure 18 presents the damage in compression and tension of concrete in RC floor slabs and deck concrete composite slabs when the maximum value is reached during earthquake. The damage in compression is focused on the surrounding of the large opening of the third floor and the slab on the both ends of truss 1, the most compressive stress of concrete are lower than 60% of compressive strength while few concrete reach the compressive strength, and obvious cracks appear over most of slabs. The steels inside the slabs remain elastic for the maximum stress value of 277 MPa (Fig. 19). The analysis results indicate that serious failures occurred only on both ends of truss 1 for most of the failures of slabs are slight. The cause for the focused failure phenomena is that, there is no slab along the span of truss 1 (spanned with 50.4 m) and the truss 1 should balance the relative displacement of tube 1 and tube 2 during earthquake, then the forces transferred by truss 1 have to be carried by the locally surrounding slabs.

4 CONCLUSIONS

1. The top displacement and the inter storey drifts of vertical components can meet the deformation requirements by code of China sufficiently. Effects from vertical earthquakes cannot be neglected for this large span structure.
2. The shear wall tubes in the four corners of the structure have carried the majority of earthquake effects. Slight failures are found in wall panels while failures occur on

coupling beams. The large span trusses connecting vertical components act as mega beams to transmit internal forces and remain unyielding during analysis. Dissipating components as RC beams and steel beams failure accompanied with concrete crushing and steel yielding.
3. The new hybrid structural system can provide rational force distribution among all components and direct way of force transmitting; the general structure and components can satisfy good seismic performances.

REFERENCES

ABAQUS Inc., *ABAQUS Theory Manual* [M]. (2006).
GB 50011-2010. *Code for seismic design of buildings*. (2010). China Architecture & Building Press, Beijing.
GB 50010-2010. *Code for design of concrete structures*. (2010). China Architecture & Building Press, Beijing.
GB 50017-2003. *Code for design of steel structures*. (2003). China Architecture & Building Press, Beijing.
Mazzoni Silvia, McKenna Frank, Scott Michael H., Fenves Gregory L., et al. *Open System for Earthquake Engineering Simulation User Command—language Manual* [S], (2006). University of California, Berkeley, Calif.

The forecasting application of Beijing Urban Waterlogging risk warning in 2012–2014

Z.C. Yin & N.J. Li
Beijing Meteorological Bureau, Beijing, China

ABSTRACT: Based on the complex terrain and large city characteristics, the geographic information of Beijing was cut into 6458 grids and corresponding channels. Focused on the urban hydrodynamic and hydrographic process, the Beijing Urban Waterlogging (BUW) numerical model was built to simulate the ponding depth. Driven by high-resolution precipitation observation, BUW could simulate the spatial distribution of "7.21" urban ponding in Beijing well, and the variation and max depth under concave bridges were close to actual condition too. In the real-time operation from 2012 to 2014, risk warning has effect in advance and can better master the spatial distribution of waterlogging, providing a reference for municipal drainage and transportation departments to launch emergent dispatch vehicles.

1 INTRODUCTION

In the context of climate warming and frequent emergent extreme weather events (Wang, H.J. et al. 2010), urban waterlogging and other secondary disasters induced by sudden short-time heavy rainfall in summer (Wang, J.L. et al. 2012, Chen, Y. et al. 2012, Sun, J. et al. 2012) severely threaten the safe operation of urban transportation network and the life security of downtown residents living in old or dilapidated buildings (Shi, Y. et al. 2009). Especially, the construction of a large number of underground parking lots, recessed overpasses and subways quickly increased new urban waterlogging risk sites. For example, due to overpass waterlogging, the traffic on several major roads in central urban districts of Beijing was interrupted on July 10, 2004. In the massive natural disaster on July 21st, 2012, overpasses once again become the Achilles' Heel of Beijing City: a total of 426 waterlogging sites were formed within the City, and 63 of them were found on the roads of central urban districts; the roads were interrupted for 39,945 times. The major causes for urban waterlogging include a large proportion of impervious underlying surface, the low-standard drainage pipe network, the three-dimensional development of urban space and the increased sudden extreme heavy precipitation. The first three factors are the inevitable results of urbanization and are difficult to make improvements in a short time. Therefore, detailed short-time heavy rainfall forecast and waterlogging risk warning become important means and breakthroughs for people to deal with urban waterlogging disasters and provide scientific decision basis for the safe operation of cities.

Currently, major methods used for numerical simulation of urban waterlogging are as follows: hydraulics, hydrology and meteorology-based numerical simulations (Xie, Y.Y. et al. 2004, 2005), hydrology and meteorology-combined statistical method (Ma, X.Q. et al. 2002), comprehensive analytical method of waterlogging causes (Liu, M. et al. 2002), meteorology and social economics-combined method (Hu, H.B. et al. 2013, You, F.C. et al. 2013), AVHRR and MODIS image-based analytical method (Huang, D.P. et al. 2008). With the development of the City, urban rainwater drainage system has gradually evolved into a network from the original linear structure, so the drainage system presents more and more hydrological and hydraulic features. Therefore, it is feasible to build a numerical simulation model for urban waterlogging and implement detailed precipitation forecast to realize early warning of urban waterlogging risks.

Beijing owns the largest number of recessed overpasses and subways among all cities in China, but these two underground space usages cause the greatest risks for urban waterlogging. Based on the urban waterlogging simulation model for Tianjin, the geographical information and physical process of recessed overpasses and subway entrances were collected (Quan, R.S. et al. 2011) to construct a Beijing Urban Waterlogging numerical model. In the meantime, based on this model, the detailed precipitation monitoring and forecasting results are used to establish an early warning system for Beijing urban waterlogging risks and carry out real-time operation.

2 DATA AND MODEL

2.1 *Data*

(1) The high-quality observed precipitation data of 49 high-quality automatic meteorological stations within the Sixth Ring Road of Beijing; (2) monitoring data of overpass waterlogging depth obtained through waterlogging monitoring stations, video inspection and traffic police inspection, etc; (3) 1:10000 geographical information layer, mainly including elevation, river channel, drainage works, architecture and roads.

2.2 *Beijing urban waterlogging numerical model*

Beijing Urban Waterlogging numerical model (BUW) is built on the complex terrain in Beijing and the features of big cities in which the spatial information from geographical information system is divided into 6,458 grids and corresponding channels (Fig. 1). On the basis of the detailed precipitation observation and forecast from Beijing Meteorological Bureau, this model simulates the changes in waterlogging depth in accordance with major urban hydrological and hydraulic physical processes of urban surface, river channels and drainage network.

3 BEIJING URBAN WATERLOGGING RISK WARNING FRAMEWORK

3.1 *Beijing urban waterlogging risk warning grade indexes*

The influence of Beijing urban waterlogging disaster is mainly reflected in traffic. Given that surface waterlogging mainly affects the operation of vehicles and that the maximum fording depth is one of the important indexes to evaluate the trafficability of vehicles, the authors collected the maximum fording depths of different types of vehicles (Fig. 2) and determines the waterlogging risk grade indexes in accordance with the degree of the influence (Table 1).

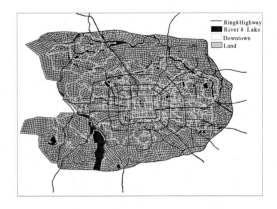

Figure 1. Simulation grids and range of BUW model, overlaid boundaries and roads.

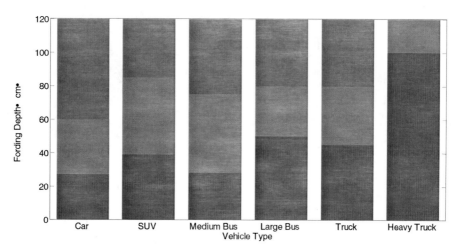

Figure 2. Maximum fording depths of different types of vehicles (unit: cm); the colors of green, orange and red represent "safe", "partially safe" and "unsafe", respectively.

Table 1. Risk warning grades.

Grades	Range (mm)	Affected reference
Blue	5 ≤ PD < 250	Pedestrian
Yellow	250 ≤ PD < 500	Car, SUV and medium bus
Orange	500 ≤ PD < 800	Large bus and truck
Red	PD ≥ 800	Heavy truck

3.2 Real-time warning system

After years of scientific research and systematic development, Beijing urban waterlogging risk warning system is established, which includes three parts: observation module, BUW forecast module and weather service module.

1. Observation module, including two parts: precipitation monitoring of radars (QPE included) and automatic meteorological station, and waterlogging monitoring of automatic water gauge and video monitor.
2. BUW forecast module, including short-time precipitation forecast, BUW module, GIS drafting module and product manufacturing module, etc.
3. Weather service module: urban waterlogging risk warning and corresponding service carriers (Apps and websites, etc).

First of all, hourly precipitation forecast within 1 × 1 km resolution is adopted to drive BUW model and predict the waterlogging depth of the grid. Then, according to the specific influence of different depths of waterlogging on Beijing traffic, a new, referable and rapid updated urban meteorological derived disaster warning is developed.

4 BUW OPERATION CASES

4.1 "7·21" torrential rainstorm

On July 21–22, 2012, a rare heavy rainfall occurred in Beijing, the heaviest citywide one since 1951. Within nearly 20 hours, the average rainfall reached 170 mm and the maximum rainfall amounted to 541 mm. In terms of urban districts, the average rainfall reached 215 mm with a maxim rainfall of 328 mm (Moshikou), and the heaviest precipitation intensity appeared

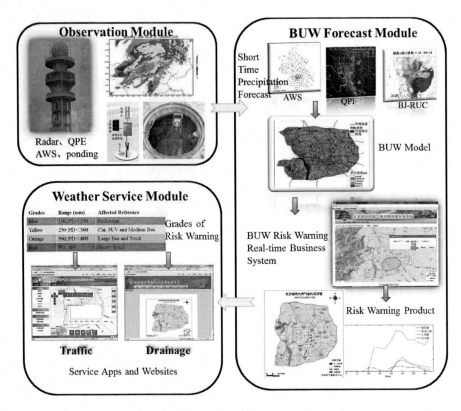

Figure 3. Beijing urban waterlogging risk warning system.

in Fengtai Sports Center, reaching 85.9 mm/h. According to the one-hour rainfall threshold stipulated in Beijing Meteorological Disaster Warning Signals and Defense Guidelines (May 2013), 93.9% of stations had a maximum rainfall of orange warning level above, and 30.6% of stations reached the red warning level.

A total of 426 waterlogging sites were formed within the whole city, and 63 of them appeared on roads of central urban districts (Fig. 4). Overall, the southern part suffered more serious waterlogging than the northern part, and the eastern part was more severe than the western part. The waterlogging sites were mostly distributed in four districts: A, B, C and D among which C and D were most plagued by waterlogging. According to the regulations of Beijing, when recessed overpasses and road sections easily exposed to waterlogging have a depth of 27 mm, they should be closed to traffic immediately and vehicle guidance work should be done. As shown in Figure 4, the waterlogging depths for red sites all exceed 30 cm, especially for Lianhua Bridge, Guangqumen Bridge, Shuangying Bridge and Xiaocun Bridge, whose waterlogging depths even reach or exceed 2 m; In District D, one person was killed due to the excessively deep waterlogging in Guangqumen Bridge. Beyond the Fifth Ring Road of Beijing, Fangshan is a severely afflicted area, and 38 people died in this torrential rainstorm: 5 in Qinglong Lake Town, 4 in Hebei Town and several around Beijing, Hong Kong and Macao Highway. Most of them were victims of urban waterlogging disasters.

BUW model was used to conduct simulation from 10 o'clock on July 21 to 1 o'clock on July 22, lasting for 16 hours. The maximum waterlogging depths are shown in Figure 5. In terms of spatial distribution, this model can not only simulate the serious waterlogging within the Fifth Ring Road of Beijing, but also displays the distribution feature of "the South and the East have more waterlogging than the North and the West, respectively". The waterlogging in A, B and C of Figure 4 can be better simulated; although a lot of waterlogging is

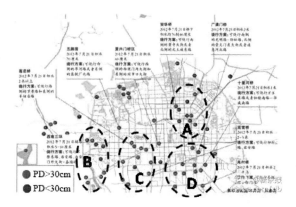

Figure 4. Distribution of urban road ponding (quoted from Beijing News).

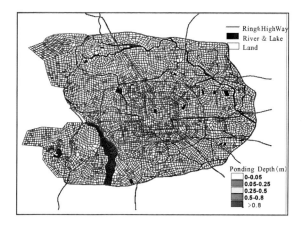

Figure 5. Maximum ponding depth in "7.21" rainstorm simulated by BUW.

Table 2. Comparison between simulation and observation of "7.21" ponding depth in bridge zone (unit: m).

Bridge	Lianhua	Wulu	Caihuying	Anhua	Fuxingmen	Shilihe
Simulation	1.77	0.65	0.49	0.62	0.61	0.69
Observation	>2	0.7	0.5	0.7–0.8	0.6	2

Bridge	Jin'an	Xiyuan	Lize	Liuli	Zhengyang	Andingmen
Simulation	0.41	0.05	0.34	0.48	0.78	0.19
Observation	0.3	0.1	0.3	0.5	0.5	0.3

Bridge	Anzhen	Fangzhuang	Muxiyuan	Zhaogongkou	Dahongmen	Xiaocun
Simulation	0.25	0.68	0.6	0.6	0.25	1.23
Observation	0.3	0.8	0.6	0.6	0.5	2

Bridge	Shuangying		Dongbianmen		Guangqumen	
Simulation	0.64		0.66		0.67	
Observation	2–3		0.5		2	

also simulated for D, it is significantly weaker than the real situation. In the meantime, the proposed model can also accurately simulate the waterlogging beyond the Fifth Ring Road of Beijing, especially Fangshan. According to the simulated results, multiple grids show a waterlogging depth of more than 1 m around Qinglong Lake and Hebei Town in Fangshan, and some grids even have a depth of more than 2 m; Serious waterlogging is also simulated around Beijing, Hong Kong and Macao Highway, which is consistent with the disaster distribution.

Recessed overpasses easily become a confluence area of surround precipitation or are prone to generate underground water, so they are high risk sites for urban waterlogging. Table 2 conducts a more detailed inspection on 21 bridges in Beijing City; these bridges are mainly concentrated in the Fifth Ring Road of Beijing, and a few of them are located beyond the Fifth Ring Road. The simulated and measured values for most of the bridges agree with each other, such as Wulu Bridge, Fuxingmen Bridge, Anhua Bridge, Lize Bridge, Liuli Bridge, Anzhen Bridge, Muxiyuan Bridge, Zhaogongkou Bridge and Caihuying Bridge. The waterlogging in these bridges generally has a depth of less than 1 m, and the model can accurately simulate the waterlogging within this depth scope. Lianhua Bridge has a measured waterlogging depth of more than 2 m and a simulated depth of 1.77 m; Xiaocun Bridge has a measured waterlogging depth of 2 m and a simulated value of 1.23 m. For other bridges with a measured depth of 2 m, all the simulated values do not exceed 1 m. This states that, the model has a limited simulation capacity of waterlogging with a depth around 2 m, which is probably attributed to insufficient consideration of large scale confluence and the steep walls on both sides of recessed overpasses.

4.2 On July 16, 2014

The sudden heavy thunder shower on July 16th, 2014 concentrated on the region from Haidian to Fengtai within the Sixth Ring Road of Beijing, forming multiple discrete waterlogging

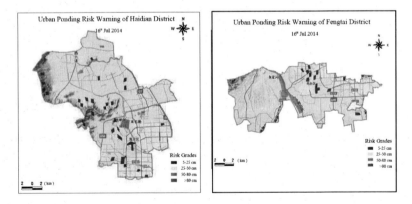

Figure 6. Risk warning for urban waterlogging in Haidian (a) and Fengtai (b) on July 16, 2014.

Table 3. Comparison between simulation and observation in 16th Jul 2014 (unit: m).

Location	Meishuguan	Dianchang Road	Yuegezhuang Northen Bridge	Puhuangyu Road
Simulation	0.26	0.302	0.198	0.074
Observation	0.1–0.2	0.1–0.2	0.1–0.2	0.1–0.2

Location	Tiancun Easten Railway Bridge	Jushan Railway Bridge	Xingong Subway	
Simulation	0.246	0.249	0.101	
Observation	1.5	1	0.1–0.2	

sites, especially the 1.5 m waterlogging under the railroad bridge of Tiancun East Road. The released Beijing urban waterlogging meteorological risk warning offered some risk prompts for the path from Haidian District to Tiancun. To be specific, for Tiancun East Road and Jushan Road Railway Bridge with a waterlogging depth of more than 1 m, a blue risk prompt was also given. Although it is lower than the actual situation, it still has certain indicating significance.

5 CONCLUSIONS AND DISCUSSIONS

BUW model can simulate the waterlogging depth by referring to major hydrological and hydraulic physical process of urban surface, river channels and drainage network on the basis of the complex terrain in Beijing and the features of big cities. With detailed precipitation monitoring as the driving conditions, BUW model can better simulate the spatial distribution of urban waterlogging for "7·21" Torrential Rainstorm, and the simulation of waterlogging depth and waterlogging process for critical bridges are also consistent with the actual situation, indicating that the model has good simulation performance. In the real-time operation from 2012 to 2014, risk warning has effect in advance and can better master the spatial distribution of waterlogging, providing a reference for municipal drainage and transportation departments to launch emergent dispatch vehicles.

It should be pointed out that, urban waterlogging numerical simulation heavily depends on the geographical features of the underlying surface and the drainage system information. Therefore, all kinds of basic information (including pumping stations, impounding reservoirs and dispatch strategies) should be constantly updated and debugged to continuously get good simulation results. The waterlogging of recessed overpasses simulated by BUW model tends to be more shallower, which may be caused by rough description of bridge structure or insufficient foreign water simulation.

ACKNOWLEDGEMENTS

This research was supported by the Science and Technology Project of Beijing (NO. Z151100002115012 and Z131100001113031).

REFERENCES

Chen, Y., Sun, J. & Xu, J., et al. 2012. Analysis and thinking on the extremes of the 21 July 2012 torrential rain in Beijing part I: observation and thinking. Meteorological Monthly (in Chinese) 38(10):1255–1266.

Hu, H.b., Xuan, C.Y. & Zhu, L.S. 2013. The pre event risk assessment of Beijing urban flood. Journal of Applied Meteorological Science (in Chinese) 24(1):99~108.

Huang, D.P., Liu, C. & Fang, H.J., et al. 2008. Assessment of waterlogging risk in Lixiahe region of Jiangsu Province based on AVHRR and MODIS image, Chin Geogr Sci 18(2):178–183.

Liu, M., Yang, H.Q. & Xiang, Y.C. 2002. Risk assessment and regionalization of waterlogging disasters in Hubei Provence. Resour Environ Yangtze Basin (in Chinese) 11(5):476–481.

Ma, X.Q., Wang, X.R. & Zhang J.D., et al. 2002. GIS-based zoning of drought and excessive rain risk at the level of city (County). Geology of Anhui (in Chinese) 12(3):171–175.

Quan, R.S., Liu, M. & Zhang, J.L., et al. 2011. Vulnerability assessment of rainstorm water-logging in subway of Shanghai. Yangtze River (in Chinese) 42(15):13–17.

Shi, Y., Shi, C. & Xu S.Y., et al. 2009. Exposure assessment of rainstorm waterlogging on old-style residences in Shanghai based on scenario simulation. Nat Hazards 53:259–272.

Sun, J., Chen, Y. & Yang, X.N., et al. 2012. Analysis and thinking on the extremes of the 21 July 2012 torrential rain in Beijing part II: preliminary causation analysis and thinking. Meteorological Monthly (in Chinese) 38(10):1267–1277.

Wang, H.J., Zhang, Y. & Lang, X.M. 2010. On the predict and of short-term climate prediction. Climatic and Environmental Research (in Chinese) 15(3):225–228.

Wang, J.L., Zhang, R.H. & Wang Y.C. 2012. Characteristics of precipitation in Beijing and the precipitation representativeness of Beijing weather observatory. Journal of Applied Meteorological Science (in Chinese) 23(3):265~273.

Xie, Y.Y., Han, S.Q. & You L.H., et al. 2004. Risk analysis of urban rainfall waterlogging in Tianjin city. Journal of the Meteorological Sciences (in Chinese) 24(3):342–349.

Xie, Y.Y., Li, D.M. & Li, P.Y., et al. 2005. Research and application of the mathematical model for urban rainstorm waterlogging. Advances in Water Science (in Chinese) 16(3):384–390.

You, F.C., Guo, L.X. & Shi, Y.S., et al. 2013. Correlation analysis and application of heavy rainfall and road waterlogging in Beijing. Meteorological Monthly (in Chinese) 39(8):1050–1056.

Bearing capacity calculation of U-shaped steel damper via assumed stress-strain curve

Hongkai Du
Key Laboratory of Earthquake Engineering and Structural Retrofit, Beijing University of Technology, Beijing, China
Collaborative Innovation Center of Energy Conservation and Emission Reduction Technology, Beijing University of Civil Engineering and Architecture, Beijing, China

Miao Han & Rongqiang Zhang
Collaborative Innovation Center of Energy Conservation and Emission Reduction Technology, Beijing University of Civil Engineering and Architecture, Beijing, China

ABSTRACT: Bearing capacity is one of the important properties of U-shaped steel dampers. A new method to calculate the bearing capacity is proposed by assumed stress-strain curves. This method has higher efficiency, because bearing capacity of U-shaped steel dampers with different radii can be calculated after one steel plate bending test. To verify the accuracy of the method, four steel plate bending tests and four U-shaped steel damper tests were completed. The calculated results obtained from the method relatively fit the tests.

1 INTRODUCTION

Damping is very important for building and bridge structures to reduce the deformation of isolation layer, so installing damper for base-isolated structures is widely used by civil engineers. The U-shaped steel damper (as shown in Fig. 1) has got the attention of scientists for its simple production process, good energy dissipation, and the advantages of installation. Most of the study for U-shaped steel dampers is based on the out of plane deformation of the bending section and arms, as shown in Figure 2. However, deformation of arms will be limited in special situations, such as the damper is mounted directly on the isolation layer between the foundation beam and structure beam. In this paper, a new method for bearing capacity calculation of U-shaped steel dampers in this situation via assumed stress-strain curves is proposed.

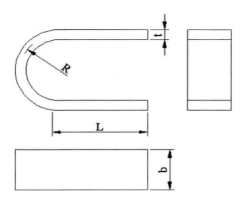

Figure 1. Diagram of U-shaped steel damper.

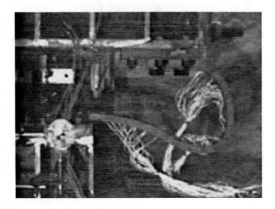

Figure 2. Deformation out of plane for U-shaped steel damper.

2 CALCULATION THEORY

The deformation process of a U-shaped steel damper consists of elastic stage, plastic stage, and crimping stage. The elastic stage is constant stiffness stage, when the load is in direct proportion with deformation. The plastic stage is stiffness decline stage, and the load still increase with the displacement in this stage. The crimping stage is zero stiffness stage, and the load keeps constant in this stage, which is called bearing capacity.

In the elastic state, the deformation of U-shaped steel damper occurs mainly at the junction of the parallel arms and semi-circular arc. Assuming the parallel arms are the fixed end, we can get the equation for the load of U-shaped steel dampers in accordance with a bending column. The equation is given by

$$F = \frac{\sigma b t^2}{6R} \quad (1)$$

where F is load; σ is maximum stress; b is width; t is thickness; and R is radius of the U-shaped steel damper.

The deformation process of the U-shaped steel damper is actually the bending process of the steel plate. When the damper comes into the crimping state, the curvature of the maximum deformation place keeps constant, as the plate bends to a certain constant curvature.

When the plate or damper bends, we assume the following equation

$$M' = \frac{\sigma' b t^2}{6} \quad (2)$$

where M' is the bending moment; σ' is the assumed stress.

According to the boundary conditions of the U-shaped steel plate, the equation relating the bending moment M' and load F' is given by

$$M' = F'R \quad (3)$$

When steel plate bending is under test, the equation relating the bending moment M' and pressure T of compression-testing machine is given by

$$M' = \frac{TS}{4} \quad (4)$$

where S is the distance between the two supporting points in steel plate bending test.

Substituting Eq. (2) in Eq. (3) and Eq. (4), the equations change to

$$F' = \frac{\sigma' b t^2}{6R} \qquad (5)$$

$$\sigma' = \frac{3TS}{2bt^2} \qquad (6)$$

As Eq. (3) and Eq. (4) are obtained under the same assumption, substituting Eq. (6) in Eq. (5) is meaningful.

The relationship between pressure T of compression-testing machine and the maximum tensile strain of steel plate ε (T–ε) can be obtained by the steel plate bending test, and it can be converted into the relationship between the assumed stress σ' and the maximum tensile strain of steel plate ε (σ'–ε) by Eq. (6). So, the curve about assumed stress-strain can be drawn.

The equation between radius of curvature and the maximum tensile strain of steel plate ε is given by

$$\varepsilon = \frac{t}{2R} \qquad (7)$$

The maximum tensile strain ε of the U-shaped steel plate with any radius R can be calculated by Eq. (7). The assumed stress σ' can be found in the assumed stress-strain curves according to the maximum tensile strain ε. Substituting σ' into Eq. (5), the bearing capacity of the U-shaped steel plate can be calculated.

Comparing Eq. (5) and Eq. (1), we can find that the form of the equations are same, except σ and σ'. So, yield load can be achieved by Eq. (5) if yield stress f_y is substituted in it.

3 TEST VERIFICATION

3.1 Steel plate bending tests

There are four Q235 steel plate samples coded SJ1 to SJ4. The width of plates is 100 mm. The thickness of SJ1 and SJ2 is 10 mm, and that of SJ3 and SJ4 is 16 mm. The bending tests were carried out in the laboratory of Beijing University of Civil Engineering and Architecture. The experimental device is 30 tons compression-testing machine of Zwick/Roell company. The distance between the two supporting points of the specimen is 300 mm. The test photo is shown in Figure 3. The load T used machine value, and the maximum tensile strain of the steel surface comes from strain gauge. In order to improve the range of strain gauge, parts of them were set at 60 degrees angle with the longitudinal direction of the plates.

The assumed stress was converted from the pressure T by Eq. (6) combining with the maximum strain of the plates; the assumed stress strain curve can be drawn, as shown in Figure 4.

As shown in Figure 4, the curves of the four plates are coincident before the steel yield, which concludes that the elastic moduli of the four steel plates are basically same. After the steel yield, test curves of the same thickness plates are coincident. For different thicknesses of the specimen, the average stress of thickness of 10 mm is almost 10% larger than that of 16 mm at the same strain.

3.2 U-shaped steel damper tests

In order to verify the accuracy of the calculation method, four different kinds of U-shaped steel dampers were designed, as shown in Table 1.

The test device of the U-shaped steel damper is shown in Figure 5. The horizontal load was provided by 50T actuator of MTS, which connected with the damper by a connecting plate and bolts. The damper was connected to the fixed plate above it by bolts to prevent the

Figure 3. Photo of steel plate bending test.

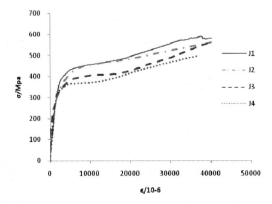

Figure 4. Assumed stress-strain curves.

Table 1. Parameters of U-shaped steel dampers.

No	R (mm)	L (mm)	b (mm)	t (mm)
U1	100	150	80	10
U2	100	200	80	10
U3	100	150	160	10
U4	100	150	80	16

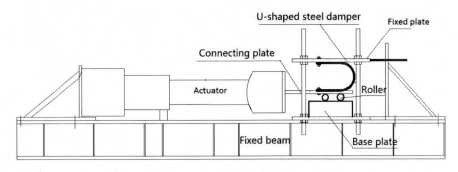

Figure 5. Installation diagram of the U-shaped steel damper test.

(a) push　　　　　　　　　　　　　　(b) pull

Figure 6. The photos of the damper during cyclic loading.

Table 2. Yield Load and Bearing capacity of U-shaped steel dampers.

	Yield load			Bearing capacity		
	Test value (F_y)	Calculation (F'_y)	F'_y/F_y	Test value (F_b)	Calculation (F'_b)	F'_b/F_b
U1	4.82	5.30	1.10	7.96	7.22	0.91
U2	4.85	5.30	1.09	7.47	7.22	0.97
U3	11.80	11.73	0.99	16.47	16.00	0.97
U4	16.28	15.02	0.92	22.19	20.48	0.92

Note: The bearing capacity values F'_b on the table were calculated by the average value of the stress-strain curves of the same thickness.

higher arm of damper moving up and down. In order to reduce the friction force, two rollers were set between the lower arm of the damper and the base plate. The test was controlled by displacement of the actuator. The photos of damper when pushed and pulled are shown in Figure 6.

Bearing capacity and yield load of the tests and calculation by Eq. (5) and Eq. (1) were given in Table 2. From the table we can see that the maximum error of bearing capacity between test value and calculation is 9%, which prove that the method fits the tests, and the maximum error of yield load is 10%, which proves Eq. (1) fist the tests too.

4 CONCLUSION

In this paper, through theoretical derivation and analysis, we propose the method for bearing capacity calculation of U-shaped steel dampers via assumed stress-strain curves. The bearing capacity of the dampers with different radii and same thickness can be calculated with one plate bending test. The accuracy of the method is proved by tests.

ACKNOWLEDGEMENTS

This work was financially supported by the National Natural Science Fund of China (51378047, 51408027), The Young Talent Program of Beijing (YETP1664), The Project of Construction of Innovative Teams and Teacher Career Development for Universities and Colleges Under Beijing Municipality (IDHT20130512).

REFERENCES

Federica Tubino, Giuseppe Piccardo. 2015. Tuned Mass Damper optimization for the mitigation of human-induced vibrations of pedestrian bridges. Meccanica. 50:809–824.
Hongkai Du, Miao Han, WeiMing Yan. 2014. Study on the calculation method of mechanical characteristics for constrained U-shaped steel plates [J]. China Civil Engineering Journal. 47(S2):158–163. (in Chinese).
Kazuaki Suzuki, Atsushi Watanabe, Eiichiro Aeki. 2005. Development of U-shaped Steel Damper for Seismic Isolation System [J]. Nippon Steel Technical Report No. 92 July.
Makoto Yamakawa, Mitsuki Nihei, Masahiko Tatibana1, etc. 2015. Modeling and simulation of spring steel damper based on parameter identification with a heuristic optimization approach. Journal of Mechanical Science and Technology. 29(4):1465–1472.
Suzuki, K., Watanabe, A. et al. 1999, 2000. Experimental Study of U-shaped Steel Damper [J]. Summary of Technical Papers of Annual Meeting, Architectural Institute of Japan, B-2.
Yao Qianfeng. 1997. Behavior of mild U-shaped steel plate restraining displacement and absorbing energy [J]. J. Xi'an Univ. of Arch. and Tech. 1997, (01):24–28. (in Chinese).
Yu Jiao, Shoichi Kishiki, Satoshi Yamada, etc. 2014. Low cyclic fatigue and hysteretic behavior of U-shaped steel dampers for seismically isolated buildings under dynamic cyclic loadings. Earthquake engineering and structural dynamics. Published online in Wiley Online Library. DOI: 10.1002/eqe.2533.

The application research of geogrids in road broadening engineering

Q.B. Tian
Ji'nan City Highway Bureau, Ji'nan, China

ABSTRACT: This paper analyzed the main types and causes of road diseases occurring in highway broadening engineering, explored the applicability of geogrids in treating soft soil foundation and preventing differential settlement in conjunct part of the new-old subgrade.

1 INTRODUCTION

With the economic development and the growth of car ownership, the increasing traffic load requires upgrading and rebuilding of existing highways. Highway broadening has been taken to meet the urgent need. However, there exist some difficulties. The original subgrade's settlement has been fixed after a long period of service. The direct broadening along the slope will cause such diseases as subgrade deformation, splitting, or subsidence due to the subsequent settlement and the differential settlement between the new and the old subgrade. Application of geogrids has been proved to be one of the preventive measures which works effectively.

2 MAIN TYPES OF ROAD DISEASES OCCURING IN ROAD BROADENING

2.1 *Destabilization of subgrade*

The remarkable differential settlement between the new and old roadbed and lateral deformation of the broadened part causes the broadened subgrade sliding along the conjunct part of the new-old subgrade, even creeping on a whole. This kind of road disease tends to occur in areas such as steep gradient topographic, soft foundation and high embankment, etc. with the minor sliding degree of the broadened subgrade; the conjunct part of the new-old subgrade becomes staggering, which causes roadbed splitting. With water infiltrating, the intensity of conjunct part decreases rapidly, the sliding degree increases due to driving load and self-weight, and then the newly broadened subgrade collapses on a whole. Overall destabilization of subgrade rarely happens due to the precautions during construction.

2.2 *Differential settlement*

Differential settlement is typical of partial sinking, pavement breaking, and severely uneven flexible pavement. The differential deformation of the new-old subgrade expresses itself in roadbed with several forms: for asphalt, surface radical splitting, breaking, binder being incompact, lateral slope changing; for cement concrete, cavity beneath slab, two-way cracking, slab staggering, slab breaking, etc. According to the survey conducted by Chen Yuliang to the road broadening engineering in Nanjing area, radical split have occurred in a few sections after the broadening engineering, shown as in Table 1 (Chen, 2003).

Table 1. Radical split in road broadening in Nanjing.

Road name	Crack length (m)	Crack width (mm)	Embankment height (m)
Ningsu road	31	2–4	7
Puzhu road	121	3–5	5–12
Tiexin intersection	28	1–3	7
Jiangdong intersection	14.5	1–3	5.5 d

*The respondent is old road.

2.3 Decrease of overall performance of subgrade

With the differential settlement and the lateral deformation becoming serious, road diseases develop, thus, road structural performance and service ability decrease remarkably. When the index of pavement condition, bearing capability, and planeness decrease to a certain level, ride comfort and traffic safety will surely be influenced negatively.

3 MAIN CAUSES OF ROAD DISEASES OCCURING IN HIGHWAY BROADENING ENGINEERING

3.1 Uncoordinated deformation of the new-old subgrade

Uncoordinated deformation of the new-old subgrade is the reflection of the spatial difference due to the settlement and deformation of foundation and embankment, which includes linear deformation and lateral deformation, whose causes mainly include: (1) differential settlement of the new-old subgrade; (2) quality problem of filling material and compaction; (3) imperfect construction of the conjunct among soft foundation, embankment, and subgrade; (4) biased subgrade geological survey and preventive measures.

3.2 Inappropriate processing of the conjunct of new-old subgrade

Inappropriate bank excavation, bar length, or layer position of bar lead to the ill combination of the new and old subgrade and the potential slide surface.

3.3 Different ability against deformation between the new and old subgrade and roadbed

The filling of a newly broadened subgrade has different time and performance from the old, and as the old roadbed has been compacted with the car load, while the new filling material still faces cumulative plastic strain, leading to the different intensity, stiffness, and the ability against deformation between the new and the old subgrade.

3.4 Imperfect subgrade drainage facility and untimely maintenances

Imperfect drainage facility and inequitable arrangement make the subgrade soak in water and become squishy, thus, its intensity decreases. Untimely maintenance of road ditch and drains makes the trapped water soak into the subgrade, thus, the speed of deformation and destabilization increases.

4 OPERATIONAL SUITABILITY TEST (OT) OF GEOGRIDS IN BROADENING ENGINEERING

4.1 Combination of geogrids and soft foundation treatment

Geogrids can be combined with the soft foundation treatment in broadening engineering. The currently used soft foundation treatment methods are shown in Table 2 (Gao, 2006).

Table 2. Current soft foundation treatment methods in road broadening engineering.

Broadening engineering	Old soft foundation treatment	Broadening soft foundation treatment
Guofuo Highway	Sacked sand drain, sand cushion	Cement injection pile, sand cushion
Huhangyong Highway	Plastic drain bar, preloading, flyash embankment	Plastic drain bar, preloading, equal-loaded precompaction, cement injection pile, embankment pile, geogrid
Shenda Highway	Plastic drain bar, preconsolidation	Plastic drain bar, preconsolidation
Haitian Island	Unsettlement or random rubble fill	Cement injection pile
Hangning Highway II	Uncharged preloading, plastic drain bar, compound geotextile	Cement injection pile, geogrid, light-weight flyash embankment
Huning Highway	Desilting and soil replacement, plastic drain bar, preloading, cement injection pile, flyash embankment	Cement injection pile, wet jetting pile, prestressed thin-wall pipe pile, EPS light-weight embankment
Nanjing City Highway	Plastic drain bar, sacked sand drain, preconsolidation, cement injection pile	Cement injection pile, wet jetting pile, CFG PILE, cast-in-place thin-wall pipe pile, low-grade concrete pile

The comprehensive comparison on the above broadening engineering shows that geogrids can be used together with commonly-used soft foundation treatment methods to enhance subgrade's intensity and stability. The following are available methods: (1) Pave geogrids between soft foundation and embankment of broadened sections to reinforce soft embankment. (2) Pave geogrids on the bottom of embankment and combine it with a plastic drain bar. (3) Pave geogrids both on the bottom of the foundation ditch and above the hardcore bed to treat small structures of soft soil foundations to increase the load of a broadened subgrade and reduce structural destruction. (4) Use solid-drawn geogrids on the combining part of a new-old subgrade and roadbed to reduce differential settlement. (5) Use geogrids to combine a new-old soil body to increase the frictional angle and form a bearing plate to enhance the overall intensity and planeness.

4.2 Interaction mechanism of geogrids and a soil body

The function of geogrids in road broadening engineering is mainly to improve stability of embankment and reduce differential settlement. A soil body itself has a weak tensile and sheer property, geogrids can be added as a tensile part, friction between geogrids and a soil body restricts lateral deformation and strengthen the shear intensity of the soil body. The specific performances are (Hu, 1996): (1) Bidirectional geogrids with equably longitudinal and transverse strength are often used to treat broadened soft soil subgrades. Interaction between geogrids and soil particles produces a stress effect, which, via the geogrid, spreads vertical deformation of upper fillings in horizontal direction. Thus, sheer deformation ability of upper fillings improves and the bearing capacity of the surface of a broadened soft soil subgrade increases remarkably. The netlike geogrid also stops fillings from sinking partially and reduces the differential settlement to a great extent. (2) Paving geogrids in the cushion course and base course of the conjunct part of a new-old subgrade restricts lateral sliding deformation of the soil body and strengthens its shear intensity and bending rigidity, improves bearing capacity of the subgrade and restricts lateral displacement of the roadbase, due to the friction, occlusion, and resistance between the geogrid and soil particles. (3) Basal cushion course has a good water permeability. Geogrid loading is good for the dissipation of pore water pressure in a subgrade and enhances the drainage consolidation.

5 TEST ABOUT USING GEOGRID ON CONJUNCT OF A NEW-OLD SUBGRADE

5.1 Relationship between geogrid paving interval and reaction modulus of a soil body

A $120 \times 160 \times 210$ cm experimental tank and $\Phi 300$ loading plate were adopted to test the reaction modulus of a soil body under different geogrid paving intervals. The result shows that the reasonable paving interval for silty clay is 30 cm–60 cm and the optimal interval is 40 cm, as shown in Figure 1 (Tongji, 2003).

5.2 *Relationship between reinforcement and uncoordinated deformation of a conjunct part of a new-old subgrade*

Simulation test was conducted about the nonlinear state of the contact interface between a soil body and geogrids via ANSYS program. The result shows: without reinforcement, the uncoordinated deformation of a conjunct part of a new-old subgrade is 28.4 cm while with reinforcement, it reduces by 5.8%, at 26.8 cm. Reinforcement can reduce the settlement and uncoordinated deformation of broadened subgrade by 11.4%, as shown in Figure 2.

5.3 *Design-construction suggestions*

Based on the test, the following should be paid attention to when using geogrids on conjunct of a new-old subgrade: (1) Geogrids can't be paved until the step along a old subgrade slope finishes being excavated and new subgrade is filled and compacted to corresponding elevation. It should be paved evenly and should extend to the inner edge of the step to prolong the passive resisting area. (2) Geogrids should meet the demand of high tensile strength, small unit extension, easy construction, and low cost. (3) Geogrid length in a new subgrade should extend to the outer margin of lane marking. Geogrids should be paved at least on the old pavement with three layers. If possible, it should be paved on every step. (4) Before large-scale usage, geogrids should be paved in an experimental area to test such indexes as flexure, modulus of resilience, etc. Reducing amplitude of uncoordinated deformation should also be analyzed.

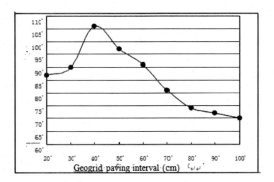

Figure 1. Relationship between geogrid paving interval and reaction modulus of a soil body.

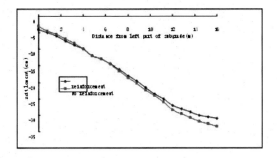

Figure 2. Relationship between reinforcement and uncoordinated deformation of a conjunct part of a new-old subgrade.

6 CONCLUSIONS

Road diseases such as longitudinal or lateral cracks often occur after road broadening engineering due to different thickness and intensity of the new and old subgrade fillings. Based on the above analysis, the paper suggests the following treatments:

Excavating steps and paving geogrids in conjunct parts of the new-old subgrade can effectively reduce its differential settlement.

Taking advantage of frictional reinforcement between geogrids and the soil body and an increasing internal frictional angle can enhance overall intensity and plainness of a roadbed and prevent reflection crack.

Geogrids should be used with reasonable design and construction.

REFERENCES

Chen, Y.L. & Yue, L.V. & Zhang, Z.N. 2003. Reasons and Causes of Roadbed Radical Split in Road Widening Engineering. East Road. (1).

Gao, X. 2006. Research on the Interaction between New and Existing Embankments and Treatment Techniques in Highway Widening Engineering [D]. Southeast University.

Hu, Y.C. 1996. Research on Geogrid Reinforcing Road Soft soil foundation. Rock and Soil Mechanics. (17).

Tongji Uni. & Changsha Uni. of Science and Technology. 2004. New Technologies of Subgrade Widening for Highways [D].

Study on behaviors of space frame structures under fire

Linbo Qiu
Central Research Institute of Building and Construction Co. Ltd., MCC Group, Beijing, China

Suduo Xue
College of Architecture and Civil Engineering, Beijing University of Technology, Beijing, China

Yun Zheng
Central Research Institute of Building and Construction Co. Ltd., MCC Group, Beijing, China

ABSTRACT: Based on the utility temperature elevation empirical formula in large space fire, the nonlinear finite element analyses of the fire performances on two kinds of space frame structures, one Kiewitt (K6) single layer reticulated dome and the other square pyramid space grids structure, are numerically simulated with the thermal parameters specified in the EUROCODE 3 and the stress-strain relationship allowing for strain-hardening of steel. The temperature fields and the displacement properties of the two structures are studied under local fire. According to the results of the analysis, it shows that most of the ultimate fire-resistant duration are smaller than 2 hours under different local fires; the influence of the fire height is the biggest on the ultimate fire-resistant duration of the structure, and the effect of the fire location is bigger than the area of the fire.

1 INTRODUCTION

In recent decades, a large number of public buildings with space frame structures, such as airport terminal, gymnasium, and museum, have been built in many cities of the world. Most of these structures are constructed by steel members and joints (Zhang 2005). However, the steel members share poor fire-resistant performances. The steel members cannot bear any loads once their temperatures reach 600 °C. In general, the temperature field of fire is about 800 °C–1000 °C, so these structures will be damaged or will even collapse under fire (Wang 2006). Although the space frame structures have been widely used, researches on their fire behaviors lag behind their developments. In the past decades, the fire-resistant researches of steel structures mainly focused on the properties of steel under fire, the temperature fields of member and space, and the fire-resistant properties of members and frame structures, etc. (Li 2000). To learn more information about the fire performances of the space frame structures, two kinds of the structures, one Kiewitt (K6) single layer reticulated dome and the other square pyramid space grids structure, were numerically simulated under fire by the nonlinear finite element analysis. The temperature fields and the displacement properties of the two structures are studied under local fire in the present paper.

2 MODEL OF NONLINEAR FINITE ELEMENT

The space frame structures include mainly the space grid structures and the space reticulated shell structures. The square pyramid space grids structure, as one style of the space grid structures, is usually used in civil engineering, and the Kiewitt (K6) single layer reticulated dome is one style of the space reticulated shell structures. So these two structures are selected to learn more information about the fire performances of space frame structures, and also the nonlinear finite element analyses of the two structures were performed.

2.1 *Sizes of structures*

The sizes of the two kinds of space frame structures are shown in Figure 1 and Figure 2. The thickness of the square pyramid space grids structure is 2.4 m, the height is 9 m. The span of the Kiewitt (K6) single layer reticulated dome is 40 m, and the span height is 8 m. The two roof systems are supported on four edges, and the yield strength of steel is 235 MPa. All steel members of the structure are seamless steel pipes in the paper.

2.2 *Finite element model*

In the present paper, the fire is assumed to be circular. The areas, coordinates, and radius of the fire are shown in Table 1 and Table 2. The fire heights, which is the distance from the fire center to the upper chord for the square pyramid space grids structure, and which is the distance from the fire center to the structural roof for the single layer reticulated dome, are

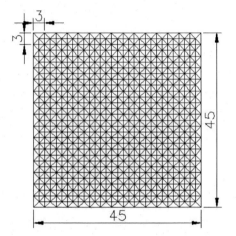

Figure 1. Plan of the square pyramid space grids structure.

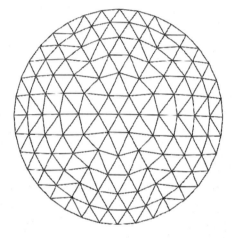

Figure 2. Plan of the single layer reticulated dome.

Table 1. Fire locations in the square pyramid space grids structure.

Fire model	Fire area (m²)	Fire radius (m)	Fire center coordinate (x,y)	Fire height (m) 1	2	3
1	50	3.99	22.5,22.5	4	6	9
2	50	3.99	18,22.5	4	6	9
3	50	3.99	13,22.5	4	6	9
4	50	3.99	9,22.5	4	6	9
5	50	3.99	4,22.5	4	6	9
6	100	5.64	22.5,22.5	4	6	9
7	150	6.91	22.5,22.5	4	6	9
8	200	7.98	22.5,22.5	4	6	9
9	250	8.92	22.5,22.5	4	6	9
10	300	9.77	22.5,22.5	4	6	9

Table 2. Fire locations in the single layer reticulated dome.

Fire model	Fire area (m²)	Fire radius (m)	Fire center coordinate (x,y)	Fire height (m) 1	2	3
1	50	3.99	(0,0)	4	6	9
2	100	5.64	(0,0)	4	6	9
3	250	8.92	(0,0)	4	6	9
4	50	3.99	(−3.67,0)	4	6	9
5	50	3.99	(−8,0)	4	6	9
6	50	3.99	(−10.77,0)	4	6	9
7	50	3.99	(−14,0)	4	6	9
8	50	3.99	(−17.18,0)	4	6	9
9	50	3.99	(−2.75,−1.59)	4	6	9
10	50	3.99	(−6.3,3.64)	4	6	9
11	50	3.99	(−10.12,−3.68)	4	6	9
12	50	3.99	(12.2,−7.04)	4	6	9
13	50	3.99	(−15.7,−6.99)	4	6	9

4 m, 6 m, and 9 m. The thermal releasing power of the fire is 25 MW, and the equation of the temperature field considering space factors is written as follows (Li 2006):

$$T_{(x,z,t)} - T_g(0) = T_z [1 - 0.8\exp(-\beta t) - 0.2\exp(-0.1\beta t)] \times \left[\eta + (1-\eta)\exp\left(-\frac{x-b}{\mu}\right) \right] \quad (1)$$

The finite element model of thermal-mechanical coupling analysis was built by the finite element software ABAQUS, the model is shown in Figure 1 and Figure 2. The pickings of the parameters of steel properties under fire are the basis and premise of the fire resistance analysis and design, and these parameters include physical and mechanical character of steel under high temperature (Li 2006). For conforming to the practical condition, the thermal elongation, the thermal conductivity, the specific heat and the stress-strain relationship allowing for strain-hardening of steel, which were varied with the environmental temperature, were chosen based on the EUROCODE 3. The Poisson ratio, the elastic modulus, and the density of steel were taken as 0.3, 206000 MPa, and 7850 Kg/m³, respectively. The geometric and material nonlinearity of steel was considered.

3 NUMERICAL ANALYSIS RESULTS OF SQUARE PYRAMID SPACE GRIDS STRUCTURE

Based on the nonlinear finite element analysis of the square pyramid space grids structure, the distribution of the temperature field about the square pyramid space grids structure under fire in Model 1 (Table 1) is shown in Figure 3, and the fire height is 9 m. Considering the symmetry of the structure, the temperature curves of the 1/4th structure are given in Figure 3 (a) and (b). The 8 curve is in the center of the upper chord, and the 9 curve in the center of the lower chord. The others are chosen in turn along the symmetrical axis, and the distance is 3 m between every two curves. The temperature of the fire center is the highest in the upper chord, and it is 580 °C (the 8 curve). The temperature decreases from the center to the edge of the structure, and the lowest temperature is 339 °C (the 1 curve). The distribution of the temperature in the lower chord is similar, in which the highest and the lowest temperatures are 403 °C (the 9 curve) and 299 °C (the 16 curve), respectively. The results in the other fire models are similar (not shown in this paper).

During the process of the numerical simulation, the fire sustained for about 2 hours. The displacement of the structure in Model 1 (Table 1) is shown in Figure 4, and the

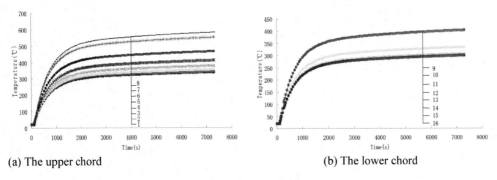

(a) The upper chord (b) The lower chord

Figure 3. Temperature curves of the points on the space grid structure.

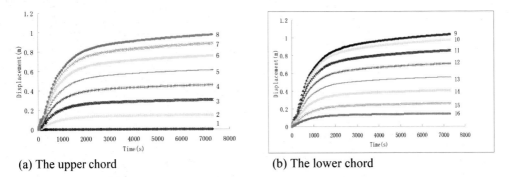

(a) The upper chord (b) The lower chord

Figure 4. Displacements of the points on the space grid structure.

points are the same as Figure 3. Figure 4 shows that the displacements are the biggest in the structure center, and there are no sudden changes in all the displacements. The biggest displacement is 1.036 m, which is smaller than 1.575 m proposed by Jin (Jin 2005). Jin suggested that the limit displacement of the structure is 3.5% of the calculation span when the whole structure collapses or cannot bear any loads. Thus in the simulation, the structure does not collapse. The results are similar in the other fire models (not shown in this paper).

However, the stresses of the members are changed suddenly within 2 hours in the simulations. The stresses of the members in the upper chord of Model 1 (Table 1) are shown in Figure 5. The sudden changes imply that the members failed. Four members of Model 1 failed in the fire center. Therefore, it shows that the failure of the structure is partial, and the ultimate fire-resistant duration can be considered as 4354s. The other models have similar results (not shown in this paper). The ultimate fire-resistant durations of the other fire models are shown in Table 3.

According to Table 3 and the analysis results, it can be drawn that at the same location of the fire, the bigger the area of the fire is, the more the failed members are, and the smaller the ultimate fire-resistant duration is correspondingly. When the area of the fire is the same, the farther the distance from the fire center to the structure center is, the more the failed members are, and the smaller the ultimate fire-resistant duration is correspondingly. When the area and location of the fire are the same, the smaller the fire height is the smaller the ultimate fire-resistant duration is correspondingly too, that is the higher of the vertical distance from the fire center to the ground or floor in the structure, the smaller the ultimate fire-resistant duration is. The influence of the fire height is the biggest on the ultimate fire-resistant duration of the structure, and the effect of the fire location is bigger than the area of the fire.

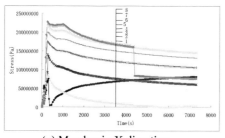 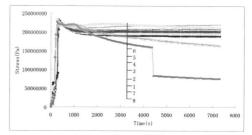

(a) Member in X direction   (b) Member in Y direction

Figure 5. Stresses of members in the upper chord.

Table 3. Ultimate fire resistant duration of different fire model (s).

Fire model	1	2	3	4	5	6	7	8	9	10
Fire height										
4 m	1112	1095	1054	934	756	1111	1051	1051	1051	1051
6 m	2732	1534	1414	1114	874	1531	1531	1531	1531	1531
9 m	4354	3274	2254	1654	1223	3571	3451	3435	3274	3211

4 NUMERICAL ANALYSIS RESULTS OF SINGLE LAYER RETICULATED DOME

Based on the nonlinear finite element analysis of the single layer reticulated dome, the distribution of the temperature field in Model 1 (Table 2) is shown in Figure 6 (a), and the fire height is 4 m. Considering the symmetry of the structure, the temperature curves of the joints along the structure radius are given in Figure 6 (b). The joints 1 and 5 are in the center of the structure, and the joint 86 at the edge of the structure. The others are chosen in turn along the structure radius. The temperature of the fire center is 655.2 °C (the curve 1 and 5). The temperature decreases from the center to the edge of the structure, and the lowest temperature is 348.0 °C (the 86 curve). The distribution of the temperature is similar to the square pyramid space grids structure. The results in the other fire models are similar (not shown in this paper).

During the process of the numerical simulation, the fire sustained for about 2 hours. The displacement curves of the structure in Model 1 (Table 2) are shown in Figure 7, and the points are the same as Figure 6. It shows that the displacements are the biggest in the structure center. There are sudden changes in all the displacements, but the times of sudden changes are not the same. When the rate of deformation about the specimen is infinite, the structure is damaged (Li 2006). The sudden changes imply that the members failed. So the whole single layer reticulated dome structure collapses or cannot bear any loads. The results are similar in the other fire models (not shown in this paper).

When the structure collapses, the sustained time of the structure under fire is the ultimate fire-resistant durations. The times of different fire models are given in Table 4. Similar to the space grid structure, at the same location of the fire, the bigger the area of the fire is the smaller the ultimate fire-resistant duration is correspondingly. As to the same area of the fire, the farther the distance from the fire center to the structure center is, the smaller the ultimate fire-resistant duration is correspondingly. When the area and location of the fire are same, the less the fire height is, the smaller the ultimate fire-resistant duration is correspondingly too, that is the higher of the vertical distance from the fire center to the ground or floor in the structure, the smaller the ultimate fire-resistant duration is. The influence of the fire height is the biggest on the ultimate fire-resistant duration of the structure, and the effect of the fire location is bigger than the area of the fire.

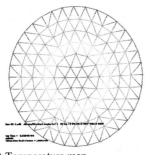 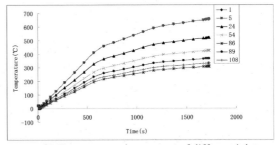

(a) Temperature map (b) Temperature-time curves of different joint

Figure 6. Temperature curves of the points on the single layer reticulated dome.

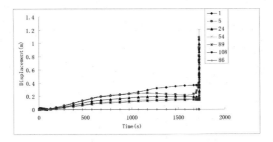

Figure 7. Displacements of the points on the single layer reticulated dome.

Table 4. Ultimate fire resistant duration of different fire model (s).

Fire model	Fire height		
	4 m	6 m	9 m
1	1728	7514	14975
2	1602	5890	8858
3	1192	2570	5289
4	1204	2676	5581
5	1196	2622	5314
6	1191	2555	5022
7	1405	3955	8124
8	1796	8621	30259
9	1359	3587	7300
10	1314	3268	6490
11	1273	3032	5924
12	1395	3825	7827
13	1571	5504	12890

5 CONCLUSIONS

Based on the analyses, some conclusions of the behaviors about the space frame structure under fire can be drawn:

1. The failed members are located in the fire center. The larger the area of the fire is, the more the failed members are. The farther the distance from the fire center to the structure center is, the more the failed members are.

2. At the same location of the fire, the bigger the area of the fire is the smaller the ultimate fire-resistant duration is correspondingly. As to the same area of the fire, the farther the distance from the fire center to the structure center is the smaller the ultimate fire-resistant duration is correspondingly. When the area and location of the fire are same, the smaller the fire height is the smaller the ultimate fire-resistant duration is correspondingly too.
3. The influence of the fire height is the biggest on the ultimate fire-resistant duration of the structure, and the effect of the fire location is bigger than the area of the fire.
4. Part of the space grid structure failed under fire within 2 hours, which could result in the whole structure collapsing. Therefore, the conclusion in reference (Jin 2005) is not suitable for the space grid structures, and the structures must be retrofitted for reusing.

REFERENCES

Li G.Q., Jiang S.C, and Lin G.X. Computation and Design of Steel Structure in Fire, Construction Press of China, 2000.

Li G.Q., Han L.H., Lou G.B., and Jiang S.C. Fire-Resistant Design of Steel Structure and Steel—Concrete Compound Structure, Construction Press of China, 2006.

Jin H. Numerical Analysis of the Fire Resistance of Pre-Stressed Composite Space Truss. Tianjin: Tianjin University, 2005, 56–75.

Wang R. Analysis and Calculation on Fire Resistance of Steel Structure. Harbin, Harbin Engineering University, 2006.12, 8, 1–12.

Zhang Y.G. Xue S.D., Yang Q.S., Fan F. Large-Span Space Structure. Mechanical Industry Press, China. 2005.

Emulational analysis of contact open-close considering transverse joints with trapeziform key hydraulic grooves

Yi-zhi Yan, Lin-jun Yu, Xiao-chen Wen & Guo-bao Wei
College of Electric Power Engineering, Kunming University of Science and Technology, Kunming, Yunnan, China

ABSTRACT: A 3-D contact element model for trapeziform key hydraulic grooves is used to simulate contact transverse joints. The contact physical equation of trapeziform key hydraulic grooves is derived; and the reasonable gradient angle is predicted. On the basis of the characteristic of the contact nonlinearity, this paper presents three contact modes that judge contact states of contact elements. Then the paper gives stress computation formula of contact element, and push over the relation between the stress and relative deformation. Finally, a typical example is analyzed. The result shows that the contact element model is able to simulate open-close state of contact and stress state of contact.

1 INTRODUCTION

In hydraulic engineering and civil engineering, there are a lot of different kinds of contact problems. The seams of the concrete dam engineering including transverse joints, bottom joints, longitudinal joints, peripheral joints, and so on. These surfaces of the joints in the loading process could be opened, closed, and undergo shearing etc. This will affect the overall structural and working state of the concrete dam, especially, influence the arch of arch dam in functioning properly (Zhang, 2005). In recent years, construction of a lot of ultrahigh concrete dam has been started, making the study of surface more and more important. There are three main methods of simulating transverse joints surfaces: the joint element method (Lau, 1998) (Malika, 2002) (Xu, 2001), the crack spread model and the dynamic contact model (Tu, 2001) (Ahmadim, 2001). The joint element method can simulate the nonlinear relation of the initial strength, space, and the normal and tangential force of conduit joint, and this element can be used to simulate the nonlinear constitutive relation between initial tensile strength, initial gap, the normal force, and tangential friction force of transverse joints; often used for the unit is the block element which is on the sides of transverse joints and conduit joint. Joint elements cannot guarantee that there is no mutual embedding between the contact bodies; to make sure the accuracy and convergence rates of the solution, it is a must, according to the experience and judgment, to select a reasonable unit spring element value or normal, tangential stiffness coefficient. The crack spread model is applied through properly modified material characteristic of conventional unit integral point to simulate the crack and joint in the average sense. This model does not need mesh, does not introduce a new degree of freedom, and can't reflect the state of contact interface. The dynamic contact model imposes constraints on the contact surface to solve the contact force that meets the conditions of the contact constraint conditions and equilibrium conditions; it is more rigorous in theory, but a lots of methods are not suitable for solving nonlinear problems. Although a lots of transverse joint simulation calculation methods are proposed, the precision and efficiency of these methods have some limitations. So the research and development of accurate and efficient algorithm is very necessary.

FneveS[7] model has been chosen on the basis of the research achievements of Goodman, Ghaboussi et al., adopting the three dimensional plane contact element for stress analysis of three dimensional nonlinearity of arch dam. The effect is admirable. Chen Houqun and the others applied FneveS model and considered the joint shearing slippage. The dynamic model of arch dam considering the effect of opening and closing of joints were studied experimentally and the three dimensional nonlinear finite element dynamic analysis method and calculation program were verified. The results of the study show that the calculation results are essentially in agreement with FenveS model using the nonlinear results; it shows that use of FenveS model to simulate the transverse joint opening, closing, shear is effective. In this paper, the three dimensional contact element model to simulate the trapeziform key hydraulic grooves transverse joint contact surface has been used. Analysis of trapeziform key hydraulic grooves parameters, B,H,M,'s influence on joint contact surface displacement and stress has been discussed.

2 CONTACT OPENING DESCRIPTION OF TRANSVERSE JOINTS WITH TRAPEZIFORM KEY HYDRAULIC GROOVES

2.1 *Transverse joints trapeziform key hydraulic grooves geometry model and the opening and closing calculation*

The purpose of put up key hydraulic grooves at transverse joints is to improve the shear capacity of transverse joint contact surface. There are two forms of key hydraulic grooves shear transfer model: partial shear and complete shear. The partial shear model propose that the transverse joints can transfer pressure and not high friction shear when closing them, and these can't transfer the pressure and shear after the opening them. The complete shear model proposes that the transverse joints can transfer pressure and shear both when opening or closing. For the contact surface shear deformation, it is necessary to consider the different key type hydraulic grooves: unidirectional key hydraulic grooves and bidirectional key hydraulic grooves, as shown in Figure 1.

As shown in Figure 2, let β be an angle between a key hydraulic grooves inclined plane and the transverse joint plane. The initial clearance for the transverse joint is d. Then the horizontal clearance of key hydraulic grooves inclined plane d' is:

$$d' = d ctg \beta \qquad (1)$$

Defining the T as the shear force of the key groove inclined plane, P and F as the component of T along the key groove slope of the normal and tangential direction, then:

$$P = T\sin\beta, F = T\cos\beta \qquad (2)$$

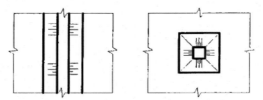

Figure 1. Unidirectional key hydraulic grooves and bidirectional key hydraulic grooves.

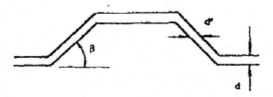

Figure 2. The plan of key hydraulic grooves.

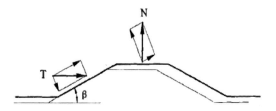

Figure 3. Trapeziform key hydraulic grooves by shear force diagram.

In order to give full play to the shear characteristics of the key groove, the required key groove should have no relative sliding under stress. Then the above two conditions meet the Coulomb friction:

$$F \leq P \cdot \tan\varphi \qquad (3)$$

The friction angle φ is that of the contact surface. The condition of key hydraulic grooves with no slip is:

$$\tan(90° - \beta) \leq \tan\varphi \qquad (4)$$

$$\text{That is: } \beta \geq 90° - \varphi \qquad (5)$$

In order to avoid the stress concentration at the root of the key groove, and to be easy to adapt to the uneven settlement of column concrete blocks, β should not be too large. The coefficient of friction between the concrete joints is about 0.8, therefore the β in reasonable trapeziform key hydraulic grooves is about 52°.

2.2 *Determination of the opening trapeziform key hydraulic grooves*

Assuming the key groove under shear stress T and the normal force N, T, and N, there can be decomposition in the normal and tangential direction along the key groove surface. Then after the key groove surface contact, the key groove may appear in the following three deformation modes:

1. Open model: $T\sin\beta - N\cos\beta \leq 0$, the key groove contact surface under normal tension, or no contact force where the key groove surface in contact have no binding effect on each other.
2. Closed model: $T\sin\beta - N\cos\beta > 0$, and $(T\sin\beta - N\cos\beta) \cdot tg\varphi < T\sin\beta + N\cos\beta$, the key groove contact surface under normal pressure, and the friction on the contact surface is larger than that of the sliding force, so the key groove surface in contact cannot undergo relative displacement.
3. Shear model: $T\sin\beta - N\cos\beta > 0$, and $(T\sin\beta - N\cos\beta) \cdot tg\varphi < T\sin\beta + N\cos\beta$, the key groove contact surface under normal pressure, and the friction on the contact surface is less than or equal to that of the sliding force, at this time, the key groove surface in contact can undergo relative displacement along the slope. The nonlinear switching problem of contact surface is transformed into contact friction problem.

3 TRANSVERSE JOINTS THREE-DIMENSIONAL CONTACT ELEMENT

Three dimensional contact element is a spatial 8 node contact element without thickness, as shown in Figure 4.

The element displacement interpolation function is:

$$u = \sum_{i=1}^{4} N_i u_i, v = \sum_{i=1}^{4} N_i v_i, w = \sum_{i=1}^{4} N_i w_i \qquad (6)$$

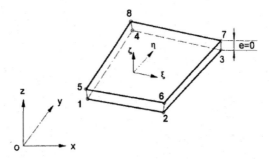

Figure 4. Schematic diagram of transverse joint three dimensional contact element model.

The interpolation function is:

$$N_i = \frac{1}{4}(1+\xi_i\xi)(1+\eta_i\eta), (i=1,4) \qquad (7)$$

Each unit node's displacement component is:

$$\{\delta\}^e = \{u_1, v_1, w_1, u_2, v_2, w_2, \ldots, u_8, v_8, w_8\} \qquad (8)$$

The contact element in the local coordinate system is $\xi\eta\varsigma$, the relative displacement of the point on the contact surface is:

$$\{\Delta\delta\}^e = \begin{Bmatrix} \Delta u' \\ \Delta v' \\ \Delta w' \end{Bmatrix} = \begin{Bmatrix} u_i - u_j \\ v_i - v_j \\ w_i - w_j \end{Bmatrix} = [N]\{\delta\}^e \qquad (9)$$

Than contact stress of any point in contact surface is:

$$\begin{Bmatrix} \tau_\xi \\ \tau_\eta \\ \sigma_\varsigma \end{Bmatrix} = [D']\{\Delta\delta\}^e + \{\sigma_0\} \qquad (10)$$

There into $[D'] = \begin{bmatrix} k_\xi & & \\ & k_\eta & \\ & & k_\varsigma \end{bmatrix}$.

Inside k_ξ, k_η are tangential stiffness coefficient of the ξ and η direction, when the isotropy $k_\xi = k_\eta$, k_ς is the coefficient of normal stiffness, $\{\sigma_0\}$ is the initial stress.

4 THE NUMERICAL EXAMPLE ANALYSIS

Figure 5 shows the column base length of 1.5 m, thickness of 1.0 m, high 10.0 m, there's a vertical joint from the left at 0.6 m. The vertical key groove is set, at the bottom of the column is the rigid constraint, the water pressure is applied on the left, and increments are in one meter.

Due to limited space, this article only gives some results of the calculation of contact hydraulic grooves. Figures 8–16 give the part value of the result chart.

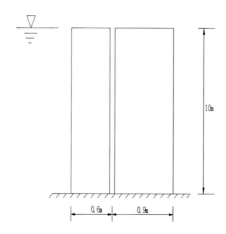

Figure 5. Schematic diagram of the contact column.

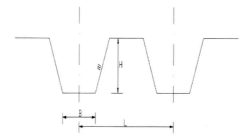

Figure 6. Schematic diagram of the trapeziform key hydraulic grooves.

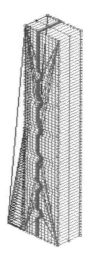

Figure 7. Column contact finite element meshes.

The key hydraulic grooves width B, height H, and slope coefficient M are three key parameters for the trapeziform key hydraulic grooves. This article is based on the three parameters for different values, respectively, to calculate and get some useful rules.

For the same slope coefficient M, with the continuous change of water depth, the relative displacement of the transverse joints key groove nodes also increases, but the key groove

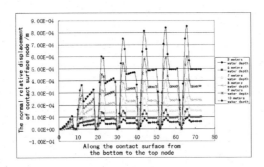

Figure 8. The normal relative displacement of trapeziform key hydraulic grooves contact surface node B = 15 H = 15 M = 1:1.

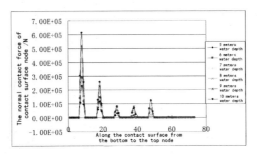

Figure 9. The normal contact force of trapeziform key hydraulic grooves contact surface node B = 15 H = 15 M = 1:1.

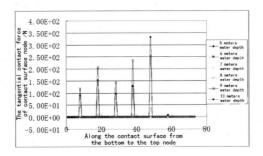

Figure 10. The tangential contact force of trapeziform key hydraulic grooves contact surface node B = 15 H = 15 M = 1:1.

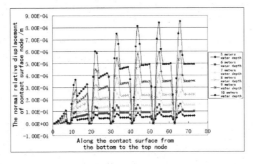

Figure 11. The normal relative displacement of trapeziform key hydraulic grooves contact surface node B = 15 H = 15 M = 1:1.5.

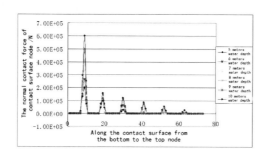

Figure 12. The normal contact force of trapeziform key hydraulic grooves contact surface node B = 15 H = 15 M = 1:1.5.

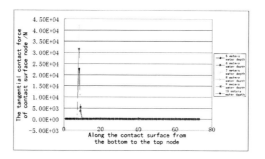

Figure 13. The tangential contact force of trapeziform key hydraulic grooves contact surface node B = 15 H = 15 M = 1:1.5.

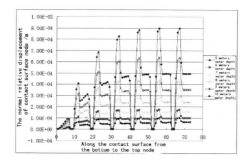

Figure 14. The normal relative displacement of trapeziform key hydraulic grooves contact surface node B = 15 H = 15 M = 1:2.

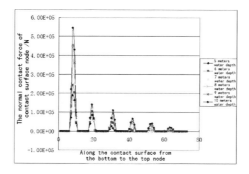

Figure 15. The normal contact force of trapeziform key hydraulic grooves contact surface node B = 15 H = 15 M = 1:2.

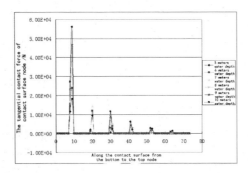

Figure 16. The tangential contact force of trapeziform key hydraulic grooves contact surface node B = 15 H = 15 M = 1:2.

contact surface normal contact force and tangential contact force decreases. With the slope coefficient increase, the relative displacement of key hydraulic grooves contact surface node increases gradually.

For the same key hydraulic grooves width B, with the continuous increase of water depth the relative displacement of the transverse joints key hydraulic grooves node increases, while the contact surface normal contact force decreases; as key hydraulic grooves width B increase, the relative displacement of key hydraulic grooves contact surface node increases, and the contact surface node normal contact force increases.

For the same key hydraulic grooves height H, with the continuous increase of water depth, the relative displacement of the transverse joints key hydraulic grooves node increases, while the contact surface normal contact force decreases, and as key hydraulic grooves height H increase, the relative displacement of key hydraulic grooves contact surface node increases, and the contact surface node normal contact force increases.

5 CONCLUSION

In this paper, the no thickness contact element model with key hydraulic grooves has been used to simulate the transverse joint contact surface. Trapeziform key hydraulic grooves contact geometry model and three trapeziform key hydraulic grooves modes on opening, closing, and cutting have been derived. Through the calculation of the column key groove contact, influence of trapeziform key hydraulic grooves width, height and slope coefficient on contact surface displacement and stress have also been analyzed.

REFERENCES

Ahmadi M.T., Izadinia M., Bachmann H. A discrete crack joint model for nonlinear dynamic analysis of concrete arch dam [J]. Computers and Structures, 2001, 79:403–420.
Fenves G.L., Mojtahedi S., Reimer R.B. ADAP 88: A computer program for nonlinear earthquake analysis of concrete arch dams [R]. Report No. UCBPEERC 89P02, Earthquake Engineering Research Center, University of California, at Berkeley, 1989.
Lau D.T., Noruziaan B. Razaqpur A.G. Modeling of contraction joint and shear sliding effects on earthquake response of arch dams [J]. Earthquake Engng. Struc. Dyn, 1998, 27:1013–1029.
Malika Azmi, Patrick P. Three-dimensional analysis of concrete dams including contraction joint nonlinearity [J]. Engineering Structure, 2002, 24:757–771.
Tu Jin, Chen Hou-qun, Du, Xiu-Ii. A. Study on the simulation scheme of joints in the nonlinear seismic response analysis of high arch dam [J]. Journal of Hydroelectric Engineering, 20010):18–25. (in Chinese).
Xu Yan-jie, Zhang Chu-han, Wang Guang-lun, et al. Nonlinear seismic response analysis of arch dam with contraction joints [J]. Chinese Journal of Hydraulic engineering, 2001 (4):68–74. (in Chinese).
Zhang Liaojun. Mechanical model for hydraulic structure contact problem and its application in the three gorges project [D]. Nanjing: Institute of civil engineering of Hohai University, 2005.

Evaluation of comprehensive treatment measures and stability property of Mayanpo of Xiangjiaba Hydropower Station

Long Jiang, Xiaogang Wang, Jinjie Zhang & Jianhui Sun
China Institute of Water Resources and Hydropower Research, Beijing, China

Huijuan Zhang
Institute of Remote Sensing and Digital Earth Chinese Academy of Sciences, Beijing, China

ABSTRACT: Most of hydropower stations were built in the mountain gorge with complex geological conditions, and so it was inevitable to involve some engineering geological problems of high slope stability in building a variety of high dams. The paper had systematically studied comprehensive treatment measures and corresponding monitoring results of Mayanpo slope, and analyzed its deformation, seepage-pressure and stress-strain characteristics, and evaluated slope stability. The results showed that: ① The comprehensive treatment scheme by using "drainage ditches + anti-slide piles + pre-stressed anchor cable" in Mayanpo slope engineering was feasible, and the two-step implementation plan of "emergency treatment and the fundamental governance" had a remarkable effect; ② Mayanpo slope was greatly influenced by the slope surface load and rainfall before completing the engineering treatment and by rainfall before and after filling the dam reservoir, and the raised invasion line caused by rising water level of the reservoir had a smaller effect on Mayanpo slope; ③ creep damage before Mayanpo slope treatment and semi-rigid destruction after treatment made the strength reduction coefficient of slope increase considerably, respectively, 1.05 and 1.72.

1 INTRODUCTION

Local cracks were found in the initial engineering construction of Mayanpo slope of Xiangjiaba Hydropower Station, and discontinuous cracks had the tendency of large-area continuous expansion, and there existed the safety hidden trouble of local or overall landslide, which could cause the project behind schedule and project cost increasing. Based on this, added geological survey was made on Mayanpo slope, and on the basis of the supplementary investigation results and engineering layout the slope comprehensive treating measures of "drainage ditches + anti-slide piles + pre-stressed anchor cable" were adopted. Through blasting excavation, rainfall and human activities during the construction, rainfall, human activities and water level increasing during initial impound period, and rainfall and water level rising and falling during operating period, no large-scale landslide phenomenon appeared so far, which effectively guaranteed the normal operation of hydraulic facilities built on the slope.

2 PROJECT OVERVIEW

2.1 *The geographical position*

Mayanpo slope located in Right Bank of Xiangjiaba Hydropower Station, and the specific location and topography was shown in Figure 1 and Figure 2.

Figure 1. The geographical position of Mayanpo slope.

Figure 2. The topography of Mayanpo slope.

Figure 3. External deformation observation layout plan of Mayanpo slope.

Figure 4. Monitoring instrument layout plan of deformation body of Mayanpo slope.

2.2 *Hydrogeology and engineering geology*

Mainly artificial slag heap, residual slope sediment, and collapse slope sediment of average thickness of 0.5 m to 6.4 m were covered on the surface of Mayanpo slope. The bedrock was sandstone and mudstone of Jurassic artesian group, the upper part was gray-white, light-yellow thick layer of fine sandstone with thin brown, brick-red sandstone, and the lower part was grey argillaceous rock.

Thick sandstone of the upper part had a relatively strong weathering resistance and a small weathering depth with the highly weathering depth of 4 m ~ 11 m and moderately weathering depth of 15 m ~ 25 m. Four weak intercalated layers exist in strata of Mayanpo slope, respectively, JC① ~ JC④ from top to bottom.

3 THE MONITORING DESIGN SCHEME OF THE PROJECT MANAGEMENT

According to the geological conditions, creep deformation characteristics, the relevant construction field distribution characteristics and scope, and the requirements of the engineering safety of Mayanpo slope, the monitoring projects of anti-slide piles stress strain, pre-stressed anchor cable force, displacement deformation, and seepage of slope soil and inner rock body were determined and took displacement deformation, the underground water-level variation, and distribution as the focus of monitoring content. The main monitoring contents are shown in Figure 3 and Figure 4.

4 ANALYSIS OF MONITORING RESULTS OF MAYANPO SLOPE

4.1 *External deformation characteristics*

By integration analysis of monitoring data of typical section, curves of horizontal and vertical displacement with time were drawn, shown in Figure 5. P is observation pier of section,

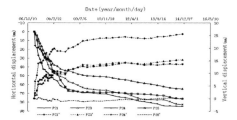

Figure 5. Curves of horizontal displacement and vertical displacement with time (one).

and for example, P08 was the curve of horizontal displacement of observation pier of 08# with time, and P08′ was the curve of vertical displacement of observation pier of 08# with time, and so on.

Figure 5 showed that horizontal and vertical displacement increases with time, which was consistent with the major time nodes of the civil engineering project construction. Displacement rapidly increased before completing the first period of treatment project (to the end of September 2007), slowly increased before completing the second period (from the month end of September 2007 to the month end of December 2008), and became gradually stable before early impoundment (from the month end of December 2008 to the month end of October 2012) in a dynamic balance of the geological environment, continued to increase slowly after impoundment (from the month end of October 2012) in another dynamic balance adjustment process after the geological environment changing.

According to the typical cross-section observation results and horizontal and vertical accumulative displacement variation, curves of horizontal and vertical displacement with time were drawn by integration analysis of data, shown in Figure 6.

Figure 6 showed that horizontal and vertical displacement rate decreases with time, consistent with the changing law of displacement. Displacement rate decreased rapidly after starting slope treatment became slowly stable after completing slope treatment, fluctuated before and after impoundment, but with a smaller increase amount in dynamic stabilization and adjustment.

4.2 *Internal deformation characteristics*

By integration analysis of monitoring data of typical inclinometer hole, deep displacement curves with depth were drawn, shown in Figure 7.

Figure 7 showed that displacement deformation of the inclinometer hole decreased with increasing depth. Dislocation displacement at a certain depth from the orifice showed that the slope had undergone slide deformation at a certain depth. Below the depth of dislocation position, displacement deformation was less and basically unchanged below a certain depth, which was more consistent with the existence of weak interlayer at a certain depth found in slope geological surveying. The orifice displacement and dislocation displacement rapidly increased before completing the first period of treatment project (to the end of September 2007), slowly increased before completing the second period (from the month end of September 2007 to the month end of December 2008), and fluctuated but basically became stable overall before early impoundment (from the month end of December 2008 to the month end of October 2012), continued to increase slowly but had a smaller increase amount after impoundment (from the month end of October 2012). The depth deformation appeared a fluctuation and slight growth trend affected by seasons and rainfall, but became basically stable overall, which showed that the deformation law was directly affected by weak interlayer of the local slope and surface rainfall, and not by the raised slope invasion line with the rising water level.

Based on the observation results of typical inclinometer hole and variation of displacement deformation with depth, as well as integration analysis of the data, curves

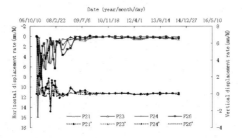

Figure 6. Curves of horizontal displacement rate and vertical displacement rate with time (two).

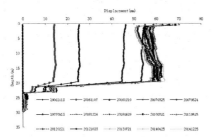

Figure 7. Displacement deformation curves of IN03 with depth.

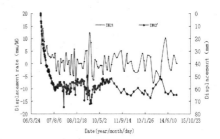

Figure 8. Curves of displacement and displacement rate of the location 1 m away from the orifice of IN03 with time.

of displacement and displacement rate of the location 1 m away from the orifice with time were drawn, shown in Figure 8. IN was inclinometer hole, IN03 was the curve of displacement rate of 3# with time, and IN03′ was the curve of displacement of 3# with time, and so on.

Figure 8 showed that displacement rate decreased with time and displacement increases with time. As a whole, displacement rate first increased and then decreased and then fluctuated until basically became stable, and local fluctuations were mainly affected by rainfall and dynamic equilibrium adjustment of the slope internal deformation.

5 SLOPE STABILITY ANALYSIS

Based on the penalty function contact of pile soil and surface of rock mass, as well as physical and mechanical parameters provided by the engineering geological survey report of Mayanpo slope, using ABAQUS finite element program, the three-dimensional numerical model of coupling of seepage and strain has been established. During analysis of the stick-slip friction model of rock mass by C3D8RP unit and anti-slide pile by C3D8R unit, the simulation of anti-slide pile position and the original rock soil shared grid unit, the original unit given their material properties, the life and death element method simulation of anti-slide pile construction, the friction coefficient in the process of analysis remains unchanged, and the application of Co. the strength reduction method for slope stability analysis.

Mayanpo slope located switch station, 110 KV substation, the temporary 500 m elevation layout pool, 6# highway, the finished material field, and other building facilities. According to the original geological conditions (before slope treatment) and setting anti-slide pile and drainage ditch (after treatment) of two kinds of working conditions, plastic strain contours with different reduction factors before and after slope treatment are shown in Figure 9 and Figure 10, lateral deformation curves in Figure 11.

Figure 9. Plastic strain contours before slope treatment.

Figure 10. Plastic strain contours after slope treatment.

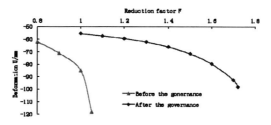

Figure 11. Curves of lateral deformation with reduction factor.

Figure 9 and Figure 10 showed that in the same load condition, plastic area was through from slope top and bottom to the middle before slope treatment, which was creep damage; through from the anti-slide pile near to both sides, which was semi-rigid destruction. The plastic region before and after treatment began from the local and gradually developed towards plastic region through with increasing reduction factor.

Figure 11 showed that lateral deformation increases with increasing reduction factor. When strength reduction factor was smaller, lateral deformation had less increment; and when strength reduction factor was larger, lateral deformation had larger increment. When reduction factor was about F = 1.05 and F = 1.72, lateral deformation sharply increased and had a huge and unlimited development of plastic deformation and displacement.

6 CONCLUSIONS

1. The comprehensive treatment scheme by using "drainage ditches + anti-slide piles + prestressed anchor cable" in Mayanpo slope engineering was feasible, and the two-step implementation plan of "emergency treatment and the fundamental governance" had a remarkable effect.
2. Surface and deep displacement law of Mayanpo slope were relatively consistent with the engineering geology, engineering treatment schedule, and reservoir conditions. Displacement rapidly increased before completing the first period of treatment project, slowly increased before completing the second period, and became gradually stable before early impoundment in a dynamic balance of the geological environment, continued to increase slowly after impoundment in another dynamic balance adjustment process after the geological environment changing.
3. Seepage pressure and anti-slide piles stress strain of Mayanpo slope were greatly influenced by the slope surface load and rainfall before completing the engineering treatment

and by rainfall before and after filling the dam reservoir, and the raised invasion line caused by rising water level of the reservoir had a smaller effect on Mayanpo slope.
4. Creep damage before Mayanpo slope treatment and semi-rigid destruction after treatment made the strength reduction coefficient of slope increase considerably, respectively 1.05 and 1.72.

REFERENCES

Beijing IWHR Technology Co., Ltd. Safety Monitoring Report of Mayanpo slope in the Right Bank of Xiangjiaba Hydropower Station in Jinsha River in 2014 [R]. 2015. (in Chinese).
Chen Zu-yu, Wang Xiao-gang, Yang Jian, et al. Stability Analysis of Rock Slope—Principle, Method, Procedure [M]. Beijing: China WaterPower Press, 2005. (in Chinese).
Chen Zu-yu, Wang Xiao-gang. High Slope Engineering in Hydropower Construction [J]. Journal of Hydropower, 1999. (in Chinese).
Lian Zhen-ying, Han Guo-cheng, Kong Xian-jing. Study on Stability of Excavated Slopes by Strength Reduction FEM [J]. Chinese Journal of Geotechnical Engineering, 2001, 23 (4). (in Chinese).
Zhang Lu-yu, Zheng Ying-ren, Zhao Shang-yi. The Feasibility Study of Strength-reduction Method with FEM for Calculating Safety Factors of Soil Slope Stability [J]. Journal of Hydraulic Engineering, 2003, (1). (in Chinese).
Zheng Ying-ren, Zhao Shang-yi, Zhang Lu-yu. Slope Stability Analysis by Strength Reduction FEM [J]. Engineering Sciences, 2002, 4 (10). (in Chinese).

Constant-ductility Residual Displacement Ratio response spectrum

Xiaobin Hu
School of Civil Engineering, Wuhan University, Wuhan, China

Huigao He
CITIC General Institute of Architectural Design and Research Co. Ltd., Wuhan, China

Weibo Jiang
School of Civil Engineering, Wuhan University, Wuhan, China

ABSTRACT: Structures subjected to violent ground shaking may be left in a displaced condition. This residual or permanent displacement may render the structures unsafe or irreparable. This paper presents the results of a parametric study aimed to construct the constant-ductility Residual Displacement Ratio (RDR) response spectrum of bilinear Single-Degree-of-Freedom (SDOF) systems subjected to strong earthquake ground motions. The RDRs, which are defined as the ratios between the residual displacement and the yield displacement, were computed using nonlinear time history analyses of bilinear SDOF systems subjected to 100 scaled ground motions. The results were statistically organized to establish the constant-ductility RDR response spectrum.

1 INTRODUCTION

The excessive residual displacement may even lead to the demolition of structures due to huge technical difficulties or cost to straighten or repair them. For example, in the 1995 Kobe Earthquake more than 20% of the over 500 minor injured bridge piers were pushed down and rebuilt due to the large permanent drifts (Kawashima 1998). Therefore, the residual deformations have been attracting considerable attentions from the earthquake engineering community and thus start to be recognized as an important parameter in performance-based seismic design and evaluation of structures (Pampanin et al. 2003; Christopoulos and Pampanin 2004).

Some researches associated with the assessment of residual displacements have been reported (Mahin et al. 1981; Macrae et al. 1997, 1998; Ruiz-Garcia et al. 2006; Li et al. 2007; Ouyang et al. 2010; Hao et al. 2013). Mahin et al. (1981) found that for some elasto-plastic systems, the mean residual displacement averaged as high as 45% of the mean peak inelastic displacement with high levels of dispersion. Macrae et al. (1997, 1998) defined Residual Displacement Ratio (RDR) as that between the residual displacement and the maximum possible residual displacement for bilinear Single-Degree-of-Freedom (SDOF) system, and established the RDR response spectrum through nonlinear time history analysis. They found that the RDRs mainly rely on the post-yield stiffness ratio of the bilinear SDOF system and have little relation to other parameters. Generally, the system with positive stiffness ratios have small residual displacements, while those with negative stiffness ratios tend to undergo little inelastic reversal of deformation and have larger residual displacements. Ruiz-Garcia et al. (2006) conducted similar research concerning both bilinear system and stiffness-degrading systems, in which RDR was expressed as the ratio between the residual displacement and the maximum elastic displacement. They concluded that RDRs exhibit large levels of record-to-record variability and sensitive to changes in local site conditions, earthquake magnitude, distance to the source range and hysteretic behavior.

It can be seen that the study methods adopted in the above literatures are basically the same, while the main difference lies in the definition of RDR. The definitions of RDR aforementioned involve using the maximum possible residual displacement or maximum elastic displacement, which are heavily dependent on the earthquake ground motions and remain unknown unless computed by time history analysis or response spectrum analysis. Therefore, a new definition of RDR is proposed for the bilinear SDOF system in this paper, which is more convenient in engineering application. Based on this, the constant-ductility RDR spectrum for the bilinear SDOF system is developed by means of time history analysis.

2 RESIDUAL DISPLACEMENT RATIOS

2.1 *Definition of RDR*

For the bilinear SDOF system, the hysteresis model is illustrated in Figure 1, where f_y and x_y are the yield force and yield displacement respectively, k is the elastic stiffness and α is the post-yield stiffness ratio. The RDR of the bilinear SDOF system is defined as

$$r = \frac{x_r}{x_y} \tag{1}$$

where x_r is the residual displacement.

2.2 *Determination of yield displacement*

The yield displacement can be computed as follows,

$$x_y = \frac{x_m}{\mu} \tag{2}$$

where x_m is the maximum inelastic displacement and μ is the target ductility factor.

It is noted that x_y could not be obtained from Eq. (2) directly since x_m is unknown. For the given natural period and earthquake ground motion, the iteration procedure of computing x_y for the target ductility factor can be executed in the following steps.

2.3 *Calculation of residual displacement*

As seen in Figure 1, it is assumed that the SDOF system rests at point a at the end of time history analysis. It is noted that the system restoring force at point a is not necessarily zero. In order to get the residual displacement, a line with the slope of k is drawn from point a and intersects the horizontal axis at point d. The displacement at the intersection point is just the residual displacement, which can be calculated as follows

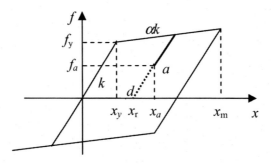

Figure 1. Hysteresis model of bilinear SDOF system.

$$x_r = x_a - \frac{f_a}{k} \tag{3}$$

where x_a and f_a are the displacement and restoring force corresponding to point a respectively, and k can be obtained as follows,

$$k = m\left(\frac{2\pi}{T}\right)^2 \tag{4}$$

2.4 *Definition of constant-ductility RDR spectrum*

Combing Eq. (1), (2), (3) and (4), it can be derived as follows

$$r = \frac{\mu\left[x_a - \frac{f_a}{m}\left(\frac{T}{2\pi}\right)^2\right]}{x_m} \tag{5}$$

For the system with specified mass and damping subjected to the given earthquake ground motion, x_a, f_a and x_m in Eq. (5) can be calculated for given μ, α and T. Therefore, Eq. (5) can be expressed in a simple form as follows,

$$r = r(\mu, \alpha, T) \tag{6}$$

It shows that r can be seen as the function of three independent variables of μ, α and T. Assuming that α is known, Eq. (6) can be further rewritten as

$$r_\mu = r_\mu(T) \tag{7}$$

The above equation describes the dependence of r on T for the specified μ, which is hereby referred as constant-ductility RDR spectrum in this study.

3 ANALYSIS PROCEDURE

3.1 *Equation of motion*

For the bilinear SDOF system under earthquake excitation, the governing equation of motion can be expressed as

$$m\ddot{x} + c\dot{x} + f(x) = -m\ddot{x}_g \tag{8}$$

where c is viscous damping coefficient; $f(x)$ is the nonlinear restoring force of the system as described in Figure 1; $\ddot{x}, \dot{x}$ and x are the acceleration, velocity and displacement of the system relative to the ground respectively; $\ddot{x}_g$ is the ground acceleration.

3.2 *Earthquake ground motions*

In this study, it is assumed that the seismic precautionary intensity is 8 degree with the design basic acceleration of ground motion being 0.3 g, the site category is type II and the design earthquake group is the second one (GB 50011-2010). The design response spectrum can be determined accordingly and used as the target spectrum, as seen in Figure 2. A total of 100 earthquake ground motion records were selected from PEER NGA database (http://peer.berkeley.edu/nga/) and scaled to be compatible with the target spectrum. These records

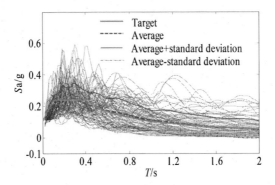

Figure 2. Elastic acceleration response spectrum of the 100 selected earthquake records.

were generated by earthquakes with moment magnitude M_w ranging between 5.5~7.8. The ensemble average and average plus/minus one standard deviation spectral acceleration over the 100 scaled records along with the target spectrum are shown in Figure 2. A good match can be seen between the ensemble average spectrum and the target spectrum.

3.3 *Analysis parameters*

Newmark-β numerical integration scheme was adopted to solve Eq. (8). The viscous damping ratio is assumed to be 0.05 and the p-Δ effect is neglected for simplification. The α values considered in this study are 0, 0.05 and 0.1. The μ values are set to be 2, 4 and 6, which are intended to be representative of three different levels of target ductility. The natural periods of the SDOF system used in this study range between 0.1 s and 2 s with an interval of 0.2 s. For the given parameters and earthquake ground motions, a total of 9000 times of time-history analyses was involved.

4 CONSTANT-DUCTILITY RDR SPECTRUM

4.1 *Typical constant-ductility RDR spectrum curve*

The ensemble average constant-ductility RDR spectrum over the suite of 100 scaled earthquake records for the system with α equal to 0 (elastic-perfectly plastic system) and μ equal to 2 is shown in Figure 3, where $\bar{r}_\mu$ denotes the ensemble average value of RDR over the 100 records. It is seen that for the period range considered in this study r is generally less than 1, which means that the residual displacement is generally smaller than the yield displacement for the system with relatively lower ductility. In addition, the spectrum curve can be divided into two distinctive regions, one is the short-period region (e.g., $T < 0.7$ s) where r decreases with the increase of T, and the other one is the long-period region (e.g., $T > 0.7$ s) where the curve tends to oscillate along the T axis.

4.2 *Simplified equation to estimate RDR*

In the context of displacement-based assessment of existing structure, it is appealing to have a simplified equation to estimate the residual displacement demand. In this section, it is further to propose the equation to estimate RDR.

Assuming that α is known, Eq. (6) can be written as

$$r = r(\mu, T) \qquad (9)$$

The following functional form is adopted for the above equation,

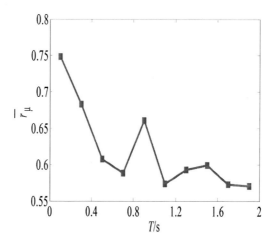

Figure 3. Typical constant-ductility RDR spectrum curve ($\alpha = 0$, $\mu = 2$).

Table 1. The expressions of a, b, c and d corresponding to different α values.

α	a	b	c	d
0	$-0.0259\,\mu^2 + 0.5855\,\mu - 0.8586$	$-0.3464\,\mu^2 + 2.5413\,\mu - 8.827$	$0.0305\,\mu^2 + 0.2696\,\mu - 0.0321$	$-0.0218\,\mu^2 + 0.1928\,\mu - 0.3463$
0.05	$-0.0358\,\mu^2 + 0.5254\,\mu - 0.5279$	$-0.3733\,\mu^2 + 2.998\,\mu - 10.506$	$-0.0124\,\mu^2 + 0.1977\,\mu + 0.091$	$-0.0254\,\mu^2 + 0.2267\,\mu - 0.3161$
0.1	$-0.0245\,\mu^2 + 0.2748\,\mu - 0.0024$	$-0.3543\,\mu^2 + 3.0555\,\mu - 11.86$	$-0.0009\,\mu^2 + 0.0286\,\mu + 0.2764$	$-0.0314\,\mu^2 + 0.2906\,\mu - 0.3928$

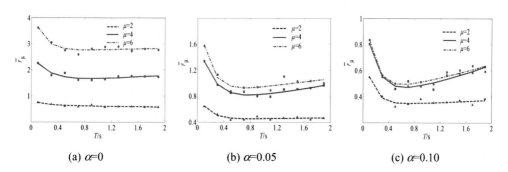

(a) $\alpha=0$ (b) $\alpha=0.05$ (c) $\alpha=0.10$

Figure 4. Comparison between the results obtained from the numerical calculation and the simplified equation.

$$r = ae^{bT} + ce^{dT} \qquad (10)$$

where a, b, c and d are all quadratic functions of the variable of μ.

The curve fitting tool of MATLAB (Mathworks Inc., 2004) was employed to fit $\bar{r}_\mu$ obtained from numerical analyses using the functional form given in Eq. (9)~(10). Table 1 lists the expressions of a, b, c and d pertaining to different α. Figure 4 plots the curves obtained from the simplified equation along with the data points from the numerical calculation, which demonstrates that a good agreement is achieved.

5 CONCLUSIONS

This paper proposed a new definition of residual displacement ratio for the bilinear SDOF system subjected to strong earthquake ground motions and established the constant RDR spectrum by elasto-plastic time history analyses. The parameters influencing the RDR spectrum were investigated thoroughly. The following conclusions can be drawn:

1. For the system with relatively smaller period, the RDR deceases with the increase of period.
2. The RDR generally increases with the rise of ductility factor or the drop of post-yield stiffness ratio. The effects of ductility factor and post-yield stiffness ratio on the RDR are basically independent of the system period.
3. For the system with relatively larger post-yield stiffness ratio, it is less pronounced to reduce the RDR by lowering the post-yield stiffness ratio.

ACKNOWLEDGEMENT

This work was supported by grants from The National Natural Science Foundation of China (51208386) and The Hubei Provincial Natural Science Foundation of China (2011CDB266) respectively. The authors are grateful to the sponsors. However, the opinions and conclusions expressed in this paper are solely those of the writers and do not necessarily reflect the views of the sponsors.

REFERENCES

Christopoulos C., Pampanin S. 2004. "Towards performance-based seismic design of MDOF structures with explicit consideration of residual deformations," *Journal of Earthquake Technology*, 41(1):53–73.

GB 50011-2010, Code for seismic design of buildings. Beijing: *China Architecture Industry Press*, 2010. (in Chinese).

Hao J., Wu G., and Wu Z. 2013. "A study on constant-relative-strength residual deformation ratio spectrum of SDOF system," *China Civil Engineering Journal*, 46(10):82–88. (in Chinese).

Kawashima K., Macrae G.A., Hoshikuma J., et al. 1998. "Residual displacement response spectrum," *Journal of Structural Engineering*, 124:523–530.

Li F., Zhu X. 2007. "Residual Displacement Ratio Spectrum of Single Degree of Freedom Bilinear Structure under Ground Motions with Pulse in Near Fault Zones," *China Railway Science*, (3):49–55. (in Chinese).

Macrae G.A., Kawashima K. 1997. "Post-earthquake residual displacements of bilinear oscillators," *Earthquake Engineering and Structural Dynamics*, 26:701–716.

Mahin S.A., Bertero V.V. 1981. "An evaluation of inelastic seismic design spectra," Journal of the Structural Division, 107(9):1777–1795.

Mathworks Inc. 2004. MATLAB: The Language of Technical Computing, Version 7.0. The Mathworks Inc.: Nattick, MA.

Ouyang C., Liu C. 2010. "Residual displacement of SDOF system under earthquakes," *World Earthquake Engineering*, (1):143–146. (in Chinese).

Pampanin S., Christopoulos C., Priestley M.J.N. 2003. "Performance-based seismic response of frame structures including residual deformations. Part II: Multidegree-of-freedom systems," *Journal of Earthquake Engineering*, 7(1):119–147.

Ruiz-Garcia J., Miranda E. 2006. "Residual displacement ratios for assessment of existing structures," *Earthquake Engineering and Structural Dynamics*, 35(3):315–336.

Numerical simulation of flow across a transversely oscillating wavy cylinder at low reynolds number

Lin Zou, Miao Wang, Can Xiong & Hong Lu
School of Mechanical and Electronic Engineering, Wuhan University of Technology, Wuhan, China

ABSTRACT: Numerical simulations of laminar, incompressible flow across a transversely oscillating wavy cylinder (the dimensionless wavelength λ/Dm = 2 and 6) have been undertaken by using the conjugating Finite-volume Method with dynamic mesh technique at a fixed Reynolds number of 100 and a fixed oscillation amplitude ratio (Ae/Dm) of 0.4. The effects of the excitation frequency ratios (fe/fs), the dimensionless wavelength λ/Dm, and the flow parameters have been studied. Compared with the straight cylinder, it is observed that the force amplifications, the lock-on range, and the near-wake vortex-shedding modes behind the wavy stay-cable depend upon the shape of wavy cylinder surface. The wavy cylinder of λ/Dm = 2 displayed a similar Lock-on phenomena at the excitation frequency ratios of 0.7 to 1.3 as the straight cylinder and have no apparent drag reduction. But for the wavy cylinder of λ/Dm = 6, the natural St frequencies are still dominant at that region, the mean drag reduction is up to 27% and the fluctuating lift coefficients is close to 0.1 at = 1.1, the well-organized wake structures are generated behind the wavy cylinder of λ/Dm = 6. Such simulation results will establish a comprehensive database to further our understanding about the physical mechanisms of this fluid–structure interaction problem.

1 INTRODUCTION

The flow around oscillating bluff bodies is an important engineering problem both from the academic and practical points of view. In addition to engineering applications, the flow problem permits investigation of the complex underlying phenomena of fluid-structure interaction and vortex formation processes. The periodic vortex shedding and the fluctuating velocity fields behind the cylindrical bodies can cause structural damages, shorten the life of the structures and even lead to severe accidents. So vortex induced vibrations has attracted numerous experimental and numerical studies. In the recent study, (Cagney and Balabani, 2013) proposed that the vortices for the symmetric S-I mode and the alternate A-II wake mode are formed at the cylinder response frequency and cause a fluctuating drag when a cylinder experiencing vortex induced vibrations in the streamwise direction. (Singh and Chatterjee, 2014) carried out one series of two-dimensional numerical on the vortex induced vibrations of an elastically mounted circular cylinder in linear shear flows at low Reynolds numbers, and observed the phenomena of hysteresis when Re is around 84 and 325. As a simplification model of vortex induced vibrations, the forced vibration is also an issued research on flow over a circular cylinder given forced transverse oscillations. (Do et al., 2007) provided an analysis of the fluctuating forces and identified the lock-on regimes for a circular cylinder oscillating longitudinal and transverse to the free stream direction with emphasis given to the case with 60° oscillations. (Placzek et al., 2009) analyzed the wake characteristics of the oscillating cylinder with the forced oscillations characterized by frequency ratio and dimensional amplitude, and suggested three modes of vortex shedding: Karman mode vortex street (2S), anti-symmetric shedding mode (2P), and asymmetric mode with one vortex pair and a single vortex produced per motion cycle (P + S).

However, it is a challenge to control the vortex shedding phenomenon and hence to reduce the potential for Flow Induced Vibration (FIV). (Lam and Lin, 2009) and (Zou and Lin, 2010) had studied several types of cylinders with a surface profile in sinusoidal curve along their spanwise direction and found the wavy cylinders could achieve the purpose of drag reduction and vibration suppression at the same time. We would expect a drag reduction by employing a wavy cylinder with an optimal wavelength for the best effect on FIV suppression at laminar flow. For the better understanding of the physical mechanisms of this fluid–structure interaction problem, in this paper, the transversely forced oscillating wavy cylinder ($\lambda/Dm = 2$ and $\lambda/Dm = 6$) have been studied at the excitation frequency ratios varied from 0.3 to 2. The purpose of our present work is to study the relationship between the force of cylinder suffered and frequency ratios, and to observe the force amplifications, the lock-on range, and the near-wake vortex-shedding modes behind the wavy cylinder. It is hoped that such an investigation will be helpful to comprehend the physical mechanisms on the control of flow induced vibration and drag reduction and applied in engineering applications such as the vibration of cables in suspension bridges and multiple risers in offshore engineering.

2 NUMERICAL METHOD

2.1 *Governing equations*

The flow field is governed by the Navier-Stokers equations, which read for a Newtonian incompressible fluid:

$$\nabla \cdot \mathbf{u} = 0, \frac{\partial \mathbf{u}}{\partial t} + \mathbf{u} \cdot \nabla \mathbf{u} = -\nabla p + \frac{\nabla^2 \mathbf{u}}{\mathrm{Re}} \tag{1}$$

where $\mathbf{u} = (u, v, w)^T$ is the velocity vector with u, v, w being respectively the streamwise, transverse, and spanwise velocity components, p is the pressure. The pressure-velocity decoupling is achieved by the SIMPLE (Semi-Implicit Method for Pressure-Linked Equations) algorithm. The second-order upwind differencing scheme is used for convective terms. The second-order central difference scheme is adopted foe diffusion terms, while the second-order implicit scheme is employed to discretize the equations in time.

2.2 *The motion equation of cylinder*

In the case of an oscillating cylinder, the prescribed cylinder motion is solved in moving mesh boundary to determine cell motion velocity at which each single boundary is moving. The cylinders impulsively perform time-dependent sinusoidal oscillations in the transverse direction only with a displacement represented by:

$$X(t) = A_e \sin(2\pi f_e t) \tag{2}$$

where t is the dimensional time, A_e is the dimensional amplitude that is equal to 0.4 Dm, and f_e is excitation frequency of cylinder displacement.

2.3 *Numerical model*

Figure 1(a) shows the main characters of the wavy cylinder, which described by the following equation:

$$D_z = D_m + 2a\cos(2\pi z/\lambda) \tag{3}$$

where D_z denotes the local diameter of the wavy cylinder and varies in the spanwise direction z. a and λ denotes the amplitude of the wavy cylinder and the wavelength of the wavy cylinder.

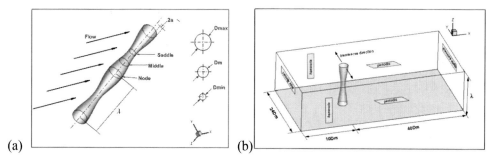

Figure 1. Geometric model of wavy cylinder and the computational domain.

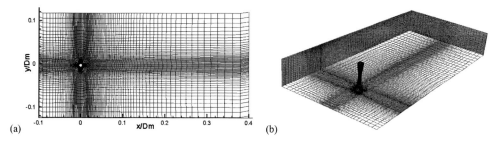

Figure 2. Computational grid distribution: (a) grid in X-Y plane; (b) grid around a wavy cylinder.

Dm is the mean diameter that is equal to the mean value of the maximum local diameter D_{max} and the minimum local diameter D_{min}. The axial location with the maximum local diameter is called the "node," while the axial location of the minimum diameter is called "saddle." The "middle" plane is located in the midpoint position between the node and saddle planes.

In the present study, the wavy cylinders with a spanwise wavelength ratio of $\lambda/Dm = 2$ and 6 and a fixed wave amplitude ratio of $a/Dm = 0.15$ were employed the computational domain is a cube that is similar to the study of (Lam and Lin, 2009). The size of the cube is set to be $50\,D_m \times 24\,D_m \times \lambda$. The upstream boundary of the computational domain is set at a distance of 10 Dm from the axis of the cylinder while the downstream boundary is 40 Dm away from the cylinder, and each lateral surface is 12 Dm away from the axis of the cylinder. The boundary conditions adopted are specified in Figure 1(b). In order to compare conveniently, in the further analysis of the article, the straight cylinder is called CY. Meanwhile, the wavy cylinder of $\lambda/Dm = 2$ is marked as WY1, the other wavy cylinder of $\lambda/D_m = 6$ is flagged as WY2.

As for the grid distributions, an unstructured hexahedral grid is employed to divide the computational domain. The circumferential direction of each cylinder uniformly distributed 80 grid points, while the distance between the first grid and the cylinder surface was 0.01 Dm. Figure 2 shows the grid distribution of the wavy cylinder.

3 RESULT AND DISCUSSION

3.1 Force characteristics

Figure 3 summarizes the variations of the mean drag coefficient Cd_{mean} ($\overline{C}_D = 2\overline{F}_D/\rho U_\infty^2 DH$) and the fluctuating lift coefficient Clr.m.s ($C'_L = 2F'_L/\rho U_\infty^2 DH$) of cylinders with different frequency ratios (fe/fs, where fe is the cylinder oscillation frequency and fs is the corresponding vortex shedding frequency for stationary cylinder). The data of two-dimensional oscillating cylinder in (Nobari and Naderan, 2006) at Re = 106 and (Anagnostopoulos, 2000) at Re = 100 are quoted to illustrate the correctness of the present results. For the straight oscillating cylinder, it can be clearly seen that the r.m.s. lift coefficient in the present study is similar to previous research result. Meanwhile, the tendency of the mean drag coefficient

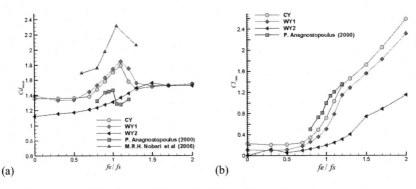

Figure 3. Mean drag coefficient (a) and fluctuating lift coefficient (b).

and lock-on region is also very consistent to the formers' work. It shows that the present gird schemes and numerical method are applicable for the present study.

From Figure 3(a), in general, for the straight cylinder (CY) and the $\lambda/Dm = 2$ wavy cylinder (WY1), Cd_{mean} has a wide peak between fe/fs = 0.7 and fe/fs = 1.3 but otherwise does not vary strongly with fe/fs. It indicated that the WY1 may be experiencing a similar transition as the CY. The values of Cdmean either side of transition are very similar. Surprisingly, for the $\lambda/Dm = 6$ wavy cylinder (WY2), the Cd_{mean} always increases moderately and is much smaller than that of CY and WY1 when the frequency ratios is smaller than 1.3, the mean drag reduction is up to 27% at fe/fs = 1.1, and then the values of Cd_{mean} is similar as that of CY and WY1 after fe/fs = 1.3. It seems to show that the natural St frequencies are still dominant at transition region. Further, from Figure 3(b), for CY and WY1, the fluctuating lift coefficient Clr.m.s have a distinct jump at fe/fs = 0.7, the frequency at which this occurs is defined as the transition frequency. It can be explained that lock-on behavior begins to occur and the flow modes will be change at this region. But for the WY2, the evident jump of Clr.m.s is not noticed at fe/fs = 0.7 and the Clr.m.s rises slowly. The fluctuating lift coefficients is close to 0.1 at fe/fs = 1.1. It indicated that the shape of WY2 has strong effects on the drag force and fluctuating lift force control.

Figure 4 shows Power spectrum of lift force for the straight or wavy cylinders at fe/fs = 0.5 and fe/fs = 1.1. When fe/fs = 0.5, the power spectrums of CY and WY1 exist two evident peaks (Fig. 4(a)), it indicates the presence of two independent frequencies and reflects the two cylinders are outside of the lock-on region. When fe/fs = 1.1, their power spectrum have only one peak (Fig. 4(b)), The vortex shedding frequency in this two cases is transferred to the cylinder oscillation frequency and therefore the flows is lock-on in nature. However, the power spectrums of WY2 in both two cases have only one smaller peak value than that of the CY and WY1 in the same frequency ratio. It is no difficult to conclude that the WY2 is not easy to be induced and the natural St frequencies are always dominant.

3.2 *Wake structure*

In order to have a further understanding of the flow physics of WY2 for reducing drag and suppressing lift, Figure 5 presents that instantaneous 3-D wake structures behind the cylinders at the frequency ratio fe/fs = 1.1. The wake structure behind straight cylinder displays a quasi-2D wake structure and the mode of vortex shedding is the typical Karman mode vortex street (2S mode) (see Fig. 5(a)). The wake structure behind WY1 displays a periodic 3-D wake structure but the mode of vortex shedding is generally still similar Karman mode vortex street (2S mode) (see Fig. 5(b)). For the WY2, The wake structure becomes more stable and the mode of vortex shedding transfers the two pairs of vortex of identical signs are shed per cycle (2C mode) (see Fig. 5(d)). It indicated that the change of the vortex-shedding modes are strongly the shape of cylinder surface dependent, and it will lead to the different drag, lift, and lock-on behavior because of the different cylinder surfaces.

Figure 6 and Figure 7 are the instantaneous 3-D wake structure behind the cylinders at fe/fs = 1.1. From Figure 6, it can be seen that the vortices are clearly apparent in the entire

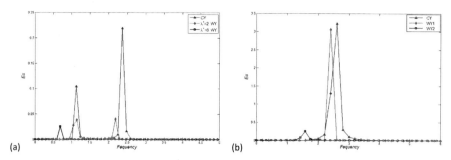

Figure 4. Power spectrum of lift force for the straight or wavy cylinders: (a) fe/fs = 0.5; (b) fe/fs = 1.1.

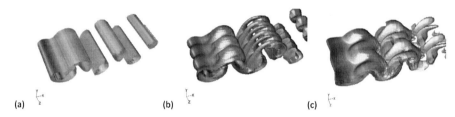

Figure 5. Instantaneous 3-D wake structure behind the cylinders at fe/fs = 1.1. (a) straight cylinder; (b) $\lambda/Dm = 2$ wavy cylinder (WY1); (c) $\lambda/Dm = 6$ wavy cylinder (WY2).

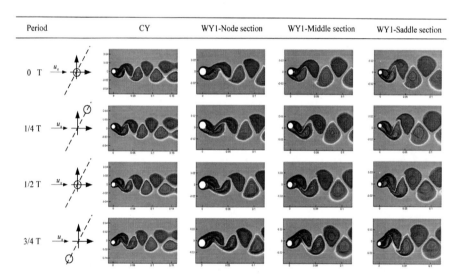

Figure 6. The instantaneous vorticity contour over one oscillating period, for CY and WY1, fe/fs = 1.1.

downstream region behind CY and WY1. Note that the shape of the vortices is circular and they move perfectly in-phase for the CY, and display the typical Karman mode vortex street (2S mode). For WY1, the wake presents a 2S mode in node section, but unconspicuous 2C mode (two pairs vortices of identical signs are shed per cycle) occurs in middle section and saddle section. In general, the flow at fe/fs = 1.1 is considered as the synchronous lock-on flow for both CY and WY2. The data of force coefficient supported this behavior (see Fig. 4). Figure 7 shows the difference of vortex structure in an oscillating period for CY and WY2. It can be seen that the wake mode for WY2 are typical 2C mode. Furthermore, the formation length of vortex behind WY2 both are longer than that of CY, and the flow wake become more stable and well organized. It is considered as the synchronous non-lock-on behavior fe/fs = 1.1 for WY2.

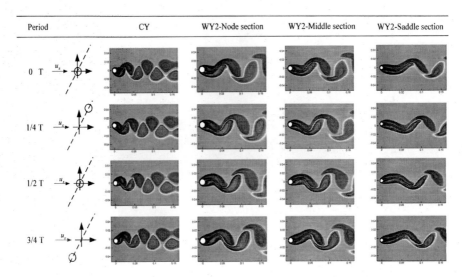

Figure 7. The instantaneous vorticity contour over one oscillating period, for CY and WY2, fe/fs = 1.1.

4 CONCLUSIONS

The paper presents a numerical study of flow across a transversely oscillating wavy cylinder (λ/Dm = 2 and 6) at low Reynolds number Re = 100. The excitation frequency ratio (fe/fs) has a very wide variation from 0 to 2. Compared with straight cylinder, we observed that the synchronous lock-on flows occur behind WY1 at 0.7 < fe/fs < 1.3, but the synchronous non-lock-on behavior occurs at this region for WY2. The more stable and the well-organized wake structures are generated behind WY2, and which lead to the mean drag reduction coming to 27%, and the fluctuating lift coefficients is close to 0.1 at fe/fs = 1.1. It indicated that the shape of WY2 has strong effects on the drag force and fluctuating lift force control.

ACKNOWLEDGMENTS

This work was financially supported by the National Natural Science Foundation of China (Grant No. 11172220 and No. 51275372).

REFERENCES

Anagnostopoulos, P. 2000. Numerical Study of the Flow Past a Cylinder Excited Transversely to the Incident Stream. Part 1: Lock-In Zone, Hydrodynamic Forces And Wake Geometry. *Journal of Fluids & Structures*, 14, 819–851.
Cagney, N. & Balabani, S. 2013. Mode competition in streamwise-only vortex induced vibrations. *Journal of Fluids & Structures*, 41, 156–165.
Do, T.T., Chen, L., Tu, J.Y., Do, T.T., Chen, L. & Tu, J.Y. Numerical Simulations of Flows over a Forced Oscillating Cylinder. 16th Australasian Fluid Mechanics Conference (AFMC), 2007. 573–579.
Lam, K. & Lin, Y.F. 2009. Effects of wavelength and amplitude of a wavy cylinder in cross-flow at low Reynolds numbers. *Journal of Fluid Mechanics*, 620, 195.
Nobari, M. & Naderan, H. 2006. A numerical study of flow past a cylinder with cross flow and inline oscillation. *Computers & Fluids*, 35, 393–415.
Placzek, A., Sigrist, J.-F. & Hamdouni, A. 2009. Numerical simulation of an oscillating cylinder in a cross-flow at low Reynolds number: Forced and free oscillations. *Computers & Fluids*, 38, 80–100.
Singh, S.P. & Chatterjee, D. 2014. Impact of transverse shear on vortex induced vibrations of a circular cylinder at low Reynolds numbers. *Computers & Fluids*, 93, 61–73.
Zou, L. & Lin, Y.F. 2010. Numerical simulation of turbulent flow around wavy cylinders at a subcritical Reynolds number and the investigation on drag reduction. *Chinese Journal of Hydrodynamics*.

Bond behavior of Near Surface Mounted Carbon Fiber Reinforced Polymer rod under temperature cycles

Heeyoung Lee, Youngmin Song & Chunhong Park
Kyung Hee University, Department of Civil Engineering, Yongin, Gyeonggi-do, Korea

Wootai Jung
SOC Research Institute of Construction Technology, Iisan, Gyeonggi-do, Korea

Wonseok Chung
Kyung Hee University, Department of Civil Engineering, Yongin, Gyeonggi-do, Korea

ABSTRACT: In this paper, a Near Surface Mounted (NSM) strengthening technique that increases the resistance of concrete structures is proposed. The bond between the NSM rods and concrete is the key component of the NSM strengthening technique. In this technique, there are two bond interfaces: one between the reinforcement composite rod and the filling material; the other between the filling material and concrete. These two interfaces must be investigated to enable the technique to efficiently perform. The aim of our experimental evaluation was to investigate the effects of filling material properties depending on the temperature variation (−15 °C~55 °C) and number of temperature cycles (0, 3, 100) for Carbon Fiber Reinforced Polymer (CFRP) rods used in the NSM technique. A total of 27 cubic specimens were fabricated, and pullout tests were conducted. The results are presented in terms of failure load, average bond stress, strains in CFRP bar, and mode of failure.

1 INTRODUCTION

The costs required to maintain worldwide aging social infrastructures have recently become a serious problem for which huge budgets are annually allocated. This has resulted in an enormous burden on users due to the effects of cost and work spent on industrial activity maintenance. Most rods used in Pre-Stressed Concrete (PSC) structures are made of steel, and many concrete structures in shore areas are exposed to inclement environments. Consequently, serious problems have occurred in these structures on account of the loss of load behavior capacity from the corrosion of the steel reinforcements. Recently, Carbon Fiber Reinforced Polymer (CFRP) has gained growing interest as an alternative material for replacing steel reinforcements, rods, pre-stressing tendons, and tensile elements owing to its non-corrosiveness, high strength, and lightweight properties. The common strengthening methods adopted for strengthening weakened concrete bridges are the bonding of steel sheets, pre-stressing of steel rods, and bonding of Fiber Reinforced Polymer (FRP). FRP rods are not likely to corrode, which is a major disadvantage of steel. In this technique, FRP reinforcement usually collapses on account of early de-bonding; thus, the full potential of FRP is not efficiently used. This disadvantage of early de-bonding can be overcome through a Near Surface Mounted (NSM) technique. In this technique, grooves are cut into the concrete cover of a Reinforced Concrete (RC) member; FRP rods are then placed in the grooves, which are filled with epoxy adhesive or mortar grout. FRP reinforcement is bonded to concrete, which ensures higher stress transfers between concrete and FRP.

The Externally Bonded (EB) CFRP techniques involve attachment to only one side. Therefore, the efficiency of EP techniques has not been fully achieved on account of the bond failure between the CFRP rods and concrete interface. Near surface mounted CFRP techniques have been developed to minimize these problems and increase the efficiency of CFRP techniques. The NSM technique has three key advantages over the EB CFRP technique. For one, the NSM technique reduces de-bonding from the concrete because the technique employs attachment to three sides. In addition, the NSM technique can be introduced at the pre-stress stage. Furthermore, in this technique, the CFRP rods are protected from damage by the concrete cover.

Even though NSM CFRP shows adequate bond performance in surrounding conditions, the bond between NSM CFRP and concrete under temperature conditions is still a serious problem because of sensitivity of filler to temperature variation. Regardless of the NSM strengthening technique, the behavior of CFRP strengthening systems at various temperatures is an issue because of the polymeric nature of the CFRP matrix and of the epoxy adhesives used to bond it to the concrete substrate. In fact, the glass transition temperatures of epoxy adhesives can be reached in specific applications.

A primary objective of this study is to improve NSM CFRP strengthening techniques with consideration of the bond characteristics under temperature variation (−15 °C~55 °C) and temperature cycles (0, 3, 100). The detailed objectives of this study are to:

1. Investigate the bond behavior of NSM CFRP techniques under temperature cycles (0, 3, 100).
2. Determine the appropriate filling material in the groove under temperature cycles (0, 3, 100).
3. Identify the sensitivity of the filling material according to temperature variation.

2 EXPERIMENTAL EVALUATION

2.1 *Test specimens*

The experimental evaluation consisted of 27 pullout tests on NSM FRP specimens at three types of temperature cycles, as shown in Table 1. Pullout tests are commonly used in the evaluation of bond performance of FRP rods in the concrete. To analyze the bond between FRP rods and filler, three types of tests were implemented. In this experiment, 27 specimens were tested using the shape of the modified pullout test. C-shaped concrete blocks of 300 × 300 mm in external dimensions, with a height of 300 mm, were used. An anchorage system bonded to the loaded end of the FRP rods was used to tie the reinforcement to the Universal Testing Machine (UTM).

2.2 *Specimen properties*

The CFRP rods adopted for the tests were fabricated through preparatory manufacture with oxide coating at surface. The oxide coating bonded at the surface was anticipated to increase

Table 1. Test parameters.

Specimen	Diameter of rod (mm)	Filler type	Water content (%)	Epoxy content	Temperature cycle number
NM-0	10.0	Mortar	14.5	–	0
EP-0	10.0	Epoxy	–	100	0
EM-0	10.0	Epoxy mortar	–	40	0
NM-3	10.0	Mortar	14.5	–	3
EP-3	10.0	Epoxy	–	100	3
EM-3	10.0	Epoxy mortar	–	40	3
NM-100	10.0	Mortar	14.5	–	100
EP-100	10.0	Epoxy	–	100	100
EM-100	10.0	Epoxy mortar	–	40	100

the bond force by increasing the frictional force of the CFRP rod. As mentioned, the experimental evaluation was comprised of pullout tests between concrete blocks and CFRP rods installed according to the NSM technique and bonded with three types of fillers: a non-shrink mortar (series NM), a conventional epoxy putty (series EP), and epoxy mortar (series EM). This filler was chosen for incorporating epoxy mortar because it was expected that its mechanical and bond properties would be less affected by temperature variation compared to the conventional epoxy one. Adhesives with cement presented good performance at elevated temperature (compared to epoxy) in a recent study (Palmieri, 2012).

The type of mortar (series NM) was applied to the specimens using non-shrink mortar in a high strength of 70 MPa as filler with water content of 14.5% and a curing time of approximately 14 days. The type of epoxy putty (series EP) used consisted of primer with epoxy putty. According to the manufacturer, a first layer of fluid "primer" must be applied to ensure good bonding between the epoxy and the concrete; this primer application is to be followed by the epoxy application 18 hours later with a curing time of approximately 7 days. The type of epoxy mortar (series EM) consisted of non-shrink mortar and fluid of epoxy. The properties of epoxy mortar were modified by adding a special epoxy.

2.3 Specimen fabrication

After casting the concrete blocks, the test specimens were prepared according to the following procedures. (1) The slits inside the concrete cover, with dimensions of 40 mm × 30 mm (depth × width), were executed using a diamond slit cutter. (2) The CFRP rod and application of the instrumentation (thermocouples and strain gauges) were clean (with acetone). (3) The CFRP rods were inserted inside the grooves. After the CFRP rod was matched to the central axial, we proceeded to the next step. (4) The fillers were prepared and introduced into the groove. The specimen was then cured and maintained at a constant temperature of 20 °C and a constant humidity of 50% for 28 days (at room condition). After preparation, the zero cycle (NM-0, EP-0, EM-0) specimens were cured at room temperature. The three cycle (NM-3, EP-3, EM-3) and 100 cycle (NM-100, EP-100, EM-100) specimens were exposed to different temperatures ranging from −15 to 55 °C for 20 hours (1 cycle). The experiments performed at the same temperature. This experimental program was established to employ a reasonable range for actual temperature (from −15 to 55 °C) on a lower bridge during construction for NSM reinforcement.

2.4 Test procedure

Pullout tests were performed at room temperature after 0, 3, and 100 cycles (three specimens for each cycle period and for each cycle type). The specimens were tested in pullout using a UTM. A UTM with a load capacity of 980 kN was used to apply the mechanical load. The test rate of speed of 0.021 mm/s until failure corresponded with the pullout test standard of ACI.

3 EXPERIMENTAL RESULTS

Table 2 presents the results of the bond test after the temperature cycles. In the case of Non-shrink Mortar (NM), the temperature cycle variation increased and its bond stress change was minimal. On the contrary, in the case of EP, the bond stress of the specimen with 3-cycle temperature variation decreased by 54.1% more than that of the specimen under no temperature variation; moreover, the bond stress of the specimen with 100-cycle temperature variation decreased by 74.3%. In the case of EP, at the time of 3-cycle temperature variation, detachment of primer occurred. In addition, at the time of 100-cycle temperature variation, primer detachment likewise occurred and bond stress was generated only by the frictional force of filler and concrete. In the case of EM, the bond stress of the specimen with three-cycle temperature variation decreased by 11.6% more than that of the specimen with no temperature variation; moreover, the bond stress of the specimen with 100-cycle temperature variation decreased by 14.8%. Although the bond stress of an EM specimen decreased after

Table 2. Test results.

Specimen	Bond stress (MPa)	Failure mode*
NM-0	10.9	FR int
EP-0	13.3	FR int
EM-0	15.5	FR int
NM-3	10.6	FR int
EP-3	5.0	FC int
EM-3	13.7	FR int
NM-100	11.8	FR int
EP-100	2.8	FC int
EM-100	13.2	FR int

* FR int: failure at the filler-bar interface.
FC int: failure at the filler-concrete interface.

100-cycle temperature variation, it was relatively greater than that of a non-shrink mortar specimen with no temperature variation.

In the test of the bond between the CFRP rod and filler after 3-cycle temperature variation, the largest average bond stress was found in EM, followed by NM, and EP, in the given order. The NM and EM filler specimens as parameters had failure mode (FR int) between the groove filler and rod. The failure mode (FR int) led to sufficient bond stress. In the cases of NM and EM, three specimens showed constant bond stress. In the case of EP, failure mode (FC int) was found between the concrete and filler after three-cycle temperature variation. Each EP specimen had a different bond stress after three-cycle temperature variation. After 100-cycle temperature variation, the largest average bond stress was found in EM, followed by NM, and EP, in the given order. The specimens of the NM and EM fillers had failure mode (FR int) between the groove filler and rod surface. It was determined that, even after the repetition of temperature variation, bond strength between the filler and concrete was relatively greater than that between the CFRP rod and filler. The EM specimens showed a deviation of bond stress after 100-cycle temperature variation. It is judged that the deviation was attributable to a reduction in bond stress of EP filler by 100-cycle temperature. As a result, in terms of failure mode after temperature variation, the specimens had more debris in pullout filler than the specimens without temperature variation and with 3-cycle temperature variation. The EP specimens had failure mode (FC int) between the concrete and filler after 100-cycle temperature variation. This result was attributable to the change in the primer used before epoxy coating, which was caused by temperature variation. After 100-cycle temperature variation, the specimens were completely detached and had a constant bond stress. The complete detachment of concrete and filler was attributable to the fact that temperature variation occurred more than glass transition temperature (50 °C) of primer. In terms of the EP specimens, the used primer was sensitive to temperature variation and thus bond behavior deteriorated. Therefore, it is determined that the specimen is disadvantageous owing to long-term behavior. The EM specimens had good early bond stress; however, the more the cycle of temperature variation increased, the more the bond stress decreased. Nevertheless, after 100-cycle temperature variation, the decreased bond stress of the EM was relatively larger than the NM bond stress. Therefore, it is judged that the specimen has appropriate bond strength even in long-term behavior.

4 CONCLUSION

In this paper, the results of an experimental study on the influence of filler type and number of temperature cycles on the bond behavior of the pullout test were briefly discussed. Based on the test results, the following conclusions can be made:

1. At room condition, the type of filler affects the bond strength and failure modes of the NSM CFRP system. The type of epoxy fillers has a greater influence on bond strength than does mortar type. It was determined that the surface treatment of the CFRP rod differently influences average bond stress depending on filler type.
2. At 3 and 100 cycles, the largest bond stress was found in epoxy mortar, followed by non-shrink mortar, and then epoxy, in the given order. Even if the cycle of temperature variation increased, the non-shrink mortar specimens showed a constant bond stress because mortar has a material property that features no significant change by temperature. With an increase in the cycle of temperature variation, the bond stress of epoxy mortar specimens decreased more. However, after the 100-cycle temperature variation, the epoxy mortar specimens had a relatively larger bond stress than non-shrink mortar specimens. Using epoxy, bond strength with a small number of temperature cycles decreased 50% compared to the case without temperature variation. The temperature change in the primer of epoxy specimens caused detachment, and thus the bond stress of epoxy specimens rapidly decreased.
3. At 3 and 100 cycles, the failure mode of the specimens with epoxy is through pullout of fillers (FC int). This is in contrast to room temperature failure mode, which is through pullout of rod (FR int). The failure mode of the specimens with epoxy mortar and mortar is through pullout of CFRP rod for all cases. Therefore, when considering the FRP strengthening technique to be used in the temperature environment, the selection of a suitable epoxy is very important.

ACKNOWLEDGEMENTS

This research was supported by Basic Science Research Program through the National Research Foundation of Korea (NRF) funded by the Ministry of Science, ICT and future Planning (No. 2014R1A2A2A01005895).

This research was supported by a grant from the Strategic Research Project (Development of Bridge Strengthening Method using Prestressed FRP Composites) funded by the Korea Institute of Construction Technology.

REFERENCES

ACI Committee 440. 2003. Guide for the Design and construction of concrete reinforced with FRP bars (ACI 440.1R-03) American Concrete Institue.

Palmieri, A., Matthys, S., & Taerwe, L. 2012. Experimental investigation on fire endurance of insulated concrete beams strengthened with near surface mounted FRP bar reinforcement. *Composites Part B: Engineering* 43(3): 885–895.

… Resources, Environment and Engineering II – Xie (Ed.)
© 2016 Taylor & Francis Group, London, ISBN 978-1-138-02894-4

Study on the wave maker stroke and the energy dissipation of SPH numerical wave flumes

Yan Xu, Yi Pan & Yongping Chen
Coastal and Offshore Engineering, College of Harbor, Hohai University, China

ABSTRACT: SPH (Smooth Particle Hydrodynamic) method is a relative new numerical solution of computational fluid dynamics. One of the usages of SPH method is to develop numerical wave flumes. However, as mentioned by many researchers, the SPH wave flume has some issues, e.g., on the unphysical wave energy dissipation and the rate of wave maker stroke to the wave height. To get an understanding on the characteristic of a SPH wave flume, a 2-D SPH numerical wave tank is set up. Scenarios with different parameters were simulated, and by analysis the results of the effects of the smooth coefficient, particle size and water depth to the wave energy dissipation, and the rate of wave maker stroke to the wave height were studied and discussed.

1 INTRODUCTION

SPH (Smooth Particle Hydrodynamic) method is a relative new numerical solution of computational fluid dynamics. It has been developed rapidly during the last decade due to its applications in engineering. In the study of Monaghan (Monaghan, 1994), the first attempt was presented to study free-surface flows. Monaghan also studied the behavior of gravity currents (Monaghan, 1996) and solitary waves (Monaghan, 1999). In the study of Cola Rossi (Cola Rossi et al., 2003), SPH method was presented to treat two-dimensional interfacial flows. In the study of Dalrymple (Dalrymple et al., 2004), a single wave generated by a dam break with a tall structure was modeled with a three-dimensional version of SPH method. In the study of Gotoh (Gotoh et al., 2004), SPH method was applied to tracking free surfaces without numerical diffusion.

However, as mentioned by many researchers, the SPH wave flume has some issues, e.g., on the wave energy dissipation and the rate of wave maker stroke to the wave height. To get an understanding on the characteristic of a SPH wave flume, this paper adopted the SPHysics open-source platform based on SPH method, and applied wave damping absorber to create a nonreflecting tank. Scenarios with different parameters were simulated, and by analysis the results of the effects of the smooth coefficient, particle size and water depth to the wave energy dissipation, and the rate of wave maker stroke to the wave height were studied and discussed.

2 NUMERICAL MODEL

2.1 *The basic theory*

In SPH, the fundamental principle is to approximate any function $f(r)$ by

$$f(r) = \int_\Omega f(r)W(r-r',h)dr \qquad (1)$$

where h is the smoothing length, and $W(r-r',h)$ is the weighting function or kernel. This approximation, in discrete notation, leads to the following approximation of the function at a particle (interpolation point) i,

$$f(r_i) = \sum_{j=1}^{N} \frac{m_j}{\rho_j} f(r_j) \cdot W_{ij} \qquad (2)$$

where the summation is over all the particles within the region of a compact support of the kernel function. The mass and density are denoted by m_j and ρ_j respectively and $W_{ij} = W(r_i - r_j, h)$ is the weight function or kernel.

Using Eq. 2 and its derivative, the original partial differential equations, etc. the control equations and boundary conditions can be rewritten as explicit equations. The transition is the basic idea of SPH method.

2.2 *Governing equation*

In this paper, momentum equation and continuity equation are

$$\frac{D\vec{v}}{Dt} = -\frac{1}{\rho}\vec{\nabla}P + \vec{g} + v_a\nabla^2\vec{v} \qquad (3)$$

$$\frac{D\rho}{Dt} = -\rho\nabla\vec{v} \qquad (4)$$

where $\vec{v}$ is the particle movement speed, p is the pressure, v_0 is the kinematic coefficient of viscosity of water. So, in SPH notation, Eq. 3 and Eq. 4 can be written as:

$$\frac{d\vec{v}_i}{dt} = -\sum_{j=1}^{N} m_j \left(\frac{p_j}{\rho_j^2} + \frac{p_i}{\rho_i^2}\right)\vec{\nabla}_i W_{ij} + \vec{g} + \sum_{j=1}^{N} m_j \left[\frac{4v_0 \vec{r}_{ij} \vec{\nabla}_i W_{ij}}{(\rho_a + \rho_b)|\vec{r}_{ij}|^2}\right]\vec{v}_{ij} \qquad (5)$$

$$\frac{d\rho_i}{dt} = \sum_{i=1}^{N} m_j \vec{v}_{ij} \vec{\nabla}_i W_{ij} \qquad (6)$$

2.3 *Wave absorbing module*

In order to avoid wave reflection, wave damping layer was usually placed at the end of the numerical flume to accelerate energy dissipation and make the water surface rapidly restore calm. Based on the wave damping absorber theory of Ni (Ni et al., 2014), this paper employed a two-part wave damping absorber, and modified the particle acceleration compulsively in the damping wave absorber layer to realize rapid absorption and avoid wave reflection. The lateral acceleration correction coefficient r_c of particles i in the damping wave absorber layer is determined by following equations:

$$r_{c1} = \begin{cases} \alpha\cos\left(\pi\frac{x_i - x_L}{x_M - x_L}\right) - \alpha + 1 & \vec{a}_x \cdot \vec{v}_x > 0 \\ 1 & \vec{a}_x \cdot \vec{v}_x \leq 0 \end{cases} \quad x_R < x_i < x_M \qquad (7)$$

$$r_{c2} = \begin{cases} \alpha\cos\left(\pi\frac{x_i - x_M}{x_R - x_M}\right) - \alpha + 1 & \vec{a}_x \cdot \vec{v}_x > 0 \\ 1 & \vec{a}_x \cdot \vec{v}_x \leq 0 \end{cases} \quad x_M < x_i < x_L \qquad (8)$$

where x_L, x_M, and x_R are the coordinates of left border, middle point and right border in the wave absorbing layer of the coordinates, respectively; x_i, $\vec{a}_x$ and $\vec{v}_x$ are particle coordinates, lateral acceleration component and lateral velocity component, respectively; and α represents adjustable parameter with an optimal value of 0.5 in this study. The longitudinal acceleration correction method is the same as the lateral acceleration.

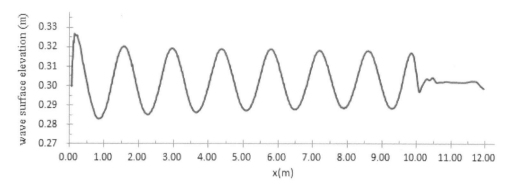

Figure 1. Wave surface in case D5 when t = 19.1T.

Figure 1 illustrates, the wave surface generated in the 12-meters-long numerical tank at t = 19.1T, with wave absorbing layer of 3 m and particle spacing of 0.012 m. As it can be seen, the effect of wave absorber is significant. When the waves propagate to the absorbing layer, it is attenuated rapidly.

3 RESULTS AND DISCUSSION

To analyze the impact of smooth coefficient, particle precision and water depth on wave maker stroke and wave energy dissipation, numerical simulation under several working conditions (shown in Table 1) were conducted. The length of the numerical tank is 12 m, and the height is 0.5 m. In addition, in order to study the wave energy dissipation, a parameter ε is defined as the rate of average wave height of $H_{7.5}$ (7.5 m from the left) to $H_{2.5}$ (2.5 m from the left).

3.1 Wave maker stroke

In this paper, the waves in the numerical flume are generated by a piston wave maker. According to the linear wave theory, the movement velocity U from wave maker can be represented by the following equation:

$$U = \frac{\omega}{W}\eta \qquad (9)$$

where ω is the circular frequency of the simulated regular wave, η abides by the wave surface equation of regular wave, $\eta = a\cos(\omega t)$ based on small amplitude wave theory. W is the function of hydraulic transmission, which represents the ratio of wave amplitude and the stroke of wave maker and can be calculated by the following equation:

$$W = \frac{2(\cosh 2k_p h_w - 1)}{\sinh 2k_p h_w + 2k_p h_w} \qquad (10)$$

where k_p is wave number and h_w is water depth.

The horizontal displacement of one time step of wave pushing plate is as follows:

$$\Delta x = U \Delta t \qquad (11)$$

However, the ratio is different in the SPH wave flume compared to that in the physical flume. In this section several cases were run to study the feature of the ratio of wave amplitude and the stroke. In all the cases, the stroke of the wave maker was 0.02 m, and the period is 1 s.

Table 1. Cases and result.

Condition	Smoothed coefficient	Particle spacing (m)	ε	H_0 (m)	H_x (m)	Depth (m)	v_0 (m²/s)
S0	0.92	0.010	0.91	0.056	0.0411	0.30	1.0×10^{-6}
D1	0.87	0.010	0.59	0.056	0.0374	0.30	1.0×10^{-6}
D2	0.97	0.010	0.60	0.056	0.0356	0.30	1.0×10^{-6}
D3	1.02	0.010	0.56	0.056	0.0342	0.30	1.0×10^{-6}
D4	0.92	0.008	0.79	0.056	0.0411	0.30	1.0×10^{-6}
D5	0.92	0.012	0.96	0.056	0.0375	0.30	1.0×10^{-6}
D6	0.92	0.014	1.05	0.056	0.0351	0.30	1.0×10^{-6}
W7	0.92	0.010	1.02	0.052	0.0362	0.25	1.0×10^{-6}
W8	0.92	0.010	0.91	0.060	0.0452	0.35	1.0×10^{-6}
W9	0.92	0.010	0.88	0.048	0.0304	0.20	1.0×10^{-6}

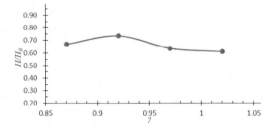

Figure 2. H/H_0 with different smooth coefficient.

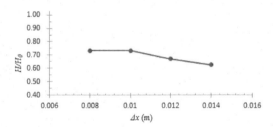

Figure 3. H/H_0 with different particle spacing.

3.1.1 Smooth coefficient

According to SPH method, the field function of particle i is a weighted average value determined by the particles present around. The radius of support region for particle i is kh, where k is defined as scale factor and can be a constant value of 2 in cubic spline function. The smoothing length is $h = 1.41\gamma\Delta x$, in which γ is smooth coefficient, representing the scale of smoothing function. Using Eq. (10), if the stroke of wave maker is known, the theoretical value of wave height can be calculated. As presented in Figure 2, when γ was 0.92, the real value of wave height could reach 73.4% of the theoretical value; when γ is less than 0.92, while only 66.8% (or 63%) of the theoretical value were obtained as γ was less (or greater) than 0.92. It is likely because when h is small, the support domain will not have enough particles acting on the given particle, which will lead to large error between the results. When h is large, the support domain will have too many particles acting on the given particle, part information or local features of the given particle will be lost, which can negatively affect the accuracy of the results as well.

3.1.2 *Particle precision*

Due to the limitation of computer hardware and SPH algorithm, 0.01 m spacing of particles is a common accuracy selection. As exhibited in Figure 2, with the same wave making stroke, when the particle spacing was greater than 0.01 m, the real wave height is smaller; when the particle spacing was 0.008 m or 0.01 m, the real wave heights approached the theoretical ones. The phenomenon can be explained for that, when particle spacing excesses the critical value, there are not enough particles to support the free surface, which will lead to the real wave height less than the theoretical one. On contrary, when particle spacing is less than the critical value, particle dependence gets smaller, that promising the wave deviation without further improvement.

3.1.3 *Water depth*

Using Eq. (10), when the actual wave height is known, the equivalent stroke of wave maker can be calculated. According to the results of condition S0, W1, W2, and W3, to achieve the same wave height, the relation between equivalent stroke W and theoretical stroke W_0 can be shown in Figures 4 and 5. Figure 4 illustrates that, the modeling result of SPH method is generally smaller than theoretical values, and thus, the model needs to overvalue the stroke of wave maker to achieve the theoretical wave height. Figure 5 show how the needed extent of overvaluation changes with water depth. It can be clearly seen that, the ratio H/H_0 (i.e. W/W_0) increases with water depth, but the increasing rate gradually decreases, which means that, it is less difficult to achieve the theoretical with larger water depth.

3.2 Wave energy dissipation

In the process of wave transmission, if there is no energy loss, the waves will keep the original waveform forward movement. However, in SPH method hypothesis of theoretical incompressible fluid is compressible in fact. This caused the particles oscillating, lead to energy dissipation. This will result in particle oscillation, and lead to energy dissipation. In this paper, we discuss wave dissipation in terms of the support region and the particle accuracy.

3.2.1 *Smooth coefficient*

According to the results of condition S0, R1, R2, and R3, the impact of smooth coefficient on wave dissipation can be depicted in Figure 6. When the smooth coefficient was less than 0.92, the support domain did not have enough particles to calculate, as a result, ε is only 0.41. When smooth coefficient became greater than 0.92, due to excessive particles in support domain, the identity of particles turned out obvious (ε is only about 0.40). When the smooth coefficient was set as 0.92, the particles in the support domain precisely avoided the above two cases (ε is as high as 0.91).

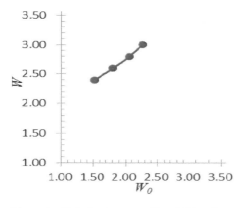
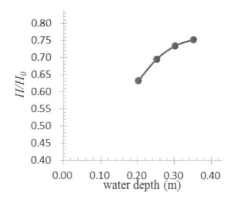

Figure 4. Relation between W and W_0 at $H_{2.5}$.   Figure 5. H/H_0 of different water depth at $x = 2.5$ m.

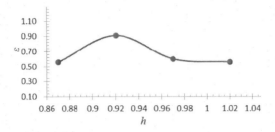

Figure 6. ε with different smooth coefficient.

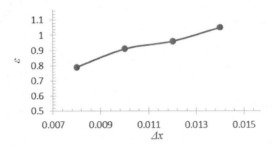

Figure 7. ε with different particle spacing.

3.2.2 *Particle precision*

In any numerical simulation method, precision of particle (or grid precision) will greatly affect the results and efficiency of calculation. In this paper, conditions were set up with water depth of 0.3 m. The particle spacing was 0.008 m, 0.01 m, 0.012, m and 0.014 m, respectively. As shown in Figure 7, when the particle spacing was 0.008 m, the wave dissipation reached 20%. With the increase in particle spacing, the dissipation dropped. As the particle spacing became greater than 0.012 m, the accuracy of the model declined due to a bad particles precision, but the wave height increased forward and could even excess the value of one.

4   CONCLUSIONS

In this paper, a 2-D SPH numerical wave tank was set up. Scenarios with different parameters were simulated, and by analysis the results of the effects of the smooth coefficient, particle size and water depth to the wave energy dissipation, and the rate of wave maker stroke to the wave height were studied and discussed. Conclusions are given as follows:

The ratio of wave amplitude to the stroke of wave maker can be influenced by the smooth coefficient, the particle size and the water depth. The effect of the smooth coefficient is relative small, however, when the smooth coefficient is 0.92 the ratios is the largest. Small particle size would lead to a larger ratio of wave amplitude to the stroke of wave maker. When the particle size is smaller than a certain value (e.g., 0.01 m in this paper), the effect is insignificant. The ratio of the SPH W to the theoretical W for the physical flume increases with water depth, but the increasing rate gradually decreases.

The wave dissipation in the SPH flume can be influenced by the smooth coefficient and the particle size. The influence of the smooth coefficient is significant, and under an optimal value of 0.92; the unphysical energy dissipation is the most insignificant. The effects of the particle size is relatively small and larger particle size has smaller dissipation, which can be explained by the definition of the parameter ε *which* is not quite appropriate. Larger particle size leads to smaller $H_{2.5}$, so the energy dissipation seems smaller, on which future study is still needed.

REFERENCES

Colagrossi, A. & A. Souto-Iglesias. 2013. *Smoothed-particle-hydrodynamics modeling of dissipation mechanisms in gravity waves.* Physical Review E 87(2): 46–56.

Colagrossi, A. & Landrini, M. 2003. *Numerical simulation of interfacial flows by smoothed particle hydrodynamics.* Journal of Computational Physics 191(2): 448–475.

Gómez-Gesteira, M. & Dalrymple, R.A. 2004. *Using a 3D SPH method for wave impact on a tall structure.* J. Waterw. Port Coast. Ocean Engineering 130 (2): 63–69.

Gotoh, H. & Shao, S. 2004. *SPH-LES model for numerical investigation of wave interaction with partially immersed breakwater.* Coastal Engineering 46 (1): 39–63.

Madsen, P.A. & H.A. Schäffer. 2006. *A Discussion of Artificial Compressibility.* Coastal Engineering 53(1): 93–98.

Molteni, D. & R. Grammauta. 2013. *Simple Absorbing layer conditions for shallow wave simulations with Smoothed Particle Hydrodynamics.* Ocean Engineering 62(4): 78–90.

Monaghan, J.J. 1994. *Simulating free surface flows with SPH*, Journal of Computational Physics 110: 399–406.

Monaghan, J.J. 1996. *Gravity currents and solitary waves.* Physica D 98: 523–533.

Monaghan, J.J. & Kos, A. 1999. *Solitary waves on a Cretan beach.* Journal of Waterway Port Coastal and Ocean Engineering 125 (3): 145–154.

Ostad, H. & S. Mohammadi. 2009. *Analysis of shock wave reflection from fixed and moving boundaries using a stabilized particle method.* Particuology 7(5): 373–383.

Xingye Ni, Weibing Feng. 2014. *Numerical wave tank based on DualSPHysics.* Journal of Waterway and Harbor, 35(2): 105–111.

Experimental study on the hydraulic performance of baffle-drop shafts

Z.G. Wang, D. Zhang, H.W. Zhang & R. Zhang
China Institute of Water Resources and Hydropower Research, Beijing, China

ABSTRACT: The baffle-drop shaft was one kind of the important hydraulic structures which were designed to convey flow vertically downward. In this paper, the hydraulic performances of the baffle-drop shafts were tested with the help of physical model tests, and the results showed that: baffle spacing was the key parameter in shaft designing. Too large baffle spacing might lead to S-shaped sweeping motion which was harmful to the structural safety and longevity, while too small values would reduce the flow conveying capacity. The baffle spacing would be appropriate only when the corresponding flow patterns in the shaft were cascade drops for its good performances on energy dissipation. In addition, the dynamic pressure measuring data indicated that the pressure characteristics of the cascade drops were also rational. Therefore, the baffle spacing should be decided to help to form cascade drops in baffle-drop shafts.

1 INTRODUCTION

The baffle-drop shaft is one typical design of drop structures. It's always of the cylindrical shape and has a dividing wall in it. In most cases, the dividing wall is centrally placed and would divide the shaft equally into two zones: the dry zone and the wet zone. The dry zone is designed to improve the air conditions and to provide access for maintenance. However, the wet zone is kept for water conveyance. In order to make the flow much more stable, variable number of baffles are commonly employed in the wet zone and arranged in a staggered form. The detailed structure of the baffle-drop structure is shown in Figure 1, in which, R is the inner radius of the shaft; B is the width of the baffle; h is the baffle spacing; H is the total depth of the shaft, l is the space between two pressure measuring points.

The baffle-drop shaft was first applied in the year of 1911 in Cleveland, Ohio, and it was mostly used in the sewer projects (Margevicius et al. 2009). However, as the concerns

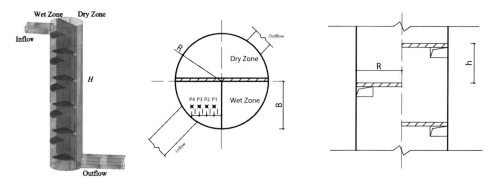

Figure 1. Sketch of the baffle-drop shaft: left figure showing the 3D view of the baffle-drop shaft; middle figure showing the plan view of the shaft; right figure showing the elevation of the shaft.

of the hydraulic engineers were focused on the flow conveying capacity and the design discharges were always small (mostly smaller than 5.0 m³/s), the design of the baffle-drop shafts were commonly based on the engineering experiences and little attention was paid to the flow details of the shaft. Thereafter, the understanding of the hydraulic characteristics of the baffle-drop shaft was nearly not improved a little for a long time. Recently, due to the influences of storm-induced floods, the design discharges of the shafts were getting larger and larger, i.e. Regional Municipality of York, Canada had employed a baffle-drop shaft with the design discharge of 16.2 m³/s (Odgaard et al. 2013b); Donghao Creek Deep Tunnel Project of Guangzhou, China prepared to construct a shaft to convey as large flow as 31.0 m³/s. Clearly, the designing of the large-discharge shafts needs a deep understanding of the detailed hydraulics to ensure its rationality. Although Odgaard et al. (2013a) had conducted some relative researches and found that two alternatives of flow patterns, namely S-shaped sweeping motion and cascade drops, might appear in the shafts, the knowledge of hydraulic characteristics of baffle-drop shafts were still not enough and systematic research were strongly needed.

In this paper, a physical model would be employed to test the hydraulic performance of the baffle-drop shafts. The flow patterns would be visual observed, and their corresponding working conditions would be analyzed as well. Besides, the dynamic pressures on baffles were measured in any of the working conditions. Accordingly, the determining principles of baffle spacing would be proposed, which could be used to facilitate the application of baffle-drop shafts.

2 PHYSICAL MODEL ARRANGEMENT

2.1 Model design and fabrication

According to the court size, the physical model was constructed with the shaft depth H being 2.14 m and the inner radius R being 50 cm. The dividing wall was centrally placed and the shaft was equally divided into the dry zone and the wet zone. The flow was supplied by an iron pipe of Φ100 mm, in which a butterfly valve and an electromagnetic flow meter were mounted to control and show the inflow discharge separately. The tail water was let out to underground tank directly through an iron pipe of Φ350 mm, in which another butterfly valve was installed to control the water elevation in the shaft. To be added, the shaft was made of Plexiglas in order to make the visual observation of flow patterns easily.

In addition, some typical shafts were chosen to test the pressure characteristics. In general, four pressure measuring points were mounted on each chosen baffle and arranged along the perpendicular bisector of the baffle edge. They were always coded in the order from 1 to 4, and the details were shown in Figure 1. The dynamic pressures were monitored by pressure sensors which covered a measuring range of 4 m. The data collecting were accomplished using Data Collecting System DJ800 which was developed by China Institute of Water Resources and Hydropower Research (IWHR).

2.2 Tested working cases

Define dimensionless baffle spacing h^* and dimensionless flow rate Q^* as Equation 1 and Equation 2

$$h^* = h/R \tag{1}$$
$$Q^* = Q/\sqrt{gR^5} \tag{2}$$

where, Q = flow rate; g = gravitational acceleration. Clearly, h^* and Q^* could well reflect the characteristics about body shape and conveying flows of baffle-drop shafts. According to the present application of the baffle-drop shafts, it was found that the dimensionless baffle spacing h^* was mostly lying in the range of 0.40~0.85, while the dimensionless flow rate Q^*

Table 1. Baffle numbers and pressure measuring arrangements at different shaft designs.

	Baffle spacing h/cm	Dimensionless baffle spacing h^*	Quantity of baffles	No. of the baffle to measure pressures
Type I	20.67	0.83	10	2rd, 7th, 8th
Type II	12.40	0.50	16	4th, 9th, 10th
Type III	10.80	0.43	18	2th, 10th, 11th

was mainly varying between 0.020 and 0.060. Accordingly, three typical baffle-drop shafts were designed to analyze the hydraulic performances. The tested flow rate was in the range of 2.0 L/s~6.0 L/s, with its corresponding dimensionless flow rate being 0.020~0.056. Besides, as the baffle spacings were different from one another, thus the number of baffles employed was not unique and the typical baffles which were chosen to measure the dynamic pressures were not the same. The details were shown in Table 1.

3 RESULTS AND DISCUSSIONS

3.1 *Flow patterns*

According to the tested results of Type I, it was found that the flow patterns under different flow conditions were similar, thereafter the results of Type I conveying the flow rate of 5.5 L/s was chosen as the example to analyze the flow patterns. Clearly, after the water flowing into the shaft, the cascade drops were soon formed. However, as the flow energy was not fully dissipated, thus the kinetic energy of the flow would get larger and larger as the flow was conveying down. Meanwhile, the velocities exiting the baffles would also get greater and greater, and eventually the flow nappe would strike the shaft wall directly instead of the next baffle, indicating the S-shaped sweeping motion was formed. This kind of flow pattern furtherly reduced the energy dissipation efficiency, which was harmful to the safety and longevity of the baffle-drop shaft. Figure 2 showed the typical flow pattern of Type I conveying a flow rate of 5.5 L/s.

The hydraulic tested results of Type II of baffle-drop shaft showed that no energy accumulation was observed on the baffles no matter what the flow rates were, including both small flow rate of 2.0 L/s and large flow rate of 6.0 L/s. The flow patterns in the baffle-drop shaft were of stable cascade drops, the velocities exiting baffles remaining the same and the horizontal length of jet nappes keeping unchanged as well. The flow details were shown in Figure 3. After testing the hydraulic performance of Type III of baffle-drop shaft, the stable cascade drops were also formed, similarly with the flow patterns of Type II. The details of the flow characteristics of Type III were also shown in Figure 3.

To deeply analyze the flow patterns of the baffle-drop shaft, the backwater characteristics at the joint of the baffle, the shaft wall and the dividing wall was tested and measured, with the results shown in Table 2. As the flow patterns in Type I was S-shaped sweeping motion and no stable backwater was formed, thus only the results for Types II and III were given.

According to Table 2, it could be seen that when the flow rate Q was between 2.0 L/s and 5.5 L/s, namely dimensionless flow rate Q^* was lying in 0.020~0.056, stable hydraulic backwater structures could be formed from the middle of the shafts. Through further analysis, it was thought that the upper part of the shaft was used to regulate the flow, while the lower part was used to convey the flow. When it arrived to the approximate middle part of the shaft after being adjusted, the flow patterns tended to be stable, and the energy released by the gravitational potential energy could be dissipated by the cascade drops.

In all, the S-shaped sweeping motion was formed in Type I of baffle-drop shaft, as the baffle spacing was too large, making the energy dissipation rate on the baffles being not enough and the kinetic energy of the flow getting larger and larger. Thus the baffle spacing should not be evaluated too large. Comparatively, the flows in the Types II and III were of the type

Figure 2. Flow patterns of Type I of baffle-drop shaft conveying a flow rate of 5.5 L/s.

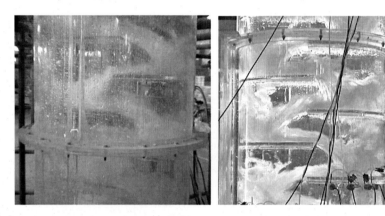

Figure 3. Flow patterns of baffle-drop shaft conveying a flow rate of 5.5 L/s: left figure for Type II and right figure for Type III.

Table 2. Maximum backwater depth and it's first appearing position for Types II & III.

	Flow rate/L/s	Dimensionless flow rate	Max. backwater depth/cm	Position for first appearing
Type II	2.0	0.020	7.0	9th
	4.0	0.041	11.0	8th
	5.5	0.056	13.0	7th
Type III	2.0	0.020	6.5	7th
	4.0	0.041	10.5	9th
	5.5	0.056	13.5	9th

of stable cascade drops, no matter what the flow rates were. After comparing the backwater depth of both types II and III, it was found that the smaller baffle spacing could reduce the backwater depth and corresponding flow velocity exiting the baffles, which was good for maintaining the safety and longevity of the shaft structure. Furthermore, the flow rate of 8.0 L/s was tested to verify the flow conveying capacity of the shafts, and the results showed that the flow patterns had become irregular and surge happened intermittently, namely too small baffle spacing would limit the flow capacity of the shafts. Therefore, the values of the baffle spacing should be neither too big nor too small. Only the proper baffle spacing could help to form stable cascade drops.

3.2 *Dynamic pressure*

The dynamic pressure characteristics could reflect the flow patterns from some angle, and could reflect the influence of the flow on the hydraulic structures, thus the dynamic pressures were tested for anyone of the three types of baffle-drop shafts. Without loss of generality, the pressure characteristics of the flow rate of 5.5 L/s were chosen as the example to make a further analysis. The details of the pressure tested results were shown in Figure 4–6.

According to the tested results shown in Figure 4, it was found that the average pressures on the 7th and 8th baffles in Type I of baffle-drop shaft were small, with the maximum value being 4.00 kPa; meanwhile, the fluctuating pressures were also small and the maximum value

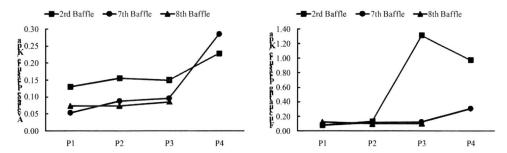

Figure 4. Pressure distribution on the baffles of shafts in Type I.

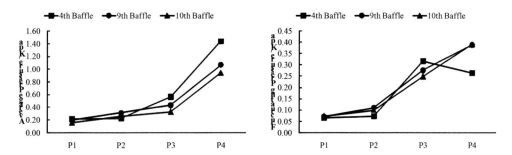

Figure 5. Pressure distribution on the baffles of shafts in Type II.

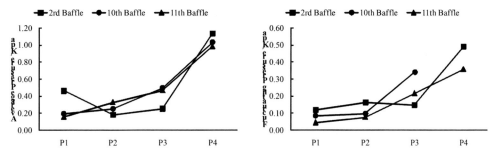

Figure 6. Pressure distribution on the baffles of shafts in Type III.

was smaller than 5.00 kPa. Besides, the dynamic pressures, no matter average pressure or fluctuating pressure, were changed in a small range as the position was changing. The pressure characteristics of Type I showed that the flow pattern of stable cascade drops was not formed, and the strong turbulent energy dissipating flow structure was not appeared either. The energy dissipation rate was relatively low.

The tested results of Figure 5 showed that: in Type II, the pressure distributions kept the same on the 9th baffle and 10th baffle. The average pressures increased from the shaft center to the shaft wall, with the maximum value approximating 15.00 kPa and the minimum value approximating 4.00 kPa. It agreed well to the flow patterns of cascade drops. In addition, the fluctuating pressure presented a similar distribution. The maximum value was about 6.00 kPa and appeared around the shaft wall, showing that the flow turbulence at the shaft wall was strong and the energy dissipation rate was high. The tested results of Type III showed that the pressure characteristics were nearly the same as those of Type II, no matter the pressure distribution or the pressure values. The pressure characteristics of both Type II and Type III reflected that after the regulation of the upper baffles, the stable cascade drops had been formed at about the middle part of the shaft.

4 CONCLUSIONS

Baffle-drop shaft is a typical hydraulic structure used for conveying flows vertically downward. As it could well adapt various of hydraulic conditions, such as unsteady inflow, multi-angle inflows, multi-elevation inflows, thus it will enjoy a good application prospect. In this paper, with the help of the physical model tests, the hydraulic performances, including flow patterns and pressure characteristics, were studied, and the main conclusions were shown as follows:

i. Too big baffle spacing would lead to the formation of S-shaped sweeping motion. This kind of flow pattern had bad energy dissipation performance and unreasonable pressure distribution. As a result, the velocity exiting the baffles would be getting larger and the flow at the bottom would be getting wavy violently. Thus S-shaped flow pattern should be avoided.
ii. Reducing baffle spacing was good for improving the flow pattern. However, too small baffle spacing might limit the flow conveying capacity.
iii. Appropriate baffle spacing should make the flow pattern to be of the stable cascade drops. In this flow pattern, the energy released by the gravitational potential energy of the flow could be fully dissipated by the corresponding drops, thus it would benefit for avoiding the velocity in the shaft getting excessively larger.

ACKNOWLEDGEMENT

The authors gratefully acknowledge the financial support of the special program for the scientific research of China Institute of Water resources and Hydropower Research (Grant No. HY0145B16201500000).

REFERENCES

Margevicius, A., Schreiber, A., Switalski, R., Lyons, T., Benton, S., and Glovick, S. 2009. A baffling solution to a complex problem involving sewage drop structures. *33rd IAHR Congress: Water Engineering for a Sustainable Environment*. IAHR, Madrid, Spain.

Odgaard, A.J., Lyons, T.C., and Craig, A.J. 2013a. Baffle-drop structure design relationships. *Journal of Hydraulic Engineering*, 139(9): 995–1002.

Odgaard, A.J., Lyons, T.C., Craig, A.J., Margevicius, A. and Servidio, D. 2013b. Baffle drop structures for division of storm water to underground tunnels. *35th IAHR Congress: Water Engineering and Civilization*. IAHR, Chengdu, China.

Residual bearing capacity of stud connectors for steel-concrete composite beam bridges subjected to fatigue damage

Xue-liang Rong
School of Civil Engineering, Shijiazhuang Tiedao University, Shijiazhuang, China

Chong-fa Song
China Road and Bridge Corporation, Beijing, China

Pin Zhao
School of Civil Engineering, Shijiazhuang Tiedao University, Shijiazhuang, China

ABSTRACT: This work tested and studied the degradation of ultimate bearing capacity under cyclic fatigue load at the junctions of a steel-concrete composite beam bridge. A fatigue life model of stud connectors was established based on fracture mechanics, taking the initial defects of components into consideration. A method to identify the relationship between residual shear bearing capacity and times of cyclic loading with constant amplitude was proposed based on parameter fitting from experimental data. Parameter analysis for various diameter, initial defect and amplitude of shear stress was conducted by using this method. The results showed that the degradation degree of the shear bearing capacity for the stud with a larger diameter is greater than that of the stud with a smaller diameter. When the initial defect rate of the stud is greater than 0.1 and the amplitude of shear stress is greater than 100 MPa, shear bearing declines rapidly and needs to be controlled during engineering design.

1 INTRODUCTION

Stud connector is widely used in steel-concrete composite beam bridges. Since the 1950s, researchers have extensively studied the static and dynamic performance of studied connector and have established a sophisticated calculation system which laid the theoretical foundation for its application. However, most of the previous research considered the static bearing capacity of stud and fatigue life separately. There is either no research on the degradation of shear bearing capacity of stud caused by cumulative fatigue damage reported. So, the bearing capacity and evaluating service life of existing stud connectors is unable to be estimated based on cumulative traffic flow or fatigue load spectrum. As the stud is embedded in concrete, regular examination and repairment of stud is impossible. Once studs crack, it may cause sudden damage of the structure. Therefore, connector, the major components of steel-concrete composite structure, needs to be assessed. And research on the inspection method of bearing capacity and prediction of residual life are necessary.

The degradation of static bearing capacity under cyclic loading for stud connectors commonly used in composite beams is discussed. Its variation with increasing times of fatigue loading and its effects on structural usage and safety throughout the structure's design life are also researched. The results may provide a scientific basis for the assessment of the stud connectors of composite beams and the estimation of residual life.

2 FATIGUE PERFORMANCE OF STUD BASED ON FRACTURE MECHANICS

Fracture mechanics is a powerful tool for studying materials with initial defects as well as the strength of the structures. It can be used to evaluate the effects of initial defects on fatigue life quantitatively. It can also be used to identify residual life and inspection cycle from structural damages based on the size and growth rate of cracks. According to fracture mechanics, the growth rate of cracks da/dN is a function of the amplitude of stress intensity ΔK, which is represented as an S-shaped curve. This curve can be divided into three zones (J.S Guo and J.Z. Sun 1999).

1. Zone I—no growth in this zone
   In this zone, $\Delta K \leq \Delta K_{th}$. ΔK_{th} is the threshold of fatigue crack growth, Because the amplitude of stress intensity at structural defect is less than this value, the crack will not grow, which means that the crack is secure.
2. Zone II—crack growth in this zone
   Crack growth rate in this zone satisfies the PARIS formula, i.e., Eqn. 1.

$$\frac{da}{dN} = C(\Delta K)^m \quad (1)$$

where a is the crack length (mm); N is the number of cyclic load; ΔK is the amplitude of stress intensity factor; $\Delta K = K_{max} - K_{min}$; C, m are the basic parameters to describe the performance of fatigue crack growth, which are determined through experiments.

3. Zone III—crack grows fast in this zone
   Crack grows very fast in this zone. Therefore, their lifetime is not included (it is not included in the lifetime)

This paper mainly studies the fatigue performance of stud connectors in Zone II. According to fracture mechanics, the estimation of fatigue life is mainly determined by the calculation of stress intensity factor K, which is affected by material, physical dimensions, welding performance, and surrounding circumstances. Deducing the analytical solution of stress intensity factor K accurately is difficult, and formula construction itself is also very complex. This paper focuses on the interaction between static bearing capacity and fatigue loading within the lifetime of the stud connector. Therefore, the detailed solution of stress intensity factor is not discussed in this paper. Stress intensity factor K is calculated in accordance with general formula, i.e., Eqn. 2.

$$K = F\sigma\sqrt{\pi a} \quad (2)$$

where F is the geometric correction factor, which presents the effects of physical dimensions of the component and crack on the stress field of the crack tip (determined by experiments); σ is the generalized load value (normal stress near the crack tip or shear stress); a is the characteristic length of the fatigue crack.

Stud connectors in steel-concrete structure mainly bear shear load. Hence, σ is the shear stress which the stud can withstand. The amplitude of stress intensity factor ΔK can be expressed by Eqn. 3.

$$\Delta K = F\sqrt{\pi}\Delta\tau a^{0.5} \quad (3)$$

where $\Delta \tau$ is the amplitude of nominal shear stress, i.e., amplitude of shear stress at the initial stage of fatigue loading.

Then, integral transform for Eqns. 1 to 3 may be performed, and the number of cyclic loadings N_c when the crack size from a_0 to a_c may be obtained; see Eqn. 4 below.

$$\int_{a_0}^{a_c} a^{-0.5m}\,da = C(F\sqrt{\pi})^m \Delta\tau^m N_c \quad (4)$$

Let $M = 1-0.5m$, $N = CM(F\sqrt{\pi})^{2-2M}$. Then, Eqn. 4 can be expressed as:

$$a_c^M - a_0^M = N\Delta\tau^{2-2M} N_c \tag{5}$$

where a_c is the characteristic length of the cross-sectional crack after N_c times of cyclic loading, which can be approximately expressed as Figure 2.

A_c denotes residual net cross-sectional area with fatigue crack length of a_c after N_c times of cyclic loading. A denotes the initial cross-sectional area of the stud. Therefore, the relationship between A_c and A can be approximately expressed as Eqn. 6 (Y.H. Wang 2009).

$$\frac{A_c}{A} = 1 - \frac{a_c}{d} \tag{6}$$

Then, the solution is

$$a_c = \left(1 - \frac{A_c}{A}\right) d \tag{7}$$

We assume that stud connector will fail as soon as the load on it reaches $P_{u,c}$, which is its static ultimate bearing capacity after N_c times of cyclic loading. Our experimental results show that $P_{u,c}$ can be calculated approximately according to Eqn. 8.

$$P_{u,c} = A_c f_u \tag{8}$$

where f_u is the ultimate strength of the steel stud
From Eqns. 7 and 8, we obtain

$$a_c = \left(1 - \frac{P_{u,c}}{A f_u}\right) d \tag{9}$$

Based on Eqn. 9, Eqn. 5 can then be written as:

$$\left(1 - \frac{P_{u,c}}{A f_u}\right)^M d^M - a_0^M = N\Delta\tau^{2-2M} N_c \tag{10}$$

Then, we obtain:

$$\left(1 - \frac{P_{u,c}}{A f_u}\right)^M = \frac{N}{d^M}\Delta\tau^{2-2M} N_c + \left(\frac{a_0}{d}\right)^M \tag{11}$$

where N_c is the number of cyclic loading with constant amplitude;
$P_{u,c}$ is the residual static ultimate bearing capacity after N_c times of cyclic loading;
$\Delta\tau$ is the amplitude of shear stress for stud (MPa);
f_u is the ultimate strength of steel stud (MPa);
a_0 is the initial crack length of stud (mm);
d, A is the cross-sectional diameter (mm) and area (mm²), respectively;
N, M are the constant parameters related to material, size, shape, etc of stud, which can be defined from a large number of experimental data fittings.

The effects of cumulative fatigue damage are considered in Eqn. 11, which indicates that $P_{u,c}$ is the residual static ultimate bearing capacity after N_c times of constant cyclic loading. The effects of initial defects, material, physical dimensions, and loading parameters are also taken into account.

3 ESTIMATION OF RESIDUAL SHEAR BEARING CAPACITY OF STUD

Eqn. 11 can be used to solve the following problems:

1. To determine the fatigue life and remaining service life of the stud connector

Based on existing research and the results of many fatigue experiments, the fatigue life of the stud is defined as: After N_e times of cyclic loading with load peak P_{max} and amplitude of fatigue load ΔP, fatigue life is reached when loading with peak load P_{max} cannot be done; during the next loading cycle, the stud will be under monotonic loading until destruction occurs. Thus, the fatigue life of is N_e. This definition indicates that the residual ultimate bearing capacity is equal to the peak fatigue load P_{max} when stud suffers fatigue failure. In other words, the ultimate bearing capacity $P_{u,c} = P_{max}$ when the times of cyclic loading N_c is equal to its fatigue life N_e. This relationship is substituted into Eqn. 11 that has

$$\left(1 - \frac{P_{max}}{Af_u}\right)^M = \frac{N}{d^M} \Delta \tau^{2-2M} N_e + \left(\frac{a_0}{d}\right)^M \qquad (12)$$

Eqn. 12 shows that the fatigue life of stud N_e is not only related to the amplitude of fatigue shear stress $\Delta \tau$ but is also affected by fatigue loading peak P_{max}, the physical dimensions of stud, material, initial defects a_0/d, and many other factors.

The undetermined parameters N and M in Eqn. 12 are generally determined through data fitting based on a large number of fatigue experiments. Extensive fatigue life and strength studies as well as the results of fatigue performance tests are available. Therefore, we adopted existing experimental data for parameter fitting. A total of 205 specimens of fatigue tests were collected from literature (Roger G. 1966; Lee P.G. et al. 2001,2005; Gerhard Hanswille et al. 2000,2007; Scott A. 2003; Ahn J.H. et al. 2007; J.G. Nie 2005; En Xie 2011). In view of different experimental designs, the specimens were screened based on the following principles: 1) the physical dimensions of specimens were consistent with Eurocode 4; 2) the concrete flange was installed with a two-way steel mesh; 3) one-way loading has a frequency in the range of 3 Hz to 5 Hz. Specimens that satisfy the above conditions were sorted, and specimens with larger errors were removed according to Chauvenet's criterion for rejection. As a result, 102 qualified specimens were selected.

Parameters N and M were determined through parameter fitting based on fatigue test data collected from the 102 specimens. Given the complexity of the formula, 1stOpt (1.5 Pro), a common mathematical optimization analysis program, was used for parameter fitting.

N and M are substituted into Eqn. 12 to estimate the fatigue life of the stud connector and the residual fatigue life. **Determining the static ultimate bearing capacity of stud connector and effects of fatigue loading**

The fatigue lifetime of the stud connector can be estimated from Eqn. 11. Parameters M and N were determined through analysis and parameter fitting from a large number of fatigue test data. Then, the following was obtained:

$$\left(1 - \frac{P_{u,c}}{Af_u}\right)^{-1.05} = -\frac{-6.19 \times 10^{-16}}{d^{-1.05}} \Delta \tau^{4.1} N_c + \left(\frac{a_0}{d}\right)^{-1.05} \qquad (13)$$

Eqn. 13 can be used to estimate the degradation of static ultimate load $P_{u,c}$ after the stud connector underwent N_c times of cyclic loading.

Fatigue resistance tests for corroded stud connectors were completed. One non-corroded piece was included in the test. After 2 million times of cyclic loading, the specimen did not break. Then, static monatomic loading test was conducted to record the loading value of destruction. Experimental data from these tests were used to verify the accuracy of Eqn. 13.

Experimental data given in Table 1 and calculation results from Eqn. 13 were compared; the initial crack length a_0 was not available. Generally, a_0 of welding structure can be determined from observation or non-destructive testing techniques. When naked eye monitoring cannot detect any defects but information confirms that weld quality reaches Class II faults, a_0 can be assumed as 2 mm (Y.H Wang and J.G. Nie 2009). Here, a_0 is temporarily given as 2 mm.

Table 2 shows the comparison between the calculation and the testing results.

3. Eqn. 13 is an approximate formula based on certain assumptions and simplification process, and the relevant parameters of the calculation model are fitted from experimental data based on standards recognized by domestic and foreign scholars. If test conditions are consistent or similar with context on which the formula relies, the results may fit better. If test conditions cannot satisfy the context, the results may be discrete. In the current research, the test results are generally used as basis for establishing standards. Compared with static load test, the fatigue test itself has a high degree of dispersion. Therefore, as a reference for bridge engineers, Table 2 and Figure 3 indicate that Eqn. 13 can be used to preliminarily determine the degradation of the ultimate bearing capacity after the stud connector experienced cyclic loading.

The calculation results based on Eqn. 13, i.e., degradation of ultimate bearing capacity with fatigue loading times for studs with different diameters, are given in Figure 4. If the stud materials are the same, the initial crack rate a_0/d in welding parts is the same under the same fatigue loading. Based on the figure, if within its fatigue life, the degradation degree of the ultimate bearing capacity for the stud with a larger diameter is obviously more serious than that of the stud with a smaller diameter. This finding indicates that in practice, a smaller stud diameter is better. A stud with a larger diameter can carry a heavier load. A larger stud may present greater risk than a smaller stud. Therefore, using a normal-sized stud instead of a larger stud connector is recommended.

The changes of ultimate bearing capacity with times of cyclic loading under various initial defect rate a_0/d and amplitude of shear stress $\Delta\tau$ are shown in Figures 5 and 6, respectively. Both figures show that both a_0/d and $\Delta\tau$ have great effects on the degree of degradation of the ultimate bearing capacity for the stud within its fatigue life. When the initial defect rate $a_0/d = 0.1$, the degradation of the ultimate bearing capacity for the stud within its fatigue life declines relatively slowly with a relative change below 10%; when $a_0/d = 0.125$, the degradation of the ultimate bearing capacity starts to accelerate; when $a_0/d = 0.15$, the ultimate bearing capacity of the stud within its fatigue life declines rapidly, especially after loading 0.8 million times the bearing capacity declines by 50%. Figure 6 shows that when $\Delta\tau = 100$ MPa, the degradation of the shear bearing capacity declines slowly. After loading 2 million times, the decrease of shear bearing capacity does not exceed 10% of the initial value. However, when $\Delta\tau = 120$ MPa, the ultimate bearing capacity declines rapidly, especially after 1.5 million times of loading, the bearing capacity declines by more than 50%, which is similar with the fatigue performance of the stud. The initial defect and amplitude of loading have greater impact.

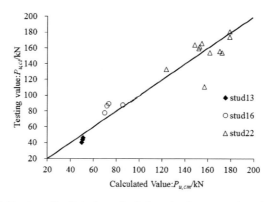

Figure 3. Residual static strength of studs under fatigue loads: comparison between calculated value and test result.

4 CONCLUSIONS

The degradation of the ultimate bearing capacity under fatigue load on the stud of a steel-concrete composite beam bridge was studied. The major conclusions are as follows:

1. The fatigue life model for stud connectors was established based on fracture mechanics with consideration of the initial component defects. The model shows that the fatigue life of the stud was not only related to the amplitude of shear stress but also to the peak of fatigue load, the physical dimensions of the stud, initial defects, and other factors.
2. The relationship between residual shear bearing capacity and times of cyclic loading with constant amplitude was proposed from the parameter fitting. The accuracy of this method was verified through previous fatigue tests and literature review. Using this method to estimate the degradation of the ultimate bearing capacity after the stud experienced cyclic loading is feasible.

REFERENCES

Ahn J.H., Kim S.H., Jeong Y.J. (2007), "Fatigue experiment of stud welded on steel plate for a new bridge deck system", Steel and Composite Structures, Vol. 7, No. 5, pp. 391–404.

Ahn J.H., Kim S.H., Jeong Y.J. (2008), "Shear behavior of perfobond rib shear connector under static and cycli loadings", Magazine of Concrete Research, Vol. 60, No. 5, pp. 347–357.

Bro M., Westberg M. (2004), "Influence of fatigue on headed stud connectors in composite bridge", PhD Thesis, Lulea University of Technology, Sweden.

Deric John Oehlers. (1989), "Deterioration in strength of stud connectors in composite bridge beams", Journal of Structural Engineering, Vol. 116, No. 12, pp. 3417–3413.

D.J. Oehlers, A. Ghosh, and M. Wahab. (1995), " Residual strength approach to fatigue design and analysis", Journal of Structural Engineering, Vol. 121, No. 9, pp. 3.1271–1279.

En Xie (2011). "Fatigue strength of shear connectors", University of Minho, Guimaraes, Portugal.

Gerhard Hanswille, Markus Porsch, Cenk Ustundag. (2000), " Resistance of stud shear connectors to fatigue", Journal of Constructional Steel Research, No.56, pp. 101–116.

Gerhard Hanswille, Markus Porsch, Cenk Ustundag. (2007), "Resistance of headed studs subjected to fatigue loading Part I: Experimental study", Journal of Constructional Steel Research, Vol. 63, No. 4, pp. 475–484.

Gerhard Hanswille, Markus Porsch, Cenk Ustundag. (2007), "Resistance of headed studs subjected to fatigue loading Part II: Analytical study", Journal of Constructional Steel Research, Vol. 63, No. 4, pp. 485–493.

Guo Jian-sheng, Sun Guo-zheng. (1999), "Fracture mechanics method for estimating the fatigue life of welded steel structure", Hoisting and Conveying Machinery, N0. 10, pp. 9–12.

Huang Qiao (2004). "Design principle of steel-concrete composite bridges", China Communications Press, Beijing, China.

Lee PG, Shim CS, Chang SP. (2005), "Static and fatigue behavior of large stud shear connectors for steel-concrete composite bridge", Journal of Constructional Steel Research, Vol. 61, No. 9, pp. 1270–1285.

Lee PG, Shim CS, Chang SP. (2001), "Design of shear connection in composite steel and concrete bridge with precast decks", Journal of Constructional Steel Research, Vol. 53, No. 7, pp. 203–219.

NIE Jian-guo (2011). "Steel-concrete composite bridges", China Communications Press, Beijing, China.

NIE Jian-guo (2005). "Steel-concrete composite beams structure: experiment, theory and application", Science Press, Beijing, China.

Roger G. Slutter, John W. Fisher. (1966), "Fatigue strength of shear connectors", Lehigh University Institute of Research.

Scott A. Civjan, M. ASCE. (2003), "Behavior of shear studs subjected to fully reversed cyclic loading," Journal of Structural Engineering, Vol. 129, No. 11, pp. 1466–1474.

Wang Yuhang, Nie Jianguo. (2009), "Fatigue behavior of studs in a composite based on fracture mechanics", J Tsinghua Univ (Sci & Tech), Vol. 49, No. 9, pp. 35–38.

Yang Xiao-hua, Yao Wei-xing, Duan Cheng-mei. (2003), "The review of ascertainable fatigue cumulative damage rule", Engineering Science, Vol. 5, No. 4, pp. 81–86.

Resources, Environment and Engineering II – Xie (Ed.)
© 2016 Taylor & Francis Group, London, ISBN 978-1-138-02894-4

Kinematics study of 6-DOF spatial mechanism

E. Novikova, V. Morozov, A. Zhdanov & I. Volkova
Vladimir State University Named Alexander and Nikolay Stoletovs, Russia

ABSTRACT: The article presents the design of 6DOF-spatial mechanism for the realization of accurate spatial displacements. Feature of the mechanism is to use a combination of three rotational and three translational pairs connected by hinges to ensure micromovings. The article presents the analytical dependence for calculating the kinematic characteristics of the proposed mechanism.

1 INTRODUCTION

The Parallel Mechanisms (PM) are mechanisms, on which the output section is connected to the base of the design by kinematic chains, which in turn contain multiple drives or lay the connexions to the movement of the output section. Mechanisms have closed kinematic chains and perceive the load as a spatial truss, which leads to an increased accuracy and carrying capacity.

The scientific research in the field of development and creation of mechatronic drives of linear motion for engineering tools with parallel kinematics are mainly concentrated in Japan and the Western and North Europe countries (Merlet, 2008, Hao, 2004, Ivanov, 2006, Bushuyev, 2001).

In the course of the study, such mechanisms as the platform of Stewart (Hauff), the spatial mechanism of Delta and the structure of Argos were examined. However, their schemata have a complex structure and a small operating zone, so there is a need to develop a new parallel kinematic mechanism with a simplified schema, bigger operating zone, high accuracy and studies of its kinematic characteristics.

As a result of the patent researches and the analysis of the structure of the kinematic chains of the PM, such mechanism has been developed, that do not change the margins as they have been set in the template: Top-2.54 cm, Right-2.54 cm, Bottom-3.05 cm, Left-2.54 cm. Do not add page numbers to your paper. We will insert these later.

2 MATERIALS AND METHODS

Copy the template file B1ProcA4.dot (if you print on A4 size paper) or B1ProcLe.dot (for Letter size paper) to the template directory. This directory can be found by selecting the Tools menu Options and then by tabbing the File Locations. When the Word program has been started, open the File menu and choose New. Now select the template B1ProcA4.dot or B1ProcLe.dot (see above). Now start by renaming the document by clicking Save As in the menu Files.

2.1 *Title, author and affiliation frame*

The mechanism with parallel kinematics type Tripod (Fig. 1) (Volkov, 2010) contains three rotational screw axis 1–3, forming in the mechanism's base four triangles, while three lineal pairs with a screw element are moving along the axes 5–7. The linear actuators 8–10 are

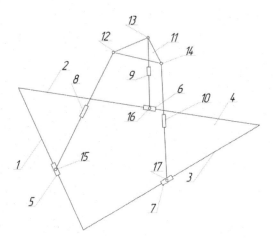

Figure 1. Schema of a spatial mechanism with three freedom degrees.

linked with advancing pairs and a slave member by 11 cylindrical hinges 12–17. Linear actuators are based on drives and units with rollerscrew mechanisms (Morozov, 1999, Morozov, 2005, Novikova, 2015), which provide high rigidity, accuracy, and smoothness.

The mechanism works as follows. During the rotation of the screw axis 1–3, assist by the drive mechanisms, the translational pairs 5–7 are driven in movement, which have helical elements and are located on these axes. The slave member 11 is connected with the translational pairs 5–7 by the linear actuators 8–10 through the cylindrical hinges 12–17. Each linear actuator is able to ensure the advance of its movable part by a predetermined distance. The total movement of the translational pairs along the axes, as well as the advance of the movable part of the linear actuators locates the slave member in a predetermined region of the space.

In the Cartesian coordinate system XYZ (Fig. 2) the points forming the base's area, are labeled as A, B, C; the translational cylindrical hinges are D, E, F; the cylindrical hinges of the advancing part of the actuators in a retracted state are H, G, I; the cylindrical hinges of the advancing part of the actuators in a intermediate state are H, G, I; the point N is the point of the bisectors' intersection of the triangle KLM, which forms the area of the movable platform; Y, T are the points which belong to the area of the tripod's base; α is the angle between the straight lines YN and NT (perpendicular from the point N to the base's area); R is radius of the region of possible values of the coordinates x, z of the point N with its centre at the intersection point of the ABC triangle's bisectors.

The coordinates of the described points are shown in Table 1.

Based on the received symbols has been drawn up a mathematical model describing the movement's points of the actuators depending on the given values of the coordinates points A, B, C, N, the lengths of the triangle's sides KLM, the lengths EH, DG, FI, the lengths LN, MN, KN and the angle α.

In the described mathematical model, the following assumptions were adopted:
$|AB|=|BC|=|AC|$; $|KL|=|LM|=|KM|$; $|EH|=|DG|=|FI|$; $|LN|=|MN|=|KN|$; the angle α is located perpendicularly to the plane of the base and the parallel straight line AB.

With the solution of the mathematical model, the lengths sides of the triangle or the coordinates of its vertices are determined.

$$|AB|=\sqrt{(x_2-x_1)^2+(y_2-y_1)^2+(z_2-z_1)^2}$$

$$|BC|=\sqrt{(x_3-x_2)^2+(y_3-y_2)^2+(z_3-z_2)^2}$$

$$|AC|=\sqrt{(x_3-x_1)^2+(y_3-y_1)^2+(z_3-z_1)^2}$$

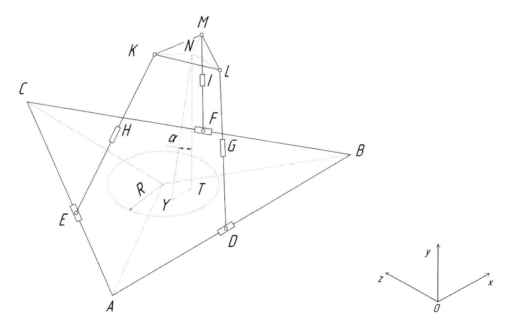

Figure 2. Design scheme for finding the motion of the mechanism.

Table 1. The coordinates of the described points.

Points	Coordinates	Points	Coordinates	Points	Coordinates
Y	x_0, y_0, z_0	E	x_5, y_5, z_5	L	x_{10}, y_{10}, z_{10}
A	x_1, y_1, z_1	F	x_6, y_6, z_6	K	x_{11}, y_{11}, z_{11}
B	x_2, y_2, z_2	G	x_7, y_7, z_7	M	x_{12}, y_{12}, z_{12}
C	x_3, y_3, z_3	H	x_8, y_8, z_8	N	x_{13}, y_{13}, z_{13}
D	x_4, y_4, z_4	I	x_9, y_9, z_9	T	x_{14}, y_{14}, z_{14}

According to the conditions of the problem $D \in |AB|; F \in |BC|; E \in |AC|$.

To find the values of the coordinates of the points H, G, I, we construct the equations of the straight lines, which includes DL, FM, and EK:

$$\text{For the line DL} \quad \frac{x_7 - x_4}{x_{10} - x_4} = \frac{y_7 - y_4}{y_{10} - y_4} = \frac{z_7 - z_4}{z_{10} - z_4}$$

$$\text{For the line FM} \quad \frac{x_8 - x_5}{x_{11} - x_5} = \frac{y_8 - y_5}{y_{11} - y_5} = \frac{z_8 - z_5}{z_{11} - z_5}$$

$$\text{For the line EK} \quad \frac{x_9 - x_6}{x_{12} - x_6} = \frac{y_9 - y_6}{y_{12} - y_6} = \frac{z_9 - z_6}{z_{12} - z_6}$$

For an unambiguous value of the position of the points H, G, I, we write the following relations:

For the point G: $|DL| = |DG| + |GL|$

$$\sqrt{(x_{10} - x_4)^2 + (y_{10} - y_4)^2 + (z_{10} - z_4)^2} = \sqrt{(x_7 - x_4)^2 + (y_7 - y_4)^2 + (z_7 - z_4)^2}$$
$$+ \sqrt{(x_{10} - x_7)^2 + (y_{10} - y_7)^2 + (z_{10} - z_7)^2}$$

231

For the point H: $|EK|=|EH|+|HK|$

$$\sqrt{(x_{11}-x_5)^2+(y_{11}-y_5)^2+(z_{11}-z_5)^2} = \sqrt{(x_8-x_5)^2+(y_8-y_5)^2+(z_8-z_5)^2}$$
$$+\sqrt{(x_{11}-x_8)^2+(y_{11}-y_8)^2+(z_{11}-z_8)^2}$$

For the point I: $|FM|=|FI|+|IM|$

$$\sqrt{(x_{12}-x_6)^2+(y_{12}-y_6)^2+(z_{12}-z_6)^2} = \sqrt{(x_9-x_6)^2+(y_9-y_6)^2+(z_9-z_6)^2}$$
$$+\sqrt{(x_{12}-x_9)^2+(y_{12}-y_9)^2+(z_{12}-z_9)^2}$$

We create a system of equations for determining the values of the coordinates points H, G, I:

$$\begin{cases} \dfrac{x_7-x_4}{x_{10}-x_4}=\dfrac{y_7-y_4}{y_{10}-y_4}=\dfrac{z_7-z_4}{z_{10}-z_4}, \dfrac{x_8-x_5}{x_{11}-x_5}=\dfrac{y_8-y_5}{y_{11}-y_5}=\dfrac{z_8-z_5}{z_{11}-z_5}, \dfrac{x_9-x_6}{x_{12}-x_6}=\dfrac{y_9-y_6}{y_{12}-y_6}=\dfrac{z_9-z_6}{z_{12}-z_6} \\ |DG|=\sqrt{(x_7-x_4)^2+(y_7-y_4)^2+(z_7-z_4)^2} \\ |EH|=\sqrt{(x_8-x_5)^2+(y_8-y_5)^2+(z_8-z_5)^2} \\ |FI|=\sqrt{(x_9-x_6)^2+(y_9-y_6)^2+(z_9-z_6)^2} \\ \sqrt{(x_{10}-x_4)^2+(y_{10}-y_4)^2+(z_{10}-z_4)^2}=\sqrt{(x_7-x_4)^2+(y_7-y_4)^2+(z_7-z_4)^2} \\ \qquad +\sqrt{(x_{10}-x_7)^2+(y_{10}-y_7)^2+(z_{10}-z_7)^2} \\ \sqrt{(x_{11}-x_5)^2+(y_{11}-y_5)^2+(z_{11}-z_5)^2}=\sqrt{(x_8-x_5)^2+(y_8-y_5)^2+(z_8-z_5)^2} \\ \qquad +\sqrt{(x_{11}-x_8)^2+(y_{11}-y_8)^2+(z_{11}-z_8)^2} \\ \sqrt{(x_{12}-x_6)^2+(y_{12}-y_6)^2+(z_{12}-z_6)^2}=\sqrt{(x_9-x_6)^2+(y_9-y_6)^2+(z_9-z_6)^2} \\ \qquad +\sqrt{(x_{12}-x_9)^2+(y_{12}-y_9)^2+(z_{12}-z_9)^2} \end{cases}$$

3 RESULTS

The calculations were carried out using a mathematical model to determine the most rational geometric parameters of the investigated spatial mechanism: $|AB|=|BC|=|AC|=600$ mm; $|LK|=|LM|=|KM|=120$ mm; R = 100 mm; NT = const = 285 mm; $-10° \le \alpha \le 10°$.

An example of calculation for the two possible positions for the mobile platform in the space is described. In the results of the calculation, the found values of the parameters ΔGL, ΔHK, ΔIM, ΔD, ΔE, ΔF allowed us to use a three-dimensional geometric model, as well as research's software package of the mechanisms' kinematics to identify the main characteristics of the kinematic (movement trajectory, velocity and acceleration points) of the investigated mechanism during its displacement from one position to another. According to the design scheme, the model of the investigated mechanism was formed and the conditions that determine the movement were obtained in the package SolidWorks. Through the research's software package of the mechanisms' kinematics, as well as the results of theoretical calculations of the mathematical model, were obtained the velocity graphics (Figs. 3–5) and the acceleration graphics (Figs. 6–8).

As a result, the computational studies, the operating zone of the table-positioner centre was defined, which is a circle with a diameter of 200 mm, relative to the cutting tool. The determine angle of inclination of the table-positioner is $-10° < \alpha < 10°$.

Due to the reciprocating movable translational pairs in the base, the operating zone of the spatial mechanism of action increases, compared with the same mechanism, if they are

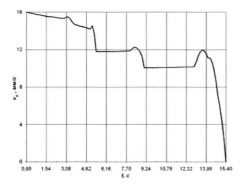

Figure 3. Graphic of the velocity p. K.

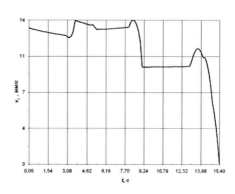

Figure 4. Graphic of the velocity p. L.

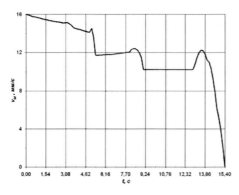

Figure 5. Graphic of the velocity p. M.

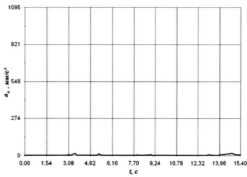

Figure 6. Graphic of the acceleration p. K.

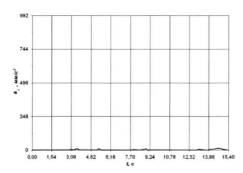

Figure 7. Graphic of the acceleration t. L.

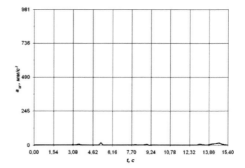

Figure 8. Graphic of the acceleration t. M.

absent in the base of such mechanism. Through the organization of the movement of the actuators in the space simultaneously with a table-positioner, the load applied on the table-positioner from the cutting tool will be more uniformly distributed on the actuators, unlike the construction of the tripods and hexapods, in which the actuators are rigidly connected to the base. As results of the research, it may be noted that the developed spatial mechanism is not inferior to the world analogues according to the basic kinematic and loading characteristics.

This research was supported by Russian Foundation of Basic Research under contract 14-08-00772.

REFERENCES

Bushuyev, V.V. & Holshev, I.G. 2001. Mechanisms of parallel structure in mechanical engineering, Machines-tools and instruments, 1, 3–6.
Hao, F. & Merlet, J.-P. 2004. Multi-criteria optimal design of parallel manipulators based on interval analysis. Mechanism and Machine Theory. 1–15.
Ivanov, A.V. 2006. Guarantee of the quality indicators of the assembly of the tool-machine-manipulator with parallel kinematics, Komsomolsk-on-Amur, 21.
Merlet, J.P. 2008. Parallel Robots, 2nd Edition. *Springer*.
Morozov, V.V. 2005. Roller screw mechanisms. Kinematic Characteristics. (monograph). *VlSU*, Vladimir, 78.
Morozov, V.V., Kosterin, A.B. & Novikova E.A. 1999. Smooth of dynamic links of the electromechanical actuators (monograph). *VlSU*, Vladimir, 158.
Novikova, E.A. 2015. Increase in accuracy and smoothness of movement of the output link actuators of linear micro motions. *Advanced mechanical and materials*. 96, 951–957.
Volkov M., Volkova I., Zhdanov A., Trefilov M. Spatial mechanism. Patent 99371, application No2010125410, priority date 21.06.2010.

Inhibiting the spontaneous combustion of sulfide ores by bacteria desulfurization

Chunming Ai, Aixiang Wu, Hongjiang Wang & Jiandong Wang
The Key Laboratory of High-Efficient Mining and Safety of Metal Mines, Ministry of Education, University of Science and Technology Beijing, Beijing, China

ABSTRACT: Acidophilus bacteria desulfurization column leaching was used to explore the effect of the desulfurization of metal sulfide ore and the feasibility of inhibiting the spontaneous combustion. The calculation method of desulfurization rate was discussed moreover. Uniform design was used in the experiment whose factors include initial pH value, the liquid arrangement intensity, and average particle size of ore. The regression analysis of desulfurization rates by liquid shows that the liquid arrangement intensity influences the desulfurization rate most, ore particle size followed, and pH the least. Liquid arrangement intensity and pH are positively correlated with the desulfurization rate, while ore particle size negatively. The result of SEM shows that the sulfur content of the ore surface is reduced from about 43% to 15% or less. The ore flammability is lower than before and the surface desulfurization rate is 65%.

1 INTRODUCTION

The fire caused by spontaneous combustion of sulfide ore is one of the major disasters in mining high sulfur content ores, which affects about 30% of metalliferous mines in China (Pan, W. et al. 2010). Ore spontaneous combustion not only results in the loss of ore, but also causes plenty of engineering abandonment and the waste of resources. Meanwhile toxic air produced by ore spontaneous combustion will pollute the underground environment and affect the health of workers (Soundararajan, R. et al. 1996, Yang, F.Q. & Wu, C. 2010). The fundamental reason for internally-caused fire is the combustion and exothermic reaction of sulfur in the ore (Pan, W. & Wu, C. 2011), while enough oxygen and constant heat concentration are the main external factors (Wang, L.L. et al. 2010, Yang, F.Q. et al. 2011). Traditional suppression measures on spontaneous combustion of sulfide ore are mainly concentrated on cutting off the air and accelerating heat dissipation, which can't fundamentally solve the problem of sulfur oxide and eradicate internally-caused fire (Wu, D.M. 2001, Li, Z.J. et al. 2009). Combined with the basic principles of solution mining and control of internally-caused fire and focusing on the sulfur content of mineral surface, the bacteria desulfurization technology uses bacteria to degrade the sulfide ore on the mineral surface, thus seeking a new method to prevent and control internally-caused fire.

The bacterial desulfurization technology mainly applied to processing sour gas and oil desulfurization and purifying the sulfur-containing wastewater is a research hotspot (Soleimani, M. et al. 2007, Sarti, A. et al. 2010). The desulfurization effectiveness of this technology is remarkable. Bacterial desulfurization on mineral mainly used in coal and pyrite slag has achieved good effect (Jorjani, E. et al. 2007, Cardona, I.C. & Márquez, M.A. 2009). The present study showed that it's feasible to use the bacterial desulfurization to prevent fire in high sulfur content metal mines (Luo, F.X. et al. 2009). Using bacteria liquid to oxidate the sulfur on the mineral surface can solve the problem about oxidation and heat-release of low sulfur so that the prevention and control of hazard change from passive to active. Through the experiment on the column desulfurization of metal sulfide ores, we discuss the influencing factors and the effect on inhibiting ore spontaneous combustion by bacterial desulfurization. It also analyzes and compares the different calculation methods to quest the desulfurization efficiency.

2 DESULFURIZATION WITH COLUMN BIOLEACHING

2.1 Experimental materials

The ore sample mainly containing colloidal pyrite ore was taken from the Anhui Tongling Xinqiao Pyrite Mine. There is a great risk of spontaneous combustion owing to sulfur mass fraction of 45%. According to the experimental requirements, crushed ore is sieved into different particle sizes and then sealed preservation.

The mesophilic sulfur-oxidizing acidophiles applied in this research, which has been isolated from the mine water of Anhui Tongling Xinqiao Pyrite Mine, is domesticated with ore after enrichment.

2.2 Experimental scheme

A three-factor and six-level uniform design scheme of six experiment groups were used, as shown in Table 1. Factors considered in our experiment include the pH value of the solution at initial stage, liquid arrangement intensity and the ore particle size.

During the 44-day experiment period, the leaching columns were sampled regularly at the interval of once every 4 days to measure pH value, electrical potential E value, and SO_4^{2-} concentration.

2.3 The computing method of desulfurization rate

Desulfuration rate, an important indicator of judging bacteria desulfuration effect, is conventionally computed by liquid or slag.

Desulfuration rate by liquid, tied to SO_4^{2-} concentration and solution volume, is calculated as follows (Eqs. (1)):

$$S_l = \frac{\alpha_i V_i}{QC} \times 100 \qquad (1)$$

where S_l = desulfuration rate by liquid (%); α_i = mass concentration of sulfur in the solution of level i (g/L); V_i = total volume of the solution of level i (L); Q = initial quantity of ore (g); and C = initial sulfur grade of ore (%).

Desulfuration rate by slag, tied to sulfur content in leaching residue and ore, is calculated as follows (Eqs. (2)):

$$S_t = \frac{QC_1 - Q'C_2}{QC_1} \times 100 \qquad (2)$$

Table 1. Scheme of bacterial desulfurization column leaching.

No.	pH	Liquid arrangement intensity/L·m^{-2}·h^{-1}	Ore particle size/mm
Column 1	2.7	240	3
Column 2	2.1	480	5
Column 3	1.5	160	7
Column 4	3	400	9
Column 5	2.4	80	11
Column 6	1.8	360	13

where S_t = desulfuration rate by slag (%); Q = initial quantity of ore (g) and Q' = total mass after desulfuration (g); C_1 = initial sulfur grade of ore (%); and C_2 = sulfur grade of ore after desulfuration (%).

Although these two easy-to-use computing methods can evaluate overall desulfuration effect, its results fail to present surface desulfuration rate. The desulfuration reaction at the ore surface is more intensive than that of ore interior. Due to this, conventional desulfuration rate which tends to be more expressive of the desulfuration effect of small particle ores or tailings desulphurization does not apply to ores of column leaching and heap leaching.

Electron-microscope scanning technology has been used in this experiment in view of the limitations of the desulfurization rate by liquid. Sulfur content on the ore surface can be obtained from energy spectrum analysis. Due to this, surface desulfuration rate is figured out, and thus the traditional method can be improved. Surface desulfuration rate is calculated as follows (Eqs. (3)):

$$S_S = \frac{W_1 - W_2}{W_1} \times 100 \tag{3}$$

where S_S = surface desulfuration rate (%); W_1 = weight ratio of sulfur at ore surfaces before the desulfuration treatment (%); and W_2 = weight ratio of sulfur at ore surfaces after the desulfuration treatment (%).

3 RESULTS AND DISCUSSIONS

3.1 *Desulfurization rate by liquid*

The changes in desulfuration rate by liquid of column leaching test against time are shown in Figure 1.

As shown in Figure 1, desulfuration rate increases with time. The rate of increment of column 2 is highest while that of column 5 is lowest. It can be concluded that the higher liquid arrangement intensity, the higher the desulfuration rate. The liquid arrangement intensity of column 2 is 480 L/(m²·h) and that of column 5 which is lowest among six columns is 80 L/(m²·h).

The quadric polynomial return analysis through Data Processing System (DPS) toward desulfurization rate by liquid gives the following relationship (correlation coefficient of R = 0.9994, p = 0.0499):

$$S = 8.1415 + 4.255 \times 10^{-5} x_2^2 - 0.06628 x_3^2 + 0.1793 x_1 x_3 + 2.85 \times 10^{-4} x_2 x_3 \tag{4}$$

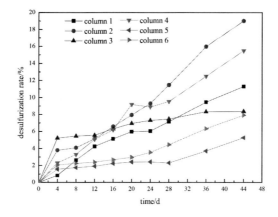

Figure 1. Changes of desulfuration rate by liquid with time.

where S = desulfuration rate (%); x_1 = pH value; x_2 = liquid arrangement intensity (L/(m²·h)); and x_3 = ore particle size (mm).

The standard regression coefficients based on the regression analysis are $r_1 = 0.3363$, $r_2 = 0.8420$, $r_3 = -0.5343$. Therefore the liquid arrangement intensity has the most significant contribution toward the desulfuration rate, with the pH value the least significant. According to the sign of correlation coefficient, the desulfuration rate is positively correlated with liquid arrangement intensity and pH value, and inversely correlated with the ore particle size.

3.2 Surface desulfuration rate

Ore samples were collected for electron-microscope scanning from column 2 before and after the desulfuration and their images are given in Figure 2 and Figure 3.

It can be seen clearly that the ore surface before the desulfuration is much more compact and smoother than it is after the oxidation, indicating intense erosion due to the desulfuration.

As shown in Figure 3, it is clear that by removing some of the sulfur at the ore surfaces, the weight gain rate by desulfuration is down to 14.79% from 43.21%. Surface desulfuration rate which is much higher than desulfurization rate by liquid of column 2 (18.9%) is 65%. The content of element O increased from 28.73% to 75.72%. The main sulfur desulfuration reaction occurs with bacteria as shown in the following equations:

$$4Fe^{2+} + 4H^+ + O_2 \xrightarrow{Bacterial} 4Fe^{3+} + 2H_2O \quad (5)$$

$$FeS_2 + 8H_2O + 14Fe^{3+} \longrightarrow 16H^+ + 2SO_4^{2-} + 15Fe^{2+} \quad (6)$$

$$2FeS_2 + 2H_2O + 7O_2 \xrightarrow{Bacterial} 4H^+ + 4SO_4^{2-} + 2Fe^{2+} \quad (7)$$

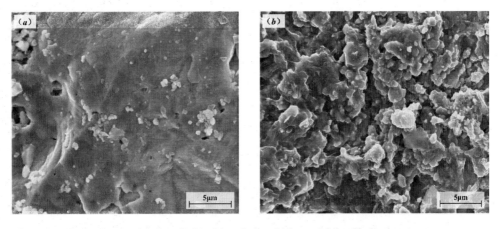

Figure 2. Ore surface morphology before (a) and after (b) bacterial desulfurization.

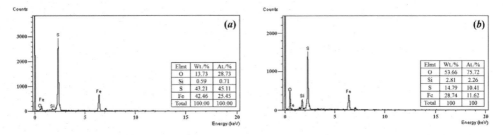

Figure 3. Energy spectrum of ore surface before (a) and after (b) bacterial desulfurization.

In the bacteria breeding stage, bacteria react with Fe²⁺ in solution and FeS₂ at particle surfaces as shown in Eqs. (5) and (6). In the rapid leaching period, the bacteria adsorb at particle surfaces and react with FeS₂ as shown in Eqs. (7). At this point, the direct reaction, indirect reaction, and combination reaction proceed simultaneously. Sulfur on the ore surface transformed into ions leach into solution in order to achieve the desulfurization bacteria.

4 EVALUATION OF SPONTANEOUS IGNITION SUPPRESSING EFFECT AND DESULFURATION RATE ANALYSES

4.1 Evaluation of spontaneous ignition suppressing effect

The autoignition of metallic sulfide ore is bound up with sulfur content of ore surface, and there is a low possibility of autoignition for ores with less than 15% of sulfide content (Zhang, H. & Zhang, C.S. 2004). Bacteria desulfuration experiment showed that though with less than 20% desulfuration rate of liquid, the desulfuration rate of ore surface is high. Sulfide content on ore surface falling below 15% demonstrate that bacteria desulfuration effectively prevent sulfur oxidation of interior ore and restrain autoignition of sulfide ore.

Autoignition cases in Xinqiao Pyrite in the 1990s are shown in Table 2 (Wu, F.C. 2002). The safety heaping time is unfixed owing to complexity and anisotropy of stopes. But based on the predicted heaping times and the autoignition times in the cases, the safety heaping time of sulfide ore is about 30–50 days.

After 44 days of the column leaching experiment, the sulfide content of ore surface decreases below the safety level by the bacteria desulfuration within the safety heaping time. And the ore has no risk of autoignition during the bacteria desulfuration reaction because of infiltration by bacteria solution.

4.2 Numerical analyses of bacteria desulfuration effect

The temperature change was simulated by using numerical simulation software Comsol Multiphysics. Based on heat transfer theory, the thermal equilibrium equation of undesulfured sulfide ore can be described as:

$$\rho_B \cdot C_B \frac{\partial T}{\partial t} = K_B \nabla^2 T + Q_p - \varepsilon \rho_g \cdot U_g C_g \nabla T \qquad (8)$$

where ρ_B = volume density of ore heap (Kg/m³); C_B = thermal capacity of sulfide ore (KJ/(Kg·K)); K_B = heat conductivity coefficient (KJ/(m·K)); ε = volume fraction; ρ_g = density of air inside ore (Kg/m³); U_g = flowing velocity of air inside ore (m/s); C_g = thermal capacity of air (KJ/(Kg·K)); Q_p = reaction heat of sulfide ore redox reaction (KJ/(m³·s)); and T = temperature of sulfide ore (K).

In the bacteria desulfurization, heat mainly comes from biological oxidation reaction heat, meanwhile, the flow of bacteria solution will cause convection and evaporation diffusion. The thermal equilibrium equation in desulfurization process can be described as:

$$\rho_B \cdot C_B \frac{\partial T}{\partial t} = K_B \nabla^2 T + Q_0 - \varepsilon \rho_g \cdot U_g C_g \nabla T - \varepsilon \rho_L \cdot U_L C_L \nabla T \qquad (9)$$

Table 2. Timetable of spontaneous combustion in Xinqiao Pyrite.

Time	1992.5	1992.10	1998.6	1999.6
Stope number	231	506II	921~922	821~822
Storage time/d	>20	50	30	>50

where ρ_L = density of bacteria solution (Kg/m³); U_L = velocity of bacteria solution (m/s); C_L = thermal capacity of bacteria solution (KJ/(Kg·K)); and Q_0 = reaction heat of desulfurization (KJ/(m³·s)).

The sulfide ore thermal equilibrium before and after desulfurization was simulated, as shown in Figure 4.

From the stimulating result of temperature field, it can be understood that the interior temperature is apparently higher than that of surface due to heat congregation and weak fluidity of air. From Figure 4, the highest temperature of an ore heap in 30 meters is 442.16 K, while that of desulfurated ore is 358.42 K. Even though desulfurization reaction is exothermic, the enthalpy change and temperature of desulfuration were less than those of natural oxidation reaction. With the same environmental factors, the pyrophorisity would decrease after bacteria desulfuration.

4.3 Comparison of different desulfuration rate analyses

Effect of inhibiting ore spontaneous combustion after the desulphurization by bacteria was obvious in the case that desulfuration rate by liquid was low in this experiment, which agrees with surface desulfuration rate.

Ore particles can be considered to be spherical based on the shrinking core model, as shown in Figure 5 (Lizama, H.M. 2004). The ore particle with diameter of r_0 has a reactionless

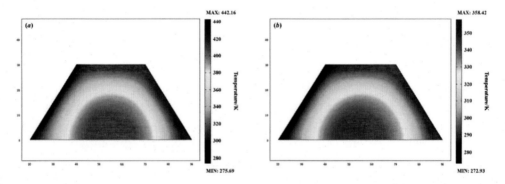

Figure 4. Temperature distribution of stockpile before (a) and after (b) bacterial desulfurization.

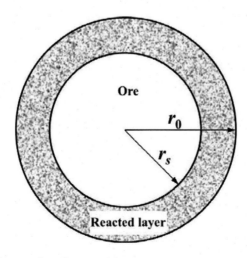

Figure 5. Desulfurization reaction of ore particles.

core whose radius is r_s. W is the amount of desulfurization in the bacterial desulphurization reaction. Desulfuration rate by liquid (S_l) and surface desulfuration rate (S_s) are calculated as follows:

$$S_l = \frac{3w}{4\pi r_0^3 \rho \cdot \gamma} \quad (10)$$

$$S_s = \frac{3w}{4\pi (r_0^3 - r_s^3) \rho \cdot \gamma} \quad (11)$$

As shown in Eqs. (10) and Eqs. (11), S_l was apparently lower than S_s. Desulfurization rate by liquid, defined as the ratio between the sulfur content in the solution and the total number of sulfur in ore, tends to be lower because the bacteria desulfurization reaction mainly occurs in the ore surface.

Sulfur on the ore surface is completely removed when S_s is at 100%, S_l is calculated as follows (Eqs. (12)):

$$S_l = 1 - \left(\frac{r_s}{r_0}\right)^3 \quad (12)$$

S_l is close to 0 when r_s approximately equals r_0. Even if the surface desulphurization rate reaches 100%, liquid desulfurization rate is almost 0 at early stage of reaction. From this both differences are obvious. After the desulphurization reaction, spontaneous combustion can be prevented altogether because the ore surface won't heat up with oxidation. Therefore, the surface desulfurization rate plays an important role in evaluating desulfurization effect.

4.4 *Optimization of desulfuration rate analyses*

Two methods of calculating desulfuration rate, including conventional desulfuration rate by liquid and surface desulfuration rate by electron-microscope scanning, have been used in this experiment. Desulfuration rate by liquid is considered to be simple, but it fails to present real surface desulfuration rate. Surface desulfuration rate obtained from the change in surface sulfur content by electron-microscope scanning tends to be more expressive of the desulfuration effect.

Surface desulfuration rate, as simple as desulfuration rate by liquid, fails to realize continuous monitoring on account of complicated procedure of SEM. Handheld ore analyzer based on XRF spectrum analysis technology is widely used in rapidly detecting metallic and nonmetallic elements on the ore surface. Therefore, exploring the accuracy of the technology contributes to promote surface desulfuration rate to heap leaching production of metal mine.

5 CONCLUSIONS

1. According to the regression analysis of desulfuration rate by liquid, factors influencing desulfuration rate are irrigation rate, ore particle size, and pH, listed in a sequence of declined significance.
2. For a processing time within sulfides' safe stocking period, bacteria desulfuration leaves less than 15% sulfur residual, risking relatively lower spontaneous ignition possibility. The numerical analysis indicates that bacteria desulfuration is able to bring down ore pile temperature. Actual desulfuration effect agrees more with surface desulfuration rate.
3. Although conventional desulfuration rate by liquid estimation facilitates online monitoring, it fails to present surface desulfuration effect. By comparison, surface desulfuration rate tends to be more expressive of the desulfuration effect of large particle ores, proving to be a feasible method to evaluate heap desulfuration leaching. The calculated surface desulfuration rate of column leaching test is 65%.

ACKNOWLEDGEMENTS

This work was financially supported by the National Natural Science Foundation of China (Nos. 51374034 and 51304011), and the National Key Technologies R&D Program for the 12th Five-year Plan (No. 2012BAB08B02).

REFERENCES

Cardona, I.C. & Márquez, M.A. 2009. Biodesulfurization of two Colombian coals with native microorganisms. *Fuel Processing Technology* 90(9):1099–1106.
Jorjani, E., Chehreh Chelgani, S. & Mesroghli, S. 2007. Prediction of microbial desulfurization of coal using artificial neural networks. *Minerals Engineering* 20(14):1285–1292.
Li, Z.J., Wang, F.S. & Li, G.X. 2009. Experimental study on the spontaneous combustion inhibitor of sulfide ores. *Journal of Safety and Environment* (3):132–134.
Lizama, H.M. 2004. A kinetic description of percolation bioleaching. *Minerals Engineering* 17(1):23–32.
Luo, F.X., Wang, H.J. & Wu, A.X. 2009. Analysis on feasibility of removing sulfur from metal sulfide ores with microorganism. *Journal of Safety Science and Technology* 5(4):23–26.
Pan, W., Wu, C. & Liu, H. 2010. Self-heating test of sulfide ore heap and numerical simulation of temperature field. *The Chinese Journal of Nonferrous Metals* 20(1):149–155.
Pan, W., Wu, C., Li, Z.J. 2011. Simulation experiment of dynamic self-heating process of sulfide ore heap. *Journal of Central South University (Science and Technology)* 42(7):2126–2131.
Sarti, A., Pozzi, E., Chinalia, F.A., et al. 2010. Microbial processes and bacterial populations associated to anaerobic treatment of sulfate-rich wastewater. *Process Biochemistry* 45(2):164–170.
Soleimani, M., Bassi, A. & Margaritis, A. 2007. Biodesulfurization of refractory organic sulfur compounds in fossil fuels. *Biotechnology Advances* 25(6):570–596.
Soundararajan, R., Amyotte, P.R. & Pegg, M.J. 1996. Explosibility hazard of iron sulphide dusts as a function of particle size. *Journal of Hazardous Materials* 51(1–3):225–239.
Wang, L.L., Wang, L.G. & Li, L.M. 2010. Study on the influencing factors of heat release of sulfide ore pile. *Mining Research and Development* (1):96–99.
Wu, D.M. 2001. The characteristics of spontaneous combustion of sulfide ores and the complex measures for preventing and extinguishing fire. *Industrial Minerals and Processing* (10):20–22.
Wu, F.C. 2002. The reason of spontaneous combustion of ores and anti-extinguishing measures in Xinqiao Pyrite. *Mining Safety & Environmental Protection* 29(3):55–56.
Yang, F.Q. & Wu, C. 2010. New test method of oxidation and self-heating properties of sulfide ore samples. *The Chinese Journal of Nonferrous Metals* 20(5):976–982.
Yang, F.Q., Wu, C. & Liu, H. 2011. Thermal analysis kinetics of sulfide ores for spontaneous combustion. *Journal of Central South University (Science and Technology)* 42(8):2469–2474.
Zhang, H. & Zhang, C.S. 2004. Principle of pyrite spontaneous combustion and its prevention. *Copper Engineering* (3):53–54.

… Resources, Environment and Engineering II – Xie (Ed.)
© 2016 Taylor & Francis Group, London, ISBN 978-1-138-02894-4

Analysis of superposition of wave load and wind load on offshore wind turbine based on load simulation

BoWen Jiang, MingJie Zhao, Pan Liu & Peng Jin
Chongqing Jiaotong University, Chongqing, China

ZiYuan Tang
Nanjing University, Nanjing, China

ABSTRACT: Based on Fifth-order Stokes regular wave theory and logarithmic law, wave load and wind load were simulated individually in accordance with nonlinear waves and static wind. Through random sampling of the probability density function of the two forces individually and calculating the superposition of each value, the probability density function of the total force is obtained. Observationally, the conclusions are summarized as follows: the relationship between the two is approximate negative correlation; the probability density function of total wind force is the superposition of five probability density function, each of them meets Gauss distribution; the general trend of the distribution of total force can be described as Gauss distribution.

1 INTRODUCTION

Compared with inland sea wind turbine, offshore wind turbine has advantages of no land occupation, high wind speed, and high efficiency. However, in order to bear the wind loads, wave load, and vertical load caused by upper platform and equipment, the foundation of offshore wind turbines are with much more structural diversities and technical difficulties. Therefore, the environmental load-induced dynamic response analysis of offshore turbine is with tremendous significance in offshore wind turbine engineering. Wave load and wind load, two of the main loads on offshore wind turbine, drew the attention of experts and scholars.

2 ANALYSIS OF WAVE LOAD AND WIND LOAD

2.1 Wave load

Focusing on the motion of the fluid particles, based on Lagrange method, two main famous theories to describe the movement characteristics of fluid particles were proposed: According to the linear wave, Airy proposed Small Amplitude Wave Theory in 1845 while Stokes proposed Finite Amplitude Wave Theory in 1847 to describe the nonlinear wave. Fifth-order Stokes regular wave theory based on Finite Amplitude Wave Theory were widely used in calculation of wave load on offshore wind turbine. Zhu Yanrong gave the following suggestions on how to select the wave theory in 1983: Small Amplitude Wave Theory should be selected when $S < 0.6$; when $0.6 < S < 10.0$, fifth-order Stokes regular wave theory suits better, see, for example, Equation 1 below:

$$S = T\sqrt{\frac{g}{h}} \qquad (1)$$

where g = acceleration of gravity (m/s^2); h = water depth (m) and T = wave period (s).

In fifth-order Stokes regular wave theory, in order to get the wave velocity potential function and wavefront rise function (see, for example, Equation 2 and Equation 3), some parameters of wave are supposed to be determined by solving transcendental equations (see, for example, Equation 4 and Equation 5). The velocity field function equals the first order derivative of velocity potential function with respect to time (s). The acceleration field function equals the second order derivative of velocity potential function with respect to time (s).

$$\Phi = \frac{L}{kT}\sum_{n=1}^{5}\Phi_n \cosh(nks)\sin(n\theta) \qquad (2)$$

where Φ = velocity potential function; k = wave number; L = wavelength (m); T = wave period (s); θ = phase angle (rad), and Φ_n = coefficient of velocity potential function.

$$\eta = \frac{1}{k}\sum_{n=1}^{5}\eta_n \cosh(n\theta) \qquad (3)$$

where η = wavefront rise function; k = wave number; T = wave period (s); θ = phase angle (rad), and η_n = coefficient of wavefront rise function.

$$\frac{\pi H}{h} = \frac{1}{h/L}\left[\lambda + \lambda^3 B_{33} + \lambda^5\left(B_{35} + B_{55}\right)\right] \qquad (4)$$

$$\frac{h}{L_0} = \frac{h}{L}\tanh(kh)\left(1 + \lambda^2 C_1 + \lambda^4 C_2\right) \qquad (5)$$

where H = significant wave height (m); L = wavelength (m); $L_0 = gT^2/2\pi$ (m); h = water depth (m); T = wave period (s); B_{ij} = coefficient B of wave and C_i = coefficient C of wave.

According to the Morison Equation (see, for example, Equation 6), total wave force on offshore wind turbine (column of small diameter, shown in Figure 1(a)) can be calculated.

When the data of a wave (H = 4.2 m, T = 6S, h = 16 m) is selected, we can get the wave velocity potential function, the wavefront rise function, the wave velocity field function, and the wave acceleration field function through calculation. The relationship between the total wave force on offshore wind turbine and time is obtained, as shown in Figure 1(b).

$$F_H = \int_{-h}^{\eta}\left(\frac{1}{2}C_D Du|u| + C_M\rho\frac{\pi D^2}{4}\frac{\partial u}{\partial t}\right)dz \qquad (6)$$

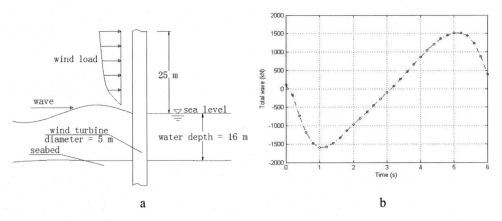

Figure 1. The model of wind turbine and the relationship between the total wave force on offshore wind turbine and time.

where F_H = total wave force (kN); η = wavefront rise function; (m); h = water depth (m); C_D = coefficient of velocity force; C_M = coefficient of acceleration force; t = time (s); z = distance from seabed (m); D = diameter of offshore wind turbine (m), and u = velocity of wave with respect to time and distance from seabed (m).

According to Figure 1(b), the total wave force gets a maximum value and a minimum value in each period, while their absolute values are roughly equaled. Through the calculation above, we can draw a conclusion that the maximum wave force is controlled by the significant wave height, period, and wavelength. Taking the significant wave height as variable, the maximum wave force is also variable. The function of the maximum wave force with respect to the significant wave height is fitted (shown in Table 1) by using Matlab.

According to the data from DongHai meteorological station, we can make an assumption that the distribution of significant wave height meets I-Shaped extreme value distribution. The mean value is 4.2 m, the standard deviation is 9.85 m².

2.2 Wind load

The mode of action of wind load has fluctuation characteristics. Therefore, wind loads can be divided into static wind and fluctuating wind. The latter is with much higher value of wind pressure contribution than former. The logarithmic law and exponential law are two common laws of wind velocity function with respect to height. Most meteorologists think logarithmic law suits better, see, for example, Equation 7, Equation 8, Equation 9, and Equation 10:

$$U(Z) = \frac{1}{k} u^* \ln\left(\frac{Z - Z_d}{Z_0}\right) \quad (7)$$

$$Z_d = H - \frac{Z_d}{k_d} \quad (8)$$

$$k_d = \left[\frac{k}{\ln(10/Z_0)}\right]^2 \quad (9)$$

$$u^* = \left(\frac{\tau_0}{\rho}\right)^{\frac{1}{2}} \quad (10)$$

where $U(Z)$ = wind velocity with respect to Z (m/s); Z = height (m); k = Von Karman's coefficient; u^* = velocity of flow hear stress (m/s); Z_d = height of zero plane (m); H = the average height of the buildings on the ground (m); Z_0 = roughness length (m); τ_0 = flow hear stress (Pa), and ρ = density of air (kg/m³).

In Equation 8, we can define H as wave height. Then a conclusion can be made that wave height has influence on the wind velocity function. As a result, the wind force on the turbine tower decreases to some extent with the increase of the wave height. The relationship between the total wind force, wave height, and standard wind velocity on the measuring point can be described as a function through calculation, which is shown in Figure 2(a). We can know that by calculating, the loss of total wind force is less more than 25% when wave height is less than 8 m (shown in Fig. 2(b)).

Table 1. The fit function of the total wave force with respect to the significant wave height.

Parameters	P_1	P_2	Model	Mean square error	Coefficient of determination
Upper limit	301.7	53.88	f(x) = P_1*x + P_2	0.008272	0.9992
Fit	299.6	48.75			
Lower limit	297.5	43.63			

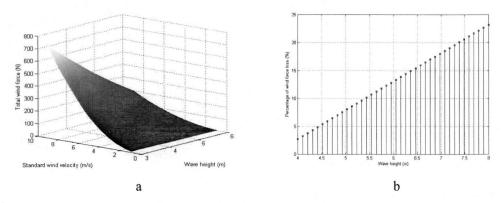

Figure 2. Total wind force with respect to wave height and standard wind velocity; the percentage of total wind force loss with respect to wave height.

Table 2. Parameters adapted to data from DongHai meteorological station.

Parameters	Roughness length (m)	Density of air (kg/m^3)	Mean value of standard wind velocity (m/s)	Standard deviation of standard wind velocity (m/s)	Height of measuring point (m)	Significant wave height (m)
Vanule	0.01	1.233	3.52	6.78	24.8	4.2

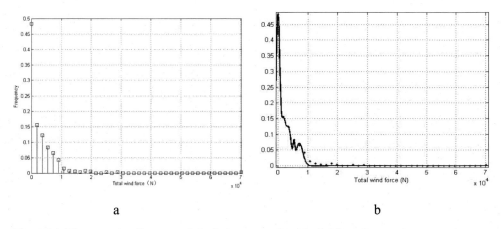

Figure 3. The stem-plot diagram and the fitting curve of total wind force frequency.

Table 3. The fit of probability density function with respect to the total wind force.

Parameters	a_i	b_i	c_i	Model	Mean square error	Coefficient of determination
i = 1	0.4751	4.85e-4	840.6	$f(x) = \sum \{a_i * \exp\{-[(x-b_i)/c_i]^2\}\}$ (i = 1~5)	0.005973	0.9947
i = 2	0.14	1800	1050			
i = 3	0.1146	3600	1155			
i = 4	0.05202	5399	415.9			
i = 5	0.07143	7215	1620			

According to the statistical data from DongHai meteorological station, the distribution of standard wind velocity on measuring point meets I-Shaped extreme value distribution. The parameters adopted to data from DongHai meteorological station are shown in Table 2.

The stem-plot diagram of total wind force frequency (shown in Fig. 3(a)) can be obtained by random sampling of the standard wind velocity function. Then we can get the distribution of total wind force by using Matlab to fit the probability density function (The fit function is shown in Table 3, the fitting curve is shown in Fig. 3(b)).

3 SUPERPOSITION OF WAVE AND WIND

Although the probability distribution of wind force is influenced by wave height, the joint probability density function of wind force and wave force is incalculable as the result of not meeting ordinary Gaussian distribution. But the loss of total wind force is limited in a certain range, and the effect of wave height change on wind force is also limited. So the wind force and the wave force are still calculated as independent random variables after the correction of wind force.

The stem-plot diagram of total force frequency (shown in Fig. 4(a)) can be obtained by random sampling of the probability density function of the two forces individually and calculating the superposition of each value. Then we can get the distribution of the total force by using Matlab to fit the probability density function (The fit function is shown in Table 4, the fitting curve is shown in Fig. 4(b)).

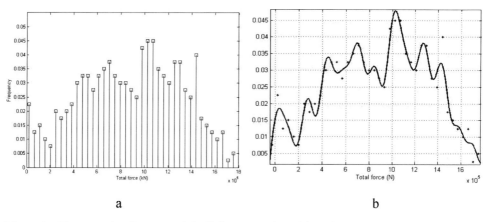

a   b

Figure 4. The stem-plot diagram and the fitting curve of total force frequency.

Table 4. The fit of probability density function with respect to total force.

Parameters	a_i	b_i	c_i	Model	Mean square error	Coefficient of determination
i = 1	0.03688	1.642e-6	0.07565	$f(x) = \sum a_i * \sin(b_i * x + c_i)$	0.004341	0.9422
i = 2	0.002964	6.06e-6	0.286	(i = 1~8)		
i = 3	0.002734	9.906e-6	1.812			
i = 4	0.004061	1.881e-5	0.8485			
i = 5	0.003303	3.184e-5	−0.3096			
i = 6	0.002551	2.46e-5	1.786			
i = 7	0.002896	1.28e-5	1.113			
i = 8	0.00208	4.376e-5	1.93			

4 CONCLUSION

Based on Fifth-order Stokes regular wave theory, this paper has studied the variation characteristics of the wave load on offshore wind turbine by simulating nonlinear waves. Through the study, an obvious linear relationship between the maximum wave force and the significant wave height is solved. Therefore, the probability distribution of maximum wave force depends on the probability distribution of significant wave height.

According to the logarithmic law, wind load was simulated in accordance with static wind due to the higher value of wind pressure contribution. The value of wind force was influenced by the wave height, the relationship between the two is approximate negative correlation. But from the perspective of load superposition, the increase of wave height means the increase of wave load. Therefore, the increasing part offsets some or all of the wind force loss. On account of that the value of wind force is also influenced by standard wind velocity, and the standard wind velocity changes with the variation of environmental characteristics, the wind force and the wave force are still calculated as independent random variables after the correction of wind force. The probability density function of total wind force is the superposition of five probability density functions, each of them meets Gauss distribution.

Observationally, the general trend of the distribution of total force can be described as Gauss distribution. But, because of the multitudinous peaks, the probability density function of total force is the superposition of nine sinusoidal function when looking for the best fit with Matlab, each of the functions is controlled by different parameters. In fact, the relationship between the accuracy of fitting and the number of sinusoidal functions is positive correlation. The complicated function is unrealistic to be popularized widely in engineering application, but simulations can be executed based on the data in a specific environment.

ACKNOWLEDGEMENTS

This project is supported by National Natural Science Foundation of Chongqing (CSTC2013JJ B30002) and Research Fund for the Doctoral Program of Higher Education of China (20125522110004).

REFERENCES

Chen, X.B., Li, J. & Chen, J.Y. Calculation of the Nonlinear Wave Force of Offshore Wind Turbine Based on the Stream Function Wave Theory. *Journal of Hunan University,* vol. 38, 2011: 22–28.
Chen, Z.Q. 2005, *Bridge wind engineering.* Beijing: China Communications Press.
Chen, K., Fu, L.B., Qian, J.H. & Jin, X.Y. Study on wind-resistant design of structures based on load effects. *Journal of Building Structures,* vol. 33, 2012: 27–34.
Claes, D. & Svend, O.H. 1996, *Wind Loads on Structures.* New York: John Wiley & Sons, Ltd. Press.
Dai, G.L., Gong, W.M., Shen, J.N. & Yang, C. Wave theory analysis of foundation of offshore wind farm near GongHai Bridge. *Chinese Journal of Geotechnical Engineering*, vol. 35, 2013: 456–461.
Han, L.Q. 2006. *Tutorial of Artificial Neural Nets.* Beijing: Beijing University of Posts and Telecommunications Press.
HydroChina Huadong Engineering Corporation, Hangzhou. 2009. *The overall planning report of offshore wind farm and intertidal zone in Rudong County of Jiangsu Province.*
JTS-145-2-2013, *Code of Hydrology for Sea Harbour.*
Meng, Y. & Lei, M.T. The Advance and Suggestion for the Study on Discharge Rate in Karst Tunnel Gushing, *Carsologica Sinica,* vol. 4, 2003. Guilin: 287–29.
Zhu, Y.R. 1991. *Wave Mechanics for Ocean Engineering.* Tianjin: Tianjin University Press.

Zr modified disordered mesoporous materials for CO$_2$ adsorption

W. Wu, Z. Tang & X. Zhang
College of Science, China University of Petroleum (East China), Qingdao, Shandong, China

ABSTRACT: The CO$_2$ adsorption capacity was dramatically enhanced by aggrandizing the acid site via introduction of Zr atom into the Disordered Mesoporous Silica (DMS) after loading amine. The physicochemical and adsorption properties of the adsorbents were characterized by XRD, TEM, UV-vis spectrophotometer, FTIR and N$_2$ adsorption-desorption techniques. The highest CO$_2$ uptake (4.55 mmol CO$_2$/g-adsorbent) was obtained for Zr(0.05)-DMS-TRPN-AN in the flow of 90% CO$_2$. Consecutive 15 adsorption-desorption runs revealed that the Zr(0.05)-DMS-TRPN-AN showed only a tiny decrease in adsorption capacity (from 1.775 to 1.758 mmol CO$_2$/g-adsorbent), providing thermal durability and adsorbent longevity.

1 INTRODUCTION

With fossil energy consumed, a large number of CO$_2$ emissions into the atmosphere, which has resulted in CO$_2$ concentration in the atmosphere increasing rapidly. Nowadays, there are many kinds of CO$_2$ capture and separation technology widely used in industry, including absorption, adsorption, membrane separation, microbial-fixation method, etc. Solid amine technology integrates the advantages of absorption and adsorption. It has been adopted as one of the promising methods in capturing atmospheric CO$_2$ (Fisher et al. 2009).

We synthesized controllable relative molecular mass and abundant surface functional groups of organic amine through modifying tri (3-aminopropyl) amine (TRPN), thus increasing the functional CO$_2$ adsorption sites on the surface of the adsorbent, which achieved a higher adsorption capacity. With the deepening of the research, we explored introduction of Zr on the surface of the Disordered Mesoporous Silica (DMS) to aggrandize the sites of electrophilic and nucleophilic. Through its reaction with CO$_2$ or amine effective site, we will achieve the improvement of CO$_2$ adsorption properties and cycle stability.

2 EXPERIMENTAL

2.1 Chemicals

All the reagents, such as Tetraethylorthosilicate (TEOS, ≥ 98%), cetyltrimethylammonium bromide (CTAB, AR), sodium hydroxide (NaOH, AR) acrylonitrile (AN, AR) and zirconyl chloride (ZrOCl$_2$·8H$_2$O, 98%) were obtained from Sinopharm and used as received.

2.2 Synthesis of adsorbents

The pure siliceous DMS material used in this paper has been synthesized according the procedure in our previous report (Zhang et al. 2012). Zr-modified DMS supports were prepared by chemical grafting. A certain amount of DMS material was added to a stand-up flask, immersed with just the right amount of ethanol, magnetic stirrer fully mixing. A certain amount of ZrOCl$_2$·8H$_2$O dissolved with ethanol added to the flask, fully stirring for 24 h at 25 °C, eliminating ethanol with vacuum rotary evaporation, and then, the flask was placed in the oven and dried overnight at 80–90 °C. The dried powder was then ground finely and calcined in the Muffle furnace at 550 °C for 6 h. Here after, we will denote Zr-containing DMS materials as Zr(x)-DMS samples, where x shows the ratio of Zr/Si in the sample.

Tri (3-aminopropyl) amine (TRPN) and the preparation of AN-modified TRPN were prepared by the method described earlier (Zhang et al. 2014). In our group, We used a wet impregnation method to prepare the TRPN-AN-modified materials. In a typical preparation, 1 g of TRPN-AN was dissolved in 5 g ethanol under stirring for about 30 min at 35 °C. 1 g of mesoporous material was then added to this solution. The resultant slurry was continuously stirred for about 3 h, and eliminating ethanol with vacuum rotary evaporation, then dried at 60 °C for 1 h under vacuum.

2.3 *Characterizations*

The crystal structures of the materials were investigated by X-ray diffraction (XRD) on a Philips X'Pert PRO SUPER X-ray diffractometer ($\lambda = 0.15418$ nm). Transmission electron microscopy (TEM) measurements were taken on a JEOL JEM-2100F 200kV. UV-vis spectra were recorded on Shimadzu UV-2550. The Fourier transform infrared (FTIR) spectra were obtained using a Nicolet MAGNA 750 Spectrometer. Specific area of the samples and structure parameter of pores were tested on ASAP2010 and win3000 multifunctional adsorption instrument.

2.4 *CO_2 Sorption and desorption measurement*

The adsorption and desorption performance of the adsorbent was tested on the instrument installed in our laboratory. According to weight change of the adsorbent measured by the microbalance, we evaluated the performance of the adsorbent materials. Before testing, the sample was heated to 100 °C for 1 h to eliminate CO_2 and water under decreased pressure. During the adsorption and desorption, about 2 g of the adsorbent was placed in a sample column, heated to 25 °C, and introducing 1% and 90% CO_2 adsorbate at a flow rate of 80 mL·min^{-1}, lasting for 60 min. After adsorption, the sample was regenerated at 60 °C for 1 h under vacuum. Adsorption capacity in mg-adsorbate/g-adsorbent and desorption in percentage were used to evaluate adsorbent and calculate the mass change in adsorption and desorption process. Cyclical adsorption and desorption processes were performed to evaluate the stability of the adsorbent.

3 RESULTS AND DISCUSSION

3.1 *Characterization*

Figure 1 presents high-angle powder XRD patterns of DMS and DMS modified Zr.

The sample DMS was observed in high-angle powder XRD patterns with the single intense peak at about 23.3° (2θ) accompanied by line broadening. This peak indicates disordered structure. After modified, when the ratio of Zr/Si was 0.01, the amount of ZrO_2 crystals formation was less, there was no change of obvious diffraction peak; While the ratio of Zr/Si increased to 0.1, Clear diffraction peaks were shown at 30.1°, 34.7°, 35.1°, 50.2°, 60.1°, 62.7° (2θ) and only four weak peaks detected for Zr(0.05)-DMS. It proved that typical tetragonal ZrO_2 crystals had formed in samples after modifying. (Ref. Code: 01–079–1767) And the carbonate species formed by CO_2 adsorption on the tetragonal ZrO_2 surface are more stable than on the monoclinic phase surface.

To investigate how the Zr distributed in the silica support, TEM micrographs of DMS, Zr (0.02)-DMS and Zr(0.1)-DMS materials were recorded in Figure 2. The structure of the irregular arrangement proved the disorder structure of DMS (Fig. 2a). As the ratio of Zr/Si is 0.02, no obvious ZrO_2 crystals were discovered (Fig. 2b). The crystalline ZrO_2 distributed uniformly over the framework of mesoporous silica (Fig. 2c). These results confirm that the increase of the Zr/Si atomic ratio in the solid products reflected as a function of the initial zirconium precursor concentration selected during post-grafting. The appearance of ZrO_2 clusters in Zr(0.1)-DMS presages that excess incorporation of Zr may hinder the dispersion of gases on the surface of silica.

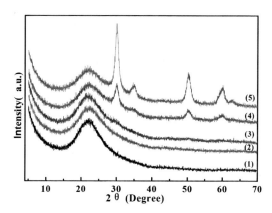

Figure 1. XRD patterns of DMS(1); Zr(0.01)-DMS(2); Zr(0.02)-DMS(3); Zr(0.05)-DMS(4) and Zr (0.1)-DMS(5).

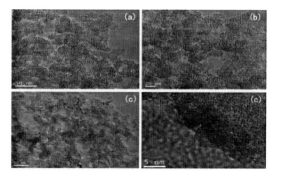

Figure 2. TEM of the DMS(a); Zr(0.02)-DMS(b); Zr(0.1)-DMS(c).

The FT-IR spectra of the Zr-containing DMS with various Zr/Si molar ratios are shown in Figure 3. The absorption peak at 1632 cm^{-1} is assigned to chemisorbed water. The peak at about 1103 cm^{-1} and 806 cm^{-1} could be ascribed to the stretching vibration of Si-O-Si in the frame structure of molecular sieve. In Figure 3 (A), the IR absorption at about 970 cm^{-1} generally was identified as the characteristic vibration of Zr-O-Si bond. After TRPN-AN impregnating, IR bands of asymmetric and symmetric stretches of CH$_2$ were located at 2940 cm^{-1} and 2830 cm^{-1}. The bands of 1467 cm^{-1} could be the stretching vibration of C-N. The TRPN-AN molecules have been successfully loaded inside the pore channels of DMS, which was confirmed by IR bands at 2247 cm^{-1} of the stretching vibration of C≡N (Chakraborty et al. 2007).

The ultraviolet spectra of a series of Zr-DMS samples and DMS are shown in Figure 4. No absorption arose from siliceous mesoporous materials of DMS material, which was transparent in the light of UV or visible. The Zr-grafted DMS samples are found to have an absorption band at 210–220 nm. The absorption band at 214 nm is usually attributed to Ligand-to-Metal Charge Transfer (LMCT) from an O^{2-} to an isolated Zr^{4+} ion in a tetrahedral configuration (Morey et al. 1999). The absorption edge in the range 230–300 nm shifts with increasing ZrO$_2$ content because of the formation of bulk zirconia (tetragonal phase) (Rodrıguez-Castellón et al. 2003).

Due to the loading and partial pores taken, a decrease of the overall pore volume and BET surface area is shown in Table 1. According to the ratio of raw materials, the expected theoretical ZrO$_2$ loading for the four samples, including Zr(0.01)-DMS, Zr(0.02)-DMS, Zr(0.05)-DMS and Zr(0.1)-DMS, would be 2.0, 3.9, 9.3 and 12.1 wt%, respectively. Results show that most of the ZrO$_2$ precursors in the solution react with the surface of DMS. With the increasing of ZrO$_2$ precursor concentration, it can't load to the pore surface completely, which affects the ultimate ZrO$_2$ content.

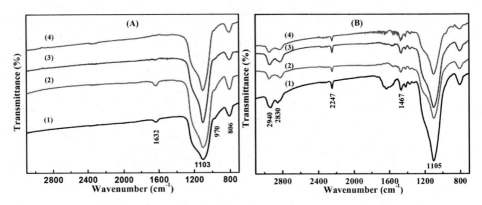

Figure 3. FT-IR spectra of Zr(0.01)-DMS (1); Zr(0.02)-DMS (2); Zr(0.05)-DMS (3) and Zr(0.1)-DMS (4) (A) before and (B) after 50 wt% impregnation of TRPN-AN.

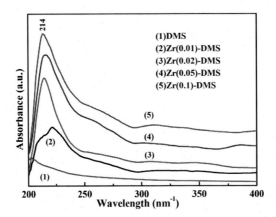

Figure 4. UV-vis spectra of calcined DMS and Zr(x)-DMS.

Table 1. Textural characteristics and chemical composition of DMS molecular sieves.

Adsorbent	Zr/Si (mol) Gel	Zr/Si (mol) Product	ZrO_2/ wt%	Zr/ atoms·nm^{-2}	Surface area/ m^2·g^{-1}	Volume/ cm^3·g^{-1}
DMS	–	–	–	–	239	1.33
Zr(0.01)-DMS	0.01	0.009	1.8	0.41	225	1.28
Zr(0.02)-DMS	0.02	0.016	3.2	0.84	220	1.21
Zr(0.05)-DMS	0.05	0.036	6.9	3.42	213	1.13
Zr(0.1)-DMS	0.1	0.067	12.1	6.40	208	1.06

3.2 *Adsorption and desorption properties*

Not accidentally, the impregnating of TRPN-AN significantly reduced the surface area and pore volume compared with the untreated DMS (239 m^2·g^{-1}) or Zr-loaded DMS (Table 2). As the Zr/Si atomic ratio increases from 0.01 to 0.1, the surface area of the sample decreases from 31.10 m^2·g^{-1} to 15.03 m^2·g^{-1}. We can also observe a similar trend in pore volume. After impregnating 50 wt.% TRPN-AN, pore volume of Zr(0.05)-DMS sharply descends, but maintaining a relatively large pore volume and surface area. Known from adsorption data in Table 2, Zr(0.05)-DMS-TRPN-AN showed the highest adsorption capacity, which is 1.3 times DMS-TRPN-AN under the same condition. Especially in low CO_2 concentration (1% CO_2), the adsorption capacity of Zr(0.05)-DMS-TRPN-AN is 2 times of DMS-TRPN-AN. Increasing the amount of zirconium, the CO_2 adsorption decreased correspondingly. This observation is similar to that reported (Kuwahara et al. 2012). In addition, the CO_2 adsorption capacity was 1.78 mmol $CO_2 \cdot$ g-adsorbent in the flow of 1% CO_2, which was much lower than that at 90% CO_2 (4.55 mmol $CO_2 \cdot$ g-adsorbent). However, compared with other reported adsorbents, Zr(0.05)-DMS-TRPN-AN showed distinguished CO_2 adsorption property under the same CO_2 concentration (Belmabkhout et al. 2009, Lee et al. 2002). Loading zirconium oxide can increase the acid sites on the surface of the adsorbent (Mei et al. 2012). When the Zr/Si atomic ratio reaches 0.05, zirconium dioxide precursor strongly interacts with silica surface by forming the bond of Zr-O-Si, evenly distributes on the surface of the adsorbent. However, when the atomic ratio increases to 0.1, only 1.33 mmol $CO_2 \cdot$ g-adsorbent. Because the moderate loading of zirconium dioxide can be evenly distributed on the surface of the DMS, excessive amounts of zirconium dioxide will result in accumulation phenomenon. It is proved that the appearance of little ZrO_2 cluster on the surface of Zr(0.1)-DMS adsorbent in TEM images(see Fig. 2(c)). Mass transfer resistance resulting in that ZrO_2 cluster blocked part of the adsorbent channel, hindered the diffusion of CO_2, which reduces the CO_2 adsorption performance (Feng et al. 2013).

3.3 Regeneration performance

The recycling performance of DMS-TRPN-AN and Zr(0.05)-DMS-TRPN-AN was tested over fifteen adsorption/desorption cycles in the flow of 1% CO_2 (Fig. 5). During the 15 cycles, the CO_2 sorption capacity of DMS-TRPN-AN adsorption decreased from 0.857 to 0.819 mmol CO_2/g-adsorbent, maintaining as much as 95% of its initial ability. After grafting acrylonitrile on TRPN, end group primary amine was effectively protected, improving the efficiency of adsorbent recycling, but the volatilization and degradation of TRPN-AN was still a problem to be solved in the process of vacuum desorption for a kind of applicable to infinite loop adsorbent preparation. After modified by Zr, the CO_2 adsorption capacity

Table 2. Organic loading and adsorption efficiencies of adsorbents (at 25 °C).

Materials	Surface area/ $m^2 \cdot g^{-1}$	Volume/ $cm^3 \cdot g^{-1}$	Organic loading/ wt%	mmolCO_2/g-adsorbent 90% CO_2	1% CO_2
DMS-TRPN-AN	31.10	0.238	48.9	3.47	0.86
Zr(0.01)-DMS-TRPN-AN	21.02	0.115	48.4	4.21	1.63
Zr(0.02)-DMS-TRPN-AN	17.07	0.150	49.1	4.52	1.75
Zr(0.05)-DMS-TRPN-AN	17.70	0.189	48.2	4.55	1.78
Zr(0.1)-DMS-TRPN-AN	15.03	0.080	48.6	3.72	1.33

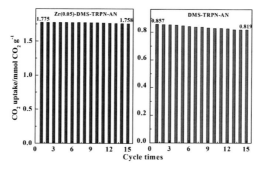

Figure 5. Multicycle CO_2 adsorption/desorption testing of DMS-TRPN-AN and Zr(0.05)-DMS-TRPN-AN.

of Zr(0.05)-DMS-TRPN-AN was 1.758 mmol CO_2/g-adsorbent during the 15 cycles, only decreased by 0.96%. The introduction of heteroatoms zirconium changed the acid or base characteristics of the silica support, which significantly increased the adsorption properties of adsorbent and circulation efficiency, the reason was probably that the existence of the surface of zirconium dioxide stabilized organic amine TRPN-AN, thus reducing the losses in the process of recycling.

4 CONCLUSION

Incorporation of Zr specie into disordered mesoporous silica has been shown to create amine-modified adsorbents with dramatically enhanced CO_2 adsorption capacities and recycling stabilities. The highest CO_2 uptakes attained 4.55 and 1.78 mmol CO_2/g-adsorbent at 25 °C under 90% CO_2 and 1% CO_2 conditions for sample with optimal Zr content, respectively. Compared with the conventional DMS-TRPN-AN materials, the introduction of Zr also enhanced the material recycling stability through creating more effective amine-stabilizing sites. The 15 adsorption/desorption cycle experiments revealed that the new adsorbent showed only a minor drop in CO_2 capture capacity, providing excellent thermal durability and adsorbent longevity.

ACKNOWLEDGMENTS

We thank the National Natural Science Foundation of China (Project No.21172264) and the Fundamental Research Funds for the Central Universities in China (15CX05028 A) (24720146039) for financial support.

REFERENCES

Belmabkhout, Y. & Serna-Guerrero, R. 2009. Adsorption of CO_2-containing gas mixtures over amine-bearing pore-expanded MCM-41 silica: application for gas purification. *Industrial & Engineering Chemistry Research* 49(1): 359–365.

Chakraborty, S. & Bandyopadhyay, S. 2007. Application of FTIR in characterization of acrylonitrile—butadiene rubber (nitrile rubber). *Polymer testing* 26(1): 38–41.

Feng, X. & Hu, G. 2013. Tetraethylenepentamine-modified siliceous mesocellular foam (MCF) for CO_2 capture. *Industrial & Engineering Chemistry Research* 52(11): 4221–4228.

Fisher, J.C. & Tanthana, J. 2009. Oxide-supported tetraethylenepentamine for CO_2 capture. *Environmental progress & sustainable energy* 28(4): 589–598.

Kuwahara, Y. & Kang, D.Y. 2012. Enhanced CO_2 Adsorption over Polymeric Amines Supported on Heteroatom-Incorporated SBA-15 Silica: Impact of Heteroatom Type and Loading on Sorbent Structure and Adsorption Performance. *Chemistry-A European Journal* 18(52): 16649–16664.

Lee, J.S. & Kim, J.H. 2002. Adsorption equilibria of CO_2 on zeolite 13X and zeolite X/activated carbon composite. *Journal of Chemical & Engineering Data* 47(5): 1237–1242.

Mei, B. & Becerikli, A. 2012. Tuning the Acid/Base and Structural Properties of Titanate-Loaded Mesoporous Silica by Grafting of Zinc Oxide. *The Journal of Physical Chemistry C* 116(27): 14318–14327.

Morey, M.S. & Stucky, G.D. 1999. Isomorphic substitution and postsynthesis incorporation of zirconium into MCM-48 mesoporous silica. *The Journal of Physical Chemistry B* 103(12): 2037–2041.

Rodrıguez-Castellón, E. & Jiménez-López, A. 2003. Textural and structural properties and surface acidity characterization of mesoporous silica-zirconia molecular sieves. *Journal of Solid State Chemistry* 175(2): 159–169.

Zhang, X. & Zhang, S. 2014. Development of TRPN dendrimer-modified disordered mesoporous silica for CO_2 capture. *Materials Research Bulletin* 56: 12–18.

Zhang, X. & Zheng, X. 2012. AM-TEPA impregnated disordered mesoporous silica as CO_2 capture adsorbent for balanced adsorption-desorption properties. *Industrial & Engineering Chemistry Research* 51(46): 15163–15169.

Resources, Environment and Engineering II – Xie (Ed.)
© 2016 Taylor & Francis Group, London, ISBN 978-1-138-02894-4

The bending of triangular plate under a concentrated bending moment

Yingjie Chen & Zhou Lei
Construction Engineering and Mechanics College of Yanshan University, Qinhuangdao, China

ABSTRACT: The issue of the bending of triangular plate under the action of a bending moment is involved in many engineering projects, but that under a concentrated bending moment is seldom discussed. On the basis of mixed variables method [1], this thesis studies the bending of triangular plate under a concentrated bending moment and different constraint conditions. With no need to consider the displacement function hypothesis, this method gets rid of the tedious procedures used in the classical solution by which the total potential energy of mixed variables can be obtained as well as the curvilinear equation of the triangular plate with one free hypotenuse and two fixed border. Besides, the unknown quantity will become clear through the corresponding curvilinear equations mentioned.

1 INTRODUCTION

Scholars and experts of various countries have ever made a great amount of studies on the inflexion problem of elastic thin plates [2.3]. However, very few people have ever studied that of triangular plate under the action of a concentrated bending moment [4.5]. With the development of national economy, more and more large scale plants and high-rise buildings are under construction. Therefore, the influence of bending moment exerted on triangular plates will be inevitably taken into consideration, such as the bending moment of large-scale plants' plates caused by motor eccentric rotor, and that caused by a sudden start of a mechanical equipment. On the basis of the mixed variables method, this thesis studies the inflexion problem of any point on a triangular plate under the action of a concentrated bending moment.

2 THE BENDING OF TRIANGULAR PLATE WITH A FREE HYPOTENUSE AND TWO FIXED SIDES STARTED

2.1 *The distribution of constraints on actual system*

The actual system is a right triangular plate with a concentrated bending moment M acted on any point (x_0, y_0) of it, as shown in Figure 1. On releasing the bending constraint on the fixed side, the constraint will be replaced by the distributed moment, as shown in Figure 2.

$$M(x)_{y=0} = \sum_{m=1}^{\infty} E_m \sin\frac{m\pi x}{a} \quad (1)$$

$$M(y)_{x=0} = \sum_{n=1}^{\infty} A_n \sin\frac{n\pi y}{b} \quad (2)$$

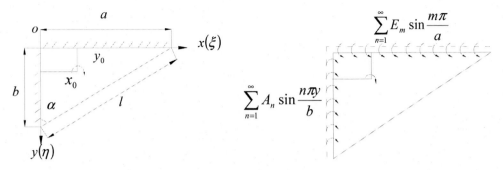

Figure 1. The actual system.

Figure 2. The actual system free from constraint.

2.2 Deflection surface equation

Suppose the deflection and the rotating angle equations of the free side are, respectively:

$$W_n = \sum_{m=1}^{\infty} a_m \sin\left(1 - \frac{x}{a}\right) m\pi \tag{3}$$

$$\theta_n = \sum_{m=1}^{\infty} b_m \cos\left(1 - \frac{x}{a}\right) m\pi \tag{4}$$

Suppose the deflection surface equation of the triangular plate is:

$$W(x,y) = \sum_{m=1}^{\infty} \sum_{n=1}^{\infty} a_{mn} \sin\alpha_m x \sin\beta_n y \tag{5}$$

Applying the mixed variables method into the actual system shown in Figure 2, the total potential energy of the mixed variables is obtained:

$$\Pi_{mp} = \int_0^a \int_0^b \frac{1}{2} D \left\{ \left(\frac{\partial^2 W}{\partial x_0^2} + \frac{\partial^2 W}{\partial y_0^2}\right) - 2(1-\nu)\left[\frac{\partial^2 W}{\partial x_0^2}\frac{\partial^2 W}{\partial y_0^2} - \left(\frac{\partial^2 W}{\partial x_0 \partial y_0}\right)^2\right] \right\} dxdy$$
$$- \left[\frac{\partial^2 W}{\partial x_0^2}\frac{\partial^2 W}{\partial y_0^2} - \left(\frac{\partial^2 W}{\partial x_0 \partial y_0}\right)^2\right] \right\} dxdy - \frac{l}{a}\int_0^{\frac{a(b-\eta)}{b}} V_n W_n dx - \frac{l}{a}\int_{\frac{a(b-\eta)}{b}}^{a} V_n W_n dx \tag{6}$$
$$- \frac{l}{a}\int_0^{\frac{a(b-\eta)}{b}} M_n \theta_n dx - \frac{l}{a}\int_{\frac{a(b-\eta)}{b}}^{a} M_n \theta_n dx - MW(x_0, y_0; \varepsilon, \eta)$$

By replace V_n, M_n, by (5) and taking the variational extremum of a_{mn}, the expression of a_{mn} is obtained. By putting a_{mn} into (5) and conducting transformation of trigonometric series and hyperbolic function, the deflection surface equation of the triangular plate is obtained:

$$\alpha_n = n\pi a/b$$

$$W = \frac{Ma^2}{b\pi^3 D}\sum_{m=1}^{\infty}\left[1 + \beta_m cth\beta_m - \frac{\beta_m(b-y_0)}{b}\cdot cth\frac{\beta_m(b-y_0)}{b} - \frac{\beta_m \eta}{b} cth\frac{\beta_m \eta}{b}\right] \tag{7}$$

$$\cdot \frac{1}{m^3 sh\beta_m} sh\frac{\beta_m \eta}{b} sh\frac{\beta_m(b-y_0)}{b}\cos\frac{m\pi x_0}{a}\sin\frac{m\pi\xi}{a} + \frac{a}{\pi}\sum_{m=1}^{\infty}\left\{\left[\frac{8a^4 + 2a^2b^2(4-\mu) + b^4}{a(4a^2+b^2)^2}\right.\right.$$

$$\begin{aligned}&\left. \cdot sh\beta_m + \frac{a(1-\mu)\beta_m}{4a^2+b^2}ch\beta_m \right]\sin\frac{2m\pi\eta}{b} - \frac{1-\mu}{b}sh\frac{\beta_m\eta}{b}sh\frac{\beta_m(b-\eta)}{b}\left[1+\frac{b^2(4a^2-b^2)}{(4a^2+b^2)^2}\right.\\ &\left.\cdot\cos\frac{2m\pi\eta}{b}\right] + \frac{1-\mu}{2b}\left(1-\frac{b^2}{4a^2+b^2}\cos\frac{2m\pi\eta}{b}\right)\left(2\beta_m sh\frac{\beta_m\eta}{b}\cdot ch\frac{\beta_m(b-\eta)}{b}+\frac{\beta_m\eta}{b}\right.\\ &\left. sh\frac{\beta_m(b-2\eta)}{b} - sh\beta_m\cdot\frac{\beta_m\eta}{b}\right) + \frac{a(1-\mu)}{4a^2+b^2}\sin\frac{2m\pi\eta}{b}\cdot\left[\frac{4b^2}{4a^2+b^2}\cdot ch\frac{\beta_m\eta}{b}sh\frac{\beta_m(b-\eta)}{b}\right.\\ &\left. -\beta_m\cdot ch\frac{\beta_m(b-2\eta)}{b}-2\frac{\beta_m\eta}{b}sh\frac{\beta_m\eta}{b}sh\frac{\beta_m(b-\eta)}{b}\right]\right\}\frac{a_m\cos m\pi}{msh\beta_m}\cdot\sin\frac{m\pi\xi}{a}+\frac{ab}{\pi^2 l}\\ &\cdot\sum_{m=1}^{\infty}\frac{4a^4+a^2b^2(4+\mu)+b^4}{(4a^2+b^2)^2}\cdot\frac{b_m\cos m\pi}{m^2}\sin\frac{2m\pi\eta}{b}\sin\frac{m\pi\xi}{a}+\frac{a^2}{2\pi^2 D}\\ &\cdot\sum_{m=1}^{\infty}\frac{E_m}{m^2}\left[-\frac{\beta_m}{sh^2\beta_m}sh\frac{\beta_m\eta}{b}+cth\beta_m\frac{\beta_m\eta}{b}\cdot ch\frac{\beta_m\eta}{b}-\frac{\beta_m\eta}{b}sh\frac{\beta_m\eta}{b}\right]\sin\frac{m\pi\xi}{a}\\ &+\frac{b^2}{2\pi^2 D}\sum_{m=1}^{\infty}\frac{A_m}{n^2}\left[-\frac{\alpha_n}{sh^2\alpha_n}sh\frac{\alpha_n\xi}{a}+cth\alpha_n\frac{\alpha_n\xi}{a}ch\frac{\alpha_n\xi}{a}-\frac{\alpha_n\xi}{a}sh\frac{\alpha_n\xi}{a}\right]\sin\frac{n\pi\eta}{b}\end{aligned}$$

(7a)

2.3 Boundary conditions

The deflection surface equation (7) has to meet the following boundary conditions:

$$\left.\frac{\partial W}{\partial \eta}\right|_{\eta=0} = 0 \tag{8}$$

$$\left.\frac{\partial W}{\partial \xi}\right|_{\xi=0} = 0 \tag{9}$$

$$-2M_{\xi\eta}\sin\alpha\cos\alpha\Big|_{\xi=a(b-\eta)/b} = 0 \tag{10}$$

$$V_n(\xi,\eta)\Big|_{\xi=a(b-\eta)/b} = V_\xi\cos\alpha + V_\eta\sin\alpha\Big|_{\xi=a(b-\eta)/b} = 0 \tag{11}$$

In order to meet these boundary conditions, four equations are obtained after putting expression (7) into (8), (9), (10), and (11). By combing those equations, an infinite system of linear equations will be obtained, by which the values of, and can be known as well as the deflection and internal forces. A specific problem will be solved, because the point on the triangular plate, which bears the action of the concentrated load, is known.

2.4 Numerical calculation conditions

Taken the following as a numerical example. If a = b = 1 m, concentrated bending moment M = 100 N·m, Elastic modulus E = 200 GPa, and Poisson's ratio = 0.3, the deflections of the hypotenuse and the bending moments of the section can be calculated, respectively, when x = 0, y = 0, while the bending moment acts on point (0.2, 0.2),(0.3, 0.3),(0.5, 0.5). There are four unknown numbers, and the boundary conditions are four liner equations. Making use of the Iterative improvement method, so that, m = n = 1, 2, 3 … 80. Then, the values of and can be obtained with the application of Matlab. When the results are put into the deflection surface equation and the numerical distribution of the deflection function will be obtained, as shown in Figures 3, 4, and 5, Tables 1, 2, and 3.

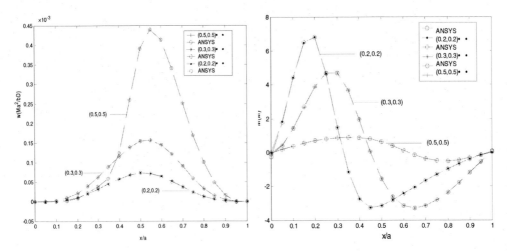

Figure 3. The distribution curve of deflection along the hypotenuse.

Figure 4. The changing curve of M_x along x axis at the point: y = 0.

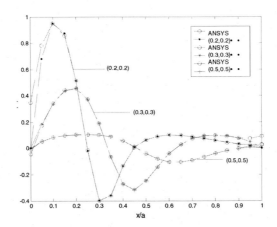

Figure 5. The changing curve of M_x along x axis at the point: y = 0.

Table 1. Data of the hypotenuse' deflection changing Ma^2/bD (×10^{-3}).

	(0.2,0.2)		(0.3,0.3)		(0.5,0.5)	
y/b	Calculation	ANSYS	Calculation	ANSYS	Calculation	ANSYS
0.0	0.00000	0.00000	0.00000	0.00000	0.00000	0.00000
0.1	0.00129	0.00129	0.00303	0.00303	0.00097	0.00097
0.2	0.01073	0.01073	0.02198	0.02197	0.00911	0.00910
0.3	0.03186	0.03186	0.06185	0.06184	0.03858	0.03857
0.4	0.05801	0.05801	0.11494	0.11492	0.12492	0.12490
0.5	0.07321	0.07319	0.15512	0.15509	0.39093	0.39089
0.6	0.06400	0.06399	0.14505	0.14503	0.41298	0.41296
0.7	0.03716	0.03716	0.09001	0.09000	0.25239	0.25238
0.8	0.01258	0.01258	0.03333	0.03332	0.10050	0.10049
0.9	0.00145	0.00145	0.00444	0.00444	0.01565	0.01565
1.0	0.00000	0.00000	0.00000	0.00000	0.00000	0.00000

Table 2. Date of M_x changing along y axis at the point: $x = 0$ (M).

y/b	(0.2,0.2) Calculation	ANSYS	(0.3,0.3) Calculation	ANSYS	(0.5,0.5) Calculation	ANSYS
0.0	0.00000	−0.24963	0.00000	−0.11039	0.00000	−0.00763
0.1	4.41701	4.41790	1.44422	1.44510	0.39438	0.39526
0.2	6.80506	6.80610	3.88001	3.88110	0.72906	0.73011
0.3	1.48612	1.48660	4.70336	4.70380	0.87202	0.87253
0.4	−2.74124	−2.74180	1.96041	1.95990	0.88389	0.88332
0.5	−3.13288	−3.13320	−1.56932	−1.56970	0.64116	0.64081
0.6	−2.43017	−2.43050	−3.14688	−3.14720	0.11304	0.11267
0.7	−1.66339	−1.66360	−3.14715	−3.14740	−0.36857	−0.36872
0.8	−0.97285	−0.97300	−2.34918	−2.34930	−0.51956	−0.51977
0.9	−0.38345	−0.38362	−1.18154	−1.18170	−0.32131	−0.32156
1.0	0.00000	0.05800	0.00000	0.07580	0.00000	0.04020

Table 3. Date of M_x changing along y axis at the point: $x = 0$ (M).

y/b	(0.2,0.2) Calculation	ANSYS	(0.3,0.3) Calculation	ANSYS	(0.5,0.5) Calculation	ANSYS
0.0	0.00000	0.34893	0.00000	−0.04190	0.00000	−0.00409
0.1	0.95023	0.95112	0.34016	0.34106	0.08180	0.08270
0.2	0.51232	0.51336	0.45479	0.45583	0.10143	0.10245
0.3	−0.39734	−0.39686	0.18992	0.19041	0.10042	0.10088
0.4	−0.13538	−0.13594	−0.27094	−0.27152	0.05683	0.05630
0.5	0.06242	0.06210	−0.24923	−0.24958	−0.03975	−0.04010
0.6	0.09933	0.09900	−0.04104	−0.04140	−0.10412	−0.10447
0.7	0.08781	0.08760	0.07150	0.07130	−0.08957	−0.08980
0.8	0.05965	0.05950	0.09813	0.09800	−0.04111	−0.04130
0.9	0.03027	0.03010	0.07795	0.07780	−0.00126	−0.00143
1.0	0.00000	0.02890	0.00000	0.07580	0.00000	0.02560

3 CONCLUSION

Based on the mixed variables method for thin plate bending, this thesis studies the triangular plate with a free hypotenuse and two fixed sides and figures out the deflection surface equation under a concentrated bending moment. On the Matlab platform, the calculation results of different problems indicate that the application of Mixed variables method lead to right results.

REFERENCES

Baolian Fu. The application of the Principle of Energy in Elastic Mechanics [M]. *Beijing:science press*, 561–585, 2004.
Zhilun Xu. Elastic Mechanics [M]. Beijing: *People's publishing House*, 1983.
Timoshenkos. et al., Theory of plates and shells [S]. *MCGRAW—HILL Book Company Inc.*, New york, 1959.
Web crippling of cold-formed steel members; *Ph. D. Thesis*, 341, 1992.
Nippon Kikai Gakkai Ronbunshu A Hen. Vol. 63, *ISSN 0387-5008*, no. 614, pp. 2167–2173, Oct 1997.

Nonlinear mechanics analysis of a thin plate under different loads in a magnetic field

Yuhong Bian
Key Laboratory of Mechanical Reliability for Heavy Equipments and Large Structures of Hebei Province, College of Civil Engineering and Mechanics, Yanshan University, Qinhuangdao, China

ABSTRACT: Based on the nonlinear magnetoelastic kinetic equations, normal Cauchy form nonlinear differential equations, which include eight basic functions in all, are obtained by the variable replacement method. Using the linearization method, the nonlinear magnetoelastic equations are reduced to the linear iteration equations, and numerical calculation method is obtained. The stresses and deformations in a thin current-carrying plate under the combined action of an electromagnetic field and mechanical loads are analyzed by considering a specific example, and the magnetoelastic effect on the plate by the side electric current and magnetic induction intensity is studied. The variation regularities of the deflection in the thin plate under the trapezoidal loads, triangular loads, and parabolic loads are respectively obtained.

1 INTRODUCTION

In modern technological fields, such as aerospace industry, magnetic suspension transportation, electromechanical power system, giant equipment, etc., there are usually that the structures, especially the current-carrying structures, are in the environment of the interaction of an electromagnetic field and a mechanical field. The stresses and deformations are influenced and controlled by an electromagnetic field and mechanical load. At present, the achievements on theoretical calculation and numerical analysis for the plates and shells are relatively complete [1–6]. However, except for the problems about the vibration and stability of the structures such as rods, beams, plates, shells, etc. in electromagnetic fields, the analysis on the stress-strain state is rarely. Therefore, the analysis on the stress-strain state in the current-carrying plates and shells in electromagnetic fields is both theoretically and practically an important topic.

The fundamental equations and a numerical algorithm of a thin current-carrying plate under combined action of an electromagnetic field and mechanical loads are developed in this paper. The stresses and deformations in a thin plate under coupling fields are calculated by this method, and the magnetoelastic effect on the plate by the side electric current and magnetic induction intensity is studied. The variation regularities of the deflection in a thin current-carrying plate under trapezoidal loads, triangular loads, and parabolic loads are respectively obtained. A method of theoretical analysis and numerical calculation is provided for devising structures in an electromagnetic field.

2 FUNDAMENTAL EQUATIONS

By satisfying magnetoelastic supposition of thin plate [6] and using Ohm's law and Maxwell equations in electromagnetic basic theories [7], at the same time, considering electromagnetic parameter relations, the nonlinear magnetoelastic kinetic equations and electrodynamics equations of thin plate under the action of a magnetic field can be derived as follows:

$$\frac{\partial (rN_r)}{\partial r} - N_\theta + r(F_r + \rho f_r) = r\rho h \frac{\partial^2 u}{\partial t^2}, \quad (1)$$

$$\frac{\partial(rQ_r)}{\partial r} + r(F_z + \rho f_z) = r\rho h \frac{\partial^2 w}{\partial t^2}, \qquad (2)$$

$$\frac{\partial(rM_r)}{\partial r} - M_\theta - rN_r\theta_r - rQ_r = \frac{r\rho h^3}{12}\frac{\partial^2 \theta_r}{\partial t^2}, \qquad (3)$$

$$-\frac{\partial B_z}{\partial t} = \frac{1}{r}\frac{\partial(rE_\theta)}{\partial r}, \qquad (4)$$

$$\sigma\mu\left[E_\theta + \frac{1}{2}\frac{\partial w}{\partial t}(B_r^+ + B_r^-) - \frac{\partial u}{\partial t}B_z\right] = -\frac{\partial B_z}{\partial r} + \frac{B_r^+ - B_r^-}{rh}, \qquad (5)$$

where h is the thickness of the plate; t is the time variable; N_r, N_θ, Q_r, M_r, and M_θ are the internal forces and moments; F_r and F_z are the mechanical loads; u and w are the displacement components; θ_r is the angle of rotation; ρf_r and ρf_z are the Lorentz force components; ρ is the mass density of the medium; σ is the electrical conductivity of the material; μ is the permeability of the material; E_θ is the electric field intensity along the θ-direction; B_z is the magnetic induction intensity along the z-direction; B_r^+ and B_r^- are the values of B_r on the upper and lower surfaces of the plate, respectively.

3 COMPUTATIONAL METHOD

Selecting $u, w, \theta_r, N_r, Q_r, M_r, B_z$, and E_θ as basic unknown functions, at the same time, considering the geometric and physical equations of thin plate [8], the magnetoelastic coupling equations, which can be written as follows boundary-value problems, are obtained.

$$\frac{\partial \mathbf{N}}{\partial r} = \mathbf{F}(r,\mathbf{N}) \quad (r_0 \le r \le r_1), \, D_1\mathbf{N}(r_0) = \mathbf{d}_1, \, D_2\mathbf{N}(r_1) = \mathbf{d}_2, \qquad (6)$$

where $\mathbf{N} = \{u, w, \theta_r, N_r, Q_r, M_r, B_z, E_\theta\}^{\mathrm{T}}$; D_1 and D_2 are given orthogonal matrixes; $\mathbf{d}_1$ and $\mathbf{d}_2$ are given vectors. For problems (6), Newmark's stable finite equidifferent formulas [6] are used to find the derivatives with respect to time in the magnetoelastic coupling equations for a time step length, thus, equations (6) can be expressed as

$$\frac{d\mathbf{N}}{dr} = \mathbf{F}(r,\mathbf{N}) \quad (r_0 \le r \le r_1), \, D_1\mathbf{N}(r_0) = \mathbf{d}_1, \, D_2\mathbf{N}(r_1) = \mathbf{d}_2. \qquad (7)$$

The problems described by Eq. (7) are nonlinear. With the linearization method, nonlinear problems can be turned into a series of linear problems, thus, we obtain a set of linear ordinary differential equations. For the corresponding initial and boundary conditions, all unknown variables can be found by the discrete-orthogonalization method.

4 EXAMPLE AND ANALYSIS

A thin annular plate made of aluminium alloy is in a magnetic field $B = \{B_r, 0, 0\}$. The density of side current in the plate is $\mathbf{J}_{\mathrm{el}} = \{0, J_{\theta\mathrm{el}}, 0\}$, the mechanical load is $\mathbf{F} = \{0, 0, F_z\}$. Let inside radius is $r_0 = 0.5$ m, outside radius is $r_1 = 1.0$ m, $h = 1 \times 10^{-3}$ m, $E = 70$ Gpa, $v = 0.33$, $\rho = 2700$ kg/m^3, $\sigma = 3.60 \times 10^7$ $(\Omega \cdot \mathrm{m})^{-1}$, $\mu = 1.25 \times 10^{-6}$ H/m, $J_{\theta\mathrm{el}} = J_\theta \sin \omega t$ A/m^2.

The boundary conditions are

$$r = r_0: u = 0, w = 0, M_r = 0, B_z = -0.1\sin\omega t \text{ T}; \quad r = r_1: u = 0, w = 0, M_r = 0, B_z = 0.$$

The initial conditions are

$$\mathbf{N}(r,t)\big|_{t=0} = 0, \dot{u}(r,t)\big|_{t=0} = \dot{w}(r,t)\big|_{t=0} = \dot{\theta}_r(r,t)\big|_{t=0} = 0.$$

In order to analyze and compare easily, the mechanical loads are chosen according to the three following schemes.

The trapezoidal loads: $P_z = -150 - (r - r_0)\tan\varphi \, \text{N/m}^2$, $\varphi = 30°$. Figure 1 shows the deflection distribution in the plate along the radius for $J_\theta = 5 \times 10^4 \, \text{A/m}^2$, $t = 10$ ms, and different magnetic induction intensity. Figure 2 shows the deflection distribution along the radius for $B_r = 0.01\,\text{T}$, $t = 10$ ms, and different electric current density. Figure 3 shows the deflection distribution along the radius for $J_\theta = 5 \times 10^4 \,\text{A/m}^2$, $B_r = 0.01\,\text{T}$, $t = 10$ ms, and different

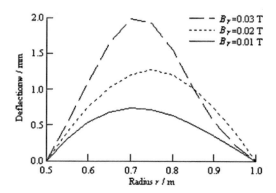

Figure 1. The w-r curves for different B_r.

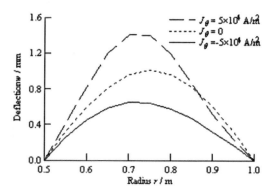

Figure 2. The w-r curves for different J_θ.

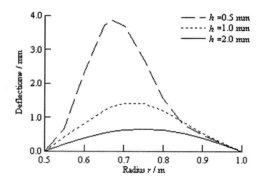

Figure 3. The w-r curves for different h.

thickness of the plate. Figure 4 shows the variation of the stress at $r = 0.75$ m with time for $J_\theta = 5 \times 10^4 \, \text{A/m}^2$ and $B_r = 0.01$ T. Curves a and b are the variation of radial stresses on the upper, lower surfaces of the plate with time, respectively; curves c and d are the variation of hoop stresses on the upper, lower surfaces of the plate with time, respectively. Figure 5 shows Lorentz force ρf_z distribution along the radius for $J_\theta = 5 \times 10^4 \, \text{A/m}^2$, $B_r = 0.01$ T, and different moment. Figure 6 shows the deflection distribution along the radius under the trapezoidal loads for $J_\theta = 5 \times 10^4 \, \text{A/m}^2$, $B_r = 0.01$ T, and different moment.

The triangular loads: $P_z = -1000 \times (r - r_0) \tan\varphi \, \text{N/m}^2$, $\varphi = 30°$. Figure 7 shows the deflection distribution in the plate along the radius under the triangular loads for $J_\theta = 5 \times 10^4 \, \text{A/m}^2$, $B_r = 0.01$ T, and different moment.

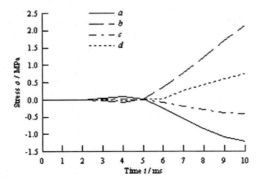

Figure 4. The σ-t curves at $r = 0.75$ m.

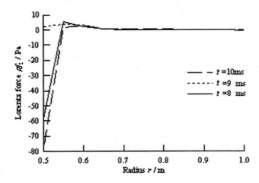

Figure 5. The ρf_z-r curves.

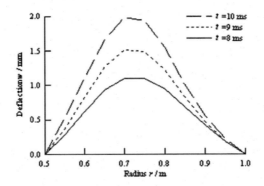

Figure 6. The w-r curves under trapezoidal loads.

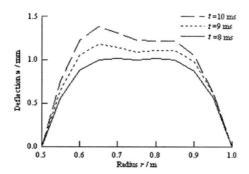

Figure 7. The *w-r* curves under triangular loads.

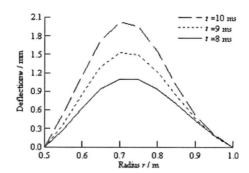

Figure 8. The *w-r* curves under parabolic loads.

The parabolic loads: $P_z = -150 - 100 \, r^2 \, \text{N/m}^2$. Figure 8 shows the deflection distribution in the plate along the radius under the parabolic loads for $J_\theta = 5 \times 10^4 \, \text{A/m}^2$, $B_r = 0.01 \, \text{T}$, and different moment.

5 CONCLUSIONS

The stress-strain state in a thin current-carrying plate can be controlled by changing the electromagnetic parameters. By analyzing the magnetoelastic effect in a thin plate under different loads, it is shown that considering electromagnetic effect is necessary when the engineering structures in electromagnetic fields are devised.

REFERENCES

Zheng, X.J. Zhang, J.P. & Zhou, Y.H. 2005. Dynamic stability of a cantilever conductive plate in transverse impulsive magnetic field. *Int. J. Solids Struct.*, 42(8): 2417–2430.

Hasanyan, D.J. Librescu, L. & Ambur, D.R. 2006. Buckling and postbuckling of magnetoelastic flat plates carrying an electric current. *Int. J. Solids Struct.*, 43(16): 4971–4996.

Zhao, B.S. & Wang, M.Z. 2006. A refined theory of magnetoelastic plates. *Engineering Mechanics*, 23(3): 82–87.

Bian, Y.H. 2015. Analysis of nonlinear stresses and strains in a thin current-carrying elastic plate. *Int. Appl. Mech.*, 51(1): 108–120.

Mol'chenko, L.V. Loos, I.I. & Fedorchenko, L.N. 2014. Influence of extraneous current on the stress state of an orthotropic ring plate with orthotropic conductivity. *Int. Appl. Mech.*, 50(6): 683–687.

Mol'chenko, L.V. 1989. *The nonlinear magnetoelasticity of thin current-carrying shells*. Kiev: Vyshch. Shkola.

Yu, F.C. & Zheng, C.K. 2004. *Electrodynamics*. Beijing: Peking University Press.

Xu, Z.L. 2006. *Elastic mechanics*. Beijing: High Education Press.

Numerical simulation on flow characteristic of Newton anti-icing fluids

Jixiang Shan
High Speed Aerodynamics Institute, China Aerodynamics Research and Development Center, Mianyang, China
Institude of System Engineering, CAEP, Mianyang, China

Shuheng Song, Liming Feng & Xin Peng
High Speed Aerodynamics Institute, China Aerodynamics Research and Development Center, Mianyang, China

ABSTRACT   The de/anti-icing technology is one of the important technologies on the flying safety in winter. In the paper, the flow of the Newton anti-icing fluids with the air flow in low speed is simulated in VOF. The flow characteristics and the boundary layer displacement thickness characteristics with time are studied in different air flow speed. It is showed that when there is Newton anti-icing fluid in the plate, the boundary layer thickness and displacement thickness increase due to the existence of the anti-icing fluids boundary layer and the transition in the air boundary layer. The boundary layer displacement thickness with anti-icing fluids changes little first and then decreases rapidly, and finally decreases slowly with the time. The boundary layer displacement thickness would decrease with increase in the air inlet velocity.

1 INTRODUCTION

In winter, the ice and snow would accumulate in the wing which would not only increase the weight of the plane, but also reduce the flying performance, even lead to an air disaster. So when there is ice in the wing and engine, the plane is not allowed to take off. To use anti-icing fluids is usually an effective way to prevent icing. There are four kinds of anti-icing fluids and the first kind is the Newton anti-icing fluids which have excellent deicing performance and stay short time. The key technology for the anti-icing fluids is stripping characteristics and the aerodynamic characteristics.

Anti-Icing Materials International Laboratory (AMIL) is set up in Canada to study the question on the deicing/anti-icing (Vanessa, 2011). There are special wind tunnel and related equipment to test the deicing performance. The stripping characteristics and the aerodynamic characteristics of the anti-icing fluids are based on the compare of the Boundary Layer Displacement Thickness (BLDT) with or without anti-icing fluids on the plate. The free stream velocity in test is same as velocity of the aircraft in taking off.

The stripping process of anti-icing fluids from the plate is a typical gas-liquid two-phase flow with a free surface. The Volume of Fluid Model (VOF method) and Level Set method are the main method to simulate the gas-liquid two-phase flow. For the VOF method, it can simulate complex free surface and is the mass conservation, but the rapid change in the free surface would be averaged (Phung, 2006). For Level Set method, it can simulate complex free surface with rapid change, but there is no mass conservation which deduces calculation errors (Murrone, 2004, Olsson, 2005).

The stripping process of anti-icing fluids from the plate is a kind of thin layer two-phase flow. In the process, the mass of the anti-icing fluid gradually decreases, the mass should be accurately simulated. So VOF method is suited to simulate this kind of thin layer flow and is used in the paper to simulate the stripping process of anti-icing fluids from the plate. The effect of air velocity on the BLDT and the boundary layer flow condition is analyzed.

2 COMPUTATIONAL METHOD AND GRID

2.1 Numerical method

In the process, VOF method is used in the paper to simulate the stripping process of anti-icing fluid from the plate. The continuity equation, convection transport, and momentum transport equation are as follows:

$$\frac{\partial F}{\partial t} + (v \cdot \nabla) F = 0 \tag{1}$$

$$\frac{\partial(\rho v)}{\partial t} + \nabla \cdot (\rho vv) = -\nabla p + \nabla \cdot \left[\mu(\nabla v) + \nabla v^T\right] + \rho g + f_g \tag{2}$$

f_g is the surface tension and is $f_g = \int \delta(x - x_f) \sigma k n ds$ which is deduced from the CSF model and VO model.

The transition turbulent model is used to simulate the flow in boundary layer.

2.2 Model description and mesh generation

The computational model and grids are shown in Figure 1. The computational zone is 1.5 m × 0.2 m. Structured grids which are 500 × 80 are used in the entire computational region. Near the wall, the dense grid is used, in which y+ is equal to 0.1 and the size of the first grid is 0.0005 mm. The velocity inlet is used in the left boundary, the pressure outlet is used in the right boundary, and the wall is used in down boundary. In the left boundary, there are special speed inlets for the air and the mixture of the air and anti-icing outflows are in the right boundary.

The anti-icing fluids with 0.002 m is initialized on the wall and the velocity of the anti-icing fluids is zero. For the anti-icing fluids, the density is 1.048×10^3 Kg/m³, and the viscosity is 1.04×10^{-5} kg·m⁻¹·s⁻¹. The effect of gravity is simulated and G is 9.81 m·s⁻². The time step is 0.005 s and it is iterated 40 times in every time step.

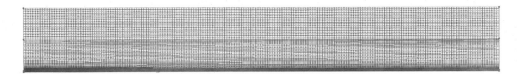

Figure 1. Computational grid.

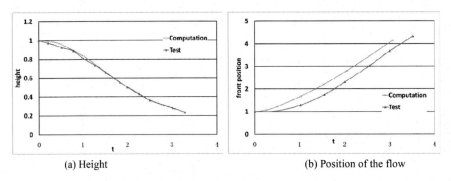

(a) Height   (b) Position of the flow

Figure 2. Comparison of the CFD results with experimental data.

2.3 Validation against experiment tests

In order to validate the present numerical method, the collapse process of the embankment and dam (Martin, 1952) is simulated and the Computational Fluid Dynamics (CFD) and experiment results are compared as shown in Figure 2. It is clear to see that the surface height of the CFD results approximately agree well with the experimental data and the change rule of the flow front position of the CFD result is same as the experimental data, but the position is advanced. In a word, the flow characteristic can be simulated accurately and the numerical method used in this paper is reliable.

3 RESULTS AND DISCUSSIONS

Boundary Layer Displacement Thickness (BLDT) corresponds to the extrapolation distance of streamline. The change of BLDT indicates effect of the anti-icing fluids on the air flow. To analyze the effect of the anti-icing fluids, the BLDT in dry surface and fluid-coated surface in v = 20 m/s is compared, which is showed in Figure 3. It can been seen from the Figure 3 that the BLDT increases lot when there is an anti-icing fluid on the plate to compare with the dry condition and the effect on the BLDT is 13.7 times thicker than the dry condition which means that the existence of anti-icing has a serious effect on the air flow. The BLDT decreases rapidly from 0.8 s to 2.0 s and then decreases gradually.

The interface shape changes with the times in v = 20 m/s are shown in Figure 4. It can be seen from Figure 4(a) and (b) that there are lots of waves in the interface which are anomalous and unordered. The size of the waves is less than the thickness of the anti-icing. In this case, the thickness of the anti-icing changes little. At t = 0.8 s, there is a big wave in the middle zone and there is no wave in other zones. In the front of the wave, the anti-icing fluids exist little and the fluid is thinner when it is further from the wave. In the back of the wave, the thickness of anti-icing fluid increases little. BLDT is rapidly decreased because of anti-icing fluid outflows. At t = 2.0 s, there is little anti-icing fluids in the plate. So the BLDT decreases little with time.

Figure 5 shows the effect of the anti-icing fluids on the velocity profiles in different sections in v = 20 m/s and t = 0.8 s. It can be seen from the Figure 5 that when there is anti-icing fluids in the plate, there are air boundary layer and anti-icing fluids boundary layer and the boundary layer thicknesses are a lot thicker in every section which leads to the increasing BLDT. In section x = 0.5 m, there is a laminar boundary layer in anti-icing fluids and a turbentunt boundary layer. To compare with laminar boundary layer in the dry condition, the transition in air boundary layer is advanced with the anti-icing fluids. In section x = 1.0 m and 1.5 m, the anti-icing fluids boundary layer is also the turbentunt boundary layer, which makes the boundary layer thicknesses increase a lot. In a word, the existence of the anti-icing fluid boundary layer and the transition in air boundary layer are the main reasons of increasing anti-icing fluid BLDT.

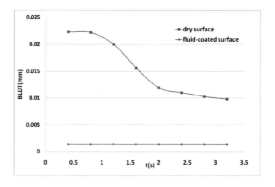

Figure 3. Effect of the anti-icing fluids on the BLDT in v = 20 m/s.

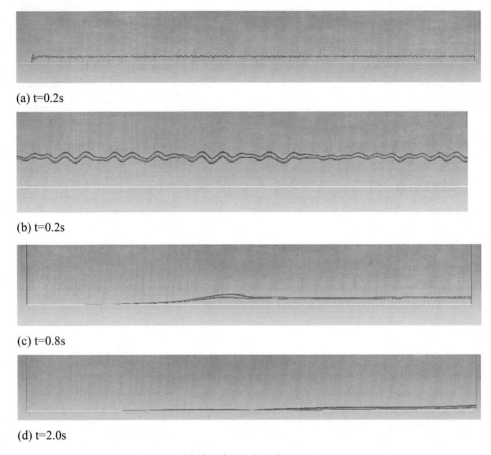

(a) t=0.2s

(b) t=0.2s

(c) t=0.8s

(d) t=2.0s

Figure 4. Interface shape changes with time in v = 20 m/s.

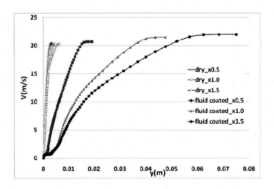

Figure 5. Effect of the anti-icing fluid on the velocity profile in different sections in v = 20 m/s and t = 0.8 s.

Figure 6 shows that BLDT changes with time in different inlet velocities when anti-icing fluids are on plate. It can be seen from the Figure 6 that the rules of BLDT with time in different inlet air velocity are same. With increase of the inlet velocity, BLDT decreases and the change of the BLDT is little ahead of time. Increase in the inlet velocity is well for the stripping of the anti-icing.

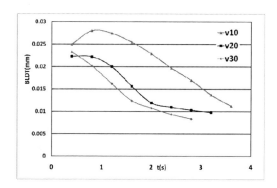

Figure 6. BLDT change with anti-icing fluid on plate in different inlet velocities.

4 CONCLUSION

In the paper, the flow of the Newton anti-icing fluids under the air flow in low speed is simulated in VOF method. It is shown that when there is Newton anti-icing fluid in the plate, the boundary layer thickness and displacement thickness increase due to the existence of the anti-icing fluids boundary layer and the transition in the air boundary layer. The boundary layer displacement thickness with anti-icing fluids changes little first and then decreases rapidly and finally decreases slowly with the time. The boundary layer displacement thickness would decrease with increase of the air inlet velocity.

ACKNOWLEDGMENTS

This work is supported by the National Natural Science Foundation of China (U1333125).

REFERENCES

Martin J.C., Moyce W.J. Part IV. An experimental study of the collapse of liquid columns on a rigid horizontal plane, *J. Philosophical Transactions of the Royal Society of London. series*, 1952, 244(4): 312–324.

Murrone A., Guillard H. A five reduced equation model for compressible two-phase flow problems [J]. *J Comput Phys*, 2004, 202: 664–698.

Olsson E., Kreiss G. A conservative level set method for two phase flow [J]. Journal of *Computational Physics*, 2005, 210(1): 225–246.

Phung D.H., Katutoshit. Verification of a VOF based two phase flow model for wave breaking and wave structure interactions [J]. *Ocean Engineering*, 2006, 33: 1565–1588.

Standard Test Method for Aerodynamic Acceptance of SAE AMS 1424 and SAE AMS 1428 Aircraft Deicing/Anti-icing Fluids.

Vanessa L. Update: wind tunnel research to support the refinement of the ice pellet allowance time table. *Marco Ruggi*. (2011).

The empirical study of hierarchical linear model of partitioned water consumption in the Yangtze River

Y.H. Zhu, P. Gao & X.F. Wu
Department of Mathematics and Physics, North China Electric Power University, Beijing, China

ABSTRACT: This paper outlines the general theory of hierarchical linear model, and based on the model, gives the empirical analysis about the changes of the provincial partitioned water consumption in the Yangtze River and Southwest Rivers from 1998 to 2012. We use the Ward clustering method to partition the water consumption into groups and determine the quantity of population and effective area of cultivated land, these as two variables with the value of the correlation coefficient. The test results show that the majority of water consumption grouped by basins and provinces presents significant linear increase, but the rate of growth is different. According to the established model, we have made an out-of-sample prediction of the partitioned water consumption in 2016. This study and the results will play a reference role in the recent planning to the development and utilization of the water resources in the Yangtze River and Southwest River Area.

1 INTRODUCTION

The planning of development and utilization of the water resources play a vital role in the sustainable development in an area. The analysis and forecast of the quantity of water resources and consumption is an important basis for water resources planning.

The Yangtze River and Southwest Rivers are two major river basins in China. There are some differences in their status, but they have strong connections in geography. Here we mainly use the model to analyze the bulletin data and related indices of the Yangtze River and Southwest Rivers in the past 15 years, to explore the regularities of the changes of the above-mentioned partitioned water consumption. Then we can make a short-term water demand forecast to provide necessary basis for development planning in the area.

2 HIERARCHICAL LINEAR MODEL

In the statistical analysis, linear regression model is a commonly used mathematical model. General linear model assumes that the samples are from the same population and they are independent. But in practical problems, many samples have a hierarchical structure or nested structure. In another word, different samples may be subordinate to different overalls (groups). In this case, using the general linear model is not appropriate, so we should use hierarchical linear model to analyze it.

2.1 *The model structure*

Following is a hierarchical linear model of two layers:

$$\text{Layer-1 (individual)} \quad Y_{ij} = \beta_{0j} + \beta_{1j}X_{1ij} + \beta_{2j}X_{2ij} + \ldots + \beta_{Qj}X_{Qij} + r_{ij} \qquad (1)$$

$$\text{Layer-2 (group)} \begin{cases} \beta_{0j} = \gamma_{00} + \gamma_{01}W_{1j} + \gamma_{02}W_{2j} + \ldots + \gamma_{0F}W_{Fj} + u_{0j} \\ \beta_{1j} = \gamma_{10} + \gamma_{11}W_{1j} + \gamma_{12}W_{2j} + \ldots + \gamma_{1F}W_{Fj} + u_{1j} \\ \vdots \\ \beta_{Qj} = \gamma_{Q0} + \gamma_{Q1}W_{1j} + \gamma_{Q2}W_{2j} + \ldots + \gamma_{QF}W_{Fj} + u_{Qj} \end{cases} \quad (2)$$

In (1) and (2), $i = 1, 2, \ldots, n_j$ Layer-1 units are subordinate to $j = 1, 2, \ldots, j$ layer-2 units. The number of independent variables in Layer-1 is Q, and in Layer-2 is F.

Formula (1) in the form of matrix is expressed as follows:

$$\text{Layer-1 (individual)} \quad Y_j = X_j \beta_j + r_j, \quad r_j \sim N(0, \sigma^2 I_{n_j}) \quad (3)$$

$$\text{Layer-2 (group)} \quad \beta_j = W_j \gamma + u_j, \quad u_j \sim N(0, T) \quad (4)$$

2.2 Model parameters estimating

After the model is given, we should estimate the unknown parameters in the model according to the actual data structure.

Firstly, the most important step is to find the optimal estimation of β_j. There are two methods: one is simply based on ordinary least squares, (OLS (general linear model)) of the data of group j, then we get:

$$\hat{\beta}_j = \left(X_j^T X_j\right)^{-1} X_j^T Y_j \quad (5)$$

The other is to forecast β_j by the characteristics of the given group represented by W_j:

$$\hat{\hat{\beta}}_j = W_j \hat{\gamma} \quad (6)$$

In (6), $\hat{\gamma}$ is estimated by the method of generalized least squares (GLS):

$$\hat{\gamma}_j = \left(\sum W_j^T \Delta_j^{-1} W_j\right)^{-1} \sum W_j^T \Delta_j^{-1} \hat{\beta}_j, \quad \Delta_j = T + V_j, \quad V_j = Var(\hat{\beta}_j) = \sigma^2 \left(X_j^T X_j\right)^{-1}$$

(5) and (6) here give the optimal combination of the two estimations:

$$\beta_j^* = \Lambda_j \hat{\beta}_j + (I - \Lambda_j) W_j \hat{\gamma} \quad (7)$$

where $\Lambda_j = T(T + V_j)^{-1}$ is the quotient of parameter discrete matrix of β_j (matrix T) and the total discrete matrix of $\hat{\beta}_j$.

Except the estimation of β_j in Layer-1, the reliability of the estimated results of these coefficients should be analyzed too. We use τ_{qq} and v_{qqj} to indicate the diagonal elements of T and V_j. The reliability is as follows:

$$\text{Reliability}\left(\hat{\beta}_{qj}\right) = \tau_{qq} / (\tau_{qq} + v_{qqj}), \quad q = 0, 1, \ldots Q \quad (8)$$

The total reliability of all the J units in Layer-2 can be summarized as follows:

$$\text{Reliability}\left(\hat{\beta}_q\right) = \frac{1}{J} \sum_{j=1}^{J} \tau_{qq} / (\tau_{qq} + v_{qqj}), \quad q = 0, 1, \ldots, Q \quad (9)$$

In practice, the coefficient estimation of Layer-1 and Layer-2 needs the information of variance components. We generally use the EM algorithm to estimate variance and covariance.

In many hierarchical linear model analysis software (such as HLM), this iterative algorithm are widely used for the estimation of variance covariance components.

2.3 Model hypothesis test

Multi-parameter test of Layer-1 coefficients is the key of the test of the model rationality. The general linear hypothesis about β is:

$$H_0: C^T \beta = 0 \tag{10}$$

If the empirical Bayes estimation of $\beta*$ is the basis for the test, then its test statistic is:

$$H_{EB} = \beta^{*T} C \left(C^T V * C \right)^{-1} C^T \beta* \tag{11}$$

where $V*$ is the $J(Q+1) \times J(Q+1)$ variance covariance matrix of all coefficients. We can use the statistics given by (11) to do the hypothesis test for H_0.

In addition to testing the coefficient, we should also do the hypothesis test for variance covariance. The null hypothesis testing for variance covariance is: $H_0: T = T_0$ and the alternative hypothesis is: $H_1: T = T_1$, where T_0 is the simplified form of T_1. In order to test such a synthesis hypothesis, we can estimate two different models, then calculate the deviation statistics D_0 and D_1 for each model, and finally we can calculate the following test statistic: $H = D_0 - D_1$. This statistic obeys a χ^2 distribution whose degree of freedom is m. Here m is the difference of the number of variance covariance components in these two models.

2 EMPIRICAL ANALYSIS

We use the data presented by Water Resources Bulletin from 1998 to 2012 as the basis to do the hierarchical linear model regression analysis for the provincial partitioned water consumption in the Yangtze River and Southwest rivers.

From the data, the provincial partitioned water consumption is approximately in a trend of linear growth. Therefore, the hypothetical hierarchical model for Layer-1 is:

$$Y_{ti} = \pi_{0i} + \pi_{1i} T_{ti} + e_{ti} \tag{12}$$

where we assume that the errors e_{ti} are independent and subject to the normal distribution with the same variance σ^2. In (10), $i(i = 1, 2, ..., 24)$, respectively, represents different Level 2 zones of the river basin. $t(t = 1998, ..., 2012)$ represents the observation point, i.e., the year.

In the hierarchical linear model, the intercept parameter π_{0i} and the growth parameter π_{1i} are allowed to change as a function of individual characteristic measurement. Therefore, the model of Layer-2 can be:

$$\begin{cases} \pi_{0i} = \beta_{00} + \sum_{q=1}^{Q_0} \beta_{0q} X_{qi} + r_{0i} \\ \pi_{1i} = \beta_{10} + \sum_{q=1}^{Q_1} \beta_{1q} X_{qi} + r_{1i} \end{cases} \tag{13}$$

where X_q $(q = 1, ..., Q)$ is the regressive independent variable of the intercept parameter π_{0i} and the growth parameter π_{1i}, which needs to be further studied.

Now we need to select independent variables X_q $(q = 1, ..., Q)$ for Layer-2. Firstly, we group them by the quantity of the water consumption in each provincial partition so that the quantities in the same group are similar and those in different groups have larger differentia. Then, these groups will be the regressive independent variable of the model for Layer-2. According to the data presented by Water Resources Bulletin, sorted by the index like: farmland irrigation water, industrial water, forestry husbandry fishery and livestock water, drinking water, urban public water, resident water, and so on, we do Ward Cluster to the data for the past 15 years and get the basically consistent clustering results as shown in Table 1.

In general, the group numbers, as the nominal variables, are not suitable to be introduced into the regression equation directly. There are six groups, so we need to introduce five dummy variables X_1, X_2, X_3, X_4, X_5.

In addition to the variation of water consumption in different groups, there are also differences between the partitions in same group. Through the correlation analysis of the data of water consumption and other social economic indicators, we found that the Pearson correlation coefficients between the quantity of population, the area of effective cultivated land and the partitioned water consumption are both more than 0.8, so there are significantly positive correlations between them. Thus, the other variables of the model for Layer-2 are as follows: X_6—the quantity of population, X_7—the area of effective cultivated land.

According to the above analysis, the hierarchical linear model of provincial partitioned water consumption is:

$$\text{Layer-1}: Y_{ti} = \pi_{0i} + \pi_{1i}T_{ti} + e_t \qquad (14)$$

$$\text{Layer-2}: \begin{cases} \pi_{0i} = \beta_{00} + \sum_{q=1}^{7} \beta_{0q} X_{qi} + r_{0i} \\ \pi_{1i} = \beta_{10} + \sum_{q=1}^{7} \beta_{1q} X_{qi} + r_{1i} \end{cases} \qquad (15)$$

By substituting the data of the related index value for the past 15 years into the model, using HLM software to do regression analysis, we get the results in Tables 2 to 4.

Table 1. Clustering results for all the provincial partitions.

Group 1	Group 2	Group 3	Group 4	Group 5	Group 6
Qinghai	Shanxi	Yunnan	Anhui	Sichuan	Hubei
Tibet	Henan	Chongqing	Shanghai	Jiangxi	Hunan
Gansu	Guangxi	Guizhou			Jiangsu
Guangdong	Tibet (SW)	Zhejiang			
Fujian		Yunnan (SW)			
Guangxi (SW)*					
Qinghai (SW)					
Xinjiang (SW)					

*Guangxi (SW) means the southwestern river basin in the territory of Guangxi, and so on.

Table 2. Parameter estimation.

π_{0i}		π_{1i}	
β_{00}	0.3332	β_{10}	0.0437
β_{01}	16.5141	β_{11}	0.3268
β_{02}	45.2424	β_{12}	1.3956
β_{03}	70.7864	β_{13}	4.7266
β_{04}	150.7422	β_{14}	5.1925
β_{05}	220.7348	β_{15}	5.3157
β_{06}	0.004612	β_{16}	−0.000613
β_{07}	0.003773	β_{17}	0.000530

Table 3. Reliability of coefficient estimation.

Layer-1 coefficient	Reliability of estimation
π_0	0.951
π_1	0.924

Table 4. The related test statistic.

Random error	Standard deviation	Variance	Random error	Free degree	χ^2 statistic	p
r_0	17.35	301.30	r_0	15	308.85	0.000
r_1	1.50	2.25	r_1	15	196.56	0.000
e	7.22	52.09	–	–	–	–

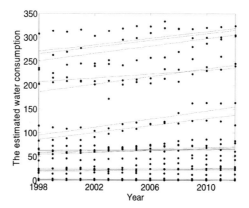

Figure 1. Regression results of the hierarchical model.

Table 5. The forecast of the provincial partitioned water consumption in 2016.

Provincial partition	Predictive value	Provincial partition	Predictive value
Qinghai	0.504	Guangxi	15.140
Tibet	1.066	Guangdong	0.509
Yunnan	52.977	Anhui	186.814
Sichuan	241.625	Jiangsu	342.973
Chongqing	90.128	Shanghai	135.476
Guizhou	69.896	Zhejiang	59.124
Gansu	2.281	Fujian	1.227
Hubei	309.713	Guangxi (SW)	1.365
Hunan	329.643	Yunnan (SW)	75.728
Jiangxi	259.692	Tibet (SW)	34.192
Shanxi	24.975	Qinghai (SW)	0.463
Henan	23.872		

From the above tables, the coefficients of Layer-1 are of high reliability. The test value of random error p is 0, far less than 0.01. So the regression effect of the model is significant. The regression results of the overall hierarchical model are as the Figure 1 below.

According to the above specific hierarchical linear model, we give the forecast of the provincial partitioned water consumption in 2016 as Table 5.

3 CONCLUSIONS

By cluster analysis, hierarchical linear model and other statistical methods, this paper gives the empirical analysis about the changes of the partitioned water consumption and a short-term forecast. The following are the main achievements and conclusions:

1. In the longitudinal view, the water consumption in each partition is mostly linearly increased over the years but significantly different. The linear changes in some partitions are strong, some changes in the process act large fluctuations. As for which partitions have a significant linear trend, it remains to be further researched and analyzed.
2. We give the forecast results for 2016, it has a reference value for the short term planning of the development and utilization of regional water resources.

Of course, due to that the selected method is single, and the model itself has some limitations, this study inevitably has some deficiencies, the specific performances are as follows:

1. In this study, we just collected the water resources bulletin data from 1998 to 2012 for a total of 15 years, based on these data, the statistical analysis often will have some obvious shortcomings. The model can be more accurate with more data.
2. In this hierarchical linear model, it is very important to select the suitable independent variables. If we can find more factors which effect the water consumption, and collect more relevant data, it will help us to establish a more rational, more perfect hierarchical linear model.

REFERENCES

Guo, Z.G. 2007. Multilevel Analysis of Sex Ratio at Birth of the 2000 Population Census [J]. *Population Studies* 31(3):20–31.

Li, Q.H. & Qian, K.X. & Xia, C.H. & Lei, J. & Deng, Y.J. 2012. The Study of Yangtze River Water Consumption Trend and the Total Control Index [J]. *Yangtze River* 43(2):13–15.

Ou, S.D. 2009. The Application of Hierarchical Linear Models in Financial Risk Measurement [J]. *Journal of Yulin Normal University* 30(5):1–4.

Roudenbush, S.W. & Bryk A.S. & Guo, Z.G. 2007. *Hierarchical Linear Models: Applications and Data Analysis Methods [M]*. Beijing: Social Sciences Academic Press.

Tai, S.C. & Sun, A.Y. & He, J.J. 2007. *The Application of Mathematical Statistics [M]*. Wuhan: Wuhan University Press.

Wan, Y. & Sui, S.L. 2006. MATLAB Regression Analysis [J]. *Journal of Qingdao Technological University* 27(4):130–132.

Wang, X.C. & Hong, W. & Wu, G.P. & Chen, J.C. & Guo, H.J. 2013. *Yangtze River and Southwest Rivers Water Resources Bulletin for 2012[R]*. Wuhan: Yangtze River Press.

Xue, W. 2001. SPSS Statistical Analysis Method and Its Application. Beijing: Electronic Industry Press.

Zhang, L. & Lei, L. & Guo, B.L. 2003. *The Application of Hierarchical Linear Model [M]*. Beijing: Educational Science Publishing House.

Bioactive properties of peptides from silkworm pupa protein

Zhenyu Zuo, Fang Yu, Hao Huang, Lingling Li & Xiaorong Qin
College of Chemical Engineering and Technology, Wuhan University of Science and Technology, Wuhan, China

ABSTRACT: Enzymatic hydrolysates of silkworm pupa (*Bombyx mori*) protein isolate were prepared by treatment with three different gastrointestinal proteases: pepsin, trypsin and α-chymotrypsin. The protein hydrolysates with a 33.6% degree of hydrolysis were fractionated by ultrafiltration showing a high concentration of short chain peptides (SPBP-1), which exhibited significantly higher antioxidant and antihypertensive capacities than fractions with higher molecular weights (SPBP-2, SPBP-3). All hydrolysates presented antioxidant properties with EC_{50} values on DPPH• from 1.7 to 5.2 mg/ml, HO• from 2.5 to 4.3 mg/ml, O_2^{-}• from 2.8 to 3.2 mg/ml, ABTS• from 2.9 to 9.5 mg/ml, respectively. All hydrolysates showed antihypertensive capacity with IC_{50} values from 0.75 to 2.48 mg/ml and SPBP-1 was the peptides with the highest ACE-inhibitory capacity. Finally, their antitumor capacities were evaluated through cell viability assay. These results verified that silkworm pupa proteins may represent a useful source of bioactive peptides after gastrointestinal enzymatic hydrolysis.

1 INTRODUCTION

Many proteins that occur naturally in raw food materials exert their physiological action either directly or upon enzymatic hydrolysis in vitro or in vivo (FitzGerald & Meisel 2003; Pihlanto & Korhonen 2003). Some peptides are inactive within the sequence of the parent protein and can be released during gastrointestinal digestion, food processing, or fermentation (Leppala 2001; Erdmann et al. 2008). Bioactive peptides, usually contain about 2–50 amino acid residues, have been defined as specific protein fragments that have a positive impact on body functions or conditions and may ultimately influence health (Kitts & Weiler 2003).

Silkworm pupa (*Bombyx mori*), an edible insect, has been used as fertilizer, animal feed, food and traditional medicines in some Southeast Asian countries where they represent a cheap source of good quality protein (Pereira et al. 2003; Mishra et al. 2003). Processing of silk industry generates annually millions of tons of low-value by-products which represent an enormous and underutilized renewable resource. New uses for these by-products are needed to increase value and create new market opportunities.

The aim of this study was to obtain bioactive peptides by gastrointestinal enzymatic hydrolysis from silkworm pupa (*Bombyx mori*) protein and to evaluate their bioactive properties: antioxidant, antihypertensive, and antitumor activities.

2 MATERIALS AND METHODS

2.1 *Materials and chemicals*

Silkworm pupa (*Bombyx mori*) protein powder was provided by Sericultural Research Institute, Hubei Academy of Agricultural Sciences (Wuhan, China). ACE (from rabbit lung) and Hippuryl-His-Leu (HHL) were purchased from Sigma–Aldrich Trading Co. (Wuhan, China). Pepsin, trypsin and α-chymotrypsin were purchased from Dingguo Biotechnology Co. (Wuhan, China). DPPH, ascorbic acid and Trinitrobenzene Sulfonic acid (TNBS) were purchased from Sigma–Aldrich Trading Co. (Wuhan, China). RPMI 1640 was from HyClone (USA); MTT (3-(4,5-dimethylthiazol-2-yl)-2,5-diphenyltetrazolium bromide) and Dulbecco's Modified Eagle's Medium (DMEM) were products of Sino-American Biotechnology Company. The Sarcoma 180 and H22 cells were obtained from China Center for Typical Culture Collection (CCTCC). All other chemicals and solvents were of analytical grade.

2.2 *Production of Silkworm Pupa protein (SP)*

Silkworm pupa powder was defatted with the azeotropic mixture of petroleum ether and Acetone (3:7) and was dried at 60°C using a drying oven. The defatted flour was dispersed in deionized water at ratio 1:60 (w/v). The suspension pH was adjusted to 9.5 by using 1 M NaOH. The suspension was centrifuged at 3000 × g for 20 min after stirring for 1 h. The proteins were precipitated after the supernatant was adjusted to pH 4.5 using 1 M HCl and centrifuged again at 3000 × g for 20 min. The precipitates were lyophilized using a freeze-dryer (FD-1D-50, Beijing Boyikang Instruments Co., Beijing, China) and stored at −20°C until used.

2.3 *Preparation of Silkworm Pupa protein Hydrolysates (SPH) and Bioactive Peptides (SPBPs)*

An aliquot (100 ml) of SPP solution (5%, w/v) was treated using an ultrasonic cell crusher (SCIENTZ-IID, Xinzhi Ultrasonic Equipment Co., Ningbo, China) with a 1.5 cm flat tip probe at 24 kHz and at a power 400 W for 20 min (Pulse durations of on-time 2 s and off-time 2 s). The solution was jacketed with ice when undergoing ultrasonic treatment and the temperature of treated solution never exceed 60°C. The hydrolysis process was carried out according to the method of Himaya with some modifications (Himaya et al. 2012). Silkworm pupae protein was dissolved in distilled water at a concentration of 35 mg/ml and hydrolyzed for 2 h using pepsin (enzyme-substrate ratio, 1500 U/mg) at pH 2.0 and 37°C with stirring. Then the reaction solution was further digested by trypsin and α-chymotrypsin (enzyme-substrate ratio of 1000 U/mg for each enzyme) for 3 h at pH 6.5 and 37°C with stirring. The reaction was stopped by heating the mixture in a boiling water bath for 15 min. Then, the solution was centrifuged at 5000 × g for 15 min and the supernatant was collected for further experimental use.

The degree of hydrolysis was calculated by determination of free amino groups by reaction with TNBS (Adler-Nissen 1979). The total number of amino groups was determined in a sample 100% hydrolyzed by incubation with 6 N HCl at 120°C for 24 h.

The collected protein supernatant was filtered sequentially using an ultrafiltration unit (Pellicon XL, Millipore, USA) through two ultrafiltration membranes with molecular mass cut-off of 5 and 3 kDa, respectively. Three fractions with MWs <3 kDa, 3–5 kDa and >5 kDa were obtained and named correspondingly as SPBP-1, SPBP-2 and SPBP-3. All fractions were lyophilized and stored at −20°C until used.

2.4 *Determination of antioxidant activity*

Four types of radical scavenging assays (hydroxyl/superoxide/DPPH/ABTS) were performed according to previously reported methods (Wang et al. 2013). The concentrations of hydrolysates and the fractions from ultrafiltration were expressed as mg protein/ml. The EC_{50} was defined as the concentration at which a sample caused a 50% decrease in the initial concentration of hydroxyl radicals, superoxide radicals, DPPH radicals, and ABTS radicals.

2.5 *Determination of Angiotensin-Converting Enzyme (ACE) inhibitory activity*

To investigate the inhibition pattern of fractionated peptides on ACE, different concentrations of fractionated peptides were added to each reaction mixture according to the method of Barbana with some modifications (Barbana & Boye 2011). The enzyme activities (absorbance at 228 nm) were measured with different concentrations of substrate (HHL). Lineweaver-Burk plots of 1/absorbance versus 1/HHL were used to determine the ACE inhibitory pattern. The IC_{50} value was defined as the Concentration of inhibitor required to inhibit 50% of the ACE activity.

2.6 Cell viability assay

The effects of the SPBPs on the viability of S180 and H22 cells were determined by an MTT-based assay. Cells were seeded in 96-well plates (5×10^3 cells/well) in DMEM with 10% FBS at 37°C for 24 h. Then the medium was removed the cells were starved for 24 h and treated with SPBPs (50–400 ug/ml) for 96 h. Cell viability was determined using the Methylthiazole Tetrazolium (MTT) method, 20 ul of MTT solution (5 mg/ml) were added to each well and incubated for 4 h. The absorbance was measured at 492 nm using enzyme-linked immunosorbent assay plate reader.

3 RESULTS AND DISCUSSION

3.1 Obtain SPBPs by using human gastrointestinal digestive proteases

A hydrolysis procedure that simulates the human gastrointestinal digestion process was developed by a two step sequential hydrolysis with pepsin, trypsin and α-chymotrypsin. Based on this procedure, a protein hydrolysate with a 33.6% of degree of hydrolysis was obtained. This protein hydrolysate was applied to an ultrafiltration unit through two ultrafiltration membranes with molecular mass cut-off of 5 and 3 kDa, respectively. Three fractions with MWs <3 kDa, 3–5 kDa and >5 kDa were collected to study the presence of bioactive peptides with antioxidant, antihypertensive, and antitumor activities.

3.2 Antioxidative activity of SPBPs

In order to evaluate the antioxidant activity of SPBPs, four radical scavenging assays (Fig. 1) were conducted, and their activities were compared with the positive controls of Ascorbic Acid (AA). Their EC_{50} values were listed in Table 1.

DPPH is a relatively stable radical that is widely used to test the ability of compounds to scavenge free radicals and therefore act as antioxidants (Tsopmo et al. 2010). As indicated in Figure 1A, DPPH radical scavenging assay indicates that SPBP-1, SPBP-2 and SPBP-3 scavenged DPPH radicals in a concentration-dependent manner with EC_{50} values of 1.724, 3.452 and 5.243 mg/ml, respectively. SPBP-1 showed the highest activity among all the samples, except for the AA positive controls. This finding indicated that SPBPs could suppress the radical chain reaction by converting DPPH radicals to less- or non-harmful substances. As indicated in Figure 1B, the activity of SPBP-1, SPBP-2 and SPBP-3 on cleaning hydroxyl radicals showed a dose-response relationship with EC_{50} values of 4.276, 2.482 and 3.853 mg/ml. SPBP-2 showed good hydroxyl radical scavenging activity, which demonstrated that it could serve as a scavenger for reducing or eliminating the damage induced by hydroxyl radicals in foods and biological systems. The superoxide radical scavenging effects of SPBP-1, SPBP-2 and SPBP-3 was investigated, and the dose-related effects was observed at peptides concentrations ranging from 0 to 2 mg/ml (Fig. 1C) with an EC_{50} of 3.175, 2.941 and 2.769 mg/ml. The ABTS assay is based on the generation of a blue/green ABTS radical that can be reduced by antioxidants, the resulting decrease in absorbance can be monitored at 734 nm. Figure 1D indicates that SPBP-1, SPBP-2 and SPBP-3 could effectively scavenge ABTS radicals in a dose-dependent manner when their concentrations ranged from 0 to 2 mg/ml. The EC_{50} of SPBP-1, SPBP-2 and SPBP-3 were 2.947, 4.782, and 9.472 mg/ml.

3.3 Inhibitory activity of ACE

ACE activity leads to an increase in blood pressure by producing the vasoconstrictor peptide angiotensin II and by degrading the vasodilator peptide bradyquinin. Inhibitors of ACE are

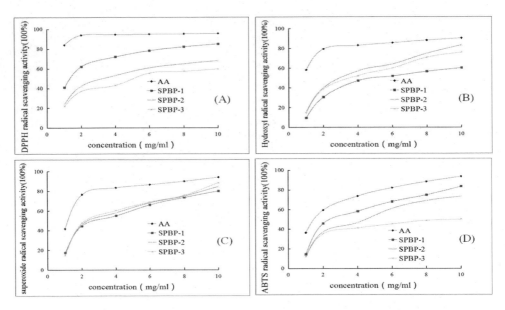

Figure 1. DPPH radical (A), hydroxyl radical (B), superoxide anion radical (C), and ABTS radical (D) scavenging activities of SPBP-1, SPBP-2, SPBP-3. All data are presented as the mean ± SD of triplicate results.

Table 1. EC_{50} values of SPBP-1, SPBP-2 and SPBP-3 on DPPH•, HO•, O_2^-• and ABTS•.

Sample	EC_{50} (mg/ml)			
	DPPH•	HO•	O_2^-•	ABTS
SPBP-1	1.724	4.276	3.175	2.947
SPBP-2	3.452	2.482	2.941	4.784
SPBP-3	5.243	3.853	2.769	9.472

All data are presented as the mean ± SD of triplicate results.

Table 2. The inhibitory rate on ACE by SPH and SPBPs.

Sample	SPI	SPH	SPBP-3	SPBP-2	SPBP-1
Weight (g)	100	33.6	6.7	10.5	16.1
IR (100%)	ND	43.5	21.4	51.3	85.7
IC_{50} (mg/ml)	ND	1.84	2.48	1.42	0.75

ND, no detected; IR, inhibitory rate; SPI, silkworm pupa protein isolate; All data are presented as the mean ± SD of triplicate results.

used in therapy against hypertension. As observed in Table 2, protein isolate did not show inhibitory activity of ACE in the amounts assayed. On the contrary, protein hydrolysate and fractions collected after ultrafiltration showed a prominent inhibitory activity of ACE. IC_{50} values (amount of sample needed to inhibit ACE by 50%) of the protein hydrolysate and fractions are shown in Table 2. SPBP-1 was the peptides with the highest ACE-inhibitory capacity with IC_{50} values of 0.75 mg/ml. These results are in agreement with the results showed by other authors in protein hydrolysates obtained with Alcalase from soybean (Wu & Ding 2002) and wheat (Matsui et al. 2000).

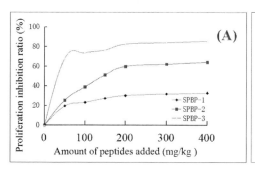 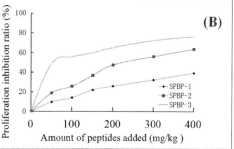

Figure 2. Anti-proliferative effect of SPBP-1, SPBP-2, SPBP-3 on S180 sarcoma (A) and H22 hepatoma (B) were performed. Each data is expressed as the mean ± SD obtained from triplicate experiments.

3.4 *Viability inhibitory effect of SPBPs on S180 and H22 cells*

MTT assays were performed to measure the viability inhibitory effect of SPBPs on S180 and H22 cells. As was shown in Figure 2, the cells growth was inhibited by SPBPs in a dose-dependent manner. Especially, incubation for 96 h with 400 ug/ml SPBP-3 induced strong escalation in inhibition rate in S180 cell lines which reached 82.3%.

4 CONCLUSIONS

Peptide fractions from silkworm pupa (*Bombyx mori*) protein hydrolysate obtained by gastrointestinal enzymatic hydrolysis display very interesting results in the biological properties studied. Oxidative stress, hypertension, and cancer are the main responsible factors in the development of certain clinical pathologies above all those related with heart and tumor diseases. In this sense, these results outstands the potential use of silkworm pupa (*Bombyx mori*) protein hydrolysates in developing functional foods for the treatment of heart and related diseases.

ACKNOWLEDGEMENTS

This work was supported by Natural Science Foundation of Hubei Province (No. 2014CFB802). We wish to thank Prof. Ming Cheng and Dr. Haodong Fan for their technical assistance.

REFERENCES

Adler-Nissen, J. 1979. Determination of the degree of hydrolysis of food protein hydrolysates by trinitrobenzenesulfonic acid. *Journal of Agriculture and Food Chemistry* 27: 1256–1262.
Barbana, C., Boye, J.I. 2011. Angiotensin I-converting enzyme inhibitory properties of lentil protein hydrolysates: determination of the kinetics of inhibition. *Food Chemistry* 127: 94–101.
Chen, J.H., Liu, J.Y. 2003. Pinelloside, an antimicrobial cerebroside from Pinellia ternate. *Phytochemistry* 64: 903–906.
Erdmann, K., Cheung, B.W.Y. & Schroder, H. 2008. The possible roles of food-derived bioactive peptides in reducing the risk of cardiovascular disease. *Journal of Nutritional Biochemistry* 19(10): 643–654.
FitzGerald, R.J. & Meisel, H. 2003. Milk protein hydrolysates and bioactive peptides. In: P.F. Fox, & P.L.H. McSweeney (Eds.), *Advanced dairy chemistry, Vol. 1: Proteins* (3rd ed.): 675–698. New York, NY, USA: Kluwer Academic/Plenum Publishers.
Himaya, S.W.A., Ngo, D.H., Ryu, B., Kim, S.K. 2012. An active peptide purified from gastrointestinal enzyme hydrolysate of Pacific cod skin gelatin attenuates Angiotensin-1 Converting Enzyme (ACE) activity and cellular oxidative stress. *Food Chemistry* 132: 1872–1882.

Kitts, D.D. & Weiler, K. 2003. Bioactive proteins and peptides from food sources. Applications of bioprocesses used in isolation and recovery. *Current Pharmaceutical Design* 9: 1309–1323.

Leppala, A.P. 2001. Bioactive peptides derived from bovine whey proteins: opioid and ace-inhibitory peptides. *Trends in Food Science and Technology* 11(10): 347–356.

Matsui, T., Li, C.H., Tanaka, T., Maki, T., Osajima, Y., & Matsumoto, K. 2000. Depressor effect of wheat germ hydrolysate and its novel angiotensin I-converting enzyme inhibitory peptide, Ile-Val-Tyr, and the metabolism in rat and human plasma. *Biological and Pharmaceutical Bulletin* 23: 427–431.

Mishra, N., Hazarika, N.C., Narain, K., Mahanta, J. 2003. Nutri-tive value of non-mulberry and mulberry silkworm pupae andconsumption pattern in Assam, India. *Nutrition Research* 23: 1303–1311.

Pereira, N.R., Ferrarese-Filho, O., Matsushita, M., de Souza, N.E. 2003. Proximate composition and fatty acid profile of Bombyx mori L. chrysalis toast. *Journal of Food Composition and Analysis* 16: 451–457.

Pihlanto, A., & Korhonen, H. 2003. Bioactive peptides and proteins. In S.L. Taylor (Ed.), *Advances in food and nutrition research, Vol. 47*: 175–276. San Diego, USA: Elsevier Inc.

Tsopmo, A., Cooper, A., & Jodayree, S. 2010. Enzymatic hydrolysis of oat flour protein isolates to enhance antioxidative properties. *Advance Journal of Food Science and Technology* 2: 206–212.

Wang, B., Li, L., Chi, C.F., Ma, J.H., Luo, H.Y., & Xu, Y.F. 2013. Purification and characterisation of a novel antioxidant peptide derived from blue mussel (Mytilus edulis) protein hydrolysate. *Food Chemistry* 138: 1713–1719.

Wu, J., & Ding, X. 2002. Characterization of inhibition and stability of soy-protein-derived angiotensin I-converting enzyme inhibitory peptides. *Food Research International* 35: 367–375.

Influences of coating process on the micro-structures of reconstituted tobacco sheets

Yi Hou, Qihui Sun, Youming Li & Songqing Hu
State Key Laboratory of Pulp and Paper Engineering, South China University of Technology, Guangzhou, Guangdong Province, China

ABSTRACT: Focused on the impacts of coating process on the micro-physical structures of reconstituted tobacco sheet, the comparison of domestic and abroad products were performed first, and the influences of viscosities of coating liquids on the micro-structures were analyzed. The results showed that coating efficiency influences the pore distribution of the surfaces directly. Coating efficiency can increased greatly with the increase of viscosity of coating liquids, 40–50 m·Pa·s viscosity of coating fluids will be helpful for the increase of basic weight and coating efficiency for domestic products. Compared with abroad products, domestic products need to resolve the problems such as wide distributions of pore size and volume porosity.

1 INTRODUCTION

Reconstituted tobacco sheets (as RT sheets), with good properties such as low density, high use efficiency and filling value, good flexibility and low discharge of tar (*Agrupis et al, 2000; Wang et al, 2011*), are products made of tobacco waste, such as tobacco stems and dusts, which cannot be utilized directly into cigarettes (*Alireza et al, 2011; Zhang et al, 2012*). In China, technology of RT sheets has developed fast since 1990's, but unfortunately quality imperfections of products limited the application of RT sheets in cigarettes (*Sun et al, 2010*), the addition in cigarettes is only 10–15% which is much lower than cigarettes manufactured abroad (almost 30%), and now focuses are mainly laid on the optimization of manufacturing technology to make up the quality gap between domestic and abroad RT sheets (*Han et al, 2002*), little attention is paid to the influences of surface microstructures and the pore distributions of tobacco sheets on the sheet qualities.

In the production of RT sheets, coating is a very important process to distribute the characteristic and flavor substances on the tobacco slices (*Gao et al, 2014*). Coating process also can improve properties of RT sheets, such as bulk property, softening property, and penetration performance, which will result directly in the mainstream smoke release quantity. This paper aimed at analyses and narrowing the quality gap between the commercial RT sheet products from China and abroad, domestic tobacco slices and sheets are researched in the influences of coating on the products properties.

2 MATERIALS AND METHODS

2.1 *Materials and instruments*

All domestic reconstituted tobacco slices (as RT slices, before coating) and reconstituted tobacco sheets (as RT sheets, after coating) were provided by a tobacco factory in GuangDong Province in China. Abroad RT sheets are samples from a famous Corporation, which is distinguished with high-class reconstituted tobacco sheets). Tobacco waste materials, such as tobacco stems and tobacco dusts, were also collected from the factory.

Water bath, viscometers (Brookfield Ltd), Rotary Evaporator (RE-522, Shanghai Yarong), Scanning Electron Microscope (SEM, Hitachi S-3700 N), and pore size analyzer (Quantum P-66) were applied in the research.

2.2 Methods

2.2.1 Preparation of coating fluids

In this paper, the water extraction solutions for tobacco stems and dusts were used as coating fluids. Every weighed 5 g tobacco stems or dusts was filled with 20 mL water in a conical flask, and then the flask was heated at 70 °C for 1.5 hr in the water bath. All the upper water liquids were collected and evaporated, and then different viscosity extraction liquids can be obtained as coating fluids.

2.2.2 Coating process

1. Coating process on RT slices
   RT slices were placed on the dried plastic board, the coating fluids were sprayed uniformly to the surfaces of slices, and then soft roller was applied to coat on the surface of the slices several times. After coating RT slices were naturally dried to get RT sheets.
2. Calculation of coating efficiency
   After balancing for 24 hrs at the standard air condition (23 °C, 50%RH), RT sheets and RT slices were weighed separately, the moistures were determined, and then the coating efficiency can be calculated as:

$$Coating-efficiency = \frac{m_2 - m_1}{m_1} \times 100\% \quad (1)$$

where m_1 = the absolute dry quantity of RT slices; m_2 = the absolute dry quantity of RT sheets.

2.2.3 Analysis of surface pores

After balancing for 24 hrs at the standards air condition (23 °C, 50% RH), RT sheets and RT slices were directly prepared for the observations of surface microstructure with SEM. After vacuum cryogenic dried for 3 hours, RT sheets and slices were researched for the surface pore distributions with pore size analyzer.

3 RESULTS AND DISCUSSIONS

3.1 Properties of commercial samples

Physical properties of RT slices and sheets from Guangdong and abroad RT sheets are analyzed as shown in Table 1. The surface observations of these samples were applied with SEM shown in Figure 1.

Compared with abroad samples, Guangdong sample is characteristic of low basic weight and ashes content. The coating efficiency of Guangdong samples can be calculated as only about 32% from Table 1. If the coating efficiency could be raised to more than 50%, the basic weight will rise to the same level as abroad samples, so coating efficiency is one of the main reasons for low basic weight of domestic samples.

The bulk and permeability of RT sheets are important factors to evaluate tobacco sheet's quality *(Liu et al, 2013)*. The greater the bulk is, the looser the conjunction of the fiber is, the higher the porosity of tobacco sheets is, the more fully tobacco sheets will burning in the process of smoking, then the harmful substances will be decreased. But the bulk can not to be too high, otherwise smoking feeling will be discounted. Compared with abroad samples, Guangdong samples are characteristic with high bulk and porosity.

Table 1. The analyses of physical properties of the samples.

	Basic weight (g/m²)	Ashes content (%)	Moisture (%)	Color	Thickness (um)	Bulk (cm⁻³/g)	Tensile strength index/ (N·m/g)	Permeability (ml/min)	Softness (mN)
RT slices from Guangdong	67.26	12.87	10.94	Grey	202	2.28	4.94	921	7.65
RT sheets from Guangdong	90.37	15.36	12.0	Brown	228	2.28	3.75	756	15.7
Abroad RT sheets	106.00	19.0	13.73	Brown	229	2.04	6.25	526	118

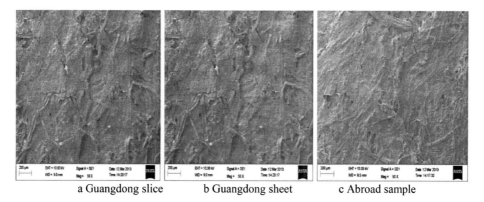

a Guangdong slice   b Guangdong sheet   c Abroad sample

Figure 1. SEM of sample's surface.

From Figure 1 it can also be easily observed that the fiber combination of Guangdong slice is very loose, the surface of the tobacco sheets is still not flat with obvious exposures of fiber after the crushing and drying process. Abroad sample has a uniform coating and we can hardly see the fiber mixed texture, the coating mass of Guangdong sheet is not enough obviously, most of the fiber are even naked.

3.2 *The coating influence on the surface structure of RT sheets*

According to the method mentioned in 1.2, coating liquids with different viscosities were applied to coat on the RT slices from Guangdong, and then RT sheets and slices were weighed separately, and the coating efficiency was calculated as shown in Figure 2. The surfaces of RT sheets were observed by SEM in Figure 3.

With the increase of viscosity of coating liquid, the coating efficiency is rising. When the viscosity is in the range of 20–50 m·Pa·s, it has a great influence on the spreading rate of fluids. Coating efficiency increased very rapidly from almost 20% to 50%, so the viscosity of coating fluids between 40–50 m·Pa·s will be helpful for the increase of basic weight of tobacco sheets as discussed above. When the viscosity reaches to more than 70 m·Pa·s, the fluids become sticky with bad mobility, and then the fluids will easily settle on the surface of fibers so the coating efficiency become extremely higher with bad porosity which is harmful for the cigarette combustion.

Figure 3 is the graphs of RT sheets coated with different viscosity fluids. The surfaces of tobacco sheets coated with low viscosity liquids are coarser and more porous, and with the increase of coating fluids viscosity, RT sheet's surface textures become smoother and more compact, and the coating layers can be easily found around the fibers to block the pores

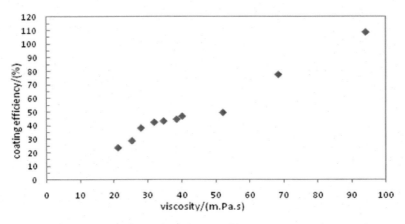

Figure 2. The coating liquid viscosity and coating rate diagram.

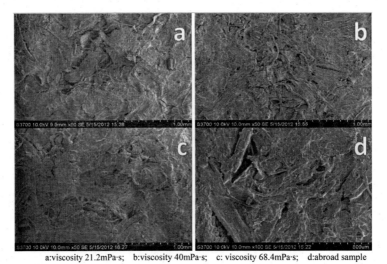

a:viscosity 21.2mPa·s;   b:viscosity 40mPa·s;   c: viscosity 68.4mPa·s;   d:abroad sample

Figure 3. Scanning micro graphs after coating with different viscosity.

between fibers. As we know the decrease of pores will limit the input of air to influence the burning of tobacco, and then more harmful substance will emit resulted from incomplete combustion, so coating liquids with higher viscosity are against to tobacco sheet quality.

3.3 *Pore distributions of tobacco sheet surface*

Since there are many pores on the surfaces of RT slices and sheets which are the important tunnels for air, the sizes and distribution of these pores will play key roles of the combustion, emission of harmful substances, the smoking feeling, and the quality of tobacco. Mercury intrusion method is applied to analyze the pores of tobacco slices and sheets with results shown in Figure 4, Figure 5 and Table 2

RT slices provided by Guangdong factory are the basic materials for the coating experiments with different viscosities fluids. From Figure 4, we can see the pore sizes of tobacco slices are between 100 nm and 200 um, the mean pore size is 1.882 um with total pore volume 1.395 mL/g and pore-throat ratio 0.026 provided by the instrument program. High pore volume and low pore-throat ratio indict that the surface structures of RT slices are very loose, and most pores are connected with others.

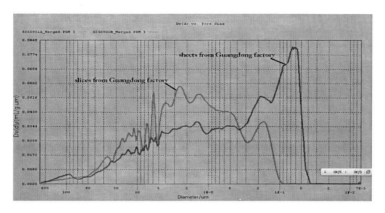

Figure 4. The pore size distribution of tobacco slice and sheet from Guangdong factory.

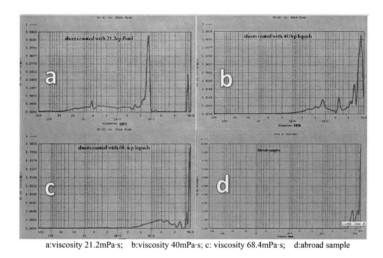

a:viscosity 21.2mPa·s; b:viscosity 40mPa·s; c: viscosity 68.4mPa·s; d:abroad sample

Figure 5. Pore size distribution of tobacco sheets with different viscosity coating fluid and Mauduits.

Table 2. The contrast of mean pore size and total pore volume about tobacco slice and sheet.

	Slices from Guangdong	Sheets from Guangdong	21.2 cp coating sample	40.0 cp coating sample	68.4 cp coating sample	Abroad samples
Mean pore size (um)	1.882	0.461	0.174	0.0369 ※	0.0133 ※	20 nm/
Total volume (mL/g)	1.3946	1.2553	1.1201	0.8179	0.3551	1.059

※For reference only.

Compared with slices, the pore size of RT sheets from Guangdong factory decreases obviously after coating, which distributed mainly between 40 nm and 200 um with mean pore size 0.461 um, total pore volume per unit mass 1.255 mL/g and pore-throat ratio 0.175, the decreased total pore volumes 10% and increased pore-throat ratio suggest that coating liquid have intruded into the pore and occupied the spaces in the coating process. This is the reason why the permeability of RT sheets is reduced compared with RT slices in Table 1.

From Figure 4, it can be easily concluded that the pore size distribution of RT slices and sheets from Guangdong are mainly in the micron range, which means the coating efficiency is relative low with big pores on the surface of tobacco sheets.

In Figure 5 and Table 2, we can find that with the viscosity increase of coating fluids, the mean pore sizes of RT sheets decreased dramatically from 1.882 um of slices down to 0.174 um of sheets coated with viscosity 21.2 cp liquid, to 0.0369 um of sheets coated with viscosity 40.0 cp liquid, and to 0.0133 um of sheets coated with viscosity 68.4 cp liquid, accompanied with the generation of a great deal of micropores (< 20 nm), the results reveal that the pore-filling effect of coating fluids is very obvious, the viscosity of coating liquid has great influence on the coating efficiency and the pore size distribution of tobacco sheets, the coating fluids with high viscosity will fill the pores and stand still because of limited flow-ability to leave small space, that is the reason for the generations of micropores.

4 CONCLUSIONS

1. A physical property comparison between the domestic and abroad reconstituted tobacco sheets was made; the results showed that domestic products are characteristic of low weight, low coating efficiency, and uneven coating.
2. Coating plays an important role for the quality of reconstituted tobacco sheets. With the increase of coating liquid viscosity, coating efficiency has a significant increase and pore-filling effect is very obvious. The 40–50 m·Pa·s viscosity of coating fluids will be helpful for the increase of basic weight and coating efficiency.
3. Compared with abrossad products, the surface of domestic reconstituted tobacco products are uneven with big holes and wide pore distributions, more attention should also be paid to the suitable total pore volumes.

Competing interests' statement: The authors declare that they have no competing financial interests.

ACKNOWLEDGMENTS

We gratefully acknowledge the assistance of research team at Jinye Reconstituted Tobacco Sheet Corporation in Guangdong Province, China. This work has been partly supported by research funds from State Key Laboratory of Pulp and Paper Engineering (2015c02) and the National Natural Science Foundation of China (21206046).

REFERENCES

Agrupis, S. & Maekawa E. & Suzuki K. 2000. Industrial utilization of tobacco stalks II: prepararion and characterization of tobacco pulp by steam explosion pulping. *The Jpn Wood Res. Soci* 46: 222–229.

Alireza, A. & Yahya, H. & Fatemeh, A. 2011. L, mon balm (melissa officinalis) stalk: Chemical composition and fiber morphology[J]. *J. Polym. and the Environ.* 19: 297–300.

Gao, W.H. & Chen, K.F. & Yang, R.D., et al. 2014. Rheological property of reconstituted tobacco coatings. *Ind. Crops and Prod* 60: 45–51.

Han, Q. & Zhang, M.Y. & Wu, Y.Y., et al. 2002. Manufacturing techniques of papermaking tobacco leaf. *J. northwest institute of light industry* 20(1): 19–22.

Liu, X.F. & Li, Y.M. & Hou Y., et al. 2014. Fiber Properties of Tobacco Stems, Powder and Rods. *Tob. Sci. & Tech* 323(6): 8–15.

Sun, X. & Su, W.Q. 2010. The research and application prospect of reconstituted tobacco by paper process. *East China pulp & paper industry* 41(4): 27–30

Wang, L. & Wen, Y.B. & Sun, D.P., et al. 2011. Study on the decrease of harmful substance in paper-process reconstituted tobacco sheet. *Adv. Mater. Res* 314: 2338–2343.

Zhang, X.Z. & Gao, H.J. & Zhang L.F., et al. 2012. Extraction of essential oil from discarded tobacco leaves by solvent extraction and steam distillation, and identification of its chemical composition. *Ind. Crops and Prod* 39: 162–169.

A new evaluation method of heavy metal contamination in soil of urban area

Yingmian Yang & Zhenjun Ye
Department of Mathematics and Physics, North China Electric Power University, Beijing, China

ABSTRACT: Firstly, a comprehensive comparison was made about the advantages and disadvantages of Nemerow pollution index method and Fuzzy Comprehensive Evaluation Method based on Membership degree function. Then in order to optimize the fuzzy comprehensive evaluation, a new method was proposed to determine the degree of membership. In the example analysis, respectively, according to the sampling data in different regions of a city, two kinds of comprehensive evaluations were calculated about the heavy metal pollution levels in soil of each region, using the following two methods. When applying the fuzzy comprehensive evaluation method, we used the new method to determine membership degree, and introduced the national secondary standard value into the calculation of weights. By calculation analysis, we got comprehensive evaluation indices of the two methods, thus the sort of the heavy metal contamination in soil of each region could be listed. Finally, we found that the sort results were consistent, which verified the feasibility of the new fuzzy comprehensive evaluation method.

1 INTRODUCTION

There are many methods of heavy metal contaminated soil rating, in which the applications of Nemerow pollution index are more common. The index can be used to evaluate urban soil contamination based on statistics. It overcomes the Inconvenience of calculation, comparison and comprehensive evaluation when describing the soil contamination with a number of indicators. However, this method only uses the maximum and average value indicators, which leads to incomplete information utilization. Fuzzy comprehensive evaluation is a comprehensive evaluation method based on fuzzy mathematics. This comprehensive evaluation method transfer qualitative evaluation into quantitative evaluation, according to the membership degree theory in fuzzy mathematics. It features clear results and strong systematicness, which properly solves the fuzzy and hard-to-quantify problems, and is suitable for a variety of non-deterministic problem solving. There are generally two methods to determine membership degree. The first one is to take the form of qualitative investigation report or expert opinion in the form of voting. And the second is to construct a reasonable expression to calculate the membership degree of evaluation factor.

This paper presents an improved method. Firstly, construct a reasonable membership degree function expression. Secondly, when seeking for the relationship matrix, we assume that classification is enough, so the pollution evaluating value of each elements in soil approximately appears normal distribution. In this way, this method combines the advantages of both methods above (the former's fitting practice and the latter's theory), and avoids the subjectivity of first method, together with the one-sidedness of second method.

2 SAMPLING AND SIMPLE ANALYSIS

The urban area can be divided into living areas, industrial zones, mountains, traffic area, green areas and park, according to their functions. Different areas suffer different impact

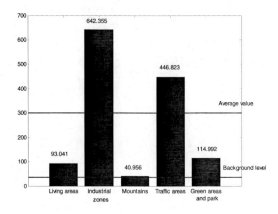

Figure 1. Histogram for concentration of *Hg* in each functional area.

Table 1. Concentration order of each element in different functional area.

	As	Cd	Cr	Cu	Hg	Ni	Pb	Zn
Industrial zones	1	1	3	1	1	1	1	1
Traffic area	4	2	2	2	2	3	3	2
Living areas	2	3	1	3	4	2	2	3
Green areas and park	3	4	4	4	3	5	4	4
Mountains	5	5	5	5	5	4	5	5

level by human activities. In some city, we sampled in grid sub-regions which space about 1 Km. We used GPS to locate each sampling point and obtained the concentration data of variety of heavy metal elements in each sample soil. After sampling, we calculated the average concentration of each heavy metal in each functional area. For instance, Figure 1 is a histogram for concentration of *Hg* in each functional area.

In addition, we sorted the concentration of functional areas, respectively according to different heavy metals, which is listed as the following table (descending order).

From Table 1, we get a preliminary conclusion: the most serious pollution area may be industrial areas, and the least may be mountainous areas.

3 QUANTITATIVE ANALYSIS OF TWO KINDS OF METHODS

3.1 *Fuzzy comprehensive evaluation*

In order to calculate the membership degree objectively, we use a reasonable function expression to solve the fuzzy relationship matrix. Taking into account that when the sample number is enough, the pollution evaluating value of each elements in soil approximately appears normal distribution $N(\mu, \sigma^2)$. We can obtain the mean value μ_j and the variance σ_j^2 of j-th column, using standardized data. In summary, the normal distribution function is:

$$f_j(x) = \frac{1}{\sqrt{2\pi}\sigma_j} e^{-\frac{(x-\mu_j)^2}{2\sigma_j^2}} \qquad (1)$$

The thought of how to determine the membership degree is that the sample has a largest membership degree in its corresponding level of contamination calculated before, and its membership degree decreases in both sides of pollution levels. So the specific practice are as follows.

In this example, we firstly divided the contamination into 5 levels: A, B, C, D, and E. Let m represents the types of heavy metal elements we detected, and n be the number of samples in the living area. Then, let the j-th column and i-th row element of an n by m matrix be a_{ij}, which is the concentration values at each sampling point. The membership degrees of a_{ij} for these five levels are respectively denoted as $e_1(a_{ij}), e_2(a_{ij}), e_3(a_{ij}), e_4(a_{ij})$, and $e_5(a_{ij})$. And each of these values is defined by the area of a certain divided region under the normal distribution curve.

$$e = \begin{Bmatrix} 0.023 & 0.057 & 0.104 & 0.193 & 0.623 \\ 0.196 & 0.198 & 0.2 & 0.202 & 0.204 \\ 0.003 & 0.012 & 0.034 & 0.094 & 0.857 \\ 0.06 & 0.107 & 0.156 & 0.228 & 0.45 \\ \vdots & \vdots & \vdots & \vdots & \vdots \end{Bmatrix}_{44 \times 5} \quad (2)$$

Here we take the first column as example (that is As element) to construct the membership degree function, according to the above way. We found the mean and variance of this column are respectively $\mu = 0.188$, $\sigma = 0.1$, then we got its normal distribution density function. The interval $(-3\sigma, 3\sigma)$ is divided into 5 equal parts, then we calculated the integral of this density function over each subinterval. The results we got were corresponding to the membership degree of 5 levels. After the actual computing, we got the membership degree matrix of As element concentration in living areas as:

$$X = w \cdot e = (w_1 \cdots w_n) \begin{pmatrix} e_{11} & e_{12} & e_{13} & e_{14} & e_{15} \\ \vdots & \vdots & \vdots & \vdots & \vdots \\ e_{n1} & e_{n2} & e_{n3} & e_{n4} & e_{n5} \end{pmatrix} = (x_1 \ x_2 \ x_3 \ x_4 \ x_5) \quad (3)$$

In the same way, we got the membership degree matrices of other heavy metal concentration in each functional area.

Assume that in the same functional area, each sampling point plays the same role in the comprehensive evaluation of the whole area. That is, each sampling point has the same weight within a common area. To simplify operation, we do not normalize the weights, but set it in the form: $w = [1 \ 1 \cdots 1]_{1 \times n}$. Right by the weights set (w) of n sampling point and the fuzzy relationship matrix (e), we can get a comprehensive evaluation matrix X, and the operation is $X = w \circ e$. Here, the operator 'o' is the composite operator of two fuzzy matrices, and we used the common matrix multiplication for simplify. Its expression is as follows.

For the As element concentration in living areas, the normalized results of comprehensive evaluation matrix is: $X = (0.199 \ 0.162 \ 0.163 \ 0.180 \ 0.296)$. This vector shows the possibility values of the level that the As concentration can be classified in, and possibility values can be regard as weight. Basing on these weight, on each kind of element, we calculated the comprehensive evaluation values in each functional area by linear weighted method.

In practice, the same concentration of different heavy metal has different effect on the soil pollution. So, we must find the comprehensive pollution index of each functional area, by assigning different weights to each kind of heavy metal element. In this paper, the national secondary standard is introduced to calculate weights. Let PI represent single pollution index, and let CI be soil pollutants found, and SI be standard index of pollutants. Then we have:

$$PI = \frac{CI}{SI} \quad (4)$$

The statistic table of heavy metal content indices within whole urban area is as shown in Table 2.

The pollution proportion can be used as the pollution effect weights of 8 kinds of element. That is, the set of weights is: $Q = (0.05 \ 0.15 \ 0.08 \ 0.16 \ 0.17 \ 0.10 \ 0.06 \ 0.23)$. Basing on the

Table 2. Heavy metal content indices within whole urban area.

	As	Cd	Cr	Cu	Hg	Ni	Pb	Zn
Average content	5.68	0.30	53.5	55.0	0.30	17.3	61.7	201
National standard	30	0.6	200	100	0.5	50	300	250
Single pollution index	0.19	0.50	0.27	0.55	0.60	0.35	0.21	0.80
Pollution proportion (%)	0.05	0.15	0.08	0.16	0.17	0.10	0.06	0.23

Table 3. Final comprehensive evaluation value of heavy metal pollution.

	As	Cd	Cr	Cu	Hg	Ni	Pb	Zn	Comprehensive value
Living areas	0.202	0.188	0.084	0.025	0.017	0.124	0.153	0.076	0.091
Industrial zones	0.227	0.223	0.062	0.042	0.051	0.135	0.172	0.079	0.107
Mountains	0.164	0.143	0.055	0.018	0.015	0.121	0.116	0.042	0.068
Traffic area	0.194	0.218	0.07	0.026	0.087	0.124	0.137	0.069	0.103
Green area and park	0.195	0.203	0.059	0.022	0.025	0.119	0.145	0.069	0.068
Weight (Q)	0.05	0.15	0.08	0.16	0.17	0.10	0.06	0.23	

Table 4. Results of Nemerow method.

	Living areas	Industrial zones	Mountains	Traffic area	Green areas and park
$(PI)_{max}$	3.74	18.35	1.31	12.77	3.29
$(PI)_{avg}$	2.47	5.43	1.19	3.84	2.04
PN	3.170	13.532	1.251	9.429	2.737
Pollution level	Severe	Severe	Mild	Severe	Moderate
Pollution degree sort	3	1	5	2	4

above weights and the comprehensive evaluation values of each element calculated, we can get the final comprehensive evaluation value within each functional area. The final results are as shown in Table 3.

From the above result, we can give the descending order of these 5 functional areas, according to their heavy metal pollution degree: Industrial zones, Traffic area, Living areas, Green areas and park, and then Mountains.

3.2 *Analyzing by Nemerow pollution index*

In Nemerow index method, we calculate the index by the expression:

$$PN = \sqrt{\frac{(PI)_{max}^2 + (PI)_{avg}^2}{2}} \qquad (5)$$

In the above expression, PN is pollution index, $(PI)_{max}$ is the maximum of single pollution index, and $(PI)_{avg}$ is the average of single pollution index. According to the expression and other related analysis, we got the results in Table 4.

4 CONCLUSION

By comparison, the results of two methods are same. The descending order of pollution degree for 5 areas is: Industrial zones, Traffic area, Living areas, Green areas and park, and then Mountains. And this order is consistent with the fact. Exhaust emissions from vehicles and industrial emissions have great impact on heavy metal pollution in industrial zones. Conversely, the mountains are away from the crowd, so the contamination is naturally minimized. And not surprisingly, the pollution degree of other areas are often in the middle.

REFERENCES

Giordano, P. P. Caputo, A. Vancheri. 2014. Fuzzy evaluation of heterogeneous quantities: Measuring urban ecological efficiency [J]. *Ecological Modelling*, 288: 112–126.

Jiabao Liu, Lu Yinan, Zou Ting. 2012. The Analysis and Evaluation on Heavy Metal Pollution of Topsoil in Chinese Large-scale Cities [J]. *Energy Procedia*, 16: 1084–1089.

Motuzova, G.V. T.M. Minkina, E.A. Karpova, N.U. Barsova, S.S. Mandzhieva. 2014. Soil contamination with heavy metals as a potential and real risk to the environment [J]. *Journal of Geochemical Exploration*, 144: 241–246.

Paustenbach D.J., Finley B.L., Long T.F. 1997. The critical role of house dust in understanding the hazards posed by contaminated soil [J]. *Int J Toxicol*, 16: 339–362.

Xiangyu Tang, Yongguan Zhu. 2004. Assessment of bioavailability of heavy metals in soil in vitro [J]. *Environment and Health*, 21(3): 183–185.

Yangwu Fu, Junsheng Qi, Shuming Chen, Jie Pan. 2009. Heavy metal pollution survey and evaluation of fluctuating soil in the Zhuxi River basin of Three Gorges Reservoir Area [J]. *Soil Science*, 40: 162–166.

Yun Zhang, Yufeng Zhang, Xin Hu. 2010. Heavy metal pollution assessment and source analysis of surface dust on the streets in different functional areas of Nanjing [J]. *Environmental Science*, 23: 1376–1381.

Yi H.M., Zhou S.L., Wu S.H., Xu K., Zhou H. 2013. An integrated assessment for regional heavy metal contamination in soil based on normal fuzzy number [J]. *Acta Scientiae Circumstantiae*, 33: 1127–1134.

Zhiyong Zhou, Xiaoju Zhang, Wenyi Dong. 2013. Fuzzy Comprehensive Evaluation for Safety Guarantee System of Reclaimed Water Quality [J]. *Procedia Environmental Sciences*, 18: 227–235.

Au and Ag adsorption from a low-concentration sulfuric acid residue leaching solution

Chunjie Zhou, Gang Zhao, Shuai Wang & Hong Zhong
College of Chemistry and Chemical Engineering, Central South University, Changsha, Hunan, China

ABSTRACT: A polystyrene-modified ethoxarbonyl thiourea resin (PSETU) was used to extract Au and Ag in a sulfuric-acid residue leaching solution. The results showed that PSETU adsorption efficiency for Au was better than other resins under the same conditions. The effects of contact time, resin dose, and adsorption temperature on Au and Ag adsorption efficiency were investigated by the batch tests. The results showed that the adsorption ratios of Au and Ag reached 75.3% and 99.3%, respectively, when the contact time was 120 min, the resin dose was 10 g/L and the temperature was 20 °C. In dynamic experiments, the adsorption ratios of Au and Ag were more than 90% and the desorption ratio of Au and Ag reached 99%, when the solution's flow velocity was 0.5 ml/min and the column height was 2 cm. In the leaching-adsorption-desorption process, the total recovery ratios of Au and Ag were 82.5% and 72.1%, respectively.

1 INTRODUCTION

Sulfuric-acid residue is the product during the roasting of pyrite ores in the sulfuric acid industry. It is commonly used in steel smelting. Sulfuric-acid residue, however, is still associated with valuable elements such as Au and Ag (Hilson, et al. 2006). Recycling the associated elements has great research significance and economic value (Atia, et al. 2014; Sun, et al. 2014).

Au and Ag leach agents from sulfuric-acid residue mainly include thiourea, sodium cyanide, ammonium thiosulfate, chloride, bromide, etc (Fu, et al. 2011; Oraby, et al. 2015). As a traditional method, cyanidation is easy to use and has low cost, but high toxicity. In recent years, thiourea has gained attention as an effective Au leach agent (Whitehead, et al. 2007; Gonen, et al. 2007). Techniques for recovering Au from acid thiourea solutions include activated carbon adsorption, replacement precipitation, solvent extraction, and ion exchange resin (Virolainen, et al. 2014). The ion exchange and adsorption methods are popular for their low costs and effectiveness.

In this paper, Au and Ag extraction in a sulfuric-acid residue leaching solution were investigated through an adsorption process using polystyrene modified ethoxarbonyl thiourea resin (PSETU) as the absorbent (Wang, et al. 2011).

2 EXPERIMENTAL

2.1 *Materials and apparatus*

Sulfuric-acid residue was acquired from Jiangxi Copper Company, China. The component analysis results of the sulfuric-acid residue were as follows: Au 1.16 g/t, Ag 2.21 g/t, and Fe 65.78%. PSETU resin was synthesized according to the reference (Wang, et al. 2011), and the other adsorbents were purchased from commercial sources. Thiourea, sodium cyanide, ammonium thiosulfate, sulfuric acid, Au wire, and Ag nitrate were of analytical grade.

Batch adsorption was performed in a SHA-C thermostatic vibrator (Jintan Medical Instrument Plant, China); Dynamic adsorption occurred in an ion exchange column with a

12 mm diameter. Metal ion concentration was determined on a WYX-9003 atomic absorption spectrometer (Shenyang Analytical Instrument Factory, China).

2.2 Preparation of the leaching solution

In the acid system, thiourea can coordinate Au to form cationic coordination compounds. The reaction is shown in Equation 1.

$$Au^+ + 2SC(NH_2)_2 \rightarrow Au[SC(NH_2)_2]^+ \qquad (1)$$

One primary sample was added in water and mixed with thiourea. The leaching solution was carried out under the following conditions: 50 °C, 8 h contact time, 15 g/L thiourea concentration, 10% sulfuric acid, and a 3:1 (V/W) liquid to solid ratio. The sample was stirred, filtered, and then dried under controlled temperatures. The Au or Ag leaching ratios were calculated by Equation 2.

$$\varepsilon_L = \frac{C_0 V_0}{10^{-3} \alpha_p m} \times 100\% \qquad (2)$$

where ε_L = Au or Ag leaching ratio; C_0 = concentration (mg/L) of the Au or Ag metal ions in the leaching solution; V_0 = solution volume (L); α_p = Au or Ag ore grade (g/t) of the sulfuric-acid residue; and m = weight (g) of the sulfuric-acid residue.

Under the above conditions, the Au and Ag concentrations in the leaching solution were 0.565 mg/L and 0.750 mg/L, respectively; the leaching ratios of Au and Ag were 87.9% and 75.3%, respectively.

2.3 Sorption experiments

2.3.1 Batch adsorption and desorption experiments

About 0.2 g adsorbent and 25 ml leaching solution were added in an iodometric flask. The flask was placed in a thermostatic vibrator and shaken at 120 r/min for 120 min. Then, the solution was filtrated and the metal ion concentrations were determined. The adsorption ratio for metal ions was calculated according to Equation 3.

$$\varepsilon_a = \frac{C_0 - C_e}{C_0} \times 100\% \qquad (3)$$

where ε_a = Au or Ag adsorption ratio; C_e = metal ion residual concentration (mg/L).

Using thiourea (15%), sulfuric acid (20%), ethanol, and distilled water as the elution solvent, a percentage of loaded resin and 25 ml elution solvent were added in an iodometric flask; this was placed in a thermostatic vibrator and then shaken at 300 r/min for 120 min at 50 °C. Then, the solution was filtrated and the metal ion concentrations were determined by atomic absorption spectroscopy. The desorption ratio was calculated according to Equation 4.

$$\varepsilon_d = \frac{C_d V_d}{V_e (C_0 - C_e)} \times 100\% \qquad (4)$$

where ε_d = Au or Ag desorption ratio; C_d = final concentration (mg/L) of metal ions in the elution solvent; V_d is the elution solvent volume (L); and V_e = adsorption solution volume (L).

2.3.2 Dynamic adsorption and desorption experiments

The leach liquid flowed through the ion exchange column at 0.5 ml/min. The effect of the resin column height on dynamic adsorption was studied under 20 °C. The adsorption ratio of Au and Ag was calculated by Equation (3).

After adsorption, PSETU resin was collected and washed with distilled water. Then, the elution solvent containing 15 g/L thiourea and 20% sulfuric acid flowed through the ion exchange column at 0.5 ml/min. The elution solvent volume was ten times that of the adsorption solution. The Au or Ag desorption ratio was calculated by Equation (4).

3 ADSORPTION AND DESORPTION EXPERIMENTS

3.1 *Batch adsorption*

3.1.1 *Effect of various adsorbents*

Leaching solution adsorption was carried out under the following conditions: 50 °C, 120 min contact time, the adsorbent dose is 8 g/L. Table 1 shows the adsorption ratio of various leaching resins solution.

As shown in Table 1, PSETU resin had higher Au and Ag adsorption ratios. Because the Au and Ag concentrations in the leaching solution were very low, the adsorption ratios were not very high. Further studies are required to improve PSETU adsorption efficiency.

3.1.2 *Effect of contact time on adsorption ratio*

The effects of contact time on adsorption of resin for Au and Ag was studied under the following conditions: 50 °C and 8 g/L resin dose. Figure 1 shows the results.

The results indicate that the resin adsorption ratio for Au and Ag increased as contact time increased. The adsorption ratio changed slowly after 120 min of contact time, indicating that the adsorption was largely complete.

3.1.3 *Effects of resin dose on adsorption ratio*

The effects of resin dose on Au and Ag adsorption were researched under the following conditions: 50 °C and 120 min contact time. Table 2 shows the results.

Table 1. Adsorption of various absorbents.

Adsorbent ratio	Activated carbon	732	XJ133	HZ-190	PSETU
Aurum/%	34.2	1.70	42.7	29.4	53.8
Silver/%	32.5	1.24	38.8	26.2	53.3

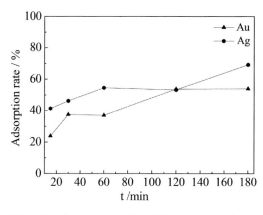

Figure 1. Adsorption ratio at different contact time t.

Table 2. Adsorption at different resin doses.

PSETU dose	Au/%	Ag/%
8 g/L	53.8	53.3
10 g/L	65.2	65.3
12 g/L	76.7	70.7

The results indicate that the adsorption ratio increased as the resin dose increased. The resin dose would need to be confirmed as 10 g/L in the static adsorption experiment, because the leaching solution concentration and flow velocity must be adjusted in actual dynamic experiment.

3.1.4 *Effect of adsorption temperature on the adsorption ratio*

The effects of temperature on the resin's Au and Ag adsorption were studied under the following conditions: 120 min contact time and 8 g/L resin dose. Figure 2 shows the results.

As shown in Figure 2, the adsorption ratio of resin for Au and Ag decreased little as the temperature rose. When the temperature was controlled at 20 °C, the adsorption ratio of resin for Au and Ag reached 75.3% and 99.3%, respectively.

3.2 Dynamic adsorption

3.2.1 *The effect of flow velocity on adsorption efficiency*

The effects of flow velocity on dynamic adsorption were studied under the following conditions: 2 cm resin column height, 20 °C, and two groups of leaching solutions flowed through the ion exchange column with different flow ratios. Table 3 shows the results.

As can be seen from the Table 3, the adsorption ratio of resin for Au and Ag increased as the flow velocity decreased. As the flow velocity increased, the Au and Ag leaching ratio decreased, and the Au adsorption reduced significantly. Based on these results, an appropriate flow velocity should be maintained during dynamic adsorption experiments.

3.2.2 *The effect of resin column height on adsorption efficiency*

The effect of resin column height on dynamic adsorption was studied under the following conditions: 0.5 ml/min flow velocity, 20 °C, and two leaching solution groups flowed through the ion exchange column when the resin column heights were 2 cm and 4 cm. Table 4 shows the results.

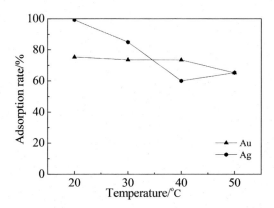

Figure 2. Adsorption at different temperatures.

Table 3. Adsorption at different flow velocities.

Flow velocity	Au/%	Ag/%
0.5 ml/min	91.2	93.3
1.0 ml/min	79.5	90.0
1.5 ml/min	72.8	85.1

Table 4. Dynamic adsorption ratio at different column heights.

Resin column height	Au/%	Ag/%
2 cm	91.2	93.3
4 cm	93.5	94.4

As can be seen from Table 4, the adsorption ratio increased with the increase in the resin amount. When the column height was 2 cm, the Au and Ag adsorption ratios were more than 90%, so the optimized column height was 2 cm.

3.3 Desorption

3.3.1 Batch desorption

Using thiourea (1.5%), sulfuric acid (20%), ethanol, and distilled water as an elution solvent, a percentage of loaded resin (8 g/L) was vibrated with the elution solvent at 300 r/min for 120 min at 50 °C. The final Au or Ag concentration in the elution solvent was measured. The PETSU desorption ratio for Au and Ag were 21.8% and 52.0%, respectively.

The results by static desorption were not so satisfied because there was distribution equilibrium of metal ions between resin and solution. Therefore, the dynamic desorption experiments were carried out through ion exchange column.

3.3.2 Dynamic desorption

The elution solvent containing 15 g/L thiourea and 20% sulfuric acid flowed through the ion exchange column at 0.5 ml/min. Under these conditions, the Au or Ag desorption ratio reached 99%.

As the above experiment showed, PSETU had superior Au and Ag adsorption ability. The desired adsorption results can be obtained through many times adsorption. Adopting this experiment's dynamic methods can further enhance the Au and Ag adsorption ratios.

3.4 Leaching-adsorption-desorption experiments

Figure 3 shows the leaching-adsorption-desorption process flow chart. The leaching experiment was carried out under the following conditions: 50 °C, 8 h contact time, 5% sulfuric acid

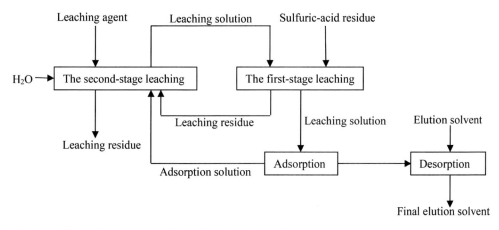

Figure 3. Flow chart of the leaching-adsorption-desorption process.

Table 5. Results of the leaching-adsorption-desorption process.

	Leaching ratio	Adsorption ratio	Desorption ratio	Total recovery ratio
Au/%	89.1	93.5	99.0	82.5
Ag/%	77.2	94.4	99.0	72.1

concentration, 15 g/L thiourea in the leaching agent, and a 3:1 (V/w) liquid to solid ratio. The dynamic adsorption was carried out under the following conditions: 2 cm column height, 20 °C, and 0.5 ml/min flow velocity. The elution solvent containing 15 g/L thiourea and 20% sulfuric acid flowed through an ion exchange column at 0.5 ml/min. To reduce expense, after adsorption, the leaching solution was used in the second-stage leaching, with 6.67 g/L thiourea and water added. Table 5 shows the results.

The Au and Ag acidic thiourea leaching ratio reached 89.1% and 77.2%, respectively. The adsorption ratio after PSETU resin adsorption reached 93.5% and 94.4%. Using acidic thiourea as an elution solvent, the desorption ratio reached 99%. The total Au and Ag recovery ratios were 82.5% and 72.1% respectively.

4 CHARACTERIZATION OF PSETU

The thiourea Au solution and thiourea Ag solution were prepared with Au wire and Ag nitrate to investigate the adsorption process and mechanism.

4.1 SEM Analysis of PSETU before and after adsorption

Figure 4 shows SEM Analysis of PSETU before and after adsorption. The smooth resin surface became coarser with granules after Au adsorption; it turned thicker and developed flakes after Ag adsorption.

Figure 4. SEM of PSETU before-and-after adsorption a-PSETU, b-PSETU + Au, c-PSETU + Ag.

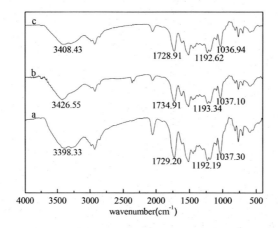

Figure 5. FT-IR spectra of PSETU before-and-after adsorption a-PSETU, b-PSETU + Au, c-PSETU + Ag.

Table 6. Selected peaks of PSETU before-and-after adsorption/cm^{-1}.

Assignment	PSETU	PSETU + Au	PSETU + Ag
$v_{C=S}$	1037.30	1037.10	1036.94
$v_{C=O}$	1729.20	1734.91	1728.91
v_{aC-O-C}	1192.19	1193.34	1192.62
v_{N-H}	3398.33	3426.55	3403.48

4.2 FT-IR Analysis of PSETU before and after adsorption

PSETU samples before and after adsorption were characterized by FT-IR. Figure 5 and Table 6 show the spectra and the peaks. In the resin spectra, C = S vibrations shifted to lower wave numbers after adsorption, and the C–O–C and N–H vibrations shifted to high wave numbers. The vibration of the C = O peak shifted from 1729.20 to 1734.91 cm^{-1} after Au adsorption, and shifted to 1728.91 cm^{-1} after Ag adsorption.

5 CONCLUSIONS

PSETU was used to extract Au and Ag from a low-concentration leaching solution of sulfuric-acid residue, whose concentrations of Au and Ag were 0.565 mg/L and 0.750 mg/L, respectively. Further results are as follows:

1. Batch adsorption showed that the Au and Ag adsorption ratios reached 75.3% and 99.3%, respectively, under a 120 min contact time, 10 g/L resin dose, and 20 °C adsorption temperature.
2. In dynamic adsorption, the adsorption ratios of Au and Ag were more than 90% when the flow velocity was 0.5 ml/min and column height was 2 cm. In dynamic desorption, the desorption ratio of Au or Ag reached 99%.
3. In the leaching-adsorption-desorption experiments, the total recovery ratios of Au and Ag were 82.5% and 72.1%, respectively.

ACKNOWLEDGMENTS

This work was financially supported by the Natural Science Foundation of China (No. 21206199), the Natural Science Foundation of Hunan Province (11JJ6014) and the Doctoral Fund of Ministry of Education of China (No. 20100162120028).

REFERENCES

Atia, A. A; Donia, A. M; Heniesh, A.M. 2014. Adsorption of Silver and Gold Ions from their Aqueous Solutions using a Magnetic Chelating Resin Derived from a Blend of Bisthiourea/Thiourea/Glutaraldehyde. *Separation science and technology* 49: 2039–2048.

Fu, F.L; Wang, Q. 2011. Removal of heavy metal ions from wastewaters: A review. *Journal of Environmental Management* 97: 219–243.

Gonen, N; Korpe, E; Yildirim, M.E; Selengil, U. 2007. Leaching and CIL processes in gold recovery from refractory ore with thiourea solutions. *Minerals Engineering* 20: 559–565.

Hilson, G; Monhemius, A.J. 2006. Alternatives to cyanide in the gold mining industry: what prospects for the future. *Journal of Cleaner Production* 14: 1158–1167.

Oraby, E.A; Eksteen, J.J. 2015. The leaching of gold, silver and their alloys in alkaline glycine–peroxide solutions and their adsorption on carbon. *Hydrometallurgy* 152: 199–203.

Sun, P.P; Song, H.I; Kim, T.Y; Min, B.J; Cho, S.Y. 2014. Recovery of Silver from the Nitrate Leaching Solution of the Spent Ag/alpha-Al$_2$O$_3$ Catalyst by Solvent Extraction. *Industrial & Engineering Chemistry Research* 53: 20241–20246.

Virolainen, S; Tyster, M; Haapalainen, M; Sainio, T. 2014. Ion exchange recovery of silver from concentrated base metal-chloride solutions. *Hydrometallurgy* 152: 100–106.

Wang, S; Zhong, H; Xia, L.Y; Wang, Z.N; Zhang, Q. 2011. Preparation and chelating properties of polystyrene modified ethoxycarbonyl thiourea resin. *Advanced Materials Research* 239–242: 781–785.

Wang, Z.N; Zhong, H; Wang, S.A; Liu, G.Y; Zhang, Q. 2011. Synthesis of 1,4-benzenedicarbonyl thiourea resins and their adsorption properties for Ag(I). *Journal of Central South University of Technology* 18: 361–366.

Whitehead, J.A; Zhang, J; Pereira, N; McCluskey, A; Lawrance, G.A. 2007. Application of 1-alkyl-3-methyl-imidazolium ionic liquids in the oxidative leaching of sulphidic copper, gold and silver ores. *Hydrometallurgy* 88: 109–120.

Swimming exercise and Caloric Restriction alter the serum cholesterol of rats

Miao Yu & Yibing Liu
Sport Human Science College, Jilin Sport University, Chang Chun City, Jilin Province, China

ABSTRACT: *Objective:* The purpose of this study was to describe changes in serum Cholesterol following Caloric Restriction (CR) intervention plus swimming training.

Methods: Eight-week-old male Wistar rats were randomly divided into six groups: normal control (NC) group, 60 minutes of swimming (S) exercise group, 20% caloric restriction, 40% caloric restriction, 20% caloric restriction, and 60 minutes of swimming exercise group and 40% caloric restriction and 60 minutes of swimming exercise group. After 12 weeks, the serums Cholesterol were measured in rats.

Results: When compared with the NC group, TG was lower significantly in S group, and HDL was higher significantly, and the rats in both 20%CR-S group and 40%CR-S had lower level of TG, and higher HDL. However, only TC in 20%CR-S group was lower. Both in 20% CR group and 40% CR group, TG were lower than in NC group, and HDL was higher. HDL was higher in 20%CR group than 40% CR group.

Conclusions: The chronic exercise was able to reduce the level of TC, TG, and LDL, and elevate high-density lipoprotein. CR alone could reduce hyperlipidemia of animals, and swimming exercise could induce elevated HDL. Additionally, the results also reinforce the idea that a combination of both strategies is better than either individually for combating hyperlipidemia.

1 INTRODUCTION

Caloric Restriction (CR) is defined as a reduction in calorie intake below the usual and libitum intake without malnutrition (Han X., et al., 2010). CR prevents or delays neurodegeneration in brain and enhances neurogenesis in animal models of Parkinson's disease and stroke (Mattson M.P., 2005). More recent studies suggest that CR protects against obesity, type2 diabetes, dyslipidemia, hypertension, inflammation, and atherosclerosis, which are major risk factors for myocardial infarction, stroke, heart failure, and chronic kidney disease, so far as to protect against cancer in humans. In nonhuman primates, CR can also increase the lifespan (Frntana L., et al., 2004; Larson-Mever D.E., et al., 2006; Weiss E.P., et al., 2006; Mohammad-Shahi M., et al., 2012; Mattison J.A., et al., 2003). CR with adequate nutrition advice that encourages consumption of a diet that is relatively lower in any one or more of fat, saturated fatty acids, cholesterol, or sodium is likely to reduce the risk of CVD (Rees K., et al., 2013). The growing prevalence of metabolic syndrome has prompted a refocus of attention on the significance of dietary macronutrients to health outcomes. The finding that restriction of energy (calorie) intake below the amount required for weight maintenance can slow the aging process and markedly extend life span was one of the most important health-related scientific discoveries of the 20th century (Connie W. Bales, et al., 2013). CR may exert antiaging by inhibiting the Growth Hormone (GH)/IGF-1 axis (Komatsu T., et al., 2011). Recent studies have reported that chronic exercise training was associated with attenuation in the response to visual food cues in brain regions known to be important in food intake regulation (Marc-Andre Cornier, et al., 2012). Low intensity swimming training may improve the coagulation and fibrinolytic activity, and help to reduce the blood lipid (Luogeng Zhang, 2009). Both CR and exercise can reduce the risks of CVD effectively, but it has been difficult to distinguish the independent effects of caloric restriction versus exercise training on serum lipid. The purpose of the study was to describe changes in lipid profile following intervention regimented supplementation program plus swimming exercise (12-week treatment intervention program).

2 MATERIALS AND METHODS

2.1 Animals

The experiment was performed on adult, 8-weeks old male Wistar rats weighing 200–220 g, all animals were obtained from the Centre for Animal Experiment of Jilin University (Changchun, China), which has been accredited by the Association for Assessment and Accreditation of Laboratory Animal Care International. All experimental animal procedures were approved by the Guidelines for the Care and Use of Laboratory Animals of the Chinese Animal Welfare Committee. Animals were kept under standard laboratory conditions (temperature 22 ± 2°C, relative humidity 50 ± 10%, 12/12 h light/dark cycle with lights turned on at 8.00 a.m.) in individual cages and had free access to tap water and received the standard-chow diet purchased from the Centre for Animal Experiment of Jilin University containing 21% kcal from protein, 68.5% kcal from carbohydrate, and 10.5% kcal from fat. The rats were weighed weekly. All animals (n = 48) were randomly divided into following groups: Normal Control group (NC group, n = 8), treated with standard-chow diet freely; treated with standard-chow diet freely plus 60 minutes of swimming exercise group (S group, n = 8), 20% calorie restriction group treated freely plus 60 minutes of swimming exercise (20%CR-S group, n = 8), 40% calorie restriction group treated plus 60 minutes of swimming exercise (40% CR-S group, n = 8), 20% calorie restriction (20% CR group, n = 8), 40% calorie restriction (40% CR group, n = 8). After 12 weeks, levels of serum cholesterol and triglyceride were measured in rats. The duration of CR was 12 weeks and allowed food intake was controlled every day. It was based on the quantity of standard-chow diet consumed and used the nutrient composition as reference. The 20% and 40% calorie restriction was designed taking standard-chow diet consumption as a reference. Body weight of investigated animals was registered. A dietary restriction group was fed a 20% and 40% standard-chow diet t; and an exercise group fed standard-chow diet that participated in 60 minutes of swimming.

2.2 Exercise protocol

Swimming was once daily per rat between 8:30 am and 10:30 am. The animals were subject to swimming sessions in a swimming system adapted for rats with water heated to 25°C. Swimming sessions began with 15 minutes in the first week and gradually increasing the length until the rats were able to swim for 60 minutes a day. The rats were subjected to swimming sessions 5 times per week for 12 weeks. Both the exercise and the dietary restriction groups were subject to their respective intervention for 12 weeks.

2.3 Analysis for blood samples

The animals were bled from the tip of the tail after been fasted overnight, then blood was collected before and after the unconditioned stimulus, and the serum was separated by centrifugation at 3000 rpm for 15 min at 4°C, then TG, HDL, LDL, and TC were measured. All experimental procedures were in full compliance with Directive of The Ethical Committee of the Faculty of Medicine, University of JiLin sports. Blood lipoprotein analysis, determination of plasma levels of TC, TG, LDL, and HDL were performed using the automatic biochemical analyzer (selectra-E, Holland).

2.4 Statistical analysis

Statistical analysis was performed on two factors: CR and swimming. All data are presented as mean ± SEM in the descriptive text and graphics. All experiments were compared using two-tailed t-test assuming equal variances with SPSS for Windows v.13. Significant differences were determined when the P value was less than 0.05 (P < 0.05).

3 RESULTS

3.1 Effects of swimming on serum lipid

Compared with NC group, TG was lower significantly in S group (P < 0.01), and HDL was higher significantly (P < 0.01). We did not observe increased levels of LDL in the S group and NC group.

3.2 Effects of CR on serum lipid

Both in 20% CR group and 40% CR group, TG were lower than in NC group (P < 0.01), and HDL was higher than in NC group (P < 0.05). There were no significant differences in LDL and TC either between NC and 20%CR group or NC and 40%CR group (P > 0.05). Further, there was no significant difference in TC, TG, and LDL between 20% CR group and 40%CR group, meanwhile HDL was higher in 20% CR group than 40% CR group (P < 0.01).

3.3 Effects of CR plus swimming on serum lipid

Compared to the NC group, the rats in both 20%CR-S group and 40%CR-S had lower level of TG, and higher HDL. However, only TC in 20%CR-S group was lower. There was no significant difference in TC and LDL between NC group and 40%CR-S group.

Table 1. Results of the effects of swimming on serum lipid (mmol/L).

	TC	TG	HDL	LDL
NC	1.23 ± 0.12	1.43 ± 0.36	0.83 ± 0.10	0.34 ± 0.04
S	1.14 ± 0.62*	0.83 ± 0.21**	1.82 ± 0.28**	0.34 ± 0.04

*P < 0.05, **P < 0.01 vs. control.

Table 2. Results of the effects CR for high fat diet on serum lipid (mmol/L).

	TC	TG	HDL	LDL
NC	1.23 ± 0.12	1.43 ± 0.36	0.83 ± 0.10	0.34 ± 0.04
S	1.14 ± 0.62	0.83 ± 0.21	1.82 ± 0.28	0.34 ± 0.04
20%CR	1.15 ± 0.48	0.76 ± 0.31**	1.11 ± 0.30*	0.38 ± 0.06
40%CR	1.15 ± 0.57	0.75 ± 0.10**	1.58 ± 0.36**&&	0.36 ± 0.04

*P < 0.05, **P < 0.01 vs. control. #P < 0.05, ##P < 0.01 vs. swimming. &P < 0.05, &&P < 0.01 vs. 20%CR.

Table 3. Results of effects of swimming and CR for high fat diet on serum lipid (mmol/L).

	TC	TG	HDL	LDL
NC	1.23 ± 0.12	1.43 ± 0.36	0.83 ± 0.10	0.34 ± 0.04
S	1.14 ± 0.62	0.83 ± 0.21	1.82 ± 0.28	0.34 ± 0.04
20%CR-S	1.07 ± 0.74	0.53 ± 0.16**##	1.95 ± 0.44**	0.40 ± 0.05*#
40%CR-S	1.26 ± 0.43	0.80 ± 0.14**&&	1.66 ± 0.17**	0.30 ± 0.06*&&

*P < 0.05, **P < 0.01 vs. control. #P < 0.05, ##P < 0.01 vs. swimming. &P < 0.05, &&P < 0.01 vs. 20%CR-S.

Compared to the S group, only TG in 20%CR-S group was lower ($P < 0.05$), nevertheless, there was no significantly difference in the other groups. Compared to 40%CR-S group, 20%CR-S group has lower TC, TG, and higher HDL.

4 DISCUSSION

We investigated the effects of swimming exercise, Caloric Restriction (CR, intake of 80% and 60% of daily energy needs), and caloric restriction plus swimming exercise (CR-S) in our study groups by evaluating the serum lipid. In accordance with our hypothesis, we observed that the differences between S group and NC group occurred for TC and TG following the 12-week swimming exercise intervention program. Although TC and TG in S group were lower significantly than NC group after the 12-week intervention program, there was no significant difference in LDL; meanwhile, HDL in S group was higher than NC group. TC and TG were the principal cardiovascular risk factors. The previous research confirmed that improvements in TC, and TG increase the risk of metabolic syndrome and associated disease states, cardiovascular disease, stroke (Garber AJ, 2004). In epidemiological studies, elevated serum cholesterol, both total and TG, induce atherosclerosis in the absence of other risk factors. This is evident in human and animal studies (Reardon CA, 2001). It means this swimming exercise intervention decrease is clinically meaningful within the context of serum lipid levels as ideal for inducing the risk of cardiovascular disease. These results agree with previous reports that regular exercise was important as nonpharmacological strategy for treating obesity as it protects against the increases in body (A.M.W. Petersen and B.K. Pedersen, 2005).

In this study, we found that both in 20% CR group and 40% CR group, TG were lower than in NC group, and HDL was higher. There was no significant difference in LDL and TC either between NC and 20%CR group or NC and 40%CR group. The accumulating evidence suggests that the promotion of diets that reduce the energy density of foods consumed may be an effective future strategy to prevent dyslipidemia (Rees K, et al., 2013). Moreover, there was no significantly difference in TC, TG, and LDL between 20%CR group and 40%CR group, meanwhile HDL was higher in 20%CR group than in 40% CR group. This is the further evidence that low-energy density diets may allow individuals to more effectively prevent chronic diseases to include cardiovascular disease (Connie W. Bales, et al., 2013).

Although chronic exercise and caloric restriction have both induced a reduction in serum lipid, the biological repercussions of both interventions were different. There was no report that which method was more effective to protect human health. Compared to the S group, both 20%CR-S group and 40%CR-S group had significantly lower level of HDL, at the same time, there was no difference between these groups. This result lend further support to suggest that chronic exercise is able to counterbalance dyslipidemia, has active effect on protecting against the development of atherosclerosis. It was previously demonstrated that swimming induces the elevated level of HDL (Frederick Wasinski, 2013). Studies have also shown that the ability of exercise to promote an increase in the HDL in HFD group may also appear to display a protective role. In concordance with previous studies (C.N. Lumeng, 2007), thus, our data regarding serum lipid showed an increased HDL in swimming group. Meanwhile we evaluated the more remarkable effect in S group than in 20%CR-S group and 40%CR-S group, we verified that the increase in HDL, an important factor for protecting cardiovascular, was reversed by exercise.

Compared to the NC group, the rats in both 20%CR-S group and 40%CR-S group had lower level of TG, and higher HDL. The observational study indicated that swimming exercise plus CR intervention could decrease the serum lipid, and increase HDL which is associated with decreased risk for cardiovascular disease. Studies in rodent species would indicate that high HDL-C levels protect against the development of atherosclerosis (Gotto AMJ, 1999). Normally, exercise intervention with dietary modifications is suggested to combat obesity; the combination of both interventions works better than exercise alone (L.H. Colbert, et al., 2004). Conversely, serum lipid reduction promoted by exercise and caloric restriction was followed by a decreased number of TC, LDL. In the High-Fat Diet (HFD), animals are subjected to high-fat diets and produce large amounts of TC, LDL. These observations suggest that exercise and caloric restriction, thought to be able to attain the same goal, proceed by different mechanisms. In our study, all the changes promoted by exercise were accompanied by control of the caloric intake. The reduction of TC, TG, and LDL induced by exercise plus caloric restriction suggests that regular swimming exercise plus CR might relieve the hyperlipidemia that have been induced by HFD. Furthermore, compared to 40%CR-S group, 20%CR-S group has lower TC, TG, and higher HDL. Higher increases in HDL levels in response to exercise have been associated with the reduction of glycogen stores (A.M.W. Petersen 2005). The study reflected a clinically meaningful question on how to control the extent of CR for more effective reduction of the risk of chronic diseases.

Therefore, as discussed above, our results suggest that chronic exercise plus appropriate CR was markedly effective. However, the changes we observed in the levels of serum lipid were strictly related to the dyslipidemia, and the significance of this finding is that CR without destination is detrimental to human health.

5 CONCLUSIONS

The chronics exercise was able to reduce the level of TC, TG, and LDL, and elevate high-density lipoprotein. CR alone could reduce hyperlipidemia of animals, and swimming exercise could induce elevated HDL. Additionally, the results also reinforce the idea that a combination of both strategies is better than either individually for combating hyperlipidemia. Our data demonstrate that both exercise and caloric restriction were able to counterbalance the deleterious effects.

REFERENCES

Ammerman A.S., Keyserling T.C., Atwood J.R., et al. A randomized controlled trial of a public health nurse directed treatment program for rural patients with high blood cholesterol. *Preventive Medicine.* 2003;36:340–51.

Anderssen S.A., Carroll S., Urdal P., Holme I., et al. Combined diet and exercise intervention reverses the metabolic syndrome in middle-aged males: results from the Oslo Diet and Exercise Study. *Scandinavian Journal of Medicine & Science in Sports.* 2007;17(6):687–95.

Baron J.A., Gleason R., Crowe B., Mann J.I., et al. Preliminary trial of the effect of general practice based nutritional advice. *British Journal of General Practice.* 1990;40(333):137–41.

Baron J.A., Gleason R., Crowe B., Mann J.I. Preliminary trial of the effect of general practice based nutritional advice. *British Journal of General Practice.* 1990;40(333):137–41.

Bishop N.A., Guarente L: Genetic links between diet and lifespan: shared mechanisms from yeast to humans. *Nat Rev Genet.* 2007, 8:835–844.

Connie W. Bales, R.D., PhD, William E. Kraus, MD, et al. Caloric Restriction Implications for Human Cardiometabolic Health. *Journal of Cardiopulmonary Rehabilitation and Prevention.* 2013; 33: 201–208.

Colbert L.H., M. Visser, E.M. Simonsick et al. Physical activity, exercise, and inflammatory markers in older adults: findings from the health, aging and body composition study. *Journal of the American Geriatrics Society.* 2004; 52(7): 1098–1104.

Danijela Vuc̆evic̆, Dus̆an Mladenovic̆, Milica Ninkovic̆, et al. The effects of caloric restriction against ethanol-induced oxidative and nitrosative cardiotoxicity and plasma lipids in rats. *Experimental Biology and Medicine.*2013; 238: 1396–1405.

Delzenne N.M., Cani P.D., et al. A place for dietary fibre in the management of the metabolic syndrome. *Curr Opin Clin Nutr Metab Care.* 2005;8:636–640.

Delzenne N.M., Cani P.D., Neyrinck A., et al. Prebiotics and lipid metabolism. In: Versalovic J., Wilson M., eds. Therapeutic MiDRobiology: Probiotics and Related Strategies. *Washington, DC: ASM Press.*2008:183–192.

Frederick Wasinski, 1 Reury F.P., Bacurau, 2 Milton R. Moraes. Exercise and Caloric Restriction Alter the Immune System of Mice Submitted to a High-Fat Diet. *Mediators of Inflammation.* 2013.

Garber AJ. The metabolic syndrome. *Med Clin North.* Am 2004;88:837–46.

Gotto A.M.J., Grundy S.M. Lowering LDL cholesterol: questions from recent meta-analyses and subset analyses of clinical trial Data Issues from the Interdisciplinary Council on Reducing the Risk for Coronary Heart Disease, ninth council meeting. *Circulation.* 1999; 99: E1–E7.

Kyu-Ho Han, Hiroaki Tsuchihira, Yumi Nakamura, et al. Inulin-Type Fructans with Different Degrees of Polymerization Improve Lipid Metabolism but Not Glucose Metabolism in Rats Fed a High-Fat Diet Under Energy Restriction. *Dig Dis Sci.* 2013;58:2177–2186.

Lumeng C.N., J.L. Bodzin, and A.R. Saltiel. Obesity induces a phenotypic switch in adipose tissue macrophage polarization. *Journal of Clinical Investigation.* 2007; 117(1): 175–184.

Marc-Andre Cornier, MD, et al. The Effects of Exercise on the Neuronal Response to Food Cues. *Physiol Behav.* 2012;105(4): 1028–1034.

Masoro E.J: Potential role of the modulation of fuel use in the antiaging action of dietary restriction. *Ann N Y Acad Sci.* 1992, 663:403–411.

Petersen A.M.W. and B.K. Pedersen. The anti-inflammatory effect of exercise. *Journal of Applied Physiology.* 2005; 98(4): 1154–1162.

Reardon C.A., Getz GS. Mouse models of atherosclerosis. *Curr Opin Lipidol.* 2001;12:167–173.

Rees K., Dyakova M., Ward K, Thorogood M, et al. Dietary advice for reducing cardiovascular risk. Published Online: 6 DEC 2013.

Rochon J., Bales C.W., Ravussin E., et al. Design and conduct of the CALERIE study: comprehensive assessment of the longterm effects of reducing intake of energy. *J Gerontol A Biol Sci Med Sci.* 2011; 66: 97–108.

Ross R., Dagnone D., Jones PJ., et al. Reduction in obesity and related comorbid conditions after diet-induced weight loss or exercise-induced weight loss in men. A randomized, controlled trial. *Ann Intern Med.* 2000;133:92–103.

Study on dynamic responses of tunnels through fracture zones

Xiaobo Yan & Hua Xiao
Fuzhou University, Fuzhou, Fujian, China

ABSTRACT: When tunnel passes through fracture zone in an orthogonal manner, the embedded depth plays a crucial role in seismic response of tunnel. This article, by numerical simulation, respectively calculates the speed and acceleration as well as the scope of plastic zone on key points of tunnel lining structure and also makes the comparison when the embedded depth is respectively 2D, 5D and 8D and when the surrounding rock grade is Grade IV and the width of fault fracture zone is 10 m. The study shows that, when the embedded depth is 8D, the dynamic response of tunnel is respectively smaller, and the influence of fault fracture zone is less; when the embedded depth is 2D, the dynamic response of tunnels through fault fracture zone is significantly greater than that without fault fracture zone, and the lining structure is dominated by shear fracture.

1 INTRODUCTION

During the excavation of a large number of mountain tunnels, the structural planes with different sizes are unavoidable. Engineering practice and theoretical research have proved that, the structural planes of different forms are the crucial factors that affect the stability of tunnel engineering (Feng et al., 2000; Li et al., 2004; Lin et al., 2007). After Wenchuan Earthquake in 2008, an enormous amount of on-site surveys show that, besides the liability of instability failure of opening, shallow-buried section, etc. other parts such as fault zone, weak fracture zone, etc. are also liable to lining facture, faulting of slab ends and other disasters. Therefore, it is of important significance for the study and analysis on dynamic responses of tunnel through fault fracture zone. This article, directed at the two main factors of width and embedded depth of fault fracture zone in design, studies the influence law of dynamic response of tunnel lining structure in earthquake.

2 CALCULATION MODEL AND PARAMETER

This paper takes the single-tunnel two-direction two-lane highway tunnel with the net width of 10.5 m as an example. To study the influence of embedded depth on characteristics of dynamic response of tunnel, the horizontal (x direction) scope of this model is chosen to be 80 m, vertical (Y direction) scope is 100 m, the distance between the bottom of the model and tunnel inverted arch is 40 m, and the width of fracture zone is 10 m. With the surrounding rock grade of Grade IV, the calculation model is respectively built in different working conditions when the embedded depth of tunnel is 2D, 5D and 8D. In the model, the thickness of primary lining is 25 m, and that of secondary lining is 40 m. Figure 1 shows the model with the embedded depth of 5D.

In terms of calculation, the parameters of surrounding rock, primary lining and secondary lining shall be chosen in accordance with Specifications for Design of Highway Tunnels (JTG D70-2004) Wang et al. (2007), and the parameters of fracture zone are shown in Table 1 Wu (2005). For seismic wave, EI Centro is adopted, and the wave is adjusted based on the acceleration peak value when the seismic intensity is magnitude 8, with the maximum peak value of 1.1 m/s². And 10 s will be chosen for baseline correction. Dynamic excitation is horizontally

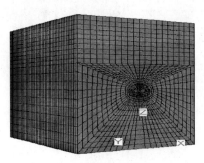 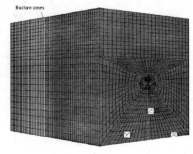

a without fracture zone   b with Fracture zone

Figure 1. Calculation model.

Table 1. Material physical mechanics parameter.

Material	Gravity/ kN·m⁻³	Poisson ratio	Modulus of elasticity/GPa	Bulk modulus/GPa	Shear modulus/GPa	Friction angle	Cohesive force MPa
Surrounding rock	21.5	0.32	4.0	3.7	1.5	33.0	0.5
Fracture zone	18.5	0.40	1.5	2.5	0.5	17	0.07
Initial lining	22.0	0.22	22.0	13.1	9.0		
Secondary lining	25.0	0.20	28.5	15.8	11.9		

and longitudinally input along the bottom of the model, and the vertical seismic wave acceleration is 2/3 of the horizontal seismic wave acceleration.

3 INFLUENCE OF EMBEDDED DEPTH ON SHEAR STRESS OF LINING

Figures 2–4 are respectively the cloud pictures about maximum shear stress σ_{xz} of secondary lining structure when embedded depth is 2D, 5D and 8D and when there is fracture zone or no fracture zone.

It can be seen from Figures 2–4 that, (1) without fracture zone, the positive shear stress σ_{xz} of tunnel lining increases with the increase of embedded depth, the maximum value is 2.6 Mpa when the embedded depth is 5D, while the negative shear stress shows the minimum value of −0.308 Mpa when the embedded depth is 5D, and reaches up to the maximum value of −4.11 Mpa when the embedded depth is 8D; with fracture zone, the positive shear stress σ_{xz} of tunnel lining basically increases with the increase of embedded depth, and it reaches up to the maximum value of 5.03 Mpa when the embedded depth is 5D, and the values of the embedded depth of 2D and 5D are almost the same; while the negative shear stress shows the minimum value of −1.3 Mpa when the embedded depth is 5D, and reaches up to the maximum value of −5.32 Mpa when the embedded depth is 8D; (2) with different embedded depths, the peak value of shear stress σ_{xz} of tunnel body changes along the longitudinal direction, and happens within the scope near the fracture zone, which is obviously different with the surrounding areas; the shear stress σ_{xz} of tunnel with fault fracture zone is larger than that without fault fracture zone, and the both show the greatest differences when the embedded depth is 2D, indicating it has the greatest influence on tunnel with or without fault fracture zone when the embedded depth is 2D.

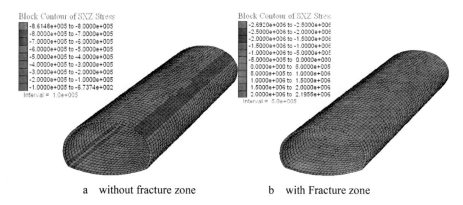

a without fracture zone b with Fracture zone

Figure 2. The max lining shear stress distribution when embedded depth is 2D.

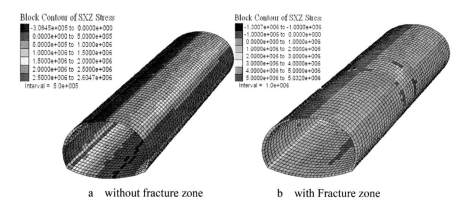

a without fracture zone b with Fracture zone

Figure 3. The max lining shear stress distribution when embedded depth is 5D.

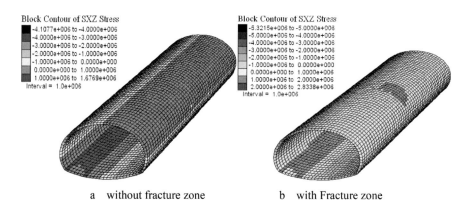

a without fracture zone b with Fracture zone

Figure 4. The max lining shear stress distribution when.

4 INFLUENCE OF EMBEDDED DEPTH ON PLASTIC ZONE

Figures 5–7 refer to the figures of the secondary lining structure and plastic zone of surrounding rock without fracture zone and with fracture zone when embedded depth is 2D, 5D, and 8D respectively.

It can be seen from Figures 5–7 that: (1) Without fracture zone and with fracture zone, plastic zone under shear is mainly distributed in the surrounding rock near the outer ring

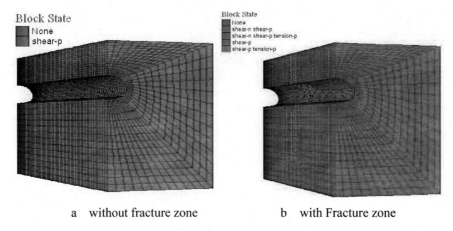

      a without fracture zone   b with Fracture zone

Figure 5.   Plastic zone distribution when embedded depth is 2D.

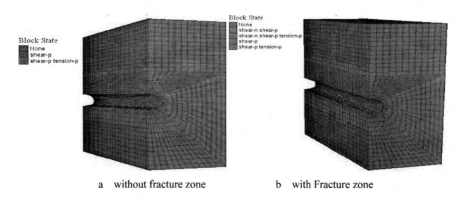

      a without fracture zone   b with Fracture zone

Figure 6.   Plastic zone distribution when embedded depth is 5D.

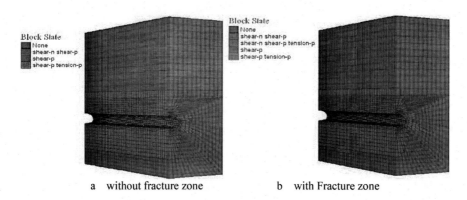

      a without fracture zone   b with Fracture zone

Figure 7.   Plastic zone distribution when embedded depth is 8D.

of lining, and increases with the embedded depth of tunnel, and its range under shear also expands continuously, longitudinally expanding to the outer-ring surrounding rock of the whole lining from the area near arch haunch and foot; (2) without fracture zone, plastic zone under shear increases with the embedded depth of tunnel, the range of tensile plastic zone also expands continuously, expanding from nonexistence and then running through the outer-ring surrounding rock of the whole lining from arch foot and vault; with fracture zone, tensile plastic zone is mainly distributed in the surrounding rock near the outer ring of lining within fracture zone and increases with the embedded depth of tunnel, and its range also expands continuously, namely, tensile plastic zone longitudinally runs through the outer-ring surrounding zone of the whole lining from arch foot and vault and tends to expand towards surrounding area and upper surrounding rock when it only exists in some area within fracture zone.

5 CONCLUSION

Under the conditions that surrounding rock grade is IV and fracture zone is 10 m wide as well as the embedded depth is 2D, 5D, and 8D respectively, through seismic numeric simulation of tunnel and after comparison and analysis on acceleration response peak, displacement response peak, bending moment response peak, shear stress response peak, and development of plastic zone, the preliminary conclusions are made:

1. The dynamic response values of different parts of tunnel lining with fracture zone are greater than those in the tunnel without fracture zone and mutation occurs near fracture zone. There are peaks within fracture zone. In case of 23 m away from fracture zone, there is no big difference between the two values. It indicates that the existence of fracture zone intensifies the dynamic response of tunnel and that the fracture zone exerts a limited influence.
2. Embedded depth is an important factor to affect the stability of tunnel. When embedded depth is 8D, tunnel has smaller dynamic response and fracture zone exerts little influence; when embedded depth is 2D, there are the greatest differences in dynamic response between the tunnels with fault fracture zone and that without fault fracture zone and fault fracture zone has the greatest influence on tunnel. Shear failure is dominated.

REFERENCES

Feng Q.M. & Guo E.D. 2000. A seismatic test for Buried Pipeline across fault. *Journal of Earthquake Engineering and Engineering Vibration* 20(1): 56–62.
Li D.W. 2004. Research on mechanical characteristics of tunnel lining with fault fracture zone. Lanzhou: China.
Lin J.Q. & Hu M.W. 2007. Research on seismic response for underground pipeline across fault. *Journal of Earthquake Engineering and Engineering Vibration.* 27(5): 129–133.
Paul B. & Ronald E.S. 1989. Centrifuge study of faulting effects tunnel. *Journal of Geotechnical Engineering* 5(7): 949–967.
Peng W. 2007. *FLAC3D Practical Course.* Beijing: China.
Wang S.C. & He F.L. 2007. *Rock Mass Classification of Tunnel Engineering.* Chengdu: China.

Study on characteristics of lipoxygenase and antioxidant enzyme changes during processing of dry-cured sausage

Ling Li
College of Life Science, Linyi University, Linyi, P.R. China

ABSTRACT: During processing and storage of meat products, lipid oxidation plays an important role in the formation of flavor. Lipid oxidation and enzyme activities were studied by evaluating the changes in lipoxygenase (LOX) activities and antioxidant enzyme activities, peroxide value (POV) and TBARS during dry-cured sausage processing. LOX activity increased significantly during 7 days ($P < 0.05$) and thereafter decreased. The antioxidant enzyme activities were significantly decreased during the whole processing ($P < 0.05$). At the end of processing, The GSH-px activity, catalase activity and SOD activity respectively decreased by 63.8%, 44.4% and 44.1% compared with 0 days (after stuffing). The correlation was extremely significant ($P < 0.01$) between the antioxidant enzyme activities. Antioxidant enzyme activities were negatively correlated with TBARS and POV ($P < 0.05$).

1 INTRODUCTION

Lipids play an important role in the development of chemical and sensory characteristics of dry-cured pork products. During processing, lipids are gradually degraded through both lipolysis and oxidation (Veiga, Cobos, Ros & Díaz, 2003). Lipoxygenase (LOX) is widely distributed in plants, animals and microorganism, and many deep studies have been performed in plants so far. LOX has been assumed to play an important role in the lipid oxidation and genesis of volatile flavor and aroma compounds in foods. In contrast, although people have increasingly realized the importance of LOX in affecting the flavor of meat products, little is known about the LOX activity in the dry-cured meat so far. Therefore, a clear correlation between LOX and lipid oxidation has not been established and needs to be further investigated.

Muscle endogenous antioxidant enzymes mainly include catalase, glutathione peroxidase (GSH-Px) and superoxide dismutase (SOD). Catalase and SOD are usually coupled enzymes, SOD scavenges superoxide anion by forming hydrogen peroxide and catalase safely decomposes H_2O_2 to H_2O and O_2. GSH-Px is a selenium-containing enzyme, catalyzing the reduction of lipid and hydrogen peroxides to less harmful alcohols and water. During dry-cured meat products processing, muscle endogenous antioxidant enzymes may play an important role in controlling the lipid oxidation. However, information on these antioxidant enzyme activities in dry-cured meat products is limited.

The main aim of this work was to manufacture a dry-cured sausage product using pig meat as materials, and to evaluate the lipolytic and oxidative changes in sausage during processing by studying the activities of LOX, antioxidant enzyme activity and the lipid oxidation parameters. Furthermore, the correlation between them was also investigated in this paper.

2 MATERIALS AND METHODS

2.1 *Sausage processing and sampling*

Sausage manufacture was carried out in a pilot plant according to industrial processing methods. The sausage mixture was made from lean pork meat and back fat in the proportions of 75:25 (weight: weight) together with 2.5% salt, 2% sugar, 0.3% black pepper and 0.015% sodium nitrite. Pork meat was ground to a 4 mm particle size, fat was chopped to a 6–8 mm particle size and divided into four portions for the different batches. Then sausage mixture was stuffed into natural casings approximately 150–200 g sausage^{-1} and diameter about 4 cm.

For ripening, all sausages were placed in a Constant Temperature and Constant Humidity Box under the same environmental conditions. Sausages were ripened as follows: 7 days at 80–85% relative humidity (RH) at 18 °C; and ripening and drying periods until day 28, at 70–80% RH at 10–12 °C. Processing was repeated three times for each formulation. Samples were obtained at various times during the maturation period; 0 (after stuffing), 7, 14, 21 and 28 days. Samples after sampling were vacuum—packaged and stored at −40°C prior to analysis.

2.2 LOX extraction and activity assay

The crude LOX was extracted according to the procedure described by Fu et al. (2009) with some modifications. About 5 g minced samples was weighed and homogenized with four volumes of 50 mM sodium phosphate buffer (pH 7.4) containing 1 mM dithiothreitol (DTT) and 1 mmol/L EDTA for 1 min at 10000 r/min in ice bath using a Polytron homogenizer. After filtration through four layers of gauze, the resulting homogenate was freeze centrifuged (4 °C) for 40 min at 10000 g. The resultant supernatant was filtered again through a layer filter paper and collected as crude LOX for analysis.

The linoleic acid substrate solution was prepared, according to the method described by Gata et al. (1996). The reactive substrate was prepared as follows: 140 mg linoleic acid was dissolved in 5 ml distilled water containing 180 μl Tween 20. The solution was clarified by adding of 2 M NaOH. Afterwards, the mixture was diluted to 50 ml with distilled water and stored under nitrogen conditions at low temperature for further use.

LOX activity was determined at room temperature by measuring the absorbance increase at 234 nm for 1 min. The reaction mixture contained 2.9 ml 50 mM citric acid buffer (pH 5.5), 200 μl substrate solution and 0.1 ml enzyme solution. The blank sample contained 3.0 ml 50 mM citric acid buffer (pH 5.5) and 200 μl substrate solution. One unit of LOX activity was defined as an increase absorbance by 1 unit per minute and per g protein at 234 nm.

2.3 Determination of antioxidant enzyme activities

The enzyme extract solution was prepared according to the method that described by G. Jin et al. (2013). A 5 g muscle sample was homogenized in 25 mL of phosphate buffer (0.05 M, pH 7) for 1 min at 12,000 rpm in an ice bath. The homogenate was centrifuged at 7000 × g for 2 min at 4 °C. The supernatant fraction was filtered through glass wool and used to determine catalase, GSH-Px and SOD activity. All the analysis was performed in triplicate.

Catalase activity assay was performed following the method of Aebi (1983) with some modifications. The supernatant of 0.1 mL was reacted at room temperature (24 °C) with 2.9 mL of 11 mM H_2O_2 in phosphate buffer (pH 7.0), and the activity of catalase was measured by recording the H_2O_2 loss by the decrease in absorbance at 240 nm during the initial 3 min. One unit (U) of catalase was defined as the amount of extract needed to decompose 1 μmol of H_2O_2/min, and the result was expressed as U/g of muscle.

GSH-Px activity was measured by a modification of the method of Paglia and Valentine (1967). The assay mixture consisted of 1 mL of 75 mM phosphate buffer (pH 7.0), 10 μL of 150 mM reduced glutathione, 10 μL of 46 U/mL glutathione reductase (Sigma G3664), 30 μL of 25 mM EDTA, 30 μL of 5 mM NADPH (Sigma N1630), 200 μL of supernatant, and 10 μL of 20% Triton X-100. The final volume of the reaction mixture was 1.5 mL. The reaction was started by addition of 50 μL of 7.5 mM H_2O_2. Conversion of NADPH to NADP+ was monitored continuously in a UV-2450 ultra-violet spectrophotometer (Shimadzu Corporation; Kyoto, Japan) at 340 nm for 3 min. The activity of GSH-Px was expressed as U/g muscle and one unit (U) was defined as the amount of extract required to oxidize 1 μmol of NADPH/min at 22 °C.

Total SOD activity was determined as proposed by Marklund and Marklund (1974), with the modifications of Gatellier, Mercier, and Renerre (2004), by measuring the inhibition of pyrogallol autoxidation. The reaction mixture contained 2850 μL of 50 mM phosphate buffer (pH 8.2), 75 μL of the supernatant and 75 μL of 10 mM pyrogallol. The absorbance at 340 nm was read as soon as the reaction begin, and the increase in absorbance during the first 2 min was recorded. One unit (U) was defined as the activity that inhibited the reaction by 50%, the result was expressed as U/g of muscle.

2.4 Analysis of lipid oxidation

The extent of lipid oxidation was evaluated by the 2-thiobarbituric acid-reactive substances (TBARS) assay and peroxide value (POV).

TBARS of sausages were determined using the method of Buege & Aust (1978). Ten grams of homogenized sausage sample was taken and TBARS were extracted twice with 15 mL of 10% trichloroacetic acid. Extracts were centrifuged and made up to 50 mL. Then 3 mL of supernatant was pipetted into a glass stopper test tube. TBA reagent (3 mL) was added and the mixture heated in a boiling water bath for 50 min. After cooling, the absorbance of each sample was read against an appropriate blank at 532 nm. A standard curve was prepared using TEP. TBARS values were expressed as mg malondialdehyde (MDA) per kg sample.

Peroxide value (POV) of lipid sample was determined following the Chinese national standard (GB/T 5538-1995). The lipid sample (1.0 g) was treated with 30 ml of organic solvent mixture (chloroform: acetic acid mixture, 2:3). The mixture was shaken vigorously, followed by the addition of 0.5 ml of saturated potassium iodide solution. The mixture was kept in the dark for 5 min; 75 ml of distilled water were then added and the mixture was mixed. To the mixture, 0.5 ml of starch solution (1%, w/v) was added as an indicator. The peroxide value was determined by titrating the iodine liberated from potassium iodide with standardised 0.01 N sodium thiosulfate solution. The POV was expressed as milliequivalents peroxide/kg lipid.

2.5 Statistical analysis

Statistical analysis and comparisons among means were carried out using the statistical package SPSS 19.0 (SPSS Inc., Chicago, IL, USA). The Duncan's multiple range was applied for comparisons of means, differences were considered significant at $P < 0.05$. In addition, the Pearson's two-tailed correlation analysis was performed to evaluate the relations between oxidative stability and antioxidant enzyme activities.

3 RESULTS AND DISCUSSION

3.1 Changes of LOX activity and lipid oxidation during processing of dry-cured sausage

Table 1 shows the evolution of POV, TBARS and LOX activity during the processing of dry-cured sausage. LOX activity increased significantly ($P < 0.05$) during the first 7 days and reached its maximum activity at the end of this stage; it then decreased ($P < 0.05$) during the drying–ripening stage. This result was similar to that reported in Jinhua ham by Zhou and Zhao (2007), in which LOX activity increased to maximum level by the end of salting stage and then decreased during the following stages. The significant increase in LOX activity during the first 7 days could be due to the activating effect of salt on LOX activity. As for the decrease of LOX activity during the drying–ripening stage, besides the effects of the process conditions on it, Fu, Xu, and Wang (2009) reported that the hydroperoxides from the lipid oxidation may cause the oxidation of thiol groups, resulting in the inactivation of LOX. Although LOX activity decreased during the drying–ripening period, it still retained higher activity during the latter period than in the 0 days (stuffing).

Table 1. Changes in lipid oxidation and LOX activity during ripening of dry-cured sausages.

Processing steps	LOX activity	TBARS	POV
0 days	11.6 ± 0.8[d]	0.41 ± 0.03[e]	0.02 ± 0.005[c]
7 days	25.4 ± 2.9[a]	0.79 ± 0.06[d]	4.1 ± 0.6[b]
14 days	21.8 ± 1.5[b]	1.02 ± 0.04[c]	8.4 ± 0.9[a]
21 days	20.4 ± 1.5[bc]	1.08 ± 0.06[b]	6.2 ± 0.5[ab]
28 days	18.0 ± 2.2[c]	1.23 ± 0.05[a]	4.5 ± 1.2[b]

Values are shown as means ± standard deviations (n = 3).
[a, b, c, d, e] Means in the same column with different superscripts are significantly different ($P < 0.05$).

The changes in TBARS in the various sausages during ripening are shown in Table 1. TBARS increased gradually ($P < 0.05$) from 0.41 mg/kg to about 1.23 mg/kg at the end of the ripening period. The rate of TBARS formation during the initial 14 days was greater than that during the latter periods. The low, but increasing rate after 14 days may result from the loss of these substances due to their involvement in further reactions. However, Bozkurt (2006) reported that TBARS values decreased at 8–14 days of ripening. The reason was that the different ripening conditions (temperature and humidity) were used in these two different dry-cured sausages.

The POV markedly increased ($P < 0.05$) during the initial 14 days and reached the peak after 14 days of drying–ripening; thereafter it decreased till the end of the drying—ripening process (Table 1). This was mainly due to the fact that the hydroperoxides are primary lipid oxidation products, and their content mainly depends on the ratio of formation to degradation. At the last drying–ripening stage, hydroperoxides could easily further oxidised to secondary oxidation products, causing the POV to decrease significantly (Jin et al., 2010).

3.2 *Changes of antioxidant enzyme activities during processing of dry-cured sausage*

For most cooked meat products, the antioxidant enzymes were active only in raw meat, and their activities was lost when meat was cooked (Mei, Crum, & Decker, 1994). However, in our present study, though the activities of all antioxidant enzymes decreased during processing, all of them still retained part of the activity until the end of processing. The GSH-px activity decreased significantly ($P < 0.05$) during 0–14 day period (Table 2). The GSH-px activity decreased by 44.1% compared with 0 days (after stuffing). Our study was similar to the study of dry-cured bacon (G. Jin et al, 2013). In our meat process conditions, the addition of $NaNO_2$ into muscle samples during salting might be an important cause for the decrease in GSH-px activity. Because, in addition to give characteristic color and flavor of cured meat, nitrate/nitrite also play an important role as antioxidant and antimicrobial agent.

Catalase is a heme-containing enzyme, which mainly decomposes the hydrogen peroxide and lipid hydrogen peroxides in meat and meat products. In our present study, the catalase activity decreased significantly ($P < 0.05$) during 0–21 day period (Table 2). Until the end of process, the catalase activity decreased by 63.8% compared with 0 days (after stuffing). This indicated that the catalase in meat was unstable as GSH-px during processing. Mei et al. (1994) found that catalase had a similar sensitivity as GSH-Px to heat. As Table 2 shows, the SOD activity was mainly decreased significantly ($P < 0.05$) during 0–21 day period. Until the end of processing, SOD activity decreased by 44.4% compared with 0 days (after stuffing). Compared with the activity decreases in SOD and GSH-px, the decrease in catalase was the lowest ($P < 0.05$), suggesting that catalase was the most unstable one among all the three antioxidant enzymes in meat during ripening of dry-cured sausages.

Table 2. Changes in antioxidant enzyme activities during ripening of dry-cured sausages.

Processing steps	GSH-px	CAT	SOD
0 days	0.68 ± 0.08[a]	373 ± 22[a]	136.3 ± 7.4[a]
7 days	0.57 ± 0.07[b]	255 ± 21[b]	113.0 ± 7.0[b]
14 days	0.42 ± 0.04[c]	195 ± 12[c]	89.7 ± 7.2[c]
21 days	0.38 ± 0.05[c]	146 ± 15[d]	76.7 ± 4.2[d]
28 days	0.38 ± 0.04[c]	135 ± 8[d]	75.7 ± 6.7[d]

Values are shown as means ± standard deviations (n = 3).
[a, b, c, d] Means in the same column with different superscripts are significantly different (P < 0.05).

Table 3. Correlation between TBARS, POV, LOX and activities of antioxidant enzymes in dry-cured sausages during processing.

	LOX	CAT	GSH-px	SOD	TBARS	POV
LOX	1	−0.47	−0.37	−0.40	0.43	0.63*
CAT		1	0.93**	0.96**	−0.96**	−0.72**
GSH-px			1	0.91**	−0.87**	−0.71**
SOD				1	−0.95**	−0.73**
TBARS					1	0.75**
POV						1

Notes: *indicating significant, P < 0.05; **indicating extremely significant, P < 0.01.

3.3 Correlations between lipid oxidation and antioxidant enzyme activities

The correlation analysis results (Table 3) showed that there were only very few correlations between antioxidant enzyme activities and oxidation indices (POV and TBARS) were significant (P < 0.05). This phenomenon was probably due to that all the antioxidant enzyme activities decreased continuously, while the TBARS increased during the processing, POV increased firstly and then decreased during the processing. The correlation was extremely significant (P < 0.01) between the antioxidant enzyme activities. Antioxidant enzyme activities were negatively correlated with TBARS and POV (P < 0.05). In addition, Table 3 results also showed that the correlation between each antioxidant enzyme and POV was stronger than their corresponding correlation with TBARS.

4 CONCLUSION

All antioxidant enzyme activities decreased significantly during dry-cured sausage processing, which reduced the oxidative stability of meat to some extent. For all dry-salted or dry-cured meat products processing, a normal rate of lipid oxidation is necessary for the development of cured flavor, while the extensive lipid oxidation can cause the rancidity. So, the antioxidant enzymes activities should be properly modulated during dry-cured meat products processing.

ACKNOWLEDGEMENTS

This study was supported by special scientific research projects for doctor in Linyi University (Grant LYDX2013BS037). The author also thanks the Chinese National Center of Meat Quality and Safety Control for their equipment assistance.

REFERENCES

Aebi, H.E. (1983). Catalase. In H.U. Bergmeyer (Ed.). *Methods of enzymatic analysis* (Vol. 3, pp. 273–286). Weinheim, Germany: Verlarg Chemie.

Bozkurt, H. (2006). Utilization of natural antioxidants: Green tea extract and Thymbra spicata oil in Turkish dry-fermented sausage. *Meat Science*, 73(3), 442–450.

Buege, J.A., & Aust, S.D. (1978). Microsomal lipid peroxidation. *Methods in enzymology*, 52, 302.

Fu, X.J., Xu, S.Y., & Wang, Z. (2009). Kinetics of lipid oxidation and off-odor formation in silver carp mince: the effect of lipoxygenase and hemoglobin. *Food Research International*, 42(1), 85–90.

Gata, J.L., Pinto, M.C., & Macías, P. (1996). Lipoxygenase activity in pig muscle: purification and partial characterization. *Journal of Agricultural and Food Chemistry*, 44(9), 2573–2577.

Gatellier, P., Mercier, Y., & Renerre, M. (2004). Effect of diet finishing mode (pasture or mixed diet) on antioxidant status of Charolais bovine meat. *Meat Science*, 67(3), 385–394.

Jin, G.F., Zhang, J.H., Yu, X., Zhang, Y.P., Lei, Y.X., & Wang, J.M. (2010). Lipolysis and lipid oxidation in bacon during curing and drying–ripening. *Food Chemistry*, 123(2), 465–471.

Jin G., He L., Yu X., et al. (2013). Antioxidant enzyme activities are affected by salt content and temperature and influence muscle lipid oxidation during dry-salted bacon processing. *Food Chemistry*, 141, 2751–2756

Marklund, S., & Marklund, G. (1974). Involvement of the superoxide anion radical in the autoxidation of pyrogallol and a convenient assay for superoxide dismutase. *European Journal of Biochemistry*, 47(3), 469–474.

Mei, L., Crum, A.D., & Decker, E.A. (1994). Development of lipid oxidation and inactivation of antioxidation enzymes in cooked pork and beef. Journal of Food Lipids, 1(4), 273–283.

Paglia, D.E., & Valentine, W.N. (1967). Studies on the quantitative and qualitative characterization of erythrocytes glutathione peroxidase. *Journal of Laboratory and Clinical Medicine*, 70(1), 158–168.

Veiga, A., Cobos, Á., Ros, C., & Díaz, O. (2003). Chemical and fatty acid composition of "Lacón gallego"(dry-cured pork foreleg): differences between external and internal muscles. *Journal of Food Composition and Analysis*, 16(2), 121–132.

Zhou, G.H., & Zhao, G.M. (2007). Biochemical changes during processing of traditional Jinhua ham. *Meat Science*, 77(1), 114–120.

Research of mine environment restoration effects and policies in Beijing since 2004

Nan Sun
Beijing Geological Engineering Design and Research Institute, Beijing, China
Department of Real Estate, Planning and Geoinformatics, Aalto University, Helsinki, Finland

Guangxin Yan
Beijing Geological Engineering Design and Research Institute, Beijing, China

Lei Tang
Beijing Geological Engineering Design and Research Institute, Beijing, China
University of Science and Technology Beijing, Civil and Environmental Engineering School, Beijing, China

Wenqing Liu
Beijing Geological Engineering Design and Research Institute, Beijing, China

ABSTRACT: Beijing as the capital of China is adjusting the industrial structures and urban functions from the beginning of twenty-first century. Based on this political background, in the past 10 years about 95.96% of the mines have been shut down since 2004. In this stage, numerous restoration projects as well as supporting policies have been carried out. In order to find out the results of the restoration projects and policies, we surveyed the 1443 mines located in the 13 districts of Beijing from the beginning of 2013 to 2014. Besides surveys on sites, we also interviewed the municipal officials who are in charge of the geological environment management of all the mines in Beijing. In this article, we try to introduce the general conditions of the 1443 mines and discuss the influences of ecological restoration to the geological environment in the mountainous mining areas since 2004. By the end of 2013, about 48.2% of them have been restored or utilized in the environmental friendly ways; about 14.5% of the mines could be restored completely by the natural powers, or only need some slightly artificial assistances. This more or less convinced us that the environmental conditions of the mining ecosystems are evolving in the sustainable direction. It is considered that the coherent policies, serious administrations, and ample deposits are the most critical guarantees to these restoration projects. However, even though the conventional technologies are maturely used in practices of restoring land and vegetation, there are very limited innovations and improvements in the water ecosystem restoration in the mountainous mining areas. The suggestion is more political focus and funds should be given to the water ecosystem restorations and technological innovations.

1 INTRODUCTION

China is one of the countries that suffered most frequently from mining geological disasters; about 50% of the disasters are caused by surface collapse, and about 33% of casualties are caused by debris flow (He, Xu, et al., 2012). As the capital of China, Beijing has rich reserves of mineral resources, and long history of mining explorations. Long-term mining has created enormous economic benefits, but caused grievous geological environment problems at the mean time, such as the deterioration of land resources and water resources, geologic hazards and the destruction of ecological landscape (Zhang, Shen, et al., 2011). With the adjust-

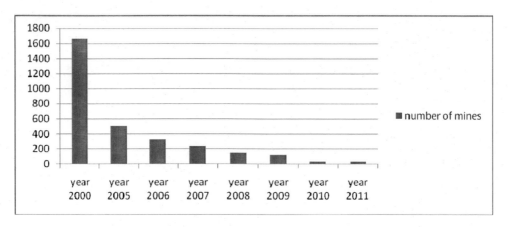

Figure 1. The number of mining enterprises in Beijing from 2000 to 2011.

ments of the industrial structures and urban functions from the beginning of the twenty-first century, a series of policies have been issued to strengthen the supervision of environment protection and restoration in Beijing. For the mining industry, the main theme is to close down and restore the mining mountainous areas in Beijing.

These policies have been implemented effectively. From 2000 to the end of 2014, the number of mines have been reduced from 1661 to 31, namely about 98.13% of the mines have been shutdown, showing the effective and powerful implementations of the municipal policies. However, two critical issues immediately related to the mining shutting down should be considered: the ecological restorations to the mines geological environments and the industrial transformations of mining industry. Each environmental management policy should be sustaining, especially considering the long term evolutions of various ecosystems. This paper will more concentrate on the first issue, while the latter one will be discussed in separate articles.

In order to rehabilitate the destroyed mine geological environment and reduce the geological disasters, Beijing Municipal Government has started a series of restoration projects as early as 2003. After the restoration projects have been carried out for about 10 years, in 2013 we spent nearly the whole year to survey the conditions of all the 1443 mines in Beijing. The main targets were to find out the current situations of the mines, the effects of restorations in the destroyed mountainous mining areas, the implementation of the related policies, and identify the key obstacles in future restoration of the mountainous ecosystem. As environmental protection and ecological restoration will continue to be the key issues of Beijing municipality in the future, these outcomes could supply important experiences and solid foundations for the future projects, and therefore make policies more effective, efficient, and focused.

All the outcomes of the projects, including the data and the pictures of the 1443 mines were edited into one database, which was named as "one map." This database is quite user friendly and integrate the vivid photographs of all the 1443 mining mountains into the management work, therefore enormously improves the efficiencies of the government managements and operations. This database is frequently used by the Municipality of Land and Resource in their daily work nowadays, making the tedious management work a bit easier.

2 METHODS AND MATERIAL

2.1 Methods

The project was implemented by 5 research groups with about 15 researchers, lasting for about 18 months in 2013 and 2014. The surveyed areas are located in the 13 districts of

Beijing, which have diversified categories of mines, including Changping, Chaoyang, Daxing, Fangshan, Fengtai, Haidian, Huairou, Mentougou, Miyun, Pinggu, Shunyi, Tongzhou, and Yanqing. The total number of mines surveyed is 1443, of which 34 mines are still in production (Fig. 2). To understand the related policies and the effects of implementations, we also interviewed officials in the two responsible government administrations—the Beijing Municipal Bureau of Land and Resources (BMBLR), and Beijing Municipal Bureau of Geology Mineral Exploration and Development (BMBGMED). In the process, about 15 officials and professionals have been interviewed for more than 20 times together. Moreover, to make the research more concrete, about 150 documents and data are collected from the National Geological Archive.

2.2 *Materials*

Beijing has a long history of mining explorations and many of the mining geological environments are seriously damaged. According to the "Beijing Mine Geological Environment Protection and Management Plan" (2006–2015) and related academic researches, the mine geological environment damages have five general categories: mine geological disasters, landscape damage, damage to the water environment, destruction and occupation of land resources, and environmental pollution (Liu, 2014). According to the plan, in Beijing there are seven most seriously damaged regions with the area of about 1021.47 km^2, eight seriously damaged regions with the area of about 2079.33 km^2, and ten less seriously damaged regions with the area of about 1024.78 km^2. Restoration projects are carried out mainly following this plan. At the end of the planned stage, we carried out the survey to find out the current stages of the abandoned mines, the results of the restoration engineering projects, and the effects of the management policies.

Besides the plan, at the beginning of 2004, the colleagues in the Geological Environmental Monitoring Station of Beijing have surveyed and evaluated the general conditions of all the mines in Beijing. In the past 10 years, their work is the most important theoretical basis of all the restoration projects and policies, and also instructive to our survey and research as well. According to their survey findings, the geo-environment problems of mines in Beijing have three main characteristics: diversified types, wide distributions, and clear division. Besides the geological disaster, the mining explorations have other more negative geological environmental effects, the resources damages and environmental pollutions. (Table 1). According to their survey, in the year of 2004, the area of land resource destroyed by mining industry is about 4,897.69 ha., while only about 11.11% of the land has been restored. It would be quite meaningful if we could make some comparisons with the situations today and 10 years before, illustrating the change and development tendency of the ecosystem in the spectrum of Beijing.

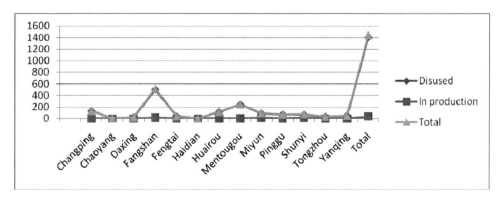

Figure 2. 1443 mines located in different districts of Beijing.

Table 1. The main geological environmental problems of mines in Beijing.

Type	Major effects
Geological disasters	Mining subsidence, ground fracture, collapse, landslides, debris flow
Resources damages	Drop of the ground water level, land resources damage, landscape damages
Environmental pollutions	Ground water pollution

3 RESULTS

3.1 General conditions of the mines in Beijing

According to the general city plan of Beijing (2004–2020), mines with low production, not meeting the safety conditions and with heavy pollution should be shut down. Since 2004, the Beijing municipality has shut down the mines in large scale. The policy implementations are powerful and effective. According to our survey, the number of mine enterprises in Beijing was reduced from 507 in 2005 to 33 in 2011 and 31 in 2014, and the production was reduced from about 30 million ton in 2005 to about 13 million ton in 2011 (Table 2). Even though most of the mines in production are state owned and in large scale, many of them are shutting down or have plan to shut down in the very near future.

The high efficiency in the mine shutting down is mainly due to the considerable funding from both central government finance and Beijing government. From 2003 to the end of 2013, 58 environmental restoration projects, with the total funds of 699.2 million Yuan, have been implemented. These projects mainly related to about 20 kinds of mines, including coal, iron, limestone, sand, and clay. At the same time, the environment pollutions and negative influences to the geological environment were avoided effectively from the sources of the mines.

Besides the precious metals ores and the related heavy metals pollutions have lately become the monitoring and treatment focus since 2013. This is mainly because of the governments and civilians have paid more and more attentions in recent years to the heavy metals pollutions of soil and ground water. However, it is rare to find the fundamental restorations to the coal mines and related water resource damage repair and pollution treatments.

3.2 General condition of the mines geological environmental restoration

According to our survey, more than half of the mines are in good conditions nowadays. Nearly 56.6% of the 1443 mines have been restored or utilized (Fig. 3). About 407 of the mines have been restored artificially, the engineering measures of which include slope cutting, earth-retaining, drainage works, and the biological measures. Around 289 of the mines have been altered into roads, parks, houses and pools, e.g., parts of the six ring road used to be sand and gravel pits, Shouyun iron ore was developed into national mine park (Fig. 4). Nearly 121 of the mines have the possibility of being restored to a favorable condition by native species, most of which are clay pits, gravel pits, or sand pits. In about 538 of them still exists kinds of problems and waiting for mass artificially assisted restorations, many of which have deep mine pits, seriously damaging the water system of mountain ecosystem and have therefore brought long negative influences to the sustainable development of the local regions; about 88 of them have been restored partially and further engineering restorations are still needed. These are applaudable improvements compared with the conditions about 10 years before.

Around 5680 acres arable land was rehabilitated. Figure 5 shows the evolutions of the ecosystem in the past decades from destroyed brown field of coal waste rock pile to nearly half natural ecosystem. In this process, honeysuckle is used as the pioneer plant in the restoration project in 2007. In 2014, when we surveyed this region, lots of native plants had grown there, showing the healthy and diversified conditions of the ecosystem. Yet, for this region, the only

Table 2. The number of mines in Beijing and the production.

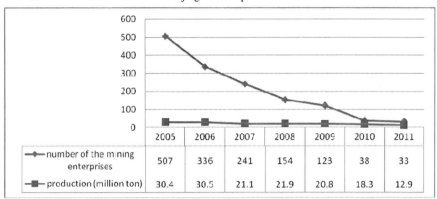

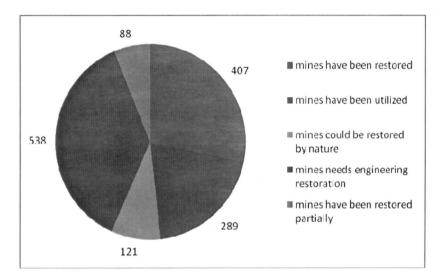

Figure 3. Current situation of the mine geological environment in Beijing.

Figure 4. Shouyun National Mine Park.

water source is rain water. Since 2000, precipitation in Beijing has diminished greatly, which will make the process of ecosystem evolution harder and slower.

The geological disasters were controlled effectively by these projects. In our survey, we found that the number of geological disasters is 114, including collapse risks in 56 places, mining subsidence in 36 places, debris flow gully in 12 places, ground fracture in 8 places,

Original Condition   Honeysuckle grown in 2007

Condition in 2009   Natural ecosystem in 2014

Figure 5. The evolution of ecosystem in restoration project of coal waste rock pile.

and landslide in 2 places. This data supplies the foundations for the restoration projects in the future. Some of the mining subsidence influenced the residential areas, which are dangerous for livings and have sustaining negative effects for the local economic developments. Government subsidies should compensate to the direct and some indirect loses.

3.3 *Related policies and funding supports*

In the past 10 years, the related policies, funding supports and administrations are valid and coherent. To fulfill the "Beijing Mineral Resources Master Plan" (2008–2015), the 13 districts in Beijing have gradually closed the medium and small mining enterprises with low productivity and heavy pollution since 2004. In 2005, Beijing Municipal People's Government Office issued the "Notice of further rectify and standardize the mineral resources development," which clarified the objectives of gradual reduction of solid mineral resources exploitations [6]. "Beijing Mine Environmental Protection and Management Plan" (2006–2015) raised the protection and renovation plan for both the disused mines and mines in production. In 2009, BMBLR has issued "Beijing Mine Geological Environment Management and National Geological Relics Protection Project Interim Measures" and "Beijing Mine Geological Environment Management Technical Guide (Trial)." Besides these policies, the two government administration—the BMBLR, and BMBGMED—have carried out a series of surveys and researches in 2005, 2008, 2009, and 2011, to assess the influences of the mines on the local ecosystem and human living environment, and the effects of restoration projects. In the October of 2014, the comprehensive database management system has been established and managed by BMBLR; all the mines' conditions could be illustrated in "one map".

What's more, the security deposits from the mining enterprises and the supports from the central finance are adequate and timely, which set important funding foundations for the restoration projects. "Beijing Mine Eco-environment Restoration and Management Deposit Interim Measures" was issued jointly by four bureaus, including the BMBLR, Beijing Municipal Commission of Development and Reform, Beijing Finance Bureau, and Beijing Municipal Environmental Protection Bureau; the implementation started from January the 1st 2009. By the end of 2012, the deposit guarantee solid mining enterprises in Beijing amounted to 251.8 million Yuan, while 104.2 million Yuan has been returned through the projects of restoration. The policy of deposit plays the critical roles for the effective mining ecological restoration. The prosperous of the mining industry in the past ten years also promote the effective implementation of this policy.

By the end of 2012, the subsidy from the central finance is about 462 million Yuan. These engineering projects solve the employment problems of the local people by supplying around 360 thousands work opportunities. Yet, after the great majority of the mines have been shut down, the influences on the mining industries and the related industries are enormous. How could local people find long term work and how could they make living in an environmental friendly way are difficult to solve at one time. Persistent concerns, patient education, and effective management of the economic-environment system in the long term are definitely required.

4 CONCLUSIONS AND SUGGESTIONS

Through the survey, interview, and research in 2013 and 2014, we deem that it is successful in shutting down, restorations, and management of the mines in Beijing in the past decade. The geological environments of the mining areas are evolving in the sustainable direction, and the geological disasters have been reduced greatly. This is mainly owing to the coherent policies, funding supports, and suitable technologies. Moreover, the restoration projects also rehabilitate arable lands and supply work opportunities to the local people as well. This partially offset the economic loss in closing down the mining industry of Beijing.

As environmental protection and ecological restoration will continue to be the key issues of Beijing municipality in the future, these outcomes could supply important experiences and theoretical foundations for the future environment protection and restoration projects, and therefore make policies more effective and focused. Even though the financial and political resources in Beijing are more beneficial than other regions in the country, this research could also offer important experiences for other ecological restoration and environment protection projects in other regions of China. What's more, in our survey we also found that water, as a critical resource, will be the key issue of restoration in the very near future. Water is not only one of the obstacles to the effects of restoration, but also one of the most limiting factors to the long term "social-economic-environment" development of the country.

In this paper, we haven't discussed the technologies in detail, while it is strongly suggested that the innovations in water ecosystem restorations and water-saving technologies are urgently needed. Through the survey, we find that compared with the historical precipitation level there is less rain fall in Beijing in the past 10 years. The condition in the mountainous areas is even worst, especially considering the water ecosystems have been seriously damaged by mining explorations. Therefore, our suggestion is to give more political focus and funding supports to the water ecosystems restoration technology innovation, as water is the critical point of both geological environment restoration and the sustainable development of the mountainous areas.

ACKNOWLEDGEMENT

The authors are grateful to the supports of the Ministry of Land and Resource and the Ministry of Science and Technology, for the research is supported by the project of International Cooperation "Research and Demonstration of the Technologies of Mine Geological

Environment Monitor and Restoration," with the code of "2012 DFA21000." This study is also funded by the Talents Project Foundation of Beijing (2013D002011000002). We are equally grateful to Beijing Municipal Bureau of Land and Resources, as this project is supported also by the project funded by them, namely the "Survey, Evaluation, and Management Studies of the Mine Geological Environment Status Quo." Special thanks also to the managers in Beijing Municipal Bureau of Geology Mineral Exploration and Development, who gave us lots of patient guidance and suggestions.

REFERENCES

Beijing Municipal Bureau of Land and Resources. Beijing mine geological environment protection and management plan (2006–2015). Visit time: 2014. October, http://www.mlr.gov.cn/kczygl/kcgh/201202/t20120224_1067174.htm.

Beijing Municipality Government. General city plan of Beijing (2004–2020). Visit time:2014. October, http://zfxxgk.beijing.gov.cn/f...O/CATA_INFO_ID = 49429–2005–04–15.

He F., Xu Y.N. Qiao G. Chen H.Q. Liu R.P. The character of Mine Geological Disaster Distribution in China. Geological Bulletin of China. 2012, March. 31(2, 3):476–486.

Liu W. The Current Station and Development Trend of Beijing Mining Geological Environment. Urban Geology. 2014, 9(1):4–8 (in Chinese).

Beijing Municipal People's Government. Notice of further rectify and standardize the mineral resources development. Visit time: 2014. October. 2005. October.19th http://zfxxgk.beijing.gov.cn/.

Zhang Z., Shen N., Wang Y., et al. Geological Environment Problems and Countermeasures of Shijiaying Mine in western Beijing. Procedia Environmental Sciences 11(2011)1245–1252.

Application of mathematical model for suspended sediment transport in mud-dumping ground of Sino-Myanmar crude oil pipeline and wharf engineering

Na Zhang, Liancheng Sun & Qixiu Pang
Tianjin Research Institute of Water Transport Engineering, Ministry of Communications, Tianjin, China

ABSTRACT: The location of spoil area, on the basis of meeting the requirements of environmental assessment, should as far as possible to reduce the distance between dredger and spoil area to save money, but at the same time, should avoid sediment from mud-dumping ground back silting to channel. Therefore, in this paper, based on analysis of natural conditions in the engineering area, such as dynamic conditions, sediment transport direction and so on, the location of spoil area is determined. Then the mathematical model is used to study suspended sediment transport in mud-dumping ground and to determine mud-dumping ground location which can satisfy the need of the construction.

1 INTRODUCTION

Sino-Myanmar crude oil pipeline and wharf engineering (Burma section) is located in Burma Rakhine state in the western of Burma, Kyaukpyu Bay waters in the eastern shore of the bay of Bengal. The wharf is located in the northeast of the Muddi Island, the waters are communicated by tidal channel and open waters. The direction of tidal channel is roughly from Northwest to Southeast. A 30-km-long channel is open along the tidal channel. The channel is deep and only the local bank need to excavate. In the paper, based on analysis of natural conditions of the engineering area, such as dynamic conditions, sediment transport direction, and so on, the location of spoil area is determined. Then the mathematical model is used to study suspended sediment transport in mud-dumping ground and to determine mud-dumping ground location which can satisfy the need of the construction.

2 PRELIMINARY ANALYSIS OF DUMPING SITE

2.1 *Current*

The excavated section can be tentatively divided into outer and inner waterway section. According for throw mud nearby principles, dumping area should also be respectively placed in open waters and the tide channel.

There is reciprocating flow in project sea area. At flood tide, the tide current from ESE enter in the tidal channel, then affected by terrain constraints, flow velocity, flow direction are adjusted. At ebb tide, the direction of the tide current is opposite to spring tide. The difference of tidal current velocity in and out Kyaukpyu Bay is not very obvious. The ebb tidal current speed is slightly larger than the flood tide. The average velocity during flood tide is 0.29~0.67 m/s and 0.44~0.71 m/s during ebb tide. The maximum velocity during flood tide and ebb tide is 1.09 m/s and 1.26 m/s respectively. Overall, current speed is in larger range, which is beneficial for the maintenance of water depth of channel (Sun, 2011).

2.2 Wind

The coastal area of Burma belongs to the monsoon climate of the tropical rain forests. The constant wind direction is NW from the annual December to April the following year, constant wind direction is the SW from May to September. The annual maximum wind speed up to 30 m/s, wind direction is SW.

2.3 Wave

According to statistical data from August 2007 to July 2009, WSW is the normal wave direction, the frequency is 28.95 percent. The strong wave is mainly from WSW and the largest wave height is 4 m. Waves with wave height larger than 3 m are also from W, SW, and WNW. With the effect of the topography, the wave on open sea is not easily transmitted to the dock wave height in front of the wharf is small and the berthing condition in the project area should be good.

2.4 Suspended sediment concentration

Measured suspended sediment concentration in project sea area is small. The measured maximum tidal period average sediment concentration is 0.125 kg/m^3. The average suspended sediment concentration is 0.080 kg/m^3. The distribution of the suspended sediment concentration is that minimum suspended sediment concentration is in the outlet, downstream section of tidal channel is slightly larger, upstream section is largest. The vertical sediment concentration distribution is that suspended sediment concentration is small at the surface and big at the bottom.

2.5 Seabed material

Sediment engineering region is more complex, sediment types, and the particle size is larger, obviously divided into three parts: The median particle diameter of clayey silt and silty clay is less 0.03 mm which mainly distributes near the inner channel and near the outer channel where water depth greater than 20 m. The median particle diameter of fine sand is greater 0.2 mm which mainly distributes near the sea of West Kyaukpyu island where water depth less than 20 m. The size distribution of seabed material showed a fine—coarse—fine pattern from the sea—gate to inner channel.

2.6 Location of spoil area

The length of design channel is about 40 km, and the excavation section is divided into several discontinuous segments along the channel. In order to reduce the dredger transport distance, improve efficiency, save money, dumping area should be provided in the section near the dredger. Meantime, dumping areas still need to meet the requirement of the water depth, and where water throwing mud cannot spread to the channel.

According to the characteristics of water depth and tidal current, dumping area in the Kyaukpyu bay is arranged in the B position, which is square and length of 1.67 km, width of 1.0 km, and area of about 1.67 km^2 and shown in Figure 1. The distance from the dredger to the north side of the channel is about 1.1 km. The dredged sediment from Muddi island waterway, crescent island waterway, 11+781~16+000 section, and 17+000~21+000 section of the outer waterway were thrown into the B dumping area.

The water depth of the dumping area should larger than the navigation water depth of the dredger which is more than 12.0 m. Meantime, considering the bed elevation of mud accumulation, selected water depth of the dumping area is greater than 14.0 m.

From the flow and the suspended sediment transport, the south of the channel is not suitable for dumping area.

Considering the constant wind direction for NW since the annual December to April next year, namely from the dumping area blow to the channel, dumping areas need to be places at the north of the channel at a certain distance.

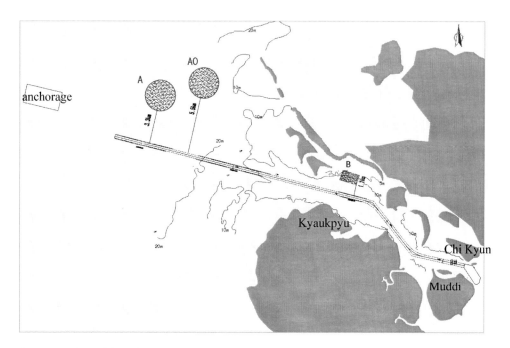

Figure 1. The position of dumping areas.

There are two dumping areas in the outer channel, named A0 and A dumping areas which are round with the radius of 1.5 km, at 3.3 km and 5.5 km, respectively, from the channel.

3 STUDY ON MATHEMATICAL MODEL OF SUSPENDED SEDIMENT

Mathematical models can be used to study the diffusion range of suspended sediment after dumping in dumping area. It can prove the rationality of the position of dumping areas.

3.1 Basic governing equation

The tidal current field at project sea area are studied by the plane two-dimensional hydrodynamic model which developed by MIKE21. The motion of tidal current is governed by the 2-D shallow water equations as follow:

$$\frac{\partial \zeta}{\partial t} + \frac{\partial p}{\partial x} + \frac{\partial q}{\partial y} = \frac{\partial d}{\partial t} \quad (1)$$

$$\frac{\partial p}{\partial t} + \frac{\partial}{\partial x}\left(\frac{p^2}{h}\right) + \frac{\partial}{\partial y}\left(\frac{pq}{h}\right) + gh\frac{\partial \zeta}{\partial x} + \frac{gp\sqrt{p^2+q^2}}{C^2 \cdot h^2} - \frac{1}{\rho_w}\left[\frac{\partial}{\partial x}(h\tau_{xx}) + \frac{\partial}{\partial y}(h\tau_{xy})\right]$$
$$-\Omega_q - fVV_x + \frac{h}{\rho_w}\frac{\partial}{\partial x}(p_a) = 0 \quad (2)$$

$$\frac{\partial p}{\partial t} + \frac{\partial}{\partial y}\left(\frac{q^2}{h}\right) + \frac{\partial}{\partial x}\left(\frac{pq}{h}\right) + gh\frac{\partial \zeta}{\partial y} + \frac{gq\sqrt{p^2+q^2}}{C^2 \cdot h^2} - \frac{1}{\rho_w}\left[\frac{\partial}{\partial y}(h\tau_{yy}) + \frac{\partial}{\partial x}(h\tau_{xy})\right]$$
$$-\Omega_p - fVV_y + \frac{h}{\rho_w}\frac{\partial}{\partial xy}(p_a) = 0 \quad (3)$$

The motion of suspended sediment is governed by the following equation:

$$\frac{\partial[(h+\zeta)S]}{\partial t} + \frac{\partial[(h+\zeta)SU_x]}{\partial x} + \frac{\partial[(h+\zeta)SU_y]}{\partial y} = -F_s + \frac{\partial}{\partial x}\left[(h+\zeta)D_x\frac{\partial S}{\partial x}\right]$$
$$+ \frac{\partial}{\partial y}\left[(h+\zeta)D_y\frac{\partial S}{\partial y}\right] \qquad (4)$$

where S is the depth-averaged sediment concentration; D_x and D_y are the horizontal diffusion coefficients of sediment; F_s is erosion and deposition function.

Bed level evolution due to suspended load:

$$\gamma_d \frac{\partial \eta_b}{\partial t} - F_s = 0 \qquad (5)$$

where η_b bed level variation due to bed load transport; γ_b is the dry volume weight of the sediment on the bed.

3.2 *Parameters*

1. D_x and D_y can be specified as constant in space which is equal to 15 m²/s.
2. Sediment setting velocity
   The setting velocity is selected as 0.045 cm/s at the flocculation state in the sea.
3. The ratio between suspended sediment and bed load (xu, 2000)
   The seabed material near Muddi Island is relatively coarse, suspension ratio of mud is smaller in Muddi island than that in the muddy coast. Due to lack of the data of sandy coast, from a security point of view, suspended sediment is calculated according to the proportion of 10%.
4. Source intensity of mud in the model is 1200 kg/s.
5. The time interval.

The time interval between the dumping consists of dredging time, the round-trip time to mud-dumping ground, and mud casting time interval (Han, 2000).

The dredging time is selected as 1.5 hour. Mud casting time interval is selected as 8 minutes. The ship for dredger travels at 10 knots per hour. By calculation, the time interval between the dumping in A and A0 dumping area are 2 hours and the time interval between the dumping in B dumping area is 1.5 hours.

3.3 *Simulation area and model verification*

Establish mathematical model for current based on the layout of the project. In order to improve the accuracy of the simulation results, two level nested models are used in hydrodynamic model. The big model range is from 92°54′14″E to 94°8′37″E and from 18°58′36″N to 20°6′30″N. The distance from E to W is approximately 129.9 km and distance from S to N is approximately 126 km. The model grid adopts quadrilateral mesh, grid spacing of the model in x and y directions are 30 m. The grid number in x and y directions are 433 and 420 respectively. The distance from E to W is approximately 86.8 km and distance from S to N is approximately 43.4 km in small model. The grid spacing in x and y direction is 100 m. The grid number in x and y direction is 868 and 433.

The water level and flux from big model is imposed at the open boundary of small model. The computational domain and grid of sediment model are identical to those of tidal current model. The field data measured in June 2007 is used to verify the present tidal current model. The simulation result of each survey point are good consistent to the local survey data. The simulation basically reflects the current movement characteristic in the project sea area and can be used to simulate suspended sediment transport.

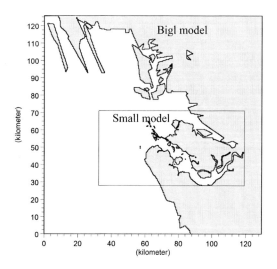

Figure 2. Computation domain.

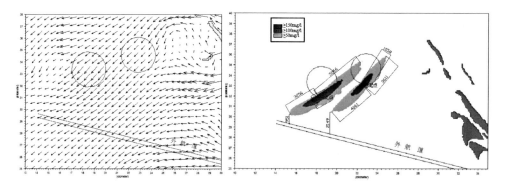

Figure 3. Flow field during ebb and Suspended sediment diffusion envelope near A and A0 mud-dumping ground.

3.4 *Results*

The mathematical model is used to simulate the whole diffusion process of the suspended sediment, Figure 3 shows flow field during ebb and the maximum range of the envelope of the total suspended sediment diffusion in a tidal cycle at A and A0. Figure 4 shows flow field during flood and the maximum range of the envelope of the total suspended sediment diffusion in a tidal cycle at B. The concentration of suspended sediment is divided into 3 intervals, greater than or equal to 50 mg/L, greater than or equal to 100 mg/L, and greater than or equal to 150 mg/L. When sediment concentration is smaller 50 mg/L, the sediment mainly exists in the form of suspension in the water and it is almost unlikely to deposition for the sediment less than 50 mg/L.

Seen from Figure 3 and Figure 4,

1. Sediment diffusion scope in 3 dumping areas is a long oval. This is mainly because the flow near spoil area is reciprocating flow, and suspended sediment diffuses mainly with the tide. The main diffusion direction of the suspended sediment is consistent with the direction of flow, and the diffusion area perpendicular to the flow direction is relatively smaller.
2. If throwing mud at the southern tip of the A dumping area, the longest diffusion distance of suspended sediment is 5.3 km with flood tide and 3.3 km with ebb tide. If throwing

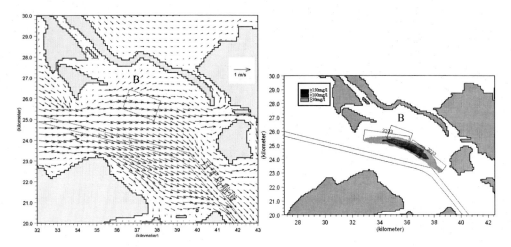

Figure 4. Flow field during flood and suspended sediment diffusion envelope near B mud-dumping ground.

mud at the southern tip of the A0 dumping area, the longest diffusion distance of suspended sediment is 3.6 km with flood tide and 4.1 km with ebb tide.
3. B mud area is located in the tidal creek where the flow is basically the reciprocating flow along the channel direction. If throwing mud at the lower left corner of B dumping area, the longest diffusion distance of suspended sediment are 4 km with flood tide and 2.76 km with ebb tide. If throwing mud at the lower right corner of B dumping area, the longest diffusion distance of suspended sediment are 3.03 km with flood tide and 3.33 km with ebb tide.
4. From the above results, throwing mud at A, A0 and B dumping area, suspended sediment did not spread to the channel. Therefore, the selection of the three dumping area is feasible.

4 CONCLUSIONS

In this paper, based on analysis of dynamic conditions of wind, wave and tidal current, sediment transport direction and water depth, and so on, the location of spoil area is determined. There are two spoil areas, A and A0, on the north side of the outer channel and one spoil area, B, in inner channel.

A 2-D mathematic model of the tidal current and sediment dispersion has been built to simulate the tidal current field near the spoil area and the diffusion range of suspended sediment from spoil area in a tidal cycle in the paper. The results show suspended sediment should not spread to the channel when throwing mud at A, A0, and B dumping area. Therefore, the three dumping area is safe, economic, feasible.

REFERENCES

Han Xi-jun, Yang Shu-sen. 2005. Influence of Dredged Material on the Basin Deposition of Gwadar Harbor. *Journal of Waterway and Harbour.* 26(3): pp.139–143.
Sun Lian-cheng, Pang Qi-xiu, Zhao Hui-min. 2011. Hydrodynamic and sediment oil Sino-Myanmar crude oil pipeline and wharf engineering. *Port & Waterway Engineering.* 5:pp. 13–17.
Xu Hong-ming. 2000. Numerical model of dredged sediment diffusion and application. *Marine Environmental Science.* 19(2): pp. 34–37.

Research on hydraulic model application in water shortage dispatch plan implementation

Xiao-yu Zhang & Yi-hui Chen
Yunnan Institute of Environmental Science, Kunming, Yunnan, China

Yue Xu
Construction Engineering College, Kunming University of Science and Technology, Kunming, Yunnan, China

ABSTRACT: Kunming is menaced by raw water shortage due to serious drought for several years. The hydraulic model is applied to multiple water reduction dispatch schemes comparison in order to resolve water shortage threat and provide basic supply service at the same time. The trials were carried out in a normal scheme, district turns water supply, and period water reduction by monitoring the flow and pressure of network. The research showed, period water reduction scheme is the best plan for considering the elements of water saving, pressure and flow stability, operability, and supply service guarantee. The scheme made a significant effort when it was implemented, the error of simulated pressure from real time pressure was neglectable and the result was high correlative. The successful case was a good reference for other regions facing raw water shortage crisis. The hydraulic model is worked as a fundamental tool for water supply dispatch supporting.

1 INTRODUCTION

Drinking water supply system is normally dispatched based on engineer's routine experience in most Chinese cities. In order to maintain the best practical range of supply pressure, the water dispatch decision is made on basis of regional water pressure distribution, number of pumping in use etc. Due to lack of prediction and accurateness, this kind of rough management can't afford the requirement of systematic water dispatch management and specific application. Drinking water hydraulic model is an effective tool that helps engineers make best decision when meeting with complicated network operation cases. The current hydraulic model researches are focusing on model building (Tao 1999, Xin et al. 2006, He et al. 2007 & Lei et al. 2009), network and pumping station design (Zhuang et al. 2011), and algorithm optimization (Tao et al. 2010), otherwise the model research on water dispatch optimization is much less.

City K is located in southwest of China. The city was disturbed by a long time raw water shortage in the past years, the available reservoir capacity was decreasing year by year, and serious contradiction between water demand and supply threaded the basic supply security. During the years 2009–2013, the volume of water impoundment experienced a downward cliff, the lowest impoundment volume was 69.6 billion m^3 in 2011, compared with high level impoundment which was 149.8 billion m^3 in 2008, reserved water decreased 80 billion m^3 which represented 53% of nominal impoundment. Under raw water shortage constraint condition, how to carefully dispatch existing impoundment and finely reduce daily production, fulfill the basic water demand of civism, and provide good water supply service to the city at a same time, has become an urgent issue to be solved. In this paper, multiple comparative dispatch schemes are established through hydraulic model to realize raw water reduction dispatching requirement. Through distribution network pressure and flow simulation, the hydraulic model provides a reliable method to well guide dispatch decision-making and secure large city water supply. Meanwhile, the research also can provide a good reference to other water shortage region in terms of water dispatch.

2 MODEL ANALYSIS

2.1 *Model establishment*

The daily production of the city is 850,000 m³/day, five Water Treatment Plants (WTPs) are worked as the main plants in use, the length of network above DN100 is 1900 km, population in service is 2.2 billion, and the service area is 200 km². The hydraulic model was built and calibrated since 2006. It includes 14,333 points and 15,299 pipelines, 35 pumps, the maximum pipe is DN1600 while the minimum pipe is DN100, the total length of network is 1200 km, the number of calibrated pressure points is 69. Under routine calibration, the error of 74% pressure points is less than 2 m, and the outlet flow error of model is less than 5%. Therefore, the hydraulic model can be applied in network master planning, dispatch optimization, in accordance with well and acceptable model accuracy; the model may play a significant role in term of network operation key decision making.

After established the hydraulic model, the paper compares the pressure and flow under the condition of normal and reduction supply to evaluate the effect of the reduction scheme, deciding whether the water reduction should be implemented or not. Secondly, when the reduction scheme is carried out, the influence of two kinds of reduction ways is explored. The water reduction ways include district turns supply scheme and period supply scheme. For the former method, the whole city is divided into several districts, and water reduction supply plan is applied to every district by turns. On the other hand, period water supply only guarantees the pipelines' pressure of peak time. From the comparison of the two methods, the paper will suggest a better way of water reduction supply. Some monitoring points are selected to conduct the research to reflect the variation of network pressure. The distribution of the pressure points is as shown in Figure 1.

2.2 *Normal operation of water supply scheme*

When the normal mode is taken, the production of the whole system will always be 850,000 m³ per day until the source of water was seriously insufficient. After that, the daily production will have to be reduced to 540,000 m³. Once the reduction scheme occurred, every WTP of the city will run according to the reduced production as shown in Table 1.

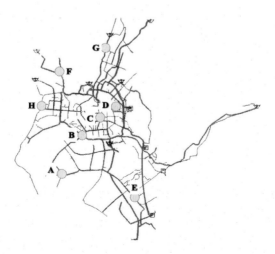

Figure 1. The layout of network pressure points.

Table 1. Daily production under normal and reduction mode.

Status	Daily production/m³						
	No. 2 WTP	No. 4 WTP	No. 5 WTP (west line)	No. 5 WTP (South line)	No. 6 WTP	No. 7 WTP	Total
Normal	67,840	45,214	124,533	103,304	78,495	430,135	849,521
Reduction	39,844	29,658	Shut down	62,393	Shut down	408,223	540,118

Table 2. Stimulated pressure values of points A–H under reduction mode.

Status	Pressure values/m								
	A	B	C	D	E	F	G	H	Average
Reduction	13.1	15.7	18.2	20.1	16.3	3.1	10.1	10.9	13.4

Table 3. Reduction water supply schedule.

WTP No.	Daily production/m³						
	Monday	Tuesday	Wednesday	Thursday	Friday	Saturday	Sunday
No. 2 WTP	51,840	51,840	51,840	51,840	51,840	Shut down	51,840
No. 4 WTP	31,622	30,587	31,148	30,571	32,760	30,382	Shut down
No. 5 WTP (west line)	Shut down	108,533	108,533	108,533	108,533	108,533	108,533
No. 5 WTP (west line)	87,304	Shut down	87,304	87,304	87,304	87,304	87,304
No. 6 WTP	62,495	62,495	62,495	Shut down	62,495	62,495	62,495
No. 7 WTP (pipe A)	258,403	236,246	Shut down	236,541	409,522	231,139	220,310
No. 7 WTP (pipe B)	259,437	237,494	410,884	237,317	Shut down	232,720	222,664
Total	751,101	727,195	752,204	752,106	752,454	752,573	753,208

On the basis of reduced water production, the network pressure is simulated by the hydraulic model. Table 2 presents the stimulated pressure values of typical monitoring points when reduction mode is run.

As shown in Table 2, after the production of every WTP is cut down, the pipeline pressure drops significantly, the average pressure is just only 13.4 m. Therefore, a sudden reduction in water supply will cause a wide shortage of water, or water will be supplied at a low pressure. The pressure of the majority of the city can't meet the basic standard. The incidence of the whole day shutdown of water supply even will occur in some parts of the city, seriously affecting the residents' daily life.

2.3 Reduction operation of water supply in advance

2.3.1 District turns water supply scheme

For this scheme, according to the balance of raw water supply and demand, the daily production will be reduced from 850,000 to 750,000 m³. The water supply shutdown region will be divided at first to determine affected area and population. The method of shutting down WTP and closing the valves is taken to conduct the scheme weekly. The reduction water supply schedule of every water plant is listed in Table 3.

On the basis of water supply schedule, the network pressure is simulated by the hydraulic model. Simulated average pressure of the district every WTP served is shown in Figure 2.

Because of the different capacities, different WTP exerted a various influence on the network. Hence, under the operation of district turns water supply, the network pressure presented a fluctuation obviously, influenced area and population floated greatly. On the other hand, the difficulty of daily network operation increased by closing valves to control shutdown district. As seen from Table 3, No. 7 WTP works as the main water supply plant in the city. Its output is above 50% of the total water volume. In Figure 2, under the condition of No. 7 WTP's Pipe A or Pipe B shut down, the pressure maintained a low water head during the night, and had a sudden drop in the peak period of the daytime. In the cases, the majority of the network pressure couldn't be guaranteed. While the coverage of No. 4 WTP was the north of the city, and the production reduction was only 30,000 m^3 per day, the average pressure could still reach 20 m after its reduction operation and the adverse impact was limited.

Although the district turns water supply program can reduce the daily water supply effectively, it still exposes a series of disadvantages, such as an unstable pressure and flow, a difficult operation and so on. Thus, the water reduction program needs a further improvement and optimization.

2.3.2 Period water reduction scheme

In order to modify the district turns water reduction scheme, a new method called period water reduction scheme is proposed to analyze the possibility of being implemented, which only ensures the pressure and water volume during the high peak. The reduced production of 750,000 m^3 per day is stimulated at first, an output of 650,000 m^3 per day later. The results show that the latter is recommended to be applied. The specific method applied is described as follows: the water supply low peak occurs from 0:00 to 6:00, and the instantaneous production is reduced to 350,000–40,000 m^3 per day; the flat peak appears at 8:00–16:00 and 22:00–24:00, the instantaneous production is kept at 450,000–650,000 m^3 per day; the high peak occurs at 16:00–22:00, the instantaneous production is increased to 850,000–1,000,000 m^3 per day in this period. Taking this program, the production is reduced as much as possible at low peak, and it should meet the minimum requirements of the pressure at flat peak. The tap water will be supplied with full capacity during the high peak period. Ultimately, the total water volume is stable at 650,000 m^3 daily, achieving the goal of saving water source

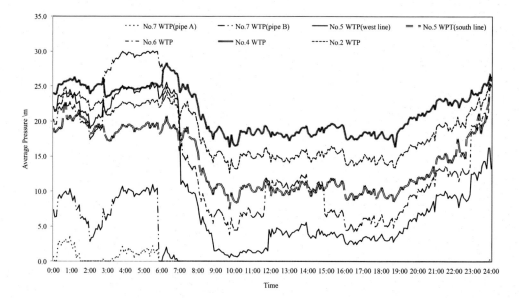

Figure 2. Daily average pressure of district turns water supply scheme.

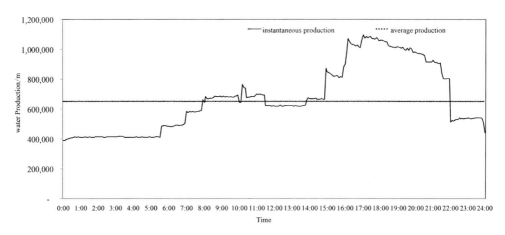

Figure 3. Daily production variation under period water reduction scheme.

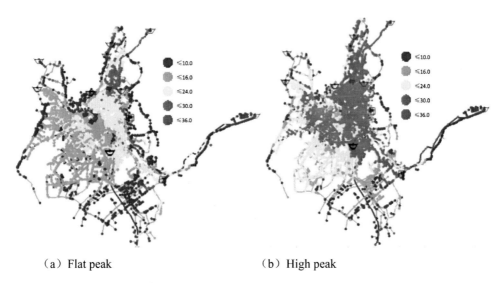

(a) Flat peak　　　　(b) High peak

Figure 4. Pressure distribution under period water supply scheme.

successfully. Figure 3 shows the variation of daily production of period water reduction program simulated by hydraulic model.

The pressure during the period of flat peak and high peak was simulated, respectively, as shown in Figure 4.

Seen from the picture (a) of Figure 4, during the flat peak period, the pressure could be kept beyond 16 m in most parts of the city in addition to some districts with high elevation located in the central and the northwestern part of the city. Particularly, the pressure of the northeast could be higher than 24 m. Seen from the picture (b) of Figure 4, during the high peak period, the pressure of the whole city could be effectively protected except the remote areas of the southeast. The pressure of the main city was kept in the range of 30–36 m.

3 RELIABILITY VERIFICATION OF THE MODEL

Among various options, period water reduction scheme was chosen to be implemented and was running well. On the other hand, SCADA system was used to monitor the network

Table 4. Accuracy evaluation of measured and simulated pressure value of points A–H.

Operation status	Pressure values/m								
	A	B	C	D	E	F	G	H	Average
Reduction	13.1	15.7	18.2	20.1	16.3	3.1	10.1	10.9	13.4

pressure when the reduction program was running for verifying the reliability of the model. The SPSS software was used to make a correlation analysis between the measured and simulated values of points A–H. The results are shown in Table 4.

As Table 4 showed, the average error of the measured and simulated pressure values is 2.05 m, and the Pearson's correlation coefficient is 0.887, which presents a highly positive correlation. Thus, the program is considered to be reliable. The simulated results reflect the distribution and variation of every district's pressure and flow, providing a scientific basis for decision-making.

4 CONCLUSIONS

Through the comparison of several water supply reduction schemes in accordance with hydraulic model, the most feasible scheme is chosen and put in practice, the method is validated and can meet the service satisfaction. This successful case in City K can be a very good reference and be applied in other water shortage regions in terms of emerging dispatch plan implementation. The model simulation successfully predicates the possible defect and decreases uncertainty of traditional and manual dispatch method. Water dispatch is becoming much systematic and efficient when supported by the hydraulic model; it improves the level of modern management of water supply companies, and achieves great social and economic benefits.

ACKNOWLEDGEMENT

This work was financially supported by Scientific Program of Yunnan Education Department (2013Z117).

REFERENCES

He fang, Li Shuping, Chen Yuhui, et al. 2007. Practical maintenance of hydraulic model for water distribution systems. Water & Wastewater Engineering 33(1): 95–98.
Lei Jingfeng, Chen Zhong, Liang Zhuoqi, et al. 2009. Setup and calibration of hydraulic model for water distribution system in Zhongshan City. China Water & Wastewater 25(17): 55–56.
Tao Jianke. 1999. Studies on dynamic model of Shanghai municipal water distribution network. China Water & Wastewater 15(4): 12–14.
Tao tao, Xia yu, Xin Kunlun, et al. 2010. First-class optimization operation of distributed complex multi-source raw water system. Journal of Tongji University (Natural Science) 38(12): 58–62.
Xin Kunlun, Liu Suiqing, Tao tao, et al. 2006. Application of pseudo-parallel genetic algorithm in optimal operation on water supply network. Journal of Tongji University (Natural Science) 34(12): 1662–1667.
Zhuang Baoyu, Yang Yufei, Zhao Xinhua. 2011. Optimization of water distribution system based on improved shuffled frog leaping algorithm. China Water & Wastewater 17(9): 54–58.

Wind stability analysis and design plan contrast of the main bridge of Pinghai Bridge

PengJun Liu
School of Architecture, Linyi University, Linyi, Shandong, China

ABSTRACT: For the auxiliary pier and not model, the dynamic mode calculation during and after construction, the flutter wind speed assessment, the critical flutter velocity calculations after construction and the actual flutter wind speed calculations of the structure in the largest cantilever state during construction are respectively conducted for the main bridge of Pinghai Bridge. The strength of the saving auxiliary pier structure was checked for the standing state of the naked tower, the largest double-cantilever state during the construction phase and under the gale loads after construction stage. The results show that the flutter stability index and the critical flutter wind speed after construction will not change with the setting of auxiliary piers; the flutter critical wind speed during the construction phase is controlled by the largest single-cantilever state and not affected because of the setting of auxiliary piers. These conclusions have some significance on the bridge wind stability analysis.

1 INTRODUCTION

In recent years, China's large-span bridge construction developed rapidly, while many problems appeared. The bridge wind resistance is an important aspect. With the improvement of the design requirements against wind, wind resistance stability analysis and design plan contrast is very important. As an example, the wind resistance stability of main bridge of Guangdong Pinghai Bridge was analyzed and the design plans were compared to provide a reference for similar bridge design.

The main bridge of Pinghai Bridge has a length of 604 m and the span is a combination of 152 + 300 + 152 m. The twin towers and single cable plane are used. It is a pre-stressed concrete cable-stayed bridge with a consolidation of pier, tower and girder. Auxiliary piers were set at 99.5 m from the centerline of the main tower at the side span.

According to the data, this basic design wind velocity of the bridge V_{10} = 43.6 m/s, and the bridge site is located in a Class A surface category. The value of design wind speed during construction stage was taken according to the return period of 30 years. According to "Wind-resistant Design Specification for Highway Bridges" (JTG/TD60-01-2004), the flutter stability and static wind load effect of the bridge were checked when there were auxiliary piers or not.

2 FLUTTER STABILITY ANALYSIS

The critical wind speed of bridge flutter is a hot topic in recent years (Ge Yaojun, 2003). The critical wind speed is an important indicator to measure the bridge flutter stability. It can be measured either by wind tunnel tests or theoretical calculations according to the measured parameters in tests.

The uncertainties impact of structural mass, stiffness and damping can be approximately estimated with flutter critical wind speed formula (Vanderput formula) which was used in a wing or a thin flat section (Xiang Haifan, 1996).

$$U_{cf} = \left[1 + (\varepsilon - 0.5)\sqrt{0.72\mu(r/b)}\right]\omega_t b \quad (1)$$

where, ε is the frequency ratio of twist and bending, μ is the density ratio of structural material and air, r is the inertia radius of the main beam section.

The 2D flutter critical wind speed calculation method which was put forward by professor Scanlan (Zhu Zhiwen, 2004) needs to be adopted to calculate the critical flutter wind speed. With the sub-state forced vibration method, for the bridge sections with vertical bending and torsional movement freedom, the corresponding critical flutter wind speed is

$$U_c = \frac{B\omega_h}{K_c} X_c = \frac{B\omega_h X_c V_{rc}}{2\pi} \quad (2)$$

where ω_h is the natural frequency of vibration of vertical bending movement, B is the bridge width, the significances of other symbols are given in Huo's literature (Huo Zhicha, 2011).

2.1 Model with auxiliary piers

2.1.1 Dynamic models calculating results of the finished state and construction process

By finite element analysis, the torsional vibration mode under a finished bridge state is given in Figure 1 (Ge Yaojun, 2001; Le Yunxiang, 2005; Honda A., 1998). The torsional vibration mode during construction stage of the structure is shown in Table 1. The torsional vibration mode diagrams of the structure are shown in Figure 2.

2.1.2 Flutter wind speed assessment

The bridge deck is located approximately 45 m above the water, and the bridge site is located in a Class A surface category, so

$$V_d = k_1 \cdot V_{10} = 1.405 \times 43.6 = 61.3 \text{ m/s}$$

In the above formula k_1 is the height variable correction factor of wind speed, which can be got according to some specifications. Taking the wind speed pulsation correction coefficient $\mu_f = 1.25$, the flutter test wind speed at the finished bridge stage:

$$[V_{cr}] = 1.2 \cdot \mu_f \cdot V_d = 1.2 \times 1.25 \times 61.3 = 92.0 \text{ m/s}$$

(a) the frequency is 0.9138Hz with auxiliary pier (b) the frequency is 0.9138Hz without auxiliary pier

Figure 1. The first order symmetrical torsion of the main beam.

Table 1. The torsional vibration mode during construction stage of the structure.

Order	Cases of auxiliary pier designed	Modal type	Natural frequency (Hz)
1	With auxiliary pier	The largest single cantilever during the construction	1.0317
2		The largest double cantilevers during the construction	1.4125
3	Without auxiliary pier	The largest single cantilever during the construction	1.0317
4		The largest double cantilevers during the construction	1.0617

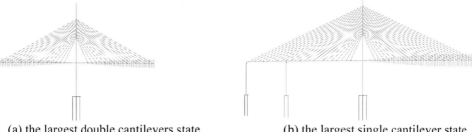

(a) the largest double cantilevers state   (b) the largest single cantilever state

Figure 2. Torsional vibration shapes diagram of the model with auxiliary piers during the construction.

Supposing the recurrence interval coefficient of wind speed during the construction phase $\eta = 0.92$, the flutter test wind speed

$$[V_{cr}^s] = 1.2 \cdot \mu_f \cdot V_{sd} = 1.2 \times 1.25 \times 0.92 \times 61.3 = 84.6 \text{ m/s}$$

2.1.3 *Flutter critical wind speed of the finished structure*

The torsional fundamental frequency of finished bridge $f_t = 0.9138$ (Hz). The flutter stability index

$$I_f = \frac{[V_{cr}^s]}{f_t \cdot B} = \frac{92}{0.9138 \times 26.9} = 3.74$$

When $2.5 < I_f < 4.0$, the critical flutter wind speed should be checked by section model wind tunnel tests. But considering the actual situation, we calculated the critical flutter wind speed according to "Wind-resistant Design Specification for Highway Bridges" (JTG/TD60-01-2004).

The density ratio of concrete girder of main span is $\mu = 63$; the inertia radius of the main beam section $r = 6.5$ m; the width of full-bridge $B = 26.9$ m; the first order symmetrical torsional frequency $f_t = 0.9138$ (Hz), so the flutter critical wind speed of the flat is

$$V_{co} = 2.5\sqrt{\mu \cdot \frac{r}{b}} \cdot B \cdot f_t = 2.5 \times \sqrt{63 \times \frac{6.5}{13.45}} \times 26.9 \times 0.9138 = 339 \text{ m/s}$$

Taking the shape factor $\eta_s = 0.6$, the effect coefficient of attack angle $\eta_\alpha = 0.8$, we can get the flutter critical wind speed:

$$V_{cr} = \eta_s \cdot \eta_\alpha \cdot V_{co} = 0.6 \times 0.8 \times 339 = 162.7 \text{m/s} > [V_{cr}] = 92.0 \text{ m/s}$$

The construction flutter stability of the finished bridge meets the regulatory requirements.

2.1.4 *Actual flutter speed in the largest cantilever state during the construction*

The torsional fundamental frequency in the largest cantilever state during the construction is $f_t = 1.0317$ (Hz). So the flutter stability index is

$$I_f = \frac{[V_{cr}^s]}{f_t \cdot B} = \frac{84.6}{1.0317 \times 26.9} = 3.05$$

The first order symmetrical torsional frequency of the mid-span concrete main beam is $f_t = 1.0317$ (Hz), other parameters are as before. so the flutter critical wind speed of the flat is

$$V_{co} = 2.5\sqrt{\mu \cdot \frac{r}{b}} \cdot B \cdot f_t = 2.5 \times \sqrt{63 \times \frac{6.5}{13.45}} \times 26.9 \times 1.0317 = 382.8 \text{ m/s}$$

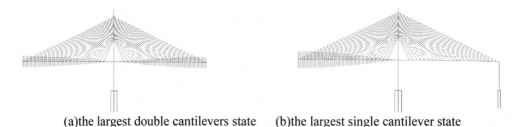

(a)the largest double cantilevers state (b)the largest single cantilever state

Figure 3. Torsional vibration shapes diagram of the model without auxiliary piers during the construction.

The critical flutter wind speed

$$V_{cr} = 0.6 \times 0.8 \times 382.8 = 183.8 \text{m/s} > [V_{cr}^s] = 84.6 \text{ m/s}$$

2.2 Model without auxiliary piers

According to finite element analysis, the torsional vibration modes under the finished bridge state are given in Table 1. The torsional vibration modes are shown in Figure 3.

The flutter testing wind speed of the structure without auxiliary pier is the same as that with auxiliary pier, the torsional fundamental frequency of the finished structure $f_t = 0.9138$ (Hz) is also the same, so the flutter speed assessment, flutter stability index and flutter critical wind speed are also the same. In addition, the torsional fundamental frequency of the largest single cantilever during the construction $f_t = 1.0317$ (Hz) is the same, so the flutter stability index and flutter critical wind speed are also the same.

3 STATIC WIND EFFECTS ANALYSIS

3.1 In the standing state of naked tower during the construction stage

When the main tower construction is completed and the rope is not hanging, takeing the effect on the main tower and main pier by the transverse and longitudinal wind loads in 30-year return period during the construction phase into consideration, we take the value of load combination as the weight of the tower and pier + wind load in construction.

3.1.1 Transverse wind load analysis

The transverse bending moment diagrams of the main tower and the main pier under combinations (1) (standard value) are given in Figure 4. The transverse internal force and cross-section check in the standing state of naked tower during the construction stage is given in Table 2.

3.1.2 Longitudinal wind load analysis

In transient situation, the stress at the edge of compressive zone in reinforced concrete members subjected to bending moment which is less than $0.8f_{ck} = 24.64$ MPa. The cross section of main tower and the main pier both meet the requirement under wind loads in longitudinal and transverse direction.

3.2 The maximum double cantilever state in construction stage

The maximum double cantilever state in construction stage is from no auxiliary pier construction structure to the time when the side span segment is closed. The effect of wind load on the structure can be analyzed from longitudinal and transverse direction both. The situation of

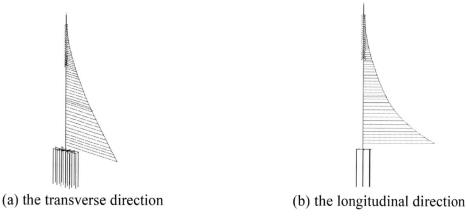

(a) the transverse direction   (b) the longitudinal direction

Figure 4. The transverse bending moment diagram of the main tower and the main pier under combinations (1) (standard value).

Table 2. The transverse internal force and cross-section check in the standing state of naked tower during the construction stage.

Wind direction	The site of direction	Standard value combination Axial force N (kN)	Shear force Q (kN)	Bending moment (kN·m)	Basic combination Axial force N (kN)	Shear force Q (kN)	Bending moment (kN·m)	Stress check (MPa) Min	Max
The transverse direction	At the bottom of main tower	34074	2752	106950	34074	3853	149730	−27.4	11.5
	At the top of main pier	40074	2825	115940	40074	3955	162316	−0.32	1.46
	In the middle of main pier	69546	3098	216800	69546	4337	303520	−0.80	3.50
	At the bottom of main pier	89387	3204	229030	89387	4486	320642	−0.20	2.05
The longitudinal direction	At the bottom of main tower	34074	818	32680	34074	1145	45752	−0.41	3.06
	At the top of main pier	40074	925	35483	40074	1295	49676.2	−0.10	1.27
	In the middle of main pier	69546	2294	79975	69546	3212	111965	−0.08	2.89
	At the bottom of main pier	89387	2807	105650	89387	3930	147910	−0.08	1.95

longitudinal wind load effect and the unbalanced load produced by cantilever construction occurs at the same time was taken into account. The bending moment diagrams of the main tower and main pier (standard value) in construction stage are given in Figure 5.

3.3 The finished bridge state

Combination items in the finished bridge state include: dead load + base displacement + temperature load + longitudinal wind load (100 years); dead load + base displacement + temperature load + transverse wind load (100 years).

Under longitudinal wind load, the longitudinal displacement at the top of main pier is 0.053 m, which is 1/1472 of the tower height. The structure displacement is shown in Figure 6a

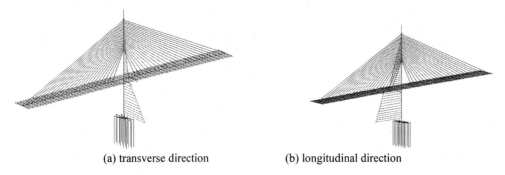

(a) transverse direction (b) longitudinal direction

Figure 5. Bending moment diagram of the main tower and main pier (standard value).

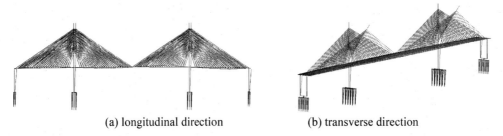

(a) longitudinal direction (b) transverse direction

Figure 6. The structure displacement diagram under wind load.

Table 3. Internal forces at the bottom of main pier caps under the basic combination of wind load in the finished bridge state.

Wind direction	Internal force	Q_{max} (kN)	N_{cor} (kN)	M_{cor} (kN·m)
The transverse direction	At the bottom of main pier caps	18897	335689	556382
The longitudinal direction	The transverse direction	15190	335700	970760
	Along the bridge direction	10508	–	310392

Table 4. Internal forces and reinforcement calculation of a single pile.

Wind direction	Axial force N (kN)	Bending moment M (kN·m)	Calculated reinforcement area (cm^2)	Calculated reinforcement ratio	Remark
The longitudinal direction	−4578	16485	395	0.65%	A
	−37383	16485	Construction reinforcement		B
The transverse direction	−400	17070	566	0.92%	A
	−41561	17070	Construction reinforcement		B

Note: A is the minimum axial force and B is the maximum axial force.

(100× magnification). Under transverse wind load, the transverse displacement at the top of main pier is 0.488 m, which is 1/160 of the tower height. The structure displacement is shown in Figure 6b (100× magnification).

The internal forces at the bottom of main pier caps under the basic combination of wind load in the finished bridge state are given in Table 3, Internal Forces and Reinforcement Calculation of a single Pile are given in Table 4.

The results showed that after removing the auxiliary pier, under the action of strong winds of 100 year frequency, the pile reinforcement of the main pier is controlled by the transverse wind load combination and the reinforcement ratio should not be less than 0.92%. For the structure in which auxiliary piers are set, auxiliary pier can share part of the transverse wind load for main pier, so the theoretical reinforcement ratio of main pier pile is not higher than 0.92%.

4 CONCLUSIONS

The following basic conclusions can be obtained from the study:

1. By the anti-wind analysis of the original structure design (with auxiliary piers) and the secondary structure without auxiliary piers of Pinghai Bridge, we find that all of the results showed the structure eliminating auxiliary pier is still safe and feasible, and the structure size will not be adjusted due to the removal of the auxiliary piers.
2. In general, the auxiliary pier setting will change the torsional fundamental frequency of structure, thereby affect the structural flutter stability index and the critical flutter wind speed. This bridge is a pier, tower, girder consolidation system. In the finished bridge stage, the first order torsional vibration mode of the structure is the torsion of the main beam in the main span. The auxiliary pier of the edge span can not achieve a restrictive effect on the torsion of the main beam in the main span, so it will not change the first-order torsional frequency.
3. In the construction phase, the critical flutter wind speed of the structure was controlled by the largest single cantilever state, so the setting of auxiliary piers would not affect the critical flutter wind speed of the structure. But the setting of auxiliary piers could reduce the cantilever length in the largest double cantilevers state, which made the statically determinate structure turn to the statically indeterminate structure as soon as possible and in favor of ensuring the the structural stability of the wind during the construction phase.

REFERENCES

Ge Yaojun, Xiang Haifan, H. Tanaka. 2003. Reliability analysis of bridge flutter under stochastic wind loading. *China civil engineering journal*. 36(6):42–46(79).

Ge Yaojun, Xiang Haifan. 2001. Probabilistic Assessment Study of Flutter Instability of Bridge Structures. *Journal of Tongji University*. 29(1):70–74.

Honda A, Miyata H, Shibata H. Aero. 1998. Dynamic stability of narrow decked suspension brigde (Aki-nada Ohashi Bridge). *Journal of Wind Engineering and Industrial Aerodynamics*. 77 & 78:409–420.

Huo Zhicha. 2011. Study on flutter of Shenzhen-nanshan bridge by CFD numerical simulation. *Harbin Institute of Technology*.

Le Yunxiang, Chang Ying, Hu Xiaolun. 2005. Analysis of Wind-resistant Stability on the Catwalk of Wuhan Yangluo Yangtze River Bridge. *Journal of Highway and Transportation Research and Development*. 22(8):40–43.

Xiang Haifan. 1996. Wind-resistant Design Guide for Highway Bridges. *China Communications Press*.

Zhu Zhiwen, Chen Zhengqing. 2004. Numerical simulations for aerodynamic derivatives and critical flutter velocity of bridge deck. *China Journal of Highway and Transport*. 17(3):44–48.

Numerical study on the street flooding in Huinan, Pudong District during the various short-duration rainstorm events

J. Huang
Shanghai Pudong New Area Hydrology and Water Resources Administration, Shanghai, China

ABSTRACT: For some years, Huinan has been suffered by the storm waterlogging disaster. To solve this problem, a numerical model by MIKE FLOOD has been set up with the rainfall runoff sub-model by MIKE URBAN, the underground network sub-model by MIKE URBAN, and the overland flow sub-model by MIKE 21. After the validation, the model was adopted to simulate the street flooding under the different drainage systems combined with the designed short-duration rainstorm events in Huinan, Pudong District. And the conclusions were obtained that: however hard it rained, Gongji Road Rain Pump would be the sustainable strategy to alleviate the urban street flooding hazard for the pipe diameters enlarged for the capacity expansion and the flow in the pipe networks drawn timely. As a result, the ponded area would always reduce, and some roads had the shallower or even no overland flood water.

1 INTRODUCTION

The flood flow in the urbanized areas constitutes a hazard with the low-lying houses got drowned and the local traffic crippled, which strongly affect the people's life and work.

Recently, the storm waterlogging in urban areas has been studied relatively despite many centuries of flood events, and a few mathematical models have been established to simulate the urban flooding (Anselmo et al. 1996, Werner 2001, Mark et al. 2004, Schmitt et al. 2004, Xia et al. 2011), e.g., Storm Water Management Model (SWMM) (Liu et al. 2007, Barco et al. 2008); InfoWorks CS (Tan 2007, Yao 2007); GIS-based urban flood inundation model (Chen et al. 2009); and MOUSE model (Huang et al. 2014).

Huinan, 121°42′~121°49′ E, 30°59′~31°5′ N, located in the south to central of Pudong District in Shanghai, has the total area of 65.78 km^2 and it is one of the major political, economic, cultural and financial centers (see Fig. 1). It has a tropical East Asian monsoon, and the oceanic climate significantly with four different seasons, plenty of good rainfall. There is Pudong Canal, Zhonggang River, Yaogou River, and Sanzaolu River (see Fig. 1), with the normal water level of 2.50~2.80 m, the maximum of 3.75 m while the minimum of 2.25 m.

At present, Huinan has the drainage system depending on the gravity flow by rainfall itself. And during the heavy rain falls, it often occurs the waterlogging with many houses, factories and croplands suffered. So it is urgent to be solved. Gongji Road Rain Pump has been planned to alleviate the waterlogging in Huinan. It will lie in the north side of Gongji Road, south to Gongbei Road, west bank of Yaogou River (see Fig. 1). Four vertical submersible axial-flow pumps will be loaded with 9.0 m^3/s total discharge. Gongji Road Rain Pump has a storage tank with the top elevation of 5.43 m and the bottom elevation of −1.70 m, a forebay with the cross-sectional area of 53.5 m^2 and the surface area of 22.5 m^2. When it works, the surroundings about 112.48 hm^2 will be served.

A numerical model of the storm waterlogging by MIKE FLOOD platform has set up with the rainfall runoff sub-model, the underground network sub-model by MIKE URBAN software and the overland flow sub-model by MIKE 21 software. In this research, the validated

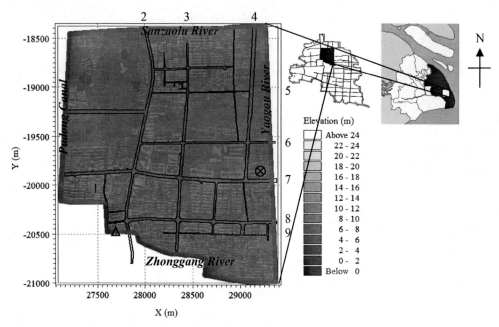

Figure 1. Location of Huinan in Pudong District ("Δ" and "⊗" show Huinan Hydrographic Station and Gongji Road Rain Pump. Numbers "1"~"9" show Meihua Road, Nanzhu Road, Jinhai Road, Chuannanfeng Road, Tanghong Road, Gongbei Road, Gongji Road, East Renmin Road, and Huayuan Road).

storm waterlogging model in Huinan, Pudong District (Huang et al. 2014) is adopted to simulate the street flooding under the different drainage systems during the various short-duration rainstorms.

2 MODEL DESCRIPTION

The numerical model by MIKE FLOOD software has been set up with coupled the rainfall runoff sub-model, the underground network sub-model by MIKE URBAN software and the overland flow sub-model by MIKE 21 software.

2.1 *Computational grids*

The research region is situated east to Yaogou River, south to Zhonggang River, west to Pudong Canal, and north to Sanzaolu River, about 5 km², including Huinan Station of Metro Line 16 and the core zone of Dongcheng District (see Fig. 1), with 642 hollows, 619 pipe lines, 22 outlets, 23.4 km of the total pipe lengths, where has divided to be 579 × 656 square grids with the uniform interval of 4 m. And 614 catchments in the study area has been generated with Thiessen Method with one water collection hollow in each catchment to make the rainfall runoff flow into the drainage pipes.

2.2 *Boundary conditions and other model parameters*

In this study, the river open boundaries are setup with the different water levels under the various short-duration rainstorms (see Table 1). And the Chicago Rainfall Pattern with the rainfall peak coefficient of 0.3 is chosen to create the diverse rainfall as the rainfall conditions (see Fig. 2).

Table 1. The rainfall and the river water levels during the various short-duration rainstorms.

Designed rainfall reappearing period (year)	Total rainfall in 120 min (mm)	River water level (m)
Once a year (P = 1)	44.52	2.50
Once every three years (P = 3)	62.01	2.85
Once every five years (P = 5)	70.14	2.90

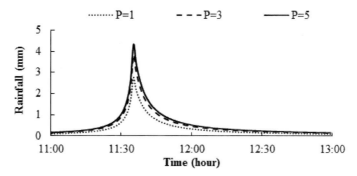

Figure 2. Graph of the rainfalls under the various short-duration rainstorms.

Table 2. Computational cases.

Designed rainfall reappearing period (year)	Gravity flow	Gongji road rain pump
Once a year (P = 1)	√	√
Once every three years (P = 3)	√	√
Once every five years (P = 5)	√	√

The rainstorm intensity in Shanghai City is computed by

$$i = \frac{17.812 + 14.6681 \lg P}{(t + 10.472)^{0.796}} \quad (1)$$

where i = the rainstorm intensity, P = the designed rainfall reappearing period, and t = the rainfall duration.

The simulated region is mainly composed of buildings, roads, green fields, and water. And the different impervious rates have been used, e.g., 90% for buildings, 85% for roads, 25% for green fields, zero for water, and 50% for the rest.

The mean surface flow velocity has adopted as 0.3 m/s, the hydraulic decay coefficient is 0.9, the initial abstraction is 0.0006 m, and the coefficient of the rainfall runoff is 0.6 in the rainfall runoff model. The roughness of the concrete drainage pipe lines is used as 0.0005 in the drainage hydrodynamic model. The manning coefficient is 32.0, and the water depth of the dry grid is below 0.002 m while the water depth larger than 0.003 m is wet in the overland flood model. And the time step to set is 1 s.

2.3 *Computational cases*

The validated model (Huang et al. 2014) is used to simulate the street flooding under the different drainage systems, i.e., gravity flow and Gongji Road Rain Pump during the designed various short-duration rainstorms as the computational cases (see Table 2).

3 SIMULATION RESULTS

When the drainage system depending on the gravity flow by rainfall itself, the maximum drowned areas under the various short-duration rainstorms are 0.53 km^2 (P = 1), 0.64 km^2 (P = 3), and 0.71 km^2 (P = 5), respectively (see Fig. 3). Correspondingly, after Gongji Road Rain Pump runs, the ponded area will reduce to 0.46 km^2 (P = 1), 0.56 km^2 (P = 3), 0.62 km^2 (P = 5), about −13.2%, −12.5%, and −12.7% (see Fig. 3).

And some roads have the shallower or even no overland flood water, especially Gongji Road (section from Jinghai Road to Yaogou River) (see Fig. 4). When the rainstorm intensity satisfies P = 3, the street flooding depth in this road section ranges from 0.10 m to 0.15 m before Gongji Road Rain Pump runs, but when the pump works, the water depth will not be more than 0.05 m (see Fig. 4).

Now, the local drainage system depends on the gravity flow by itself. Under this saturation, it is often easier to generate the surface runoff when the rainfall exceeds the full capacity of the drain lines or the rain arrives more quickly than the local pipe networks can absorb it (see Fig. 5(a)). However, it will be hard for the street flooding after Gongji Road Rain Pump works with the pipe diameters enlarged for the capacity expansion and the flow in the pipe networks drawn timely (see Fig. 5(b)).

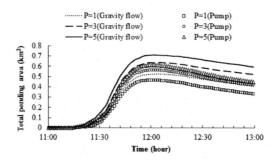

Figure 3. Changes of the total ponding areas under the different conditions during the various storms.

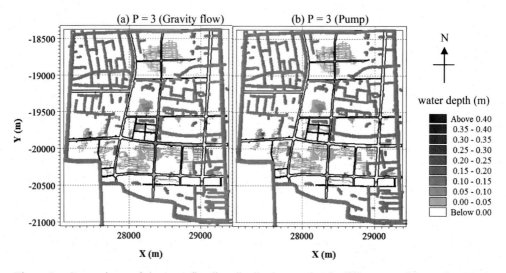

Figure 4. Comparisons of the street flooding distributions under the different conditions when P = 3.

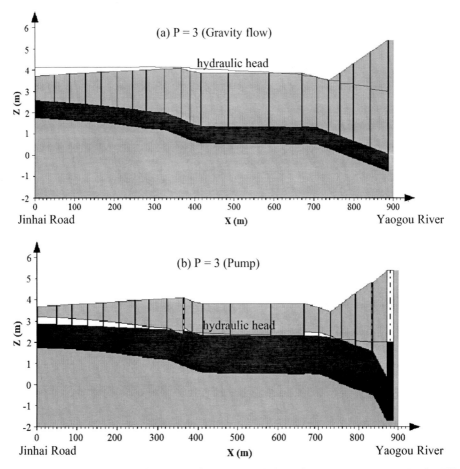

Figure 5. Pipe profiles at Gongji Road (section from Jinghai Road to Yaogou River) under the different conditions when P = 3 (Pipe flow in black).

4 CONCLUSIONS

The storm waterlogging model has been set up by MIKE FLOOD coupled with the rainfall runoff sub-model by MIKE URBAN, the underground network sub-model by MIKE URBAN, and the overland flow sub-model by MIKE 21 to simulate the street flooding under the different drainage systems during the various short-duration rainstorm events in Huinan, Pudong District. And the results show that no matter how much it rains, after Gongji Road Rain Pump works, the ponded area will always reduce in the varying degrees, and some roads have the shallower or even no overland flood water. So, it will be hard for the street flooding with the pipe diameters enlarged for the capacity expansion and the flow in the pipe networks drawn timely. Hence, the sustainable urban drainage system is the effective strategy in Huinan, Pudong District.

REFERENCES

Anselmo, V., Galeati G., Palmieri S., et al. 1996. Flood risk assessment using an integrated hydrological and hydraulic modeling approach: a case study. *Journal of Hydrology*: 533–554.

Barco, J., Wong, K.M. & Stenstorm, M.K. 2008. Automatic calibration of the U.S. EPA SWMM Model for a large urban catchment. *Journal of Hydraulic Engineering* 134: 466–474.

Chen, J., Hill, A.A. & Urbano, D.L. 2009. A GIS-based model for urban flood inundation. *Journal of Hydrology* 373: 184–192.

Huang, J., Wang, S.Z., Deng, S.Z., et al. 2014. Numerical study on the impact of Gongji Road Rain Pump on the waterlogging in Huinan, Pudong District. *Journal of Geoscience and Environment Protection* 2: 52–58.

Liu, X.P., Liu, S.Q., Li, S.P., et al. 2007. Simulation of drainage networks based on SWMM. *Water and Wastewater Engineering* 33(4): 105–108. (in Chinese).

Mark, O., Weesakul, S., Apirumanekul, C., et al. 2004. Potential and limitations of 1D modeling of urban flooding. *Journal of Hydrology* 35: 159–172.

Schmitt, T.G., Thomas, M., Ettrich, N., 2004. Analysis and modeling of flooding in urban drainage systems. *Journal of Hydrology* 299: 300–311.

Tan, Q. 2007. Application of drainage model for urban storm water quantity management. *Doctor Dissertation of Tongji University*. (in Chinese).

Werner, M.G.F. 2001. Impact of grid size in GIS based flood extent mapping using a 1D flow model. *Physics and Chemistry of the Earth, Part B: Hydrology, Oceans and Atmosphere* 26 (7–8): 517–522.

Xia, J.Q., Falconere, R.A., Lin, B.L., et al. 2011. Numerical assessment of flood hazard risk to people and vehicles in flash floods. *Environmental Modeling & Software*: 1–12.

Yao, Y. 2007. Study on the application of urban drainage networks modeling based on GeoDatebase. *Master Dissertation of Tongji University*. (in Chinese).

Spatial distribution and pollution characteristics of Ammonia-N, Nitrate-N and Nitrite-N in groundwater of Dongshan Island

H.Y. Wu, S.F. Fu, X.Q. Cai & K.Z. Zhuang
The Third Institute of Oceanography, State Oceanic Administration, Xiamen, China

ABSTRACT: In this study, the spatial variability and pollution characteristics of NH_4^+-N, NO_3^--N, and NO_2^--N in the groundwater of Dongshan Island were analyzed. The results showed that the NH_4^+-N and NO_2^--N in groundwater of Dongshan Island were at a low level on the whole, but their spatial variability were high, and their autocorrelation were poor; however, NO_3^--N was general high, but its spatial variability was moderate, and the autocorrelation was much good. The high concentration areas of NH_4^+-N, NO_3^--N, and NO_2^--N were all located in the coastal land. The domestic pollutants and human and animal wastes from towns and villages were the main sources of nitrogen pollution. Land use type, soil type, groundwater depth, pH, dissolved oxygen, season, and Fe^{2+} all influenced the distribution and transformation of NH_4^+-N, NO_3^--N, and NO_2^--N in groundwater.

1 INTRODUCTION

In natural groundwater, the main nitrogen compounds were ammonia-N (NH_4^+-N), nitrate-N (NO_2^--N), and nitrite-N (NO_3^--N), in which, NO_3^--N is the most general compound, due to the good solubility of nitrate and the difficulty of NO_3^--N to be absorbed by soil. Because of the chemical property and the higher toxic, NO_2^--N was considered as an important nitrogen pollution indicator. With the increasing use of chemical fertilizer, NO_3^--N and NO_2^--N became universal groundwater pollutants all over the world (Umezawa et al. 2008; Dragon 2013).

In many developing countries and regions, especially in remote rural areas and water resource shortage areas, groundwater is the main resource of the local residents' drinking water, domestic water, and agriculture irrigation. Thus, the groundwater quality is closely related to people's production and living. An excess of nitrogen in groundwater may be toxic to human and other animals (Salvador et al. 2009; Reddy et al. 2009; Xie et al. 2011; Liu et al. 2013), thus the World Health Organization (WHO), European Union (UN) and many countries specify the maximum allowable concentration of ammonia-N, nitrate-N, and nitrite-N in drinking water. Also too much of nitrogen in groundwater may result in eutrophication of related water body (Wang et al. 2014); therefore, to study their spatial distribution characteristic was meaningful for the controlling, prevention and treatment of nitrogen pollution in groundwater.

Islands are surrounding by seawater, most of them are shortage of freshwater, groundwater was the most important resource for the residents. In this study, we analyzed the spatial distribution characteristic and its variation pattern of ammonia-N, nitrate-N, and nitrite-N in the groundwater of Dongshan Island, as well as their pollution source and influence factors. The results would give advice to the management of groundwater resource of the island.

2 MATERIALS AND METHODS

2.1 *Study area*

Dongshan Island is located in 23°34′~23°47′N, 117°18′~117°35′E. It is surrounded by sea, with an area of 194 km². With southeast subtropical oceanic climate, the mean annual

Figure 1. Location of the sampled wells in Dongshan Island.

precipitation is 1256.0 mm and mean annual temperature is 20.9°C. No big surface water system is found. The annual mean water resources of Dongshan Island is 9.06 million m^3 and the water resource of per capita was only 450 m^3 (Shi 2004). The groundwater is quaternary pore water and mainly comes from atmospheric precipitation. Groundwater is the main resource of the local residents' drinking water, domestic water, agriculture irrigation, and freshwater aquaculture.

2.2 Sampling time and method

A total of 71 wells were chosen for analysis, in light of the distribution of town, village and farmland, as well as the terrain and soil type, in 2014 (Fig. 1). Dry season (from January to March) and wet season (from June to September) were sampled separately and compared. In order to reduce the influence of rain on groundwater quality, sampling usually took 7 days after rain.

2.3 Analysis method

Temperature, salinity, pH, and Dissolved Oxygen (DO) were measured *in situ* by a portable multi-parameter analyzer (MULTT3430); groundwater depth was measured *in situ* by using Solinst 101 water gauge; NH_4^+-N, NO_3^--N, and NO_2^--N were measured by a continuous flow analyzer method (ALLIANCE console future), and the Geo-statistic method was used to analysis the spatial variation of them; chloridion was analyzed by using silver nitrate capacity method.

3 RESULTS AND DISCUSSION

3.1 Statistical analysis

As shown in Table 1, the average concentration of NO_3^--N and NH_4^+-N in dry season was higher than in wet season, while NO_2^--N was in contrast. NO_3^--N showed moderate spatial variability, while NO_2^--N and NH_4^+-N showed high spatial variability.

According to Chinese quality standard for groundwater (for drinking water, NO_3^--N < 20 mg/L, NH_4^+-N < 0.2 mg/L, and NO_2^--N < 0.02 mg/L), among the 71 samples

Table 1. Statistical analysis of ammonia-N, nitrate-N, and nitrite-N in the groundwater of Dongshan Island.

	Time	Number of samples	Minimum	Maximum	Average value	SD	Variable coefficient
NO_3^--N	Dry season	71	1.72	72.5	23.316	14.993	0.643
	Wet season	64	0.307	49.33	18.587	11.310	0.608
NO_2^--N	Dry season	71	0	0.768	0.0237	0.0984	4.146
	Wet season	64	0	0.758	0.0598	0.175	2.935
NH_4^+-N	Dry season	71	0.0234	35.246	0.727	4.320	5.947
	Wet season	64	0.004	17.49	0.386	2.211	5.725

in dry season, 57.8%, 9.9%, and 7.0% were exceeding the standard of NO_3^--N, NO_2^--N, and NH_4^+-N, respectively; while among the 64 samples in wet season, the percentages were 42.2%, 12.5%, and 7.8%, respectively. Comparing the survey data of 2004 and 2007 (Ma et al. 2008) with this study, the concentration of NO_3^--N and NH_4^+-N were obviously increased, and NO_2^--N was decreased.

3.2 Spatial structure features and distribution

3.2.1 Spatial trend analysis

Figure 2 shows the spatial trend of NO_3^--N, NO_2^--N, and NH_4^+-N for wet season. It is seen that NO_2^--N and NH_4^+-N have no obvious spatial trend as high randomness was found in spatial distribution. NO_3^--N showed "S" in general trend, which meant the 3rd order trend effect, i.e., lower concentration in central areas and higher concentration in southern and northern areas.

3.2.2 Spatial variation analysis

According to fitted curves calculated by geostatistics software GS+9 (Fig. 3), semivariance of NH_4^+-N and NO_2^--N were less fitted to liner model, which meant poor spatial correlation and high spatial variability; while semivariance of NO_3^--N was well fitted to spherical model (fitting degree was 0.71) which meant good spatial correlation.

3.2.3 Spatial distribution characteristics and pollution assessment of ammonia-N, nitrate-N, and nitrite-N

According to the semivariogram model, using Kriging method to estimate the non observation points, then calculating the spatial distribution of ammonia-N, nitrate-N, and nitrite-N all over the island. Figure 4 shows the pollution areas of ammonia-N, nitrate-N, and nitrite-N in the wet and dry season, respectively. Obvious variations between wet and dry season were found.

In the dry season, a total of 44.7 km² areas was polluted by NH_4^+-N, mainly located in south, mid-east, and north. In the wet season, 39.1 km² was polluted and mainly located in mid-east and north. NH_4^+-N is not a direct risk to human health; however, the existence of NH_4^+-N means the groundwater is subjecting to new pollutions (Xie et al. 2011), and high concentration of NH_4^+-N means serious domestic pollution or agriculture pollution (Umezawa et al. 2008).

The pollution of NO_2^--N in wet season was heavier than in dry season as 70.5 km² areas was polluted in wet season, which was nearly two times than in dry season, it may because more rain in wet season promoted the denitrification reaction.

In dry season, the high concentration of NO_3^--N was found all over the island except for high attitude area, and the highest value was found in the southern most, which reached 72.5 mg/L. In wet season, the average concentration, maximum concentration and pollution area were all

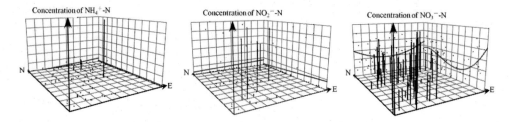

Figure 2. Trend analysis of ammonia-N, nitrite-N and nitrate-N in groundwater of Dongshan Island for wet season, respectively.

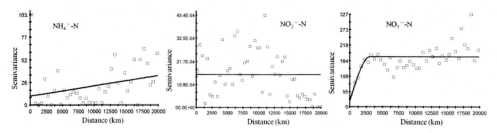

Figure 3. Semivariogram model of ammonia-N, nitrate-N, and nitrite-N in the groundwater of Dongshan Island.

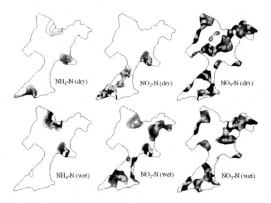

Figure 4, Pollution areas of ammonia-N, nitrate-N and nitrite-N exceeding the Third class standard of Chinese quality status for groundwater in dry and wet season.

lower than in dry season. It was attributed to the rain in wet season diluted the concentration of NO_3^--N, and denitrification reaction also reduced the concentration of NO_3^--N.

4 POLLUTION SOURCE OF AMMONIA-N, NITRATE-N, AND NITRITE-N AND THEIR TRANSFORMATION INFLUENCE FACTORS

4.1 *Pollution source of ammonia-N, nitrate-N, and nitrite-N*

Knowing the pollution source is very important for the controlling of nitrogen pollution of groundwater (Zhao et al. 2010). As exhibited in Table 2, NO_3^--N was significantly positive correlated to Cl^-, and so were NO_2^--N and NH_4^+-N. This meant the domestic pollutants and human and animal wastes were the main sources of nitrogen pollution (Wang et al. 2013).

Table 2. Correlation analysis between ammonia-N, nitrate-N, nitrite-N, and other environmental factors.

	NO_2^--N	NH_4^+-N	Depth	pH	DO	Cl$^-$
NO_3^--N						
Correlation coefficient	0.040	−0.103	0.167	0.354**	0.186*	0.308**
Sig.	0.642	0.236	0.053	0.000	0.031	0.000
Number of samples	135	135	135	111	135	135
NO_2^--N						
Correlation coefficient		0.466**	−0.193*	0.256**	−0.181*	0.183*
Sig.		0.000	0.025	0.007	0.035	0.034
Number of samples		135	135	111	135	135
NH_4^+-N						
Correlation coefficint			−0.401**	0.185	−0.184*	0.233**
Sig.			0.000	0.052	0.032	0.007
Number of samples			135	111	135	135

**Means significant at $p < 0.01$; *means significant at $p < 0.05$.

The land use types of Dongshan Island contain country, town, farmland (for vegetables), and woodland. The concentration of NO_3^--N, NO_2^--N and NH_4^+-N in each land use type were compared and nitrogen pollution in country and town was found the heaviest. The NO_3^--N was the highest in town, then country, farmland and woodland, and the NH_4^+-N in country was far higher than in other areas. This was consistent with the conclusion that was analyzed above. The domestic pollutants and human and animal wastes polluted groundwater attributed to the lack of sewage pipe networks and the high content of sand in soil.

4.2 *Transformation influence factors*

As listed in Table 2, NO_2^--N and NH_4^+-N were significantly negative correlated to groundwater depth, this meant the higher depth the lower concentration of NO_2^--N and NH_4^+-N. It was because the higher groundwater depth (i.e., thickness of unsaturated zone), the less groundwater was susceptible to pollution (Ma et al. 2012).

pH influenced the transformation among NH_4^+-N, NO_3^--N, and NO_2^--N as shown in Table 2. pH was significantly positive correlated to NO_3^--N and NO_2^--N. We compared the NO_3^--N and NO_2^--N in acidic groundwater (54 samples) and alkaline groundwater (56 samples). It was found that in alkaline groundwater, NO_2^--N was two times more than in acidic groundwater. Two reasons could account for this phenomenon. One reason was that the lower pH, the higher activity of nitrifying bacteria, whereas the higher pH, the higher activity of denitrifying bacteria (Chen et al. 2015); the other reason was relating to Fe^{2+}. Fe^{2+} is an important factor influencing the transformation between NO_3^--N and NO_2^--N. Due to the high reducibility of Fe^{2+}, in neutral or alkaline solution, NO_3^- would be reduced to NO_2^-. The concentration of Fe^{2+} in the groundwater of Dongshan Island was 0.03~12.0 mg/L (Shi 2008; Ma et al. 2008). Thus, the high concentration of Fe^{2+} and the alkaline environment resulted in the serious NO_2^--N pollution.

DO was also important as it was found significantly positive correlated to NH_4^+-N and negative correlated to NO_3^--N and NO_2^--N. It was because, the higher concentration of DO is benefit for nitration reaction that promotes NH_4^+-N transformed to NO_3^--N (Gomez et al. 2002); and at the same time, denitrification rate reduced, NO_3^--N is difficult to transform to NO_2^--N and N_2.

5 CONCLUSIONS

The nitrogen pollution of the groundwater of Dongshan Island was serious that about half of Dongshan Island was polluted by NO_3^--N, and the pollution became more serious in recent years. Similar distribution characteristic of NO_3^--N, NH_4^+-N, and NO_2^--N was found that the high concentration areas were all located in coastal land. Domestic pollutants and human and animal wastes from towns and villages were the main sources of nitrogen pollution, that this would be the first step to control the nitrogen pollution of Dongshan Island. Land use type, groundwater depth, pH, DO, season, and Fe^{2+} were all the factors that influence the distribution and transformation of ammonia-N, nitrate-N, and nitrite-N in groundwater. Thus, they were considerable factors in the control of nitrogen pollution in groundwater of Dongshan Island.

REFERENCES

Chen, J.P. & Mao, H.T. & Ding, J.Y. et al. 2015. Response mechanism of nitrogen fertilization to ammonia-nitrite-nitrates in groundwater. *Journal of Liaoning Technical University* 34(1): 118–123. (in Chinese).

Dragon, K. 2013. Groundwater nitrate pollution in the recharge zone of a regional Quaternary flow system (Wielkoposka region, Poland). *Environment Earth Science* 68(7): 2099–2199.

Gómez, M.A. & Hontoria, E. & González-López, J. 2002. Effect of dissolved oxygen concentration on nitrate removal from groundwater using a denitrifying submerged filter. *Journal of Hazardous Materials* 90(3): 267–278.

Jalali M. 2011. Nitrate pollution of groundwater in Toyserkan, western Iran. *Environment Earth Science* 62(5): 907–913.

Liu, C.W. & Wang, Y.B. & Jang, C.S. 2013. Probability-based nitrate contamination map of groundwater in Kinmen. *Environment Monitor Assessment* 185(12): 10147–10156. (in Chinese)

Ma, H.B. & Li, X.X. & Hu, C.S. 2012. Status of nitrate nitrogen contamination of groundwater in China. *Chinese Journal of Soil Science* 43(6): 1532–1536. (in Chinese)

Ma, R.X. & Zhang, Y.Z. & Lin, Z.F. & Duan, Y. 2008. Assessment on the groundwater quality of Dongshan Country and the pollution control measure. *Straits Science* 6: 29–32. (in Chinese)

Reddy, A.G.S. & Niranjan Kumar, K. & Subba Rao, D. et al. 2009. Assessment of nitrate contamination due to groundwater pollution on north eastern part of Anantapur District A.P. India. *Environment Monitor Assessment* 148(1–4): 463–476.

Salvador, P.-H. & Manuel, P.-V. Andres, S. 2009. A hydro-economic modeling framework for optimal management of groundwater nitrate pollution from agriculture. *Journal of Hydrology* 373(1–2): 193–203.

Shi, W.Y. 2004. The management and protection measures of groundwater in Dongshan Island, Fujian Province. *Technical Supervision in Water Resources* 5: 30–33. (in Chinese)

Umezawa, Y. & Hosono, T. Onodera, S. et al. 2008. Sources of nitrate and ammonium contamination in groundwater under developing Asian megacities. *Science of the Total Environment* 404(2–3): 361–376.

Wang, Q.S. & Gu, Y. & Sun, D.B. 2014. Spatial and seasonal variations of nitrate-N concentration in groundwater within Chao Lake watershed. *Acta Ecologica Sinica* 34(15): 4372–4379. (in Chinese)

Wang, X.M. & Wang, L.L. & Wu, B.R. et al. 2013. Distribution features and polluting source analysis of nitrate in shallow groundwater in the Huaibei plain, Anhui. *Geology of Anhui* 23(2): 142–145. (in Chinese)

Xie, J.H. & Liu, H.J. & Wang, A.W. 2011. Study on the transformation of ammonia-N, nitrate-N and nitrite-N and the role of ammonia-N in assessment and control of water pollution. *Inner Mongolia Water Resources* 5: 34–36. (in Chinese)

Zhao, J.C. & Li, Y.C. & Yamashita, I. et al. 2010. Summary on deduction and trace the source methods for ground water nitrate contamination. *Chinese Agricultural Science Bulletin* 26(18): 374–378. (in Chinese)

Rapid quantify HSPs mRNA in *Sebastiscus marmoratus* exposed to crude oil by LAMP

R. Chen, Z.P. Mo & Z.Z. Li
College of the Environment and Ecology, Xiamen University, Xiamen, Fujian, China

ABSTRACT: Loop-Mediated Isothermal Amplification (LAMP) is a novel and rapid nucleic acid amplification method. We used this method to develop a simple and rapid way to quantify the Heat Shock Protein (HSP) genes of *Sebastiscus marmoratus*. After exposure to different concentrations of the water-soluble fraction of crude oil for 1, 5 and 10 days, the levels of HSP60, HSP70 and HSP90 mRNA in the liver of *S. marmoratus* were measured using Real-Time PCR (RT-PCR) and Real-Time LAMP (RT-LAMP). The change patterns of all three gene levels measured using RT-LAMP were the same as those measured using RT-PCR, indicating that the RT-LAMP is an effective method for monitoring HSP mRNA. The HSP mRNA levels in the liver of *S. marmoratus* were all inhibited after exposure, which is different from the results of various other studies concerning HSPs.

1 INTRODUCTION

Heat Shock Proteins (HSPs) are a family of proteins that help organisms to modulate stress responses and to protect the organisms from environmentally induced cellular damage. These proteins are classified into several families based on their apparent molecular mass, such as HSP90 (85–90 kDa), HSP70 (68–73 kDa), HSP60, HSP47, and low molecular mass HSPs (16–24 kDa). HSPs are indicators of stress or exposure to toxicants (Mallouk et al., 1999). Several studies report the induction of fish HSPs by virus (Dong et al., 2006) and bacterial infections (Fu et al., 2011), including HSP70 s, HSP90 s and HSP60 s, and expression of HSP70 at high intensity in the tissues of fish collected from a metal polluted site compared to a less polluted site is noted (Rajeshkumar et al., 2011).

Loop-Mediated Isothermal Amplification (LAMP) is a novel method for the amplification of DNA (Notomi et al., 2000). The real-time monitoring of LAMP reactions through the use of a thermal cycler or turbidimeter allows for the potential to quantify the original amount of lambda DNA or RNA (Nagamine et al., 2002; Mori et al., 2004). A real-time turbidimetry-based method is cost-effective when compared with the present techniques of RT-PCR. However, few attempts have been made to quantitatively detect biomarker gene and thus, little is known concerning the effectiveness of RT-LAMP in quantitatively monitoring environmental pollution.

The objective of this study was to establish a convenient and accurate method (RT-LAMP) to measure the biomarkers gene in rockfish (*Sebastiscus marmoratus*) for monitoring the WSFs of crude oil.

2 MATERIALS AND METHODS

2.1 Fish

The *S. marmoratus*, weighing 20.89 ± 0.80 g, were obtained from market and transferred to glass aquaria (250 L in volume) filled with filtered seawater for 7 d to acclimate. During the experimental period, the water temperature and salinity were maintained at 18 ± 1 °C and 22–24.

2.2 Treatments and sampling

The Water-Soluble Fraction (WSF) of Arabia AM crude oil was prepared using the following procedure: the oil and clean sand filtered seawater were mixed at a volume ratio of 1:10, and continuously stirred for 4 h, followed by standing for 16 h. The lower aqueous phase was separated as the WSF mother liquor, stored at 4 °C.

The concentration of WSF was measured using a fluorescence spectrophotometry following "The Specification for Marine Monitoring-Part 4. Seawater Analysis" (GB 17378.4-2007). The excitation wavelength was 310 nm and the emission wavelength was 365 nm, and the National Marine Environmental Monitoring Center 20-3 oil was used as a standard.

The fish (n = 25 of both sexes for each concentration of WSF) were exposed to three concentrations of WSF (shown in Table 1) with a control group. The seawater was changed every day, and the fish were given feed for two hours before replacing the water. The fish were sampled after exposure on days 1, 5 and 10. The livers were isolated from six fish at random and immediately frozen in liquid nitrogen until analysis.

2.3 Construction of primers

The primers for LAMP assay were designed by the online LAMP primer designing software PrimerExplorer V4 (http://primerexplorer.jp/e/). Primer Premier 5.0 was used to design the primers for RT-PCR assay.

2.4 RT-PCR analysis

Total RNA collected from the liver of *S. marmoratus* were extracted with Trizol (Takala, Japan), and 2 μg of total RNA was used to synthesize first-strand cDNA. The fluorescence RT-PCR assay was performed using a Mx3000P (Stratagene, USA). The GAPDH gene was used as the internal control. The thermal profile included the following: 95 °C for 2 min followed by 40 cycles of 95 °C for 20 s, 63 °C for 20 s, 72 °C for 20 s. Dissociation curve analysis of the amplification products was performed at the end of each PCR reaction for confirmation of the amplification of a single PCR product. The Relative Expression Software Tool (version 2) was used to calculate the relative expression of mRNA target genes in the RT-PCR.

2.5 RT-LAMP analysis

The RT-LAMP reaction was carried out in a 25 μL reaction mixture containing 2.4 μM of each of the Forward Inner Primer (FIP) and the Backward Inner Primer (BIP), 0.4 μM of each of the forward outer primer (F3) and the backward outer primer (B3), 8 U of Bst DNA polymerase, 1.4 mM of each of the dNTPs, 10 mM of KCl and $(NH_4)_2SO_4$, 20 mM of Tris–HCl buffer (pH 8.8), 0.1% of Triton X-100, 0.4 M of Betaine, 4 mM of $MgSO_4$, and 2 μL of target DNA. The reaction mixture was incubated at 64 °C for 60 min in a Loopamp real-time turbidimeter (LA-320; Eiken, Japan), and heated at 80 °C for 2 min to terminate the reaction.

2.6 Statistical analysis

All the data are expressed as means ± Standard Error (SE). The data were analyzed with one-way analysis of variance via SPSS 17.0. A value of $P < 0.05$ was used to indicate significant difference between the control and exposed group.

Table 1. WSF concentration of crude oil in the experiment.

Experimental group	1	2	3	4
Preparation concentration (μg L^{-1})	0	20	60	180
Actual concentration (μg L^{-1})	7.6	18	54	144.5

3 RESULTS

The results of RT-LAMP were consistent with those of RT-PCR for the expression HSPs (Fig. 1), in that the HSPs were all inhibited to a certain extent, with HSP60 being the most notable. On day 1, the expression of HSP60 was severely inhibited while the expression of HSP70 and HSP90 did not show a significant change. The expression of HSP70 was inhibited on day 5 in the 60 $\mu g \cdot L^{-1}$ group and day 10 in the 180 $\mu g \cdot L^{-1}$ group. Distinct reduction in the HSP90 level was observed only on day 10 in the 20 $\mu g \cdot L^{-1}$ group.

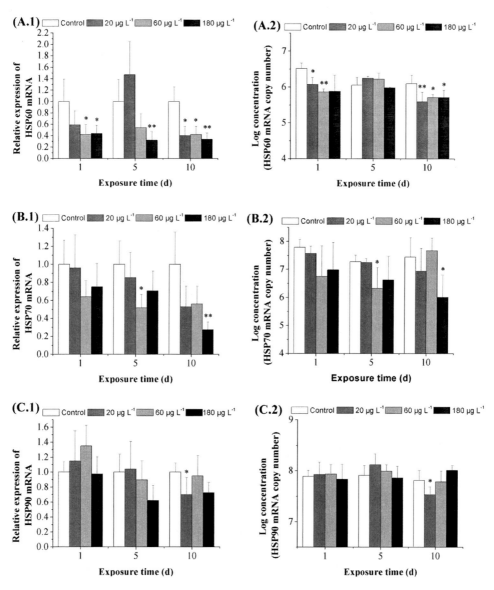

Figure 1. The mRNA expression of hepatic HSPs in *Sebastiscus marmoratus* exposed to WSF. Asterisks (*) indicate that the values are significantly different from that of the control at $P < 0.05$ (*), $P < 0.01$ (**) and $P < 0.001$ (***). (A.1) the RT-PCR results of HSP 60; (B.1) the RT-PCR results of HSP 70; (C.1) the RT-PCR results of HSP 90; (A.2) the RT-LAMP results of HSP 60; (B.2) the RT-LAMP results of HSP 70; (C.2) the RT-LAMP results of HSP 90.

4 DISCUSSION

The expression levels of HSPs mRNA in samples were quantified using both RT-PCR and RT-LAMP assay, and a similar trend might be visible. These results indicated that turbidity-based RT-LAMP is applicable to the evaluation and monitoring of oil pollution.

Compared to RT-PCR, RT-LAMP has the potential to be widely used due to its simplicity and cost-effectiveness. Many studies use LAMP in detecting target genes from bacteria, viruses or parasites (Gadkar et al., 2008; Buates et al., 2010; Wang et al., 2012; Lin et al., 2012; Du et al., 2012). Attempts have also been made to quantitatively detect functional gene targets. The real-time Reverse-Transcription LAMP (rRT-LAMP) method is also used for quantifying human astrovirus (Wei et al., 2013). However, the application of this method to the field monitoring of environmental pollution has not been reported.

HSPs are also considered to be a potential petroleum hydrocarbon target although their exact role in protection from the effects of petroleum hydrocarbons is not fully understood (Downs et al., 2001; Lüchmann et al., 2011). Many PAHs induce HSP expression in bivalves (Porte et al., 2001; Monari et al., 2011), however, the WSF inhibited the expression of HSP60, HSP70 and HSP90 in our study.

HSP60 is primarily known as a mitochondrial protein important for folding key proteins after import into the mitochondria and has been extensively studied under pathological conditions (Hwang et al., 2009; Castilla, et al., 2010). HSP60 is thought to be strongly induced in marine invertebrates by petroleum hydrocarbons (Wheelock et al., 2002; Lüchmann et al., 2011). However, Stegeman et al. (1992) noted that HSP60 levels do not remain permanently elevated after heat shock but rather decline with time. So it is not surprising to find that the expression of HSP60 mRNA was strongly inhibited in our study.

The levels of HSP90 and HSP70 mRNA were significantly inhibited on day 5 in the 60 $\mu g \cdot L^{-1}$ group and on day 10 in the 180 $\mu g \cdot L^{-1}$ group. This might have occurred for two main reasons. First, the decrease might have occurred because of a decrease in the metabolic capacity of the organism caused by the strong toxicity of WSF. Lai et al. (1984) report that HSP90 has basic physiological roles, regardless of exposure to stress, so it is less likely to be induced by pollutants. Weber & Janz (2001) suggested that at 96 h post-injection, β-naphthoflavone and dimethylbenz[a]anthracene significantly decrease HSP70 expression in juvenile catfish ovaries. Rhee et al. (2009) also found that 4-nonylphenol and 4-t-octylphenol caused down-regulation of HSP70 expression in *Tigriopusjaponicus*. Whereas defense mechanisms function under weak oxidative stress, organisms cannot achieve appropriate levels of metabolic function under strong oxidative stress (Zhang et al. 2004). Strong stress would be generated when the WSF accumulates in the tissues beyond a certain threshold concentration. Therefore, the metabolic capacity and HSP70 and HSP90 mRNA expression level would decrease accordingly. Secondly, the decrease in HSP90 expression might be due to the WSFs of crude oil inhibiting the synthesis of HSP90 by some alteration at a transcriptional or translational level, or by inhibition of the signal-transduction cascade for the induction of HSP90 (Downs et al., 2001). Lüchmann et al. (2011) also report that diesel fuel lead to a significant dose-dependent inhibition of HSP90 levels in the gill and digestive gland of *Crassostreabrasiliana*.

In conclusion, we established RT-LAMP for the analysis of the expression of HSPs. In order to ultimately judge the RT-LAMP method as a tool to measure biomarker genes, future studies are needed.

ACKNOWLEDGMENTS

This work was financially supported by the Ocean Public Welfare Scientific Research Special Appropriation Project of China (201005016).

REFERENCES

Buates, S., Bantuchai, S., Sattabongkot, J., et al. 2010. Development of a Reverse Transcription-Loop-Mediated Isothermal Amplification (RT-LAMP) for clinical detection of *Plasmodium falciparum* gametocytes. *Parasitology International* 59(3): 414–420.

Castilla, C., Congregado, B., Conde, J.M., et al. 2010. Immunohistochemical expression of Hsp60 correlates with tumor progression and hormone resistance in prostate cancer. *Urology* 76(4): 1017.e1–1017.e6.

Dong, C.W., Zhang, Y.B., Zhang, Q.Y., et al. 2006. Differential expression of three *Paralichthysolivaceus* Hsp40 genes in responses to virus infection and heat shock. *Fish and Shellfish Immunology* 21(2): 146–158.

Downs, C.A., Dillon, J.R.T., Fauth, J.E., et al. 2001. A molecular biomarker system for assessing the health of gastropods (*Ilyanassaobsoleta*) exposed to natural and anthropogenic stressors. *Journal of Experimental Marine Biology and Ecology* 259(2): 189–214.

Du, F., Feng, H.L., Nie, H., et al. 2012. Survey on the contamination of *Toxoplasma gondii* oocysts in the soil of public parks of Wuhan, China. *Veterinary Parasitology* 184(2–4): 141–146.

Fu, D.K., Chen, J.H., Zhang, Y., et al. 2011. Cloning and expression of a Heat Shock Protein (HSP) 90 genes in the haemocytes of *Crassostreahongkongensis* under osmotic stress and bacterial challenge. *Fish and Shellfish Immunology* 31(1): 118–125.

Gadkar, V., Rillig, M.C., 2008. Evaluation of Loop-Mediated Isothermal Amplification (LAMP) to rapidly detect arbuscularmycorrhizal fungi. *Soil Biology and Biochemistry* 40: 540–543.

Hwang, Y.J., Lee, S.P., Kim, S.Y., et al. 2009. Expression of heat shock protein 60 kDa is upregulated in cervical cancer. *Yonsei Medical Journal* 50(3): 399–406.

Lai, B.T., Chin, N.W., Stranek, A.E., et al. 1984. Quantitation and intracellular localization of the 85 k heat shock protein by using monoclonal and polyclonal antibodies. *Molecular Cell Biology* 4(12): 2802–2810.

Lin, Z.B., Zhang, Y.L., Zhang, H.S., et al. 2012. Comparison of Loop-Mediated isothermal amplification (LAMP) and real-time PCR method targeting a 529-bp repeat element for diagnosis of toxoplasmosis. *Veterinary Parasitology* 185(2–4): 296–300.

Lüchmann, K.H., Mattos, J.J., Siebert, M.N., et al. 2011. Biochemical biomarkers and hydrocarbons concentrations in the mangrove oyster *Crassostreabrasiliana* following exposure to diesel fuel water-accommodated fraction. *Aquatic Toxicology* 105(3–4): 652–660.

Mallouk, Y., Vayssier-Taussat, M., Bonventre, J.V., et al. 1999. Heat shock protein 70 and ATP as partners in cell homeostasis (Review). *International Journal of Molecular Medicine* 4: 463–537.

Marcheselli, M., Azzoni, P., Mauri, M., 2011. Novel antifouling agent-zinc pyrithione: Stress induction and genotoxicity to the marine mussel *Mytilusgalloprovincialis*. *Aquatic Toxicology* 102(1–2): 39–47.

Monari, M., Foschi, J., Rosmini, R., et al. 2011. Heat shock protein 70 response to physical and chemical stress in *Chameleagallina*. *Journal of Experimental Marine Biology and Ecology* 397(2): 71–78.

Mori, Y., Kitao, M., Tomita, N., et al. 2004. Real-time turbidimetry of LAMP reaction for quantifying template DNA. *Journal of Biochemical and Biophysical Methods* 59(2): 145–157.

Nagamine, K., Hase, T., Notomi, T., 2002. Accelerated reaction by loop-mediated isothermal amplification using loop primers. *Molecular and Cellular Probes* 16(3): 223–229.

Notomi, T., Okayama, H., Masubuchi, H., et al. 2000. Loop-mediated isothermal amplification of DNA. *Nucleic Acids Research* 28(12): 63.

Porte, C., Biosca, X., Solé, M., et al. 2001. The integrated use of chemical analysis, cytochrome P450 and stress proteins in mussels to assess pollution along the Galician coast (NW Spain). *Environmental Pollution* 112(2): 261–268.

Rajeshkumar, S. & Munuswamy N., 2011. Impact of metals on histopathology and expression of HSP 70 in different tissues of Milk fish (*Chanoschanos*) of Kaattuppalli Island, South East Coast, India. *Chemosphere* 83(4): 415–421.

Rhee, J.S., Raisuddin, S., Lee, K.W., et al. 2009. Heat shock protein (*Hsp*) gene responses of the intertidal copepod *Tigriopusjaponicus*to environmental toxicants. *Comparative Biochemistry and Physiology-Part C: Toxicology & Pharmacology* 149(1): 104–112.

Stegeman, J.J., Brouwer, M., Di Giulio, R.T., et al. 1992. Molecular responses to environmental contamination: enzyme and protein systems as indicators of chemical exposure and effect. In: Huggett, R.J., Kimerle, R.A., Mehrle, P.M., Bergman, H.L. (Eds.), *Biomarkers, Biochemical, Physiological, and Histological Markers of Anthropogenic Stress*. Lewis Publishers, USA, pp. 235–335.

Wang, X., Zhu, J.P., Zhang, Q., et al. 2012. Detection of enterovirus 71 using Reverse Transcription Loop-Mediated isothermal amplification (RT-LAMP). *Journal of Virological Methods* 179(2): 330–334.

Weber, L.P. & Janz, D.M. 2001. Effect of β-naphthoflavone and dimethylbenz[a]anthracene on apoptosis and HSP70 expression in juvenile channel catfish (*Ictalurus punctatus*) ovary. *Aquatic Toxicology* 54(1–2): 39–50.

Wei, H.Y., Zeng, J., Deng, C.L., et al. 2013. A novel method of real-time reverse-transcription loop-mediated isothermal amplification developed for rapid and quantitative detection of human astrovirus. *Journal of Virological Methods* 188(1–2): 126–131.

Wheelock, C.E., Baumgartner, T.A., Newman, J.W., et al. 2002. Effect of nutritional state on HSP60 levels in the rotifer *Brachionusplicatilis* following toxicant exposure. *Aquatic Toxicology* 61(1–2): 89–93.

Zhang, J.F., Wang, X.R., Guo, H.Y., et al. 2004. Effects of water-soluble fractions of diesel oil on the antioxidant defenses of the goldfish, *Carassiusauratus*. *Ecotoxicology and Environmentai Safety* 58(1): 110–116.

Analysis and evaluation of nutrition composition of *Clinacanthus nutans*

Qun Yu & Zhen-hua Duan
College of Food Science and Technology, Hainan University, Haikou, Hainan, China

Wei-wen Duan & Fei-fei Shang
Hezhou University, Hezhou, Guangxi, China

Guo-xian Yang
Hainan Wuzhishan Wanjiabao Science and Technology Company limited, Wuzhishan, Hainan, China

ABSTRACT: In order to understand the main chemical constituents, such as amino acids and trace elements of *Clinacanthus nutans*, the leaves of *C. Nutans* were analyzed in this research. The results showed that the content of protein, crude fiber, moisture, fat, vitamin C, and vitamin B1 in the leaves of *C. Nutans* were 5.73%, 2.71%, 78.3%, 0.5%, 1.57, and 0.27 mg/100 g respectively. Compared with the common vegetables, such as lettuce, the *C. Nutans* have more protein. The value of EAA/TAA and EAA/NEAA both met the FAO/WHO standards. In addition, the *C. Nutans* are a good source of zinc and iron. Thus, we conclude that the *C. Nutans* is a kind of wild vegetable, which has a high nutritional value and medicinal value.

1 INTRODUCTION

Clinacanthus nutans is also known as sabah snake grass. It is called as "bone grass" in Wuzhishan city of Hainan in China. *C. Nutans* is a kind of tall grass that grows in the thin forests of low altitude near the equator or thickets in sandy soil. It is widely distributed in south-east Asia, and there are wild *C. Nutans* in Hainan, Guangdong, Guangxi, and Yunnan Province (Lin et al. 1983; Xu et al. 2014). It has effect on clearing heat, diuresis detumescence with wet, invigorating the circulation of hydrophobic, clearing moisture, and anti-tumor. The C. Nutans is often used as medicine from the whole plant or leaf. In Thailand, they are used to prevent insects or snake and become a kind of widely used medicinal plants (Wang et al. 2013).

We have learned that the *C. Nutans* usually is used as vegetable and tea drinking in Wuzhishan city. Related products are mainly in Hainan, and they include *C. Nutans* powder, *C. Nutans* tea, and *C. Nutans* leaves. According to the literature, the *C. Nutans* contains triterpenes, flavone glycoside, glucoside and thioglycoside, and so on (Yi et al. 2012). Most of these ingredients show anti-tumor effects in the pharmacological experiments. It contains abundant nutrients, that is, the basis for food processing. The nutrient content of *C. Nutans* was analyzed in this paper, and the amino acid composition and trace element were studied. Its aim is to promote the processing and utilization of *C. Nutans*.

2 MATERIALS AND METHODS

2.1 *Materials and chemicals*

Clinacanthus nutans, Wuzhishan Wanjiabao science and technology company limited in Hainan; Waters AccQ·Fluor boric acid buffer, the Waters reagent company in America;

AQC (new amino acid derivatization reagent, Waters AccQ·Fluor derivative agent powder 2A and Waters AccQ·Fluor derivative agent diluent 2B), the Waters reagent company in America; the standard amino acid reagents, the Waters reagent company in America; hydrochloric acid, nitric acid and perchlorate acid, Longhui trading company limited in Haikou; methanol, acetone and chloroform, Longhui trading company limited in Haikou.

2.2 Instruments preparation

HP2695 high performance liquid chromatography system (with Waters-2475 fluorescence detector), the Waters company in America; PHS-3C laboratory pH meter, Weiye instrument company in Shanghai; VP50 vacuum pump, Laibo Taike instrument company limited in Beijing; hydrolysis tube (200 ml), Longhui trading company limited in Haikou; vortex generator, Germany IKA vortex generator; trace pipetting device, YE3 K116260 YE4 A175630; Half a type high efficiency Liquid chromatography, Shimadzu company in Japan; Electrospray mass spectrometer API 2000 LC/MS/MS, Applied Biosystems company in America; and Superconducting magnetic resonance imaging apparatus, Bruker company in Swiss.

2.3 The conditions of High Performance Liquid Chromatography (HPLC)

The chromatographic column is Waters AccQ·Tag amino acid analysis column. The mobile phase A is acetate phosphate buffer. The mobile phase B is acetonitrile and high pure water. The flow velocity is 1.0 ml/min. The column temperature is 37 °C. The detection wavelength is 395 nm.

2.4 Methods

2.4.1 Determination method of conventional nutrients

The determination of moisture is based on GB5009.3-2010; the determination of fat is based on the soxhlet extraction method of GB/T5009.6-2003; the determination of crude fiber is based on GB/T5009.10-2008; the determination of protein is according to the jeldah method of GB5009.5-2010; the determination of vitamin C is based on GB6195-1986; and the determination of vitamin B_1 is based on GB/T5009.84-2003.

2.4.2 Determination method of the amino acids

The amino acids of the sample were determined using a Waters HPLC system. According to the atlas, we take an external standard method to calculate the amino acids content of every 100 g sample: 200 mg of the *C. Nutans* powder were taken into the hydrolysis tube, and then 10 mL, 6 mol/L of hydrochloric acid was added into it. After vacuuming a period of time, it was put into the oven to 110 °C after 22 h hydrolysis. After vortex mixing, 70 ul buffer 1 and 20 ul AQC derivative agent were added. Vortex mixing about 15 min, to seal the sample tube by paraffin wax membrane. Take the water bath in 55 °C for 10 min, and transfer it to into the low volume of tube-Insert. Finally, determinate the sample in high performance liquid chromatograph with a 2690 system. The standards and samples were subjected to HPLC analyses with a linear gradient system of solvents A and B. The gradient elution procedure was 100%A + 0%B at 0 min, 99%A + 1%B at 0.5 min, 95%A + 5%B at 18 min, 91%A + 9%B at 19 min, 83%A + 17%B at 29.5 min, 0%A + 60%B + 40%C at 35 min, 100%A at 38 min, and 100%A at 47 min.

2.4.3 Determination method of the trace elements

The determination of plumbum is based on the Graphite furnace atomic absorption spectrophotometry method of GB5009.12-2010; the determination of cadmium is based on the Graphite furnace atomic absorption spectrophotometry method of GB/T5009.15-2003; the determination of chromium is according to GB/T5009.123-2003; the determination of mercury is according to GB/T5009.17-2003; the determination of arsenic is based on the Hydride atomic fluorescence spectrophotometry method of GB/T5009.11-2003; the determination of zinc is based on the Flame atomic absorption spectrophotometry method of GB/T5009.14-2003; and the determination of iron is based on the Flame atomic absorption spectrophotometry method of GB/T5009.90-2003.

2.5 Statistical analysis

All the tests were carried out in triplicate. The experimental data were presented as means ± Standard Deviations (SD). The statistical analysis was performed using one-way analysis of variance (ANOVA). The significant difference was determined with 95% confidence interval ($P < 0.05$).

3 RESULTS AND DISCUSSION

3.1 Chemical composition of C. Nutans

Proximate compositions in the leaves of *C. Nutans* are shown in Table 1, and the other five plants' data are from references (Yang et al. 2002). The protein content of *C. Nutans* was around 5.73%, which is about five times than in celery and Chinese cabbage. What's more, the protein content of *C. Nutans* is also about two times than in spinach, flowering cabbage, and lettuce. Protein is necessary for human, and the sticky protein among them can prevent the fatty deposits in cardiovascular and the atrophy of connective tissue in liver and kidney. It keeps arteries elastic and maintains the lubrication of respiratory tract.

The fat content of 0.5% was observed in *C. Nutans*, and that was similar than that found in lettuce and flowering cabbage. However, it was higher than that in celery and spinach. The results also showed that *C. Nutans* has higher crude fiber content (2.71%) when compared to the crude fiber content of other five plants. Cellulose itself has no nutrition; however, it is the indispensable nutritional ingredients. It can promote gastrointestinal peristalsis and increase the function of alimentary canal. At the same time, it also can prevent the absorption of cholesterol by combination or releasing the fatty acids produced by fermentation, and the crude fiber can decrease the heat energy, sugar and fat of food, so as to inhibit the occurrence of tumors, cerebral arteriosclerosis, ischemic heart disease, diabetes, and other diseases (Gong et al. 2010).

The *C. Nutans* moisture content was 78.3%; however, the moisture content of the other five plants was above 90%. Besides, vitamin C was found to be 1.57 mg/100 g, and vitamin B_1 was found to be 0.27 mg/100 g. The vitamin C content of lettuce, celery, spinach, and Chinese cabbage is higher than that in *C. Nutans*, but their vitamin B_1 content is lower. Vitamin B_1 can inhibit cholinesterase for the hydrolysis of acetylcholine. Acetylcholine plays an important role in promoting gastrointestinal peristalsis. Therefore, lacking vitamin B_1 leads to gastrointestinal peristalsis slowly, the decrease of glands and loss of appetite. The lacking of vitamin B_1 also can cause beriberi (Public Nutritionist 2012). *C. Nutans* can make up for the shortcomings of other five vegetables in the vitamin imbalance. So, the *C. Nutans* is a new kind of wild vegetable that contains high protein and balanced vitamin.

Table 1. The main chemical constituents of six kinds of vegetables.

Composition	Protein %	Fat %	Crude fiber %	Moisture %	Vitamin C mg/100 g	Vitamin B_1 mg/100 g
C. Nutans	5.73 ± 0.14	0.50 ± 0.02	2.71 ± 0.05	78.30 ± 0.29	1.57 ± 0.07	0.27 ± 0.04
Lettuce	1.60 ± 0.04	0.40 ± 0.01	1.10 ± 0.01	96.70 ± 1.05	13.00 ± 0.13	0.06 ± 0.00
Celery	0.80 ± 0.01	0.10 ± 0.00	1.40 ± 0.03	95.00 ± 0.94	12.00 ± 0.23	0.08 ± 0.00
Spinach	2.60 ± 0.07	0.30 ± 0.01	2.10 ± 0.02	92.60 ± 1.21	32.00 ± 0.54	0.11 ± 0.01
Chinese cabbage	1.20 ± 0.02	0.20 ± 0.00	0.90 ± 0.01	95.90 ± 1.53	19.00 ± 0.37	0.03 ± 0.00
Flowering cabbage	2.20 ± 0.01	0.50 ± 0.02	0.80 ± 0.01	92.50 ± 0.87	Tr	–

3.2 Amino acids analysis

The amino acid content of *C. Nutans* is shown in Table 2. Seventeen kinds of amino acids were tested in *C. Nutans*. Methionine, Valine, Lysine, Isoleucine, Phenylalanine, Leucine, and Threonine are necessary the amino acids for human body. Histidine is an essential amino acid for baby body. Tryptophan was completely destroyed because of the 24-hour acid hydrolysis, so it did not be detected. From Table 2, we can know that EAA/TAA has a value of 0.430 and EAA/NEAA has a value of 0.754. They are bigger than the ideal mode of FAO/WHO (EAA/TAA's value is 0.4; EAA/NEAA's value is 0.6).

In measured amino acids, essential amino acid content is rich, and the Lysine content is the highest, accounting for 27.4% of all amino acids. Lysine is the first limiting amino acid in some place that people only eat rice flour for food (Wang 1988), and it plays an important role in protein metabolism and inhibiting viral infection. The content of Valine, Leucine and Isoleucine is also higher. They work together to repair muscles, control blood sugar and provide energy for the body. They are essential for heavy manual labor and physical exercisers (Meng et al. 2011). Usually people can get it by beans and meat. It seems that *Clinacanthus nutans* can be used as a new food that provides essential amino acids.

In addition to abundant essential amino acids, there are some medical amino acids such as Aspartate, Glutamate, and Arginine. Glutamate is the neurotransmitter precursors and it could remove the toxicity of ammonia to the brain, and promote the generation of gastrointestinal mucosa mucin. It can be regarded as the medicine against peptic ulcer. Arginine can help to defense against male infertility due to the lack of sperm, contribute to growth and development, enhance immunity and promote wound healing (Duan et al. 2006). Besides the medicinal effects, it was abundant in seasoning amino acids such as Alanine. From Table 2, the content of Alanine is the highest, accounting for 28.82%. It can be added to food as nutritional supplements and improve the artificial flavoring taste. In a word, the *C. Nutans* has an irreplaceable role in developing pro-environmental seasoning.

3.3 Minerals and trace elements contents of C. Nutans

Table 3 shows the determination results of trace elements. Lettuce and Flowering cabbage's data were taken from references (Jiang et al. 2012). Table 3 illustrates that the zinc and iron contents of *C. Nutans* are very high, accounting for 13.6 mg/kg and 11.50 mg/kg. Compared with lettuce and flowering cabbage, the zinc content of *C. Nutans* is 6 times than lettuce and 2 times than flowering cabbage. The zinc plays an important role in maintaining appetite, enhancing immunity, and promoting wound healing. The iron content of *C. Nutans* is 2 times than lettuce. Iron is closely associated with iron deficiency anemia, known as "Blood element". It is a necessary element for body to make blood and it is a suitable activator in the synthesis of hemoglobin. It seems that *C. Nutans* can serve as good source of zinc and iron.

As shown in Table 3, the *C. Nutans* contains trace amounts of plumbum, chromium, cadmium, mercury, and arsenic. This may be associated with the soil of *C. Nutans* grow.

Table 2. The amino acid content of *Clinacanthus nutans*.

Amino acid	Content/mg	Amino acid	Content/mg
Aspartic	248.32 ± 0.24	Isoleucine	103.68 ± 0.38
Threonine	57.60 ± 0.17	Leucine	136.96 ± 0.52
Serine	84.48 ± 0.32	Tyrosine	42.24 ± 0.14
Glutamate	203.52 ± 1.13	Phenylalanine	60.16 ± 0.12
Glycine	162.32 ± 0.94	Lysine	942.08 ± 2.54
Alanine	989.44 ± 2.79	Histidine	152.32 ± 0.85
Cysteine	11.02 ± 0.09	Arginine	90.88 ± 0.75
Valine	152.32 ± 0.77	Proline	135.68 ± 1.29
Methionine	23.04 ± 0.13		

Table 3. The trace elements content of three kinds of plants.

The element (mg/kg)	Zn	Fe	Pb	Cr	Cd	Hg	As
C. Nutans	13.60 ± 0.18	11.50 ± 0.94	0.045 ± 0.001	0.077 ± 0.007	0.057 ± 0.009	0.0023 ± 0.0003	0.026 ± 0.005
Lettuce	2.20 ± 0.07	4.90 ± 0.07	0.060 ± 0.001	–	0.018 ± 0.002	–	<0.01
Flowering cabbage	5.06 ± 0.12	14.10 ± 1.69	0.050 ± 0.006	–	0.035 ± 0.003	–	<0.01

Although the plumbum content is 0.045 mg/kg, it is lower than that in lettuce and flowering cabbage. According to GB4935-94, it is lower than the national standard (Tang et al. 2005). In a conclusion, it is safe to eat *C. Nutans*.

4 CONCLUSION

1. The *C. Nutans* is an important and nutritional wild vegetable in Hainan. The result shows that the contents of protein, crude fiber, moisture, fat, vitamin C and vitamin B1 in the leaves of *C. Nutans* were 5.73%, 2.71%, 78.3%, 0.5%, 1.57 and 0.27 mg/100 g, respectively. So, the *Clinacanthus nutans* is a new kind of wild vegetable that contains high protein and balanced vitamin.
2. The amino acids analysis result shows that seventeen kinds of amino acids were tested in *C. Nutans*. The value of EAA/TAA and the value of EAA/NEAA are bigger than the ideal mode of FAO/WHO. The Lysine and Alanine are abundant.
3. The determination result of trace elements shows that *C. Nutans* has higher contents of zinc and iron and it can serve as a good source of zinc and iron.

REFERENCES

China Employment Training Technical Instruction Center Organization. 2012. *Public Nutritionist*. Beijing. China Labor Social Security Publish.

Duan, Z.H. et al. 2006. Analysis and evaluation of Hainan sea eel swim bladder. *Food and Nutrition in China* 11:43–45.

Gong, J. et al. 2010. The new progress in the study of *celllose*. *Journal of Cellulose Science and Technology* 18:62–71.

Jiang, Y.Y. et al. 2012. Analysis and evaluation of Common Vegetables in Guangxi. *Food and Nutrition in China* 18:71–74.

Lin, J.T. et al. 1983. Studies about the chemical ingredients of Torsional flowers. *Chinese Herbal Medicine* 14:337.

Liu, X. et al. 2014. Inhibitive effect of Clinacanthus nutans (Burm. f.) Lindau n-butanol extracts on Heps hepatoma in mice. *Journal of Jiangsu University (Medicine Edition)* 24:211–215.

Meng S.P. et al. 2011. Physiological function and separation and purification technology of L-leucine. *Science and Technology of Food Industry* 32:441–443.

Tang Q.F. et al. 2005. Element content determination and its nutritional value evaluation of 7 kinds of wild vegetables in Dan Zhou ShangHai Vegetables 3:81–83.

Wang X. 1988. Lysine improve protein nutritional value of flour. *Journal of nutrition* 4:368–370.

Wang X. et al. 2013. Chemical composition ananlysis about *Clinacanthus nutans*'s antitumor effect. *Chinese pharmacy* 24:4104–4107.

Yang, Y.X. et al. 2002. *Chinese food ingredients*. Beijing:Beijing medical university publish.

Yi B. et al. 2012. Analysis of Amino Acids, Trace Elements and Chemical Constituents from the Leaves of *Clinacanthus nutans*. *Pharm J Chin PLA* 28:396.

Extraction by microwave-ultrasonic assisted enzymatic hydrolysis and functional properties of Insoluble Dietary Fiber from soy sauce residue

Wen Li & Tao Wang
Jiangsu Key Construction Laboratory of Food Resource Development and Quality Safe, Xuzhou Institute of Technology, Xuzhou, China

ABSTRACT: Insoluble Dietary Fiber (IDF) was extracted from soy sauce residue by microwave-ultrasonic assisted enzymatic hydrolysis for the first time, after single factor experiment and orthogonal experiment, the optimal extraction condition was obtained as follows: NaOH concentration of 5%, ratio of solvent to material of 25:1, ultrasonic power of 300 W, extraction time of 100 s, after hydrolyzed by 0.3% papain at 65 °C for 20 min, the extraction rate of IDF was 73.86%. As to functional properties, the extracted IDF had good SC (swelling capacity), WHC (water-holding capacity), OHC (oil-holding capacity) and Nitrite ion-absorbing capacity.

1 INTRODUCTION

Dietary Fiber (DF) means carbohydrate polymers with ten or more monomeric units, which are not hydrolyzed by the endogenous enzymes in the small intestine of humans (Cui et al., 2011). As it possesses different health benefits including improving postprandial glucose response, reducing caloric intake, and decreasing total and Low Density Lipoprotein (LDL) levels (Mengmei et al., 2015), it has been used to prevent heart disease, obesity, and cancers (Elleuch et al., 2013; Huang, Ye, Chen, & Xu, 2013). The dietary fiber ingredients will be categorized into three groups: Insoluble Dietary Fiber (IDF), Soluble Dietary Fiber (SDF), and resistant starch. IDF are plant or food materials that are metabolically inert, absorbing water throughout the digestive system and easing defecation (Cui et al., 2011). It was mainly composed of cellulose, hemicellulose, lignin, original pectin, and chitosan (Zheng, 2001).

In the past years, fiber source research has focused on cereals, tubers, fruits, vegetables, and algae (Kaushik & Singh, 2011; Meyer, Dam, & Lærke, 2009; Raghavendra et al., 2004). Currently, there is an interest for alternative DF sources. Soy sauce is a conventional condiment in Asia, using soybean and wheat as material. With the production of soy sauce, a large number of soy sauce residue is also retained, which is considered as a potential cheap biomass resource for being rich in water, protein, oil, and soy isoflavones (Yamamoto et al., 2006).

Unfortunately, most of soy sauce residue is not fully utilized and is discarded as industry waste. Because of the environmental problems, the treatment and utilization of Soy Sauce Residue (SSR) have been paid more and more attention. But except for protein, many kinds of ingredients of soy sauce residue were not made full use of. Therefore, a novel food processing technique is suggested to manufacture valuable additional products, such as IDF. IDF was extracted by various processes such as alkali dissolution and enzymatic hydrolysis assisted alkali dissolution (Zhu & Hu, 2010; Wang, 2009).

To our knowledge, there is no report on IDF extraction from soy sauce residue by microwave-ultrasonic assisted enzymatic hydrolysis method. In this study, IDF was extracted from soy sauce residue by microwave-ultrasonic assisted enzymatic hydrolysis. The optimal microwave-ultrasonic extraction parameters were obtained by using single factor experiment and orthogonal experiment. Furthermore, the functional properties of the exacted IDF were also examined.

The objective of this study was to provide a theoretical basis for the industrial extraction of IDF from soy sauce residue and assess its potential applications in functional foods.

2 MATERIALS AND METHODS

2.1 Materials and chemicals

Soy sauce residue with a moisture content of 14.26 (w/w, %) was obtained from Xuzhou Hengshun Wantong food Brewing Co. Ltd. (Jiangsu, China). It was dried at 60 °C, smashed and passed through a 60-mesh sieve. 10 ml acetic ether was added to 2 g of soy sauce residue powder in a 100 ml Bunsen beaker for 3 h, which was duly covered to avoid solvent loss. After filtration, the residue was air-dried overnight, when deoiled residue was obtained.

Papain (enzyme activity 3000 U/g) was purchased from Sinopharm Chemical Reagent Co., Ltd. (Beijing, China). All other chemicals used were of analytical grade.

2.2 IDF extraction by Microwave-ultrasonic assisted enzymatic hydrolysis

2 g of deoiled residue was added to 60 ml of 3% (m/v) NaOH solution in a 250 ml extraction flask, and placed in a microwave-ultrasonic synergistic extraction instrument (CW-2000, Shanghai Tuoxin Analytical instrument Co., Ltd., China). The mixture was treated with ultrasonic at 600 W and 400 W of microwave power for 100 s with intervals. After that, the mixture was centrifuged at 4000 rpm for 10 min. The sediment was mixed with 10 ml deionized water, adjusted the pH to 6.0 with 0.1 mol/L HCl, added 0.3% (w/w) papain, hydrolyzed at 65 °C for 20 min, and 100 °C for 5 min to inactivate the enzyme. Then the mixture was centrifuged at 4000 rpm for 10 min. Residues were washed with deionized water for several times to a neutral pH, dried at 60 °C for 5 h. Then it was whitened by H_2O_2, washed with deionized water, and vacuum dried. IDF extraction rate was determined by the following equation (1):

$$\text{Extraction rate (\%)} = \frac{m_1}{m_0} \times 100\% \quad (1)$$

where m_1 is the weight (g) of IDF after treatment; m_0 is the weight of deoiled soy sauce residue.

2.3 Optimization of the microwave-ultrasonic extraction conditions

2.3.1 The analysis of single factor experiment

2.3.1.1 *Effect of NaOH concentration on IDF extraction rate*
The extraction parameters were as follows: ratio of NaOH solution to material (v/m) 30:1, ultrasonic power 600 W and 400 W of microwave power, extraction time 100 s, hydrolyzed by 0.3% (w/w) papain at 60 °C for 20 min. The selected NaOH concentrations (m/v) were 3%, 4%, 5%, 6%, and 7%.

2.3.1.2 *Effect of ratio of solvent to material (v/m) on IDF extraction rate*
Based on the optimal NaOH concentration, the ratio of solvent to material (v/m) were set at 20:1, 25:1, 30:1, 35:1, 40:1. The other steps were the same as described above.

2.3.1.3 *Effect of microwave power on IDF extraction rate*
Based on the optimal NaOH concentration and ratio of solvent to material (v/m), microwave powers of 200 W, 300 W, 400 W, 500 W, 600 W, 700 W, 800 W were chosen, the other steps were the same as described above.

2.3.1.4 *Effect of extraction time on IDF extraction rate*
Based on the optimal NaOH concentration, ratio of raw material to solvent (m/v) and microwave power, the chosen extraction time were 50 s, 100 s, 150 s, 200 s, 250 s, the other steps were the same as described above.

2.3.2 Orthogonal experiment

On the basis of the single factor experiment, the optimal microwave-ultrasonic extraction parameter was determined by $L_9(3^4)$ orthogonal experiments to evaluate the combination effects of the four factors on IDF extraction rate. Four factors including NaOH concentration, ratio of solvent to material, microwave power, extraction time at three levels were listed in Table 1.

2.4 Functional properties of IDF

2.4.1 Swelling capacity (SC)

SWC of IDF was analyzed according to the method described by Zhang et al. with minor modifications (Zhang, Bai, & Zhang, 2011). 0.5 g of IDF samples was placed in a 10 ml measuring cylinder, the volumes (ml) of the samples were recorded, then 5 ml of distilled water was added. The mixtures were vigorously stirred, the suspension was maintained at room temperature for 24 h, the final volume attained by IDF was measured. The SC was determined by the following equation (2):

$$SC\ (ml/g) = \frac{V_2 - V_1}{m} \qquad (2)$$

where V_1 is the original volume occupied by SDF sample, V_2 is the final volume occupied by SDF sample; m is the weight of SDF sample.

2.4.2 Water-Holding Capacity (WHC)

The WHC of IDF samples was determined according to the method described by Mateos-Aparicio et al. with minor modification (Mateos, Mateos-Peinado, & Ruperez, 2010). 0.5 g IDF was added with 20 ml distilled water, stirred and left at room temperature for 24 h, then they were centrifuged for 30 min at 3500 r/min, the supernatant was discarded, and the residue was collected and weighed, WHC was expressed as g of water bound per g of dry sample.

2.4.3 Oil-Holding Capacity (OHC)

The OHC of IDF samples was determined according to the method described by Chau with minor modification (Chau, 2003). 0.5 g IDF was added with 10 ml vegetable oil, stirred for 30 min and left at room temperature for 24 h, centrifuged for 30 min at 3500 r/min, OHC expressed as g of oil held by 1 g of dry sample. The density of the oil was 0.92 g/ml.

2.4.4 Nitrite ion-absorbing capacity

Accurately measured 2.0 ml of $NaNO_2$ solution (5.0 μg/ml) was mixed with IDF, placed in 37 °C water bath for 30 min, centrifuged for 30 min at 3500 r/min, the concentration of NO_2^- existing in the supernatant was measured by Nai ethyldiamine hydrochloride colorimetry (Hu, 1997). Blank samples were prepared by adding ammonium citrate buffer (pH 3.0) into IDF to a constant volume of 10 ml. The nitrite ion-absorbing capacity (y) was determined by the following equation (3):

$$y(\%) = \frac{5 \times 2 - (E_s - E_b)}{5 \times 2} \times 100\% \qquad (3)$$

Table 1. Factors and levels of orthogonal experiment.

Level	A, NaOH concentration (%)	B, ratio of solvent to material (ml/g)	C, microwave power (W)	D, extraction time (s)
1	3	25:1	200	50
2	4	30:1	300	100
3	5	35:1	400	150

where E_s is the NO_2^- concentration (μg/ml) of IDF sample, E_b is the NO_2^- concentration (μg/ml) of blank sample.

2.5 Statistical analysis

All tests were carried out in triplicate. Data were expressed as mean ± Standard Deviation (SD). Experimental design and data analyses were performed with orthogonal design assistant V3.1 and SPSS 14.0 (SPSSInc., USA). One-way Analysis of Variance (ANOVA) was conducted to determine significant differences, $P < 0.05$ was considered to be statistically significant.

3 RESULTS AND DISCUSSION

3.1 Effect of NaOH concentration on IDF extraction rate

The effect of NaOH concentration on IDF extraction rate was examined by varying NaOH concentration at 3%, 4%, 5%, 6%, 7%, the other factors were fixed. The results shown in Figure 1A reveal that the IDF extraction rate increased with the increasing of NaOH concentration and reached maximum when NaOH concentration was 3%, then decreased afterward. The main reason of this phenomenon was maybe that when the NaOH concentration was too high, the cellulose in the soy sauce residue reacted with NaOH and was destroyed, so the extraction rate decreased. Thus, 3% NaOH solution was chosen as the best extraction solvent.

3.2 Effect of ratio of solvent to material on IDF extraction rate

The effect of ratio of solvent to material (namely, 20:1, 25:1, 30:1, 35:1, 40:1) on IDF extraction rate was also investigated. The results are shown in Figure 1B. The extraction rate increased rapidly with an increase of ratio of solvent to material and reached the peak when

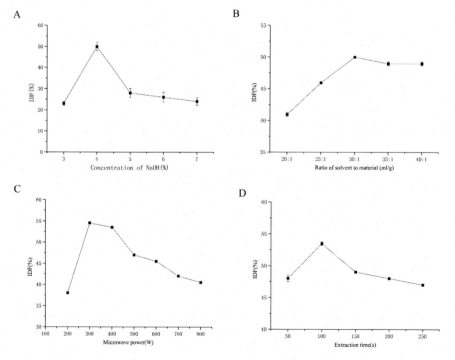

Figure 1. Effect of NaOH concentration (A), ratio of solvent to material (B), microwave power (C), and extraction time (D) on IDF extraction rate.

the ration was 30:1, but when the ratio continued to increase, there was no obvious increase or decrease in the extraction rate of IDF. One possible reason responsible for this phenomenon was that increasing solvent to material ratio could promote more ingredients to be dissolved in NaOH solution until they were fully extracted, but if ratio of solvent to raw material was too high, it would lead to a higher process cost. To avoid wasting solvent, 30:1 was set as the optimal ratio.

3.3 Effect of microwave power on IDF extraction rate

To evaluate the effect of microwave power on IDF extraction rate, microwave power was set from 200 W to 800 W, the results shown in Figure 1C suggest that the extraction rate was originally increased when microwave power varied from 200 W to 300 W, but decreased when the power exceeded 300 W, that was maybe because the extraction temperature increased with the increasing microwave power, and then more ingredients dissolved in NaOH solution; moreover, the hydrogen bond between cellulose and hemicelluloses broke, when hemicellulose also dissolved. Therefore, the optimal microwave power was 300 W.

3.4 Effect of extraction time on IDF extraction rate

Additionally, the effect of extraction time on IDF extraction rate was also examined. IDF was extracted for 50, 100, 150, 200, 250 s. As shown in Figure 1D, the extraction rate increased rapidly when extraction time varied from 50 to 100 s, and reached the maximum at 100 s, then decreased with the increasing time. A possible explanation was that with the increasing contact time, more ingredients dissolved in NaOH solution, so, the optimal extraction time was 100 s.

3.5 Orthogonal experiment

In order to obtain the best condition for IDF microwave-ultrasonic extraction, $L_9(3^4)$ orthogonal experiment design was carried out to obtain the optimal concentration of NaOH: ratio of solvent to material, microwave power, and extraction time. Based on the factors and levels listed in Table 1, nine groups of experiments were carried out and the results are shown in Table 2. According to the magnitude order of R (Max Dif), the order of effect of four factors on IDF microwave-ultrasonic extraction rate was B > D > C > A. According to the K values, the optimum formulation was $A_3B_1C_2D_2$, but the optimal group among the nine groups was $A_3B_1C_3D_2$; under such conditions, the extraction rate was 71.48%, which was not in accordance with $A_3B_1C_2D_2$ obtained. So a verification experiment was carried out to check the result of optimal formulation of $A_3B_1C_2D_2$. When IDF was extracted under condition shown

Table 2. Results of orthogonal experiment.

Number $L_9(3^4)$	A	B	C	D	IDF (%)
1	1	1	1	1	68.24
2	1	2	2	2	61.48
3	1	3	3	3	54.47
4	2	1	2	3	69.46
5	2	2	3	1	50.49
6	2	3	1	2	60.04
7	3	1	3	2	71.48
8	3	2	1	3	52.47
9	3	3	2	1	60.54
K1	61.397	69.727	60.250	59.757	
K2	59.997	54.813	63.827	64.333	
K3	61.497	58.350	58.813	58.800	
R	1.500	14.914	5.014	5.533	

in $A_3B_1C_2D_2$, the extraction rate of the IDF was up to 73.86%, which was higher than that of $A_3B_1C_3D_2$ (71.48%). So, the optimal condition was as follows: NaOH concentration of 5%, ratio of solvent to material of 25:1, ultrasonic power of 300 W, extraction time of 100 s.

The variance analyses were performed as shown in Table 3. It was clearly observed that ratio of solvent to material acted as significant factor for IDF extraction rate ($p < 0.05$). This indicated that the effect of ratio of solvent to material was more important than other conditions for IDF extraction.

3.6 Functional properties of IDF

3.6.1 Swelling Capacity (SC), Water-Holding Capacity (WHC) and Oil-Holding Capacity (OHC)

SC was defined as the ratio of the volume occupied when the sample was immersed in excess of water. WHC and OHCs were defined by the quantity of water and oil bound to the fiber without the application of any external force (except for gravity and atmospheric pressure). As shown in Figure 2A–C, SC, WHC, OHC of the extracted IDF increased obviously with

Table 3. Results of variance analysis.

Factor	Square of deviance	Freedom	F ratio	Critical value	Significance
A	4.220	2	1.000	19.000	
B	364.344	2	86.337	19.000	*
C	39.990	2	9.476	19.000	
D	52.479	2	12.436	19.000	
Error	4.22	2			

*Represent significant different ($p < 0.05$).

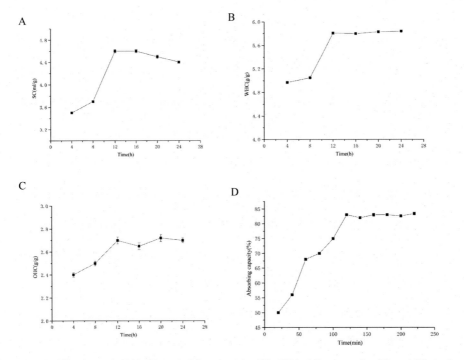

Figure 2. Swelling capacity (A), Water-holding capacity (B), Oil-holding capacity (C) and Nitrite ion-absorbing capacity (D) of IDF.

the increase of time, and reached the maximum at 12 h, when the SC, WHC, OHC was 4.62 ml/g, 5.81 g/g, and 2.72 g/g, respectively. After that, they remained relatively stable. The results showed that IDF has good capacity of swelling, holding water and oil, which was helpful in decreasing total and Low Density Lipoprotein (LDL) and forming bigger excrement (Cao, & Huang, 1997).

3.6.2 *Nitrite ion-absorbing capacity*

Nitrite is a reactive ion and can react with secondary amines and amides under acid conditions to form N-nitroso compounds, many of which have been shown to be carcinogenic in animals (Lijinsky, 1987). Nitrite ion-absorbing capacity may be a contributing factor in the possible role of dietary ingredients for protecting against gastric cancer development (Moller, Dahl, & Bockman, 1988). If IDF has good capacity of swelling and holding water, it also has strong hydrophilic and hydrophobic interactions and good absorbing capacity correspondingly. Nitrite ion-absorbing capacity at different time is shown in Figure 2D. At the beginning, the absorbing capacity increased rapidly, saturated at 120 min, when 83% nitrite ion was absorbed, then kept constant afterward.

4 CONCLUSIONS

In this paper, IDF was extracted by microwave-ultrasonic assisted enzymatic hydrolysis from soy sauce residue for the first time. After single factor experiment and orthogonal experiment, the optimal microwave-ultrasonic extraction condition was established as follows: NaOH concentration of 5%, ratio of solvent to material of 25:1, ultrasonic power of 300 W, extraction time of 100 s; after hydrolyzation by 0.3% papain at 65 °C for 20 min, the corresponding extraction rate of IDF was 73.86%. The microwave-ultrasonic assisted enzymatic extraction technology was an effective method to extract IDF. The application of microwave-ultrasonic could enhance the extraction rate and shorten the extraction time. The extracted IDF had good SC, WHC, OHC, and nitrite ion-absorbing capacity. It has a great application potential in food and pharmaceutical field.

ACKNOWLEDGMENTS

This work was funded by National Natural Science Foundation of China (No. 81273004), Qing Lan Project, the Open Project Program of Jiangsu Key Laboratory of Food Resource Development and Quality Safety, Xuzhou Institute of Technology (No. SPKF201311, SPKF201412) and Science and Technology Planning Project of Xuzhou (XF13C034).

REFERENCES

Cao, S.W., & Huang, S.H. (1997). Study on preparation technology of several kinds of dietary fiber. Food Science, 6, 41–45.

Chau, C.F. (2003). Comparison of the chemical composition and physicochemical properties of different fibers prepared from the peel of Citrus sinensis L. Cv. Liucheng. Journal of Agricultural and Food Chemistry, 9, 2615–2618.

Cui, S.W., Nie, S., & Roberts, K.T. (2011). Functional properties of dietary fiber. Guelph Food Research Centre, Guelph, ON, Canada, 517–525.

Elleuch, M., Bedigian, D., Roiseux, O., Besbes, S., Blecker, C., & Attia, H. (2013). Dietary fibre and fibre-rich by-products of food processing: characterisation, technological functionality and commercial applications: a review. Food Chemistry, 2, 411–421.

Hu, G.H. (1997). Study on rice bran dietary fiber binding NO_2^- in vitro. Cereals Oils, 2001,11, 2–3.

Huang, Z., Ye, R., Chen, J., & Xu, F. (2013). An improved method for rapid quantitative analysis of the insoluble dietary fiber in common cereals and some sorts of beans. Journal of Cereal Science, 3, 270–274.

Kaushik, A., & Singh, M. (2011). Isolation and characterization of cellulose nanofibrils from wheat straw using steam explosion coupled with high shear homogenization. Carbohydrate Research, 1, 76–85.

Lijinsky, W. (1987). Structure-activity relations in carcinogenesis by N-nitroso compounds. Cancer and Metastasis Reviews, 3, 301–356.

Mateos, A.I., Mateos-Peinado, & C., Ruperez, P. (2010). High hydrostatic pressure improves the functionality of dietary fibre in okara by-product from soybean. Innovative Food Science and Emerging Technologies, 3, 445–450.

Mengmei, M., Taihua, M., Hongnan, S., Miao, Z., Jingwang, C., & Zhibin, Y. (2015). Optimization of extraction efficiency by shear emulsifying assisted enzymatic hydrolysis and functional properties of dietary fiber from deoiled cumin (Cuminum cyminuml.). Food chemistry, 2015, 179, 270–277.

Meyer, A.S., Dam, B.P., & Lærke, H.N. (2009). Enzymatic solubilization of a pectinaceous dietary fiber fraction from potato pulp: Optimization of the fiber extraction process. Biochemical Engineering Journal, 1, 106–112.

Moller, M.E., Dahl, R., & Bockman, O.C. (1988). A possible role of the dietary fibre product, wheat bran, as a nitrite scavenger. Food Chemistry and Toxicology, 26, 841–845.

Raghavendra, S.N., Rastogi, N.K., Raghavarao, K.S.M.S., & Tharanathan, R.N. (2004). Dietary fiber from coconut residue: Effects of different treatments and particle size on the hydration properties. European Food Research and Technology, 6, 563–567.

Wang, Z.H. (2009). Preparation and characterization of insoluble dietary fiber from soy sauce residue from soy sauce residues. China Brewing, 2, 105–108.

Yamamoto, H., Takeuchi, F. Nagano, C., Sasaki, A, & Shibata, J. (2006). Separation of flavonoids and salt in bean cake disposed from soy sauce manufacturing process. Journal of chemical engineering of Japan, 7, 777–782.

Zhang, M., Bai, X., & Zhang, Z.S. (2011). Extrusion process improves the functionality of soluble dietary fiber in oat bran. Journal of Cereal Science, 1, 98–103.

Zheng, C. (2001). low-energy diet. Beijing: Light Industry Press.

Zhu, L., & Hu, Z.H. (2010). Study on extracting water-insoluble dietary fiber from soy sauce residues, Cereals Oils, 6, 20–22.

Determination of Cd, Pb and Cu in chia seeds by microwave digestion-HR-CS GFAAS

Y. Liu, S.L. Chen, Y. Li, Y.X. Zhu & S.Y. Jiang
Jiangsu Key Laboratory of Food Resource Development and Quality Safe, Xuzhou Institute of Technology, Xuzhou, China

ABSTRACT: A new application of high resolution continuum source graphite furnace atomic absorption spectrometry has been developed for the determination of Cd, Pb and Cu in chia seeds. The samples were digested by microwave. Under the selected determination conditions, the correlation coefficients better than 0.998 and the recoveries between 95.8% and 102.5% were obtained for Cd, Pb and Cu. The detection limits were 0.11 µg/L (Cd), 1.09 µg/L (Pb) and 0.88 µg/L (Cu), respectively. The results showed that the contents in chia seeds were 1.74 ± 0.06 µg/kg (Cd), 6.57 ± 0.29 µg/kg (Pb) and 19425 ± 325 µg/kg (Cu), respectively. Therefore, the proposed method was accurate and stable with a high practical value. It provided scientific basis for determination of metal elements in food.

1 INTRODUCTION

Several sample preparation techniques are used for the determination of metal elements in food, such as inductively wet digestion, dry digestion, incomplete digestion and microwave digestion. Wet digestion and dry digestion are two traditional ways, but both of them have obvious disadvantages, for instance, some dense mass of strong acids are used in wet digestion, while dry digestion is a time-consuming process (Afridi, et al., 2007; Gao, 2007). The objective of incomplete digestion is uniform and transparent digestive solutions, which do not need of complete destruction and colorless liquors. As a consequence of that, the sample-process time just need less than 20 min (Liu, et al., 2004). Nevertheless, incomplete digestion also requires a good deal of strong acids and the microemulsification stability after that is an important parameter for accurate analysis. Closed-vessel microwave-assisted digestion need only 5 mL of HNO_3 and 4 mL H_2O_2 in this study, and its main advantages are the high relative speed, good reproducibility, low blank, low possibility of contamination and the minimum loss of volatile elements (Ren, et al., 2011).

The objective of this study was to develop a simple and robust method for the fast sequential multi-element determination of Cd (Li, et al., 2012), Pb (Liu, et al., 2009) and Cu (Gao, et al., 2007) in chia seeds by high resolution continuum source graphite furnace atomic absorption spectrometry (HR-CS GFAAS) (Nunes, et al., 2011; Borges, et al., 2011; Lepri, et al., 2010; Baysal, et al., 2011) using new features.

2 EXPERIMENTAL

2.1 *Instrumentation*

An Analytik Jena ContrAA 700 High Resolution Continuum Source Atomic Absorption Spectrometer (Analytik Jena, Berlin, Germany) had been used for all measurements in this work. This spectrometer consists of a high-intensity xenon short-arc lamp, a high-resolution Double Echelle Monochromator (DEMON) and a Chargecoupled Device (CCD) array detector (Resano, et al., 2009). All absorption lines in the range from 185 nm to 900 nm are provided by the high-intensity xenon short-arc lamp as the radiation source. The highest resolution of about 2 pm is carried out by DEMON, including a pre-dispersing prism monochromator and a high-resolution echelle grating monochromator. 200 pixels of the linear CCD array detector are used for monitoring all spectral informations on both sides of center wavelength. In addition, this instrument is also equipped with an automated sampling accessory (MPE) (Liu, et al., 2014). High purity argon was used as the carrier gas in the process of determination. The sampling volume of 20 µL and the chemical modifier volume of 5 µL were used for determination. The optimal determination parameters of HR-CS GFAAS and graphite furnace temperature programs were shown in Table 1 and Table 2, respectively.

2.2 Reagents and standards

The reagents were of guaranteed reagent. Ultrapure water with a resistivity of 18.2 MΩ·cm was obtained from a Milli-Q system (Millipore, Billerica, USA). Calibration solutions of Cd, Pb and Cu were prepared in the ultrapure water with 0.5% (v/v) HNO_3 by serial dilution of the stock solutions with 100 mg/L respectively (National Chemical Reagent Company, Beijing, China). All glasswares were previously soaked overnight in dilute HNO_3 (5% v/v) for cleaning and were rinsed with abundant ultrapure water prior to avoid contamination.

2.3 Microwave digestion

The chia seeds were purchased from the supermarket (Xuzhou, China) during 2015 and analyzed of their Cd, Pb and Cu contents. They were crushed to powder. Approximately a 0.5 g of the crushed sample was preprocessed with 5 mL of HNO_3 and 2 mL H_2O_2 in the PTFE jar, which was the process of heating to 120 °C for 30 min in an air-ventilated oven. The compound (sample and acids) was digested in the intelligent microwave digestion system (Xin-tuo, Shanghai, China) after adding again 2 mL H_2O_2. A five stage program (Table 3) with a maximum pressure of 2.0 Mpa was chosen for achieving complete digestion of the crushed sample within the shorter time. The digestive liquor was diluted to 10 mL with the ultrapure water with 0.5% (v/v) HNO_3 when its volume was less than 3.0 mL. Soon after that, it was hand shaken resulting in a visually homogeneous system. A blank digest was carried out in the same way. Three independent aboved treatments of the sample were performed for obtaining the average of repetitive determinations of Cd, Pb and Cu.

Table 1. Optimal determination parameters of HR-CS GFAAS.

Element	Wavelength (nm)	Chemical modifier
Cd	228.8018	$NH_4H_2PO_4$ (10 g/L)
Pb	283.3060	$NH_4H_2PO_4$ (10 g/L)
Cu	324.7540	$Mg(NO_3)_2$ (0.5 g/L)

Table 2. Optimized graphite furnace temperature programs.

Element	Pyrolysis temperature (°C)	Atomization temperature (°C)	Ramp (°C/s)
Cd	600	1200	1400
Pb	800	1500	1500
Cu	1100	2000	1500

Table 3. Microwave digestion program.

Stages	Pressure (Mpa)	Hold (min.)	Power (W)
1	0.2	60	500
2	0.5	60	1000
3	1.0	120	1000
4	1.5	120	1000
5	2.0	60	1000

3 RESULTS AND DISCUSSION

3.1 Optimization of analytical lines

The first sensitive line was usually chosen as the analytical line in the determination of metal elements by Atomic Absorption Spectrometry (AAS), such as the analytical lines of Cd and Cu in this work were the respective first sensitive lines as shown in Table 1. There were some other absorption lines around the first sensitive lines of Pb, which produced lots of interferences for the determination of Pb, as shown in Figure 1. The choice of the second sensitive line is a simple and effective method for eliminating these interferences. But this method is not applied widely in the analysis of Line Sources Atomic Absorption Spectrometry (LS AAS), because of the hollow cathode lamp as the radiation source. The radiation intensities of other sensitive lines provided by this lamp, not the first sensitive line, are low in general. A high-intensity xenon short-arc lamp is used as the radiation source in High Resolution Continuum Source Atomic Absorption Spectrometry (HR-CS AAS), whose radiation intensity is 10–100 times higher than ordinary xenon lamp and energy is relatively stable. Thus, the respective second sensitive lines of Pb were chosen as the analytical lines in this work, as shown in Figure 2.

According to Figure 2, there was not any other absorption lines around the second sensitive lines of Pb.

3.2 Analytical performance

The analytical characteristic data of HR-CS GFAAS were shown in Table 4. The calibration curves used to determine of Cd, Pb and Cu with HR-CS GFAAS were built-up by measuring the absorbance of the calibration solutions in the selected determination conditions, as shown in Figure 3–Figure 5.

The correlation coefficients were 0.9997 (Cd), 0.9984 (Pb) and 0.9990 (Cu), respectively. The detection limits were 0.11 μg/L (Cd), 1.09 μg/L (Pb) and 0.88 μg/L (Cu), respectively. As can be seen, the proposed method represented one of more sensitive methodologies for determination of Cd, Pb and Cu.

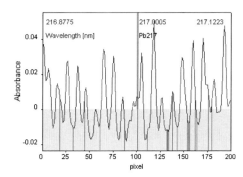
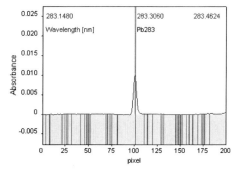

Figure 1. First sensitive line of Pb.

Figure 2. Second sensitive line of Pb.

Table 4. Analytical characteristic data of HR-CS GFAAS.

Element	Calibration function (C in μg/L)	Correlation coefficient (R2)	Upper linear range (μg/L)	Detection limit (μg/L)
Cd	A = 0.0505 × C + 0.0186	0.9997	0.25–4	0.11
Pb	A = 0.0045 × C + 0.0203	0.9984	5–50	1.09
Cu	A = 0.0069 × C + 0.0124	0.9990	2–40	0.88

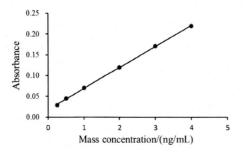

Figure 3. Calibration curve of Cd.　　　　Figure 4. Calibration curve of Pb.

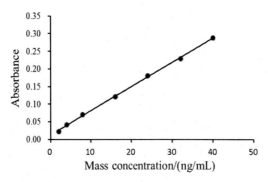

Figure 5. Calibration curve of Cu.

Table 5. Determination of Cd, Pb and Cu in chia seeds and recoveries (Avg. ± SD of three trials).

Element	Content (μg/kg)	Spiked (μg/kg)	Recoveries (%)
Cd	1.74 ± 0.06	2	97.1 ± 2.1
Pb	6.57 ± 0.29	5	95.8 ± 2.8
Cu	19425 ± 325	10000	102.5 ± 0.6

The proposed means, using microwave digestion-HR-CS GFAAS, had been applied to determination of Cd, Pb and Cu in chia seeds and was verified through spike recovery tests (Li, et al., 2012). The results, obtained as the average of three replicates of each element, were shown in Table 5. The contents in chia seeds were 1.74 ± 0.06 μg/kg (Cd), 6.57 ± 0.29 μg/kg (Pb) and 19425 ± 325 μg/kg (Cu), respectively. The recoveries were 97.1 ± 2.1 (Cd), 95.8 ± 2.8 (Pb) and 102.5 ± 0.6 (Cu), respectively. Therefore, the proposed method represented one of more accurate methodologies for the determination of the four metal elements.

4 CONCLUSIONS

The developed means provided a exact and sensitive procedure for the determination of Cd, Pb and Cu in chia seeds by HR-CS GFAAS after microwave digestion. Under the selected determination conditions, the correlation coefficients better than 0.998 and the recoveries between 95.8% and 102.5% were obtained for Cd, Pb and Cu. The detection limits were 0.11 μg/L (Cd), 1.09 μg/L (Pb) and 0.88 μg/L (Cu), respectively. The results showed that the contents in chia seeds were 1.74 ± 0.06 μg/kg (Cd), 6.57 ± 0.29 μg/kg (Pb) and 19425 ± 325 μg/kg (Cu), respectively. Low detection limits, good correlation coefficients and precisions, and high recoveries showed that the proposed method is accuracy, reliability and stability.

ACKNOWLEDGEMENTS

This work was financially supported by the University Scientific Research Projects of Xuzhou Institute of Technology (No. XKY2013325 and XKY2014214) and Spark Plans of China (No. 2013GA690417, 2013GA690418, 2014GA690103 and 2014GA690105).

REFERENCES

Afridi H.I., Kazi T.G., Arain M.B., et al. 2007. Determination of cadmium and lead in biological samples by three ultrasonic-based samples treatment procedures followed by electrothermal atomic absorption spectrometry. *Journal of AOAC International* 90(2): 470–478.

Baysal A., Akman S. 2011. A practical method for the determination of sulphur in coal samples by high-resolution continuum source flame atomic absorption spectrometry. *Talanta* 85(5): 2662–2665.

Borges A.R., Becker E.M., Lequeux C., et al. 2011. Method development for the determination of cadmium in fertilizer samples using high-resolution continuum source graphite furnace atomic absorption spectrometry and slurry sampling. *Spectrochimica Acta Part B: Atomic Spectroscopy* 66(7): 529–535.

Gao S.Y. 2007. The determination of element content in burdock by flame atomic absorption spectrophotometry. *Journal of Xuzhou Institute of Technology* 6(22): 32–35.

Gao S.Y., Cai S.J., Qu D.H. 2007. Determination of Trace Elements (Cu, Se, Cd, Pb) in Garlic by Method of Furnace Atomic Absorption. *Journal of Xuzhou Institute of Technology* 10(22): 61–64.

Lepri F.G., Borges D.L.G., Araujo R.G.O., et al. 2010. Determination of heavy metals in activated charcoals and carbon black for Lyocell fiber production using direct solid sampling high-resolution continuum source graphite furnace atomic absorption and inductively coupled plasma optical emission spectrometry. *Talanta* 81(3): 980–987.

Li Y., Chen S.L., Wang S.L., et al. 2012. Study on determination methods of cadmium content in alcoholic drink. *Journal of Xuzhou Institute of Technology (Natural Sciences Edition)* 4(27): 16–19.

Liu H., Chen S.L., Li C., et al. 2014. Sequence determination of Cd and Pb in honey by incomplete digestion-high resolution continuum source graphite furnace atomic absorption spectrometry. *Applied Mechanics and Materials* 511: 22–27.

Liu H., Wu Q.J. 2009. Determination of Pb in food sample by nano-titanium dioxide preconcentration and separation-furnace atomic absorption spectroscopy. *Journal of Xuzhou Institute of Technology (Natural Sciences Edition)* 1(24): 38–41.

Liu L., Yu M. 2004. Determination of calcium and magnesium in gelatin by noncomplete digestion-flame atomic absorption spectrometry. *Metallurgical Analysis* 5: 015.

Nunes L.S., Barbosa J.T.P., Fernandes A.P., et al. 2011. Multi-element determination of Cu, Fe, Ni and Zn content in vegetable oils samples by high-resolution continuum source atomic absorption spectrometry and microemulsion sample preparation. *Food chemistry* 127(2): 780–783.

Ren T., Zhao L.J., Zhong R.G. 2011. determination of aluminum in wheat flour food by microwave digestion-high resolution continuous source graphite furnace atomic absorption spectrometry. *Spectroscopy and Spectral Analysis* 31(12): 3388–3391.

Resano M., Briceño J., Belarra M.A. 2009. Direct determination of Hg in polymers by solid sampling-graphite furnace atomic absorption spectrometry: a comparison of the performance of line source and continuum source instrumentation. *Spectrochimica Acta Part B: Atomic Spectroscopy* 64(6): 520–529.

Determination of vitamin C in intact *Actinidia Arguta* fruits using Vis/NIR spectroscopy

Guang Xin, Bo Zhang & Shuqian Li
College of Chemistry and Life Science, Anshan Normal University, Anshan, Liaoning, China

Jingjing Mu, Changjiang Liu & Xianjun Meng
College of Food Science, Shenyang Agricultural University, Shenyang, Liaoning, China

ABSTRACT: In this paper, the visible-near infrared (Vis/NIR) spectroscopy in the wavelength from 570 nm to 1848 nm was employed to establish a calibration plot for determination of vitamin C in *Actinidia arguta* fruits. It was demonstrated that the working plot established using the Modified Partial Least Squares (MPLS) model, the first order derivatives spectrum $D_1 \log(1/R)$ and Detrend only, presented a satisfying prediction performance for vitamin C in *Actinidia arguta*. The correlation coefficient of cross validation (R_{CV}) and correlation coefficient of prediction (R_p) were 0.9584 and 0.9521, respectively, and the Root-Mean-Square Error of Cross Validation (RMSECV) and Root-Mean-Square Error of Prediction (RMSEP) were 5.9795 mg/100 g, and 5.4760 mg/100 g, respectively. This work demonstrated an alternative practical application of VIS/NIR spectroscopy for non-destructive determination of vitamin C in *Actinidia arguta* fruits.

1 INTRODUCTION

Actinidia arguta originates from China, and it is one of the most widely distributed fruit trees in Chinese region (Matich, A.J. et al., 2003; Matich, A.J. et al., 2013). *Actinidia arguta* is not only rich in all kinds of amino acids and vitamins, but also have the efficacy of health care, especially the content of vitamin C is several or ten times more than in other fruits (Krupa, T. et al., 2011). Vitamin C has an antioxidant function, so it is the good source of vitamin C among a large segment of the fruits (Latocha, P. et al., 2010). Also Vitamin C can judge maturity of *Actinidia arguta*. As fruit mature, vitamin C content changes obviously (Connie, L.F. et al., 2006; Connie, L.F. et al., 2008).

Most measurement methods of fruits physicochemical properties are mainly measured through chemical method and these methods are usually destructive, at the same time the method are rather time-consuming, which needs sample preparation tediously and high cost. Ultraviolet spectrophotometry and titration method (2,6-Dichlorophenolindophenol) (DPI) are commonly used to measure the vitamin C content in *Actinidia arguta*. They are also destructive and involve a considerable amount of manual work, which cannot satisfy the meet for reality of the fruit sorting fast grading requirements. Therefore, there is a demand for non-destructive and rapid techniques for assessing fruit quality in the picking and production fields.

Near-infrared spectrum is mainly produced by the no-resonance of molecular vibration when the molecular vibrations from the ground state to a higher level. Mainly record the vibration frequency multiplication and frequency absorption of H-C, O-H, N-H, S-H, and P-H (Cristina, P. et al., 2008). Different perssads (such as methyl, methylene, benzene ring, etc.) or the same perssads have obvious difference of near-infrared absorption wavelength and intensity in different chemical environment (Ding, X.X. et al., 2015).

Mathematical model is combined with mathematics and practical application, and now it is applied to physics and chemistry (Mariya, G. et al., 2013). Near-Infrared Spectroscopy (NIR) technique analysis focuses on using mathematical method to solve the spectral peak overlap, measuring the information of high background and lowing intensity, spectra determination of the instability caused spectrum distortion (Lorente, D. et al., 2015).

Near-Infrared Spectroscopy (NIR) technique has the advantages of accuracy, rapid-response, lower cost, chemical-free, pollution-free, and non-destructive. Hence, it has been widely adopted in the inspection of various agricultural products (Tsai, C.Y. et al., 2008). In recent years, researchers in fruit nondestructive testing have made a lot of researches. The indicators involve soluble solids content (Lu, H.S. et al., 2006), acidity, firmness (Patricia, P. et al., 2008), dry material (Lu, Q. et al., 2010), internal quality of fruit (Camps, C. & Christen, D., 2009), storage disorders (Clark, C.J. et al., 2004), etc. V. Andrew, M. et al (2002) developed the NIR methods for measurement of Kiwifruit dry matter and soluble solids content. However, few reports were available in the development of vitamin C measurement by using the near-infrared diffuse reflectance (NIR) method. In this study, we tried to apply the NIR technique to analyze the vitamin C content.

2 MATERIALS AND METHODS

2.1 *Sample preparation*

The *Actinidia arguta* samples used in this study were purchased from Han Jiayu, Anshan. All of them were picked on August 26, 2011. The selected samples were required that the mature was consistent (about eight mature), with no plant diseases and mechanical damage on the fruits.

The analysis was performed on 10,000 *Actinidia arguta* fruits, they were stored at a 24-hour cold (−0.5–0.5 °C) storage. The experiment was sampled on three occasions across different storage stages (harvest time, storage 12 days, and storage 24 days) and every stage took 180 samples randomly, in which 150 samples were taken as the calibration set and 30 samples were taken as the prediction set.

Before the experiment, *Actinidia arguta* were acclimatized to equilibrium for about 3 hours in the controlled temperature condition room. And then in the equator two metering slide parts of every fruit, draw about 1 cm in diameter of the circle, marking and sequencing. Put the sample on the tray, click the "run" on the computer, the tray use one minute to rotate a circle. To collect the spectral and determine the chemical value.

2.2 *Spectra collection*

Visible and Near-Infrared Diffuse Reflection (Vis/NIR) Spectroscopy (Infra XactTM Lab, Foss, Danmark), its spectral system is the holographic grating, the detector consists of silicon (570~1 098 nm) and indium gallium arsenic (1 100~1 848 nm). Configuration is composed of the ISI scan analysis software and Win ISI calibration software, resolution is 7 nm, removed of spectral data is 2 nm, and the accuracy of wavelength is less than 0.5 nm. When we collect the spectrum, scan the mark parts and storage of the spectral data through the ISI scan analysis software (Arthit, P., Siwalak P., Anupun, T., & Kaewkarn, P., 2012).

2.3 *Determination of Actinidia arguta vitamin C*

Measure the fruit vitamin C content on the scanning spot after scanning. Titration method was used to measure the *Actinidia arguta* of vitamin C content (DPI).

Wash the fruit clean and wipe off the water. Weigh 5 g sarcocarp and mix 5 ml of 2% oxalic acid to grind in the mortar. And 2% of oxalic acid is used to constant volume in the 100 ml volumetric flask. After filtrate, take 10 ml of leachate and use the 2,6-Dichlorophenolindophenol (calibration) to titration.

$$Vitamin\ C\ (mg/100g) = \frac{(V-V_0) \times T \times A \times 100}{M} \quad (1)$$

where V = the volume of the titration sample using 2,6-Dichlorophenolindophenol, ml; V_0 = the volume of the titration blank use 2,6-Dichlorophenolindophenol, ml; T = titer of the 2,6-Dichlorophenolindophenol, mg/ml; A = dilution multiple; and W = sample weight, g.

2.4 *Statistical treatment of data*

Use the Win ISI III software to filter and smooth the spectrum. Aimed at removing the noise and extracting the effective information. Use the different regression techniques, different derivative treatments and different scatter and standard treatments at the same time, the purpose is to make sure the prediction model of Vitamin C of *Actinidia arguta*, and then use the sample which did not participate in the calibration to verify this model. Then evaluate the feasibility of the model *Actinidia arguta* of vitamin C content (DPI).

In this study, the quality of prediction model should be evaluate as follows: the correlation coefficient of cross validation, RCV; correlation coefficient of prediction, RP; root-mean-squared error of cross validation, RMSECV; root-mean-square standard error of prediction; RMSEP; A good model should be with low RMSECV and RMSEP value, as well as the high R value, in addition RMSECV and RMSEP value differences should also be relatively small (Zhang, P. et al. 2011).

3 RESULTS AND DISCUSSION

3.1 *Original spectrum of Actinidia arguta and standard ascorbic acid spectra*

Figure 1 shows the original absorption spectrum of *Actinidia arguta* in 570~1848 nm, we could see that in 676, 976, 1176 and 1428 nm, there were the significant absorption peaks. Figure 1 was the original spectrum of *Actinidia arguta* contrasted with standard ascorbic acid spectra. The visible region had a clear absorption peaks (676 nm), it may due to the absorbing light of peel chlorophyll (Cayuela J.A., 2008). Williams & Norris (2001) found that the moisture absorption of the near-infrared spectra was 960–990 nm; 1198–1270 nm. Cayuela J.A. (2008) claimed that moisture and carbohydrates cause the absorption peaks in 978 nm and 1454 nm. Generally speaking, the absorption peaks in 976, 1176 and 1428 nm was caused by C-H and O-H. Thus indicated that vitamin C compounds contribute to the appearance of absorption peaks. In other words, it was feasible to use the VIS/NIR in 570~1848 nm wavelength range to build vitamin C of *Actinidia arguta* prediction model.

3.2 *Eliminate the outlier values*

Using the Principal Component Analysis (PCA) way and calculating the score in the cluster analysis to statistically find the differences between the samples. The Mahalanobis Distance (GH) was calculated. The GH of more than 3.0 is considered as the supernormal samples,

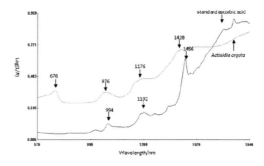

Figure 1. Original absorption spectrum of *Actinidia arguta* contrast with standard ascorbic acid spectra.

and then the samples were eliminated from the file. Figure 2 shows the three-dimensional space figure of before and after eliminating the outlier value of sample set.

3.3 *Distribution characteristics of calibration and validation*

The quality of model was to some extent depends on the real value of the fruit precision and sample to detect parameters coverage. Table 1 showed that the calibration and verification collection of chemical value set coverage was wide, and the content of the sample test sets range in calibration set range, we could accurately evaluate the quality of the model.

3.4 *The choice of mathematical modeling algorithm*

Table 2 shows the results of MPLS, PLS, and PCR model calibration. Based on regression analysis, the result of RCV of MPLS was significantly higher than in the other two algorithms. Cause Analysis, MPLS fully considered the influence of chemical data when the Principal Component Analysis (PCA) during the process of the whole spectrum. The results showed that the model using the MPLS was the best.

3.5 *Derivative deal with the choice of methods*

Based on the application of MPLS, compared and analyzed the results about the vitamin C content in *Actinidia arguta* calibration modeling, which is used for different derivative

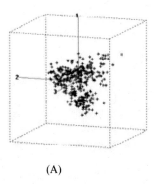
(A)

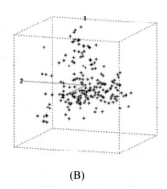
(B)

Figure 2. The three-dimensional space figure of before and after eliminate the outlier value of sample set. A-Before eliminate; B-After eliminate.

Table 1. Distribution characteristics of calibration and validation.

Sample set	Amount	Rang (mg/100 g)	Mean (mg/100 g)	SD (mg/100 g)
Calibration	150	41.95–182.99	111.35	28.46
Validation	30	61.95–165.60	106.11	24.94

Table 2. Statistical results of models established by different regression techniques.

Regression techniques	R_{CV}	RMSECV
Modified Partial Least Squares (MPLS)	0.9555	6.5079
Partial Least Squares (PLS)	0.9268	7.7.97
Principal Component Regression (PCR)	0.8528	10.7674

treatments. Table 3 shows the calibration results about the absorbance of original spectra log (1/R), the first order derivatives spectrum $D_1 \log(1/R)$, the second order derivatives spectrum $D_2 \log(1/R)$, the third order derivatives spectrum $D_3 \log(1/R)$, and the forth order derivatives spectrum $D_4 \log(1/R)$ calibration results. Through the comparison, $D_1 \log(1/R)$ of the calibration model was a better treatment. Cause Analysis, $D_1 \log(1/R)$ has improved the precision of the spectrum, reduced the baseline instability, eliminated baseline drift and did best at reducing the influence of the spectrum when particles changed. Figure 3 showed the spectrum that was processed by $D_1 \log(1/R)$.

3.6 Scattering and standardized management options

Comparative analysis of the calibration modeling results of scattering and standardized treatments of vitamin C content in *Actinidia arguta* by using the MPLS and $D_1 \log(1/R)$. Table 4 shows None, SNV, DET, MSC, SNVD, IMSC, and WMSC model calibration results. Compared with other treatments, the model that used the no scattering and standardized (None) treatment on vitamin C worked best, and it can strengthen the characteristics of spectrum, at the same time it can reduce the noise and the effect of drift. By using the MPLS, $D_1 \log(1/R)$ and None established, the model calibration $R_{CV} = 0.9584$, RMSECV = 5.9795 mg/100 g.

3.7 Vitamin C analysis model predicts evaluation

In order to predict the reliability and veracity of calibration model, we forecasted analysis of the vitamin C content of 30 fruits which did not participate in the selection of this calibration model, the results were as shown in Figure 4. The prediction results indicated that the text

Table 3. Statistical results of models established by different derivative treatments.

Derivative treatments	R_{CV}	RMSECV
Log (1/R)	0.9555	6.5079
$D_1 \log (1/R)$	0.9584	5.9795
$D_2 \log (1/R)$	0.8852	9.5366
$D_3 \log (1/R)$	0.8693	10.1942
$D_4 \log (1/R)$	0.8755	9.9900

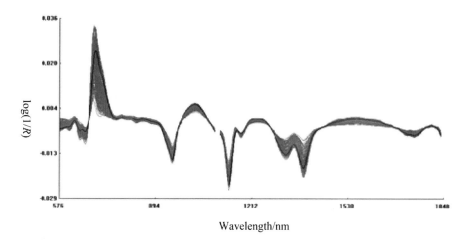

Figure 3. Processed by the $D_1 \log(1/R)$ spectrum of *Actinidia arguta*.

Table 4. Statistical results of models established by different scattering and standard treatments.

Scatter and standard treatments	R_{CV}	RMSECV
None	0.9584	5.9795
SNV	0.9290	7.5623
DET	0.9481	6.5199
MSC	0.9480	6.5326
SNVD	0.9472	6.5605
IMSC	0.9478	6.5376
WMSC	0.9327	7.4719

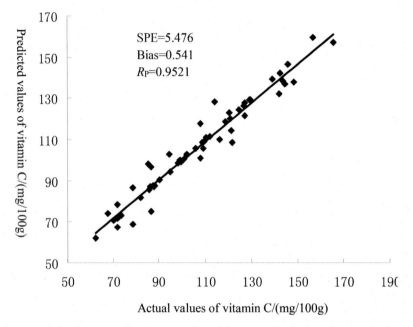

Figure 4. Correlation between using the optimal model and actual values of the vitamin C in *Actinidia arguta*.

results are successful, Root-Mean-Square Error of Prediction (RMSEP) = 5.4760 mg/100 g, and correlation coefficient (R_p) = 0.9521.

4 CONCLUSIONS

In conclusion, this method could be used to classify the maturity during harvest time and ripening time. In the future, the Vis/NIR technique can be used to predict the vitamin C content in *Actinidia arguta*, and thus the edible quality of the *Actinidia arguta*.

ACKNOWLEDGMENTS

We thank Anshan Normal College for providing the fruits used in this study, Institute of Forestry and Pomology Beijing Academy of Agriculture and Forestry Sciences for providing technical advice on the NIR instrument and providing access to the Vis/NIR analyzer.

REFERENCES

Andrew, V., M., Robert, B.J., Richard, S., Paul, J.M. 2002. Comparing density and NIR methods for measurement of Kiwifruit dry matter and soluble solids content. *Postharvest Biology and Technology* 26(2002): 191–198.

Arthit, P., Siwalak P., Anupun, T., & Kaewkarn, P. 2012. Evaluation of internal quality of fresh-cut pomelo using Vis/NIR transmittance. *Journal of Texture Studies* 43(2012): 445–452.

Camps, C. & Christen, D. 2009. Non-destructive assessment of apricot fruit quality by portable visible-near infrared spectroscopy. *Food Science and Technology* 42(2009): 1125–1131.

Cayuela, J.A. 2008. Vis/NIR soluble solids prediction in intact oranges (*Citrus Sinensis* L) cv. Valencia Late by reflectance. *Postharvest Biology and Technology* 47(1): 75–80.

Clark, C.J., McGlone, V.A., Silva, H.N. De., Manning, M.A., Burdon, J. & Mowat, A.D. 2004. Prediction of storage disorders of kiwifruit (*Actinidia chinensis*) based on visible-NIR spectral characteristics at harvest. *Postharvest Biology and Technology* 32(2004):147–158.

Connie, L.F., Mina, R.M., Bernadine, C.S., et al. 2006. Physicochemical, Sensory, and nutritive qualities of hardy kiwifruit (*Actinidia arguta 'Ananasnaya'*) as affected by harvest maturity and storage. *Journal of Food Science* 71(3): 204–210.

Connie, L.F., Alissa, M.S., Bernadine, C.S., et al. 2008. Postharvest quality of hardy kiwifruit (*Actinidia arguta 'Ananasnaya'*) associated with packaging and storage conditions. *Postharvest Biology and Technology* 47(3): 338–345.

Cristina, P., Gerard, D., Ian, M., Inigo, Z., Balbino, G.C. & Antonia, G.C. 2008. Direct classification of related species of fungal endophytes (Epichloe spp.) using visible and near-infrared spectroscopy and multivariate analysis. *Federation of European Microbiological Societies* 284(2008): 135–141.

Ding, X.X., Ni, Y.Q., Kokot, S. 2015. NIR spectroscopy and chemometrics for the discrimination of pure, powdered, purple sweet potatoes and their samples adulterated with the white sweet potato flour. *Chemometrics and Intelligent Laboratory Systems* 144(15): 17–23.

Krupa, T., Latocha, P., Liwi´nska, A. 2011. Changes of physicochemical quality, phenolics and vitamin C content in hardy kiwifruit (*Actinidia arguta* and its hybrid) during storage. *Scientia Horticulturae*, 130(2): 410–417.

Latocha, P., Krupa, T., Wolosiak, R., et al. 2010. Antioxidant activity and chemical difference in fruit of different *Actinidia* sp. *International Journal of Food Sciences and Nutrition* 61(4): 381–394.

Lorente, D., Escandell-Montero, P., Cubero, S., Gómez-Sanchis, J., Blasco, J. 2015. Visible–NIR reflectance spectroscopy and manifold learning methods applied to the detection of fungal infections on citrus fruit. *Journal of Food Engineering* 163(10): 17–24.

Lu, H.S., Xu, H.R., Ying, Y.B., Fu, X.P., Yu, H.Y. & Tian, H.Q. 2006. Application Fourier transform near infrared spectrometer in rapid estimation of soluble solids content of intact citrus fruits. *Journal of Zhejiang University* 7(10): 794–799.

Lu, Q., Tang, M.J., Cai, J.R. & Lu, H.Z. 2010. Long-term prediction of Zhonghua kiwifruit dry matter by near infrared spectroscopy. Scienceasia1 36(2010): 210–215.

Matich, A.J., Young, H., Allen, J.M., et al. 2003. *Actinidia arguta*: volatile compounds in fruit and flowers. *Phytochemistry* 63(3):285–301.

Matich, A.J., Bunn, B.J., Hunt, M.B., et al. 2006. Lilac alcohol epoxide: A linalool derivative in *Actinidia arguta* flowers. *Phytochemistry* 67(8):759–763.

Matich, A.J., Bunn, B.J., Comeskey, D.J., et al. 2007. Chirality and biosynthesis of lilac compounds in *Actinidia arguta* flowers. *Phytochemistry* 68(13):1746–1751.

Matich, A.J., Young, H., Allen, J.M., et al. 2013. *Actinidia arguta*: volatile compounds in fruit and flowers. *Phytochemistry* 63(3): 285–301.

Mariya, G., Ivana, N., Kiril, M., Nikolina, Y., Jasenka, G.K., Želimir, K. 2013. Application of NIR spectroscopy and chemometrics in quality control of wild berry fruit extracts during storage. *Croatian Journal of Food Technology, Biotechnology and Nutrition* 8(3–4): 67–73.

Nishiyama, I., Yamashita, Y., Shimohashi, A., et al. 2004. Varietal difference in vitamin C content in the fruit of kiwifruit and other *Actinidia* species. *Journal of Agricultural and Food Chemistry* 52(17): 5472–5475.

Patricia, P., Maria, S., Dolores, P., Jose, G. & Ana, G.J. 2008. Nondestructive determination of total soluble solid content and firmness in plums using near-infrared reflectance spectroscopy. *Journal of Agriculture and Food Chemistry* 56: 2565–2570.

Teye, E., Huang, X.Y., Sam-Amoah, L.K., Takrama, J., Boison, D., Botchway, F. & Kumi, F. 2015. Estimating cocoa bean parameters by FT-NIRS and chemometrics analysis. *Food Chemistry* 176(6):403–410.

Tsai, C.Y., Sheng, C.T., Chen, S., Chen, H.J., Chiu, Y.C., Jiang, J.A., Hsieh, K.W. & Hung, R.C.H. 2008. Development of a Continuous Online Detecting System for Fruits Using Near Infrared Technology. Proceedings of the 4th International Symposium on Machinery and Mechatronics for Agriculture and Biosystems Engineering (ISMAB) 27–29 May 2008, Taichung, Taiwan.

Williams, P., Norris, K.H. 2001. Vuriable Agriculture and Food Industrics. Second ed. St. *The Association of Cereal Chemist*: 171–185.

Zhang, P., Li, J.K., Meng, X.J., Zhang, P., Wang B.G. & Feng, X.Y. 2011. Nondestructive Determination of Soluble Solid Content in Mopan persimmon by Visible and Near-infrared Diffuse Reflection Spectroscopy. *Food Science*. 32(6): 191–194. (in China)

Analysis and control of bacteria flora found in the dish-marinated bitter melon and cucumber

Qinghua Yu
Inner Mongolia Finance University, Hohhot, Inner Mongolia, China

Yunsheng Jiang & Xi Liu
Yangzhou University, Yangzhou, Jiangshu, China

ABSTRACT: First, a basic formula and the traditional preparing method of Marinated Bitter Melon and Cucumber were decided based on experiments. Then the bacteria reducing rate between the main raw Materials and finished products and the change of bacterial flora the two samples stored in the refrigerator is tested. Results show that total bacteria count found in the dish prepared by traditional method is 5.4×10^4 CFU/g. 69.5% of the bacteria comes from bitter melon, 29% from cucumber, and 1.5% from onion; while the total bacteria count found in the dish prepared by scalping method is 2.3×10^3 CFU/g, reduced by 57.4%. The finding is: the contamination of Marinated Bitter Melon and Cucumber prepared by the traditional method is severe; however, the situation improved greatly using scalping technics and the shelf life of the dish is prolonged significantly. Therefore, the risks of food poisoning can be lowered. The finding is meaningful in terms of safe handling of this kind of foods.

1 INTRODUCTION

In recent years, Catering food safety, the end of the food chain, is often overlooked in China, resulted in frequently occurred bacteria-caused food poisoning. Cold dishes are the primary cause of most of these cases (Zhang, 2005), for vegetables, the main Materials of cold dishes, are commonly grown in unprocessed manure containing great amount of bacteria. Peppers, bitter melon, cucumber, onion and lettuce are especially at high risks of being contaminated. Traditional food preparing methods have proved to be less effective to decontaminate (Wang, 2003). Marinated Bitter melon and Cucumber, enjoying the reputation of easy-cooking, nice-looking and fresh-tasting, is one the classic dishes of Manchu Han Imperial Feast and has turned into a common cold dish often consumed in hotels, restaurants and at homes. In this essay, the authors will discuss the possible hurdle factors to provide tangible data for improving food safety of similar dishes.

2 MATERIALS AND METHODS

2.1 *Materials*

Bitter melon, cucumber, onion are raw and fresh; spices and seasonings, such as salt, monosodium glutamate and pepper are all bottled or bagged products, bought from Oushang supermarket, Yangzhou.

2.2 *Reagents and culture medium*

Nutrient AGAR, PSA, MRS, VRBGA and MSA, prepared in our lab, were the culture mediums used (Jiang, 2007). Beef extract, bacto-tryptone, yeast extract, and sweet red were the bio-reagents used. Gram Stain was prepared according to relevant literatures (Liu, 2009).

2.3 Instruments and equipments

XS-18 biological microscope, manufactured by Jiangnan Optical Instruments Company; pHS-3C pH meter, HG101 electric blast drying oven, manufactured by Jiangnan Optical Instruments Company; HG303 electric drying incubator, manufactured by Changshu Shuangjie Testing Instruments Company; LDZX-40B2 Automatic electric pressure vertical steam sterilizer, manufactured by Shanghai Shenan Medical Instruments Company; SW-CJ-1F clean bench, manufactured by Suzhou Cleaning Equipment Company.

2.4 Methods

2.4.1 Deciding the basic formula of marinated bitter melon and cucumber

After referencing relevant literatures and conducting experiments, the final basic formula of the dish is decided as in Table 1.

2.4.2 Preparing marinated bitter melon and cucumber using traditional method

Clean and remove the seeds of the bitter melon, then cut it into thin slices. After that, salt them for about 10–15 minutes to reduce the bitterness. Next, wash by cool boiled water and drain. Wash, peel and remove the seeds of the cucumber, then cut it into thin slices and salt them. Wash the onion, and chop it into small pieces. Put prepared bitter melon, cucumber and onion into a bowl, and add salt, vinegar, and black pepper to be ready for the test (Zhang, 1994).

2.4.3 The investigation and analysis of the quantity and source of bacteria tainted the dish-marinated bitter melon and cucumber

We test the total bacteria count found in the main-Materials and ready-to-served dish respectively (GB4789.2, 2010) and employ weighted mean to identify the source of bacteria, using the number of bacteria found in the main-Materials as the variance, and the weight of the main-Materials as the weight (Jiang, 1997).

2.4.4 Bacteria control and setting up the new method of preparing marinated bitter melon and cucumber

Wash and peel the bitter melon, remove the seeds and cut it into thin slices; wash and remove the seeds of the cucumber and cut it into thin slices; wash and chop the onion into small pieces. Put the three Materials into a container, scalping for five 5 minutes and then test the number of bacteria to form the new method of making Marinated Bitter melon and Cucumber. The bacteria testing results of the new method is compared to the traditional one to count the percentage of bacteria reduced under the new method.

2.4.5 The change of hygiene condition of marinated bitter melon and cucumber while cold stored

Put the ready-to-serve Marinated Bitter melon and Cucumber prepared by new and traditional method into refrigerator at 4 °C and take out to test the total bacterial count, the number of Pseudomonas, enterobacter, and lactic acid bacteria every 24 hours, respectively. At the same time, do the taste tests, evaluate their hygiene conditions and thus to define the shelf life of the dish prepared by new method (Huang, 1991).

Table 1. The basic formula of Marinated Bitter Melon and Cucumber.

Ingredients	Onion	Bitter Melon	Cucumber	Vinegar	Salt	Monosodium glutamate	Pepper powder
Weight (g)	20	200	200	20	4	2	1.5

3 RESULTS AND ANALYSIS

3.1 *Contamination investigation and the source analysis of marinated bitter melon and cucumber*

Table 2 shows that the number of bacteria carried by bitter melon tops the list, 6.0×10^3 CFU/g; and cucumber comes the second, 2.5×10^3 CFU/g; onion bottoms the list, 1.3×10^3 CFU/g. Using total bacteria count as variance and the usage as weight, it can be concluded that the theoretical number of bacteria in the finished product made by traditional method is 3.9×10^3 CFU/g. As far as the source of bacteria is respectively: 1.5% of the bacteria come from onion, 69.5% from bitter melon, and 29% from cucumber. Vinegar carries little bacteria and can be ignored.

The total bacterial count of the finished dish is as high as 5.4×10^3 CFU/g, which indicates that the number of bacteria also increases in kitchen operation process. Factors, such as the variety of raw Materials, contacting times of raw Materials with bacteria tainted knives and chopping boards, length of food exposed to room temperature, contacting length of food with bacteria-carrying aerosols, all contribute to the dramatic rise in the actual number of bacteria in the finished salad.

3.2 *Bacteria control of the dish-marinated bitter melon and cucumber*

Table 3 clearly indicates that scalping kills 57.4% of bacteria contaminated the finished dish and reduces nearly half of the bacteria found in all main-Materials. Therefore, scalping is a very effective method to decontaminate, can best control the number of bacteria tainted cold dishes.

3.3 *The change of mircobial commnuities tainted the dish-marinated bitter melon and cucumber while cold stored at 4 °C*

3.3.1 *The change of bacteria quantity tainted the dish-marinated bitter melon and cucumber while cold stored at 4 °C*

For the observing result of the change of Bacterial flora, please refer Figures 1 and 2.

Table 2. Testing result of the number of bacteria found in Marinated Bitter melon and Cucumber prepared by traditional method.

Items	Weight (g)	Operating duration (min)	Total bacterial count (CFU/g)
Onion	20	6	1.3×10^3
Bitter melon	200	10	6.0×10^3
Cucumber	200	8	2.5×10^3
Vinegar	20	1	<10
Finished dish prepared by traditional method	440	25	5.4×10^3

Table 3. Comparison of total bacteria count in ingredients prepared by traditional and updated methods.

Items	Total bacterial count of unprocessed ingredients (CFU/g)	Total bacterial count of scalped ingredients (CFU/g)	Percentage of bacteria reduced (%)
Onion	1.3×10^3	6.5×10^2	50.0
Bitter melon	6.0×10^3	3.1×10^4	48.0
Cucumber	2.5×10^3	1.1×10^3	56.0
Finished dish	5.4×10^3	2.3×10^3	57.4

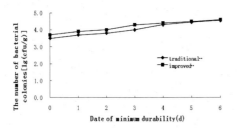 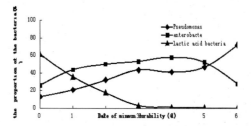

Figure 1. The change of total bacteria count in the dish-Marinated Bitter Melon and Cucumber while cold stored at 4 °C.

Figure 2. The observing result of the change of bacterial flora found in traditional method prepared Marinated Bitter Melon and Cucumber.

Figure 1 shows that the number of bacteria found in traditional method prepared dish is steadily but slowly increasing and peaks at day 6 thanks to the low temperature. In contrast, the total bacteria count in updated method prepared dish is much less than that in the initial stage, which indicates the effectiveness of cold shock treatment.

3.3.2 *The change of percentage of different bacteria tainted the dish-marinated bitter melon and cucumber while cold stored at 4 °C*

Table 4 and Figure 2 indicate that the microbial flora in Marinated Bitter Melon and Cucumber Prepared by traditional method initially consist of Lactic Acid Bacteria (61%), Pseudomonas (26%) and Enterobacter (13%). Cold stored at 4 °C, microbial flora in Marinated Bitter Melon and Cucumber Prepared by traditional method changed dramatically. Characteristics of these changes are: first, the number of Lactic acid bacteria is the largest initially but drop dramatically in first 3d to less than 3% till finally reaches zero at the end of shelf life. Second, Enterobacter is a sub-primary bacteria initially but jumps to be the primary only in 1 d and keeps the status till day 5 when it outnumbered by Pseudomonas. Third, Enterobacter grows so fast in the first 3 d and its number nearly equals that of Enterobacter in day 4 and 5, finally in day 6 it greatly outnumbered Pseudomonas. Putrefactive bacteria among the microbial flora in Marinated Bitter Melon and Cucumber are Pseudomonas, account for 72% and Enterobacter, account for 28%. The primary bacterium is Pseudomonas. The rotten process of the dish is co-conducted by these two kinds of bacteria.

Table 4 and Figure 3 indicate that the microbial flora in Marinated Bitter Melon and Cucumber Prepared by new method initially consisted of Pseudomonas (18%), Enterobacter (62%) and Lactic Acid Bacteria (20%). Cold stored at 4 °C, microbial flora in Marinated Bitter Melon and Cucumber prepared by traditional method has changed. Characteristics of these changes are: first, the number of Enterobacter remained the biggest in the whole 6 d and increases steadily. In the peak time, it accounted for 95% of the whole. Second, the percentage of Lactic acid bacteria accounted for drops quickly to 6% in day 1 and leveled for 2 d till nearly bottomed to zero in day 4. Third, the percentage of Pseudomonas accounted for increases to 27% in day 1 and then sharply dropped to 9% and stabled there. The results show that Enterobacter is the primary bacteria and Enterobacter and Lactic Acid Bacteria were responsible for the rotten process.

To comparing the two findings, we can see that the similarities are: the number of Lactic Acid Bacteria dropped sharply and became zero gradually; Enterobacter and Lactic Acid Bacteria were responsible for the rotten process. The difference is that Enterobacter was the primary bacteria initially and then outnumbered by Pseudomonas when testing the dish made by traditional method; while Enterobacter was the primary bacteria right through the end when testing the dish made by new method.

The overall statistic shows that Pseudomonas and Enterobacter were primary bacteria relatively and were responsible for the rotten process of the dishes stored at 4 °C together. Scalping can clearly lower the percentage of Pseudomonas, effectively suppressed the growth of Pseudomonas and Lactic Acid Bacteria and greatly reduced the number of

Table 4. The change of Percentage of different bacteria tainted the dish-Marinated Bitter Melon and Cucumber while cold stored at 4 °C.

Shelf life (d)	Percentage of different bacteria (%)					
	Pseudomonas		Enterobacter		Lactic acid bacteria	
	Traditional method	Updated method	Traditional method	Updated method	Traditional method	Updated method
0	12.9	18.4	25.8	61.7	61.3	19.9
1	20.7	26.8	43.6	67.1	35.7	6.1
2	32.2	8.1	50.2	85.4	17.6	6.5
3	43.5	5.4	53.4	89.6	3.1	5.0
4	41.1	4.4	57.9	95.2	1.0	0.4
5	46.8	3.6	52.5	96.1	0.7	0.3
6	72.2	9.1	27.7	90.6	0.1	0.3

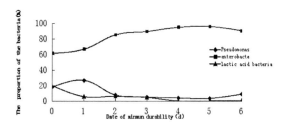

Figure 3. The observing result of the change of bacterial flora found in Marinated Bitter Melon and Cucumber prepared by the new method.

these two kinds of bacteria. Therefore, it has a satisfactory effect on bacteria suppressing. However, scalping has little effect on Enterobacter. The percentage of Enterobacter in finished dish prepared by new method is bigger than that in traditional prepared one. For the dishes stored at 4 °C, scalping has not changed the composition of Putrefactive bacteria flora.

3.4 *The sensory quality changes of marinated bitter melon and cucumber stored at 4 °C*

Sensory test shows that the taste of Marinated Bitter Melon and Cucumber, stored at 4 °C. Made by traditional method changed slightly and the freshness of it decreased on day 4. Therefore, the shelf life of the dish is 3 d. While, the taste of the dish made by new method didn't change until day 5, which indicated the shelf life of this dish should be 5 d. Comparing with the traditional method, the new method can make the shelf life of Marinated Bitter Melon and Cucumber 40% longer. The result tells that the new method involving scalping technic can restrain the growth of Pseudomonas and Enterobacter which greatly benefits food safety improving of Marinated Bitter Melon and Cucumber and prolongs its shelf life.

4 DISCUSSIONS

4.1 *Source of harmful bacteria tainted marinated bitter melon and cucumber*

The analysis shows that most of the Materials carrying bacteria and the bacterial flora are complex, for the microorganism mainly comes from soil, air, water, wild or tamed animals, insects, birds and machines and has different preference in terms of host vegetables. Main bacteria types are Lactic acid bacteria, Enterobacter, Pseudomonas, Corynebacteriumsome, Proteus, Micrococcus, enterococcus, and Bacillus (Ray, 2014). The bacteria carrying raw Materials without being decontaminated by scalping before made into finished dish will definitely make further contamination. Therefore, there will most likely stay various kinds of bacteria in finished dishes. Thus increases food poisoning risks. Around 35% of vegetable and fruit rotten are caused by bacteria; therefore, bacteria control should be prioritized (Bai, 2010).

4.2 *Analysis of the microbial flora in marinated bitter melon and cucumber*

The microbial flora in Marinated Bitter Melon and Cucumber mainly consists three kinds of bacteria: Lactic acid bacteria, Pseudomonas and Enterobacter. Lactic acid bacteria likes warm temperature, and best situated to live at the temperature between 30–37 °C. Some bacteria can live in temperature as low as 1 °C and as high as 45 °C. The habitable pH reading for this kind of bacteria is between 4.0~6.5. Lactic acid bacteria grows slowly at 4 °C and finally is supressed by advantage bacteria. Pseudomonas enjoys cold temperature, can grow comfortably at 4 °C. The habitable pH reading for this kind of bacteria is between 6.0–8.0. while, Enterobacter can only live in the environment with a 7.4 pH reading. Lactic acid bacteria is the primary bacteria in the dish made by traditional method initially; while when the micro environment is too acid to host Enterobacter on day 5, Enterobacter was outnumbered by Pseudomonas. However, in the dished made by new method, Lactic acid bacteria accounts for only a small percentage of the total bacteria count and thus produced less acid. This fact again make the Ph reading of the finished dish within the habitable range of Enterobacter. That is the reason why Enterobacter has been the primary bacteria right through the end.

4.3 *Microorganism control of marinated bitter melon and cucumber*

The experimental results show that the bacterial contaminations of the cold dish mostly originate from the kitchen. Bacteria originally attached to vegetables and fruits are spread to other places while the raw Materials are washed, peeled, chopped and seasoned. Once the dish stored, bacteria multiplies and the risks increase. Therefore, bacterial control of raw Materials is essential to the safety of cold dishes. Scalping technic used in our experiment proves to be effective to reduce total bacteria count. Meanwhile, the use of vinegar also has satisfactory results (Zhang, 2007). Besides, the employing of hurdle factors, such as preheating, lowering pH reading by adding vinegar and cold storage also clearly contributes to the lowered bacteria count found in the dish prepared by updated method. Bacterial hazard control cannot succeed without routine hygiene practices such as staff health check, hand washing, dedicated containers, cold storing of food and raw Materials and expiration dates labeling. Other factors, especially consciousness of sterile operation, no direct hand contact with food, increased machining operating level will all achieve in bacteria control (Guan, 2004). Studying possible updated cold dish preparation methods, setting up the maximum tolerance of total bacteria count, perfecting food preparation (such as storing conditions and shelf life) regulations will have positive effects on the setting up of food safety standards of cold dishes, regulating catering operation, strengthening government monitoring and protecting consumer rights.

5 CONCLUSION

The new method involving scalping the raw Materials can effectively reduce the number of colonies in raw dishes, such as marinated bitter melon and cucumber, prolong their shelf life while refrigerated and lower the risks of food poisoning.

The essay is supported by Science and Technology Development Program of China (no. 2013GA690252), and is supported. by Inner Mongolia social situation and Public Opinion Research Center of China (no. SQ1310).

REFERENCES

Bai, X.P. 2010. Hazards of Unsafe Foods and Their Control Measures. Beijing: China Metrology Publishing House.
Guan, X.Q. 2004. Microbial Control in Freshly Cut Fruits and Vegetables. Journal of Xinjiang Chemical Industry. 3:51–53.
GB4789.2.2010. People's Republic of China National Standard on Food Safety Microbiotical Testing-Total Bacterial Count. Beijing: China Standard Publishing House.
Huang, X..Z. 1991. The Storage Conditions and Shelf-life of Some Vegetables. North Gardening. 9:32.
Jiang, Y.S. 2007. Cooking and Microorganism. Beijing: China Light Industry Press.
Jiang Y.S. & Chen. J. 1997. A Study on the Hygiene of the Cold Dish-Sanmian. China Cooking Studies. 3:26–30.
Liu, H. 2009. Modern Experiment Technology on Food Microbiology. Beijing: China Light Industry Press.
Ray, B. & Bhuni, A. 2014. Fundamental Food Microbiology. H Jiang Translated. Beijing: China Light Industry Press.
Wang, H.R. 2003. Safety and Hygiene of Raw Vegan Recipes. China Food and Nutrition. 12:21.
Zhang, C.Y. & Lu, X.Q & Teng, H.S. 2007. Vinegar bacteria killing nature and its effectiveness. Qilu Medical Journal. 22(3):196–198.
Zhang, S.W. 2005. Hot Issues Studied on the Connotation of Food Safety and Setting up Related Regulations. Food Technologies, 9:1–6.
Zhang, Z.H & Yu, J. 1994. A Cookbook on Raw Food Recipes. Shanghai: Shanghai Far East Publishing House.

Efficient asymmetric synthesis of (S)-3-phenyllactic acid by using whole cells of recombinant E.coli

Yibo Zhu, Zhouqun Huang, Limei Wang & Bin Qi
Changshu Institute of Technology, Changshu, Jiangsu, P.R. China

Ying Wang
Jilin Agricultural University, Changchun, Jilin, P.R. China

ABSTRACT: (S)-3-phenyllactic acid [(S)-PLA], an excellent building block widely applied in the pharmaceutical and chemical industries, has recently been discovered to be a potential antiseptic agent. However, compared with (R)-PLA, there have been few reports on the production of (S)-PLA. Moreover, the product yield and concentration of reported (S)-PLA synthetic processes remain unsatisfactory. Recombinant E.coli BL21(DE3)/pET28a-*ldh*L overexpressing L-lactate dehydrogenase (*l-ldh*) from *Lactobacillus plantarum* subsp. *plantarum* ATCC 14917 was constructed. In the asymmetric reduction of PPA by recombinant E. coli cells in aqueous system, the yield of (S)-PLA reached 77.26 mmol L^{-1} with an enantiomeric excess of 99.08% from total 149.25 mmol L^{-1} of PPA after 2.5 h of fed batch bioconversion with intermittent PPA feeding. Given the high product enantiomeric excess and productivity (30.90 mmol L^{-1} h^{-1}), (S)-PLA was efficiently produced by the one-pot biotransformation system. Thus, a novel biocatalysis process was developed as a promising alternative for (S)-PLA production.

1 INTRODUCTION

3-Phenyllactic acid (2-hydroxy-3-phenylpropanoic acid or β-phenyllactic acid, PLA), a specialty intermediate widely used in the pharmaceutical and chemical industries, has recently emerged as a potential antiseptic agent with broad spectrum and effective activity against both bacteria and fungi and as feed additive to replace antibiotics in livestock feeds (Dieuleveux et al. 1998, Lavermicocca et al. 2003, Wang et al. 2009). 3-PLA possesses a chiral carbon atom in its molecule, and occurs as two enantiomers, (S)-PLA and (R)-PLA, which both are versatile chiral building blocks but different in biological functions and the applications in pharmaceutical and chemical industries (Zheng et al. 2011). Therefore, the preparation of optically pure chiral isomers has been one of the major tasks and the centered subjects in chiral chemical industry.

Various approaches for the preparation of PLA have been reported in recent years, including traditional chemical methods, enzymatic routes and biological methods (Kawaguchi et al. 2014). Chemical processes for chiral isomers production result in the racemic mixture of both stereospecific forms. On the other hand, biocatalytic asymmetric synthesis has highly attracted attentions from industry and academia recently because of its high conversion rate and environment-friendly nature, especially outstanding stereo selectivity (Mu et al. 2012). Furthermore, difficult processes for separation and purification of enzyme can be avoided by utilizing microbial whole cells as a biocatalyst. Therefore, microbial whole-cell asymmetric biocatalysis has become the most promising and first choice method in the asymmetric synthesis.

So for, PLA has been reported to be produced by many microorganisms, mostly lactic acid bacteria, e.g. *Lactobacillus* (Lavermicocca et al. 2000, Li et al. 2008, Prema et al. 2010, Ström et al. 2002), *Enterococcus* (Ohhira et al. 2004), and *Leuconostoc* (Valerio et al. 2004). It has not been reported much about (*S*)-PLA production by means of the asymmetric synthesis of microorganism, compared with (*R*)-PLA. A *Pseudomonas* sp. BC-I8 strain could convert racemic 3-phenyllactonitrile to (*S*)-PLA with only 75% enantiomeric excess (Hashimoto et al. 1996). The fed-batch fermentation of *Lactobacillus* sp. SK007 produced 17.4 g L^{-1} PLA with a productivity of 0.24 g L^{-1} h^{-1} (Mu et al. 2009). Whole cells of *Bacillus coagulans* SDM converted PPA to 37.3 g L^{-1} (PLA) with a productivity of 2.3 g L^{-1} h^{-1} (Zheng et al. 2011). Rubrivivax benzoatilyticus JA2 yielded a maximum of 0.92 mmol L^{-1} (*S*)-PLA from L-phenylalanine (1 mmol L^{-1}) when fructose served as carbon source (Prasuna et al. 2012). Therefore, the product yield and concentration of reported (*S*)-PLA synthetic processes remain unsatisfactory.

Lactate dehydrogenases are of considerable interest as sterospecific catalysts in the chemical preparation of enantiometrically pure pharmaceutical intermediates (Kallwass et al. 1992, Leonida et al. 2003). L-Lactate dehydrogenase (L-LDH) exhibits a wide variety of catalytic properties including catalyzing the conversion of PPA to (*S*)-PLA with the concomitant oxidation of NADH. In this study, the L-lactate dehydrogenase gene *l-ldh* of *L. plantarum* ATCC14917 was cloned and then introduced into *E.coli* BL21(DE3) through vector pET-28a(+). Then, L-LDH over-expressing recombinant *E.coli* was constructed to enhance the production of (*S*)-PLA. The engineering *E. coli* BL21(DE3) was constructed as new biocatalyst for the asymmetric production of (*S*)-PLA in this study which optical purity exceeded 99%.

2 MATERIALS AND METHODS

2.1 *Materials*

Restriction enzymes *Nde*I and *Xho*I, T4 DNA ligase, Ex Taq DNA polymerase and pMD19-T (Simple) cloning vector were purchased from TaKaRa Biotechnology Co. The expression vector pET-28a(+) was obtained from Novagen. PCR related products, Isopropyl-B-D-Thiogalactopyranoside (IPTG), kanamycin were purchased from Sangon Biotech Co., China. (*S*)-PLA, (*R*)-PLA and phenylpyruvic acid in the chemical form of sodium Phenylpyruvate (PPA), were purchased from Sinopharm Chemical Reagent Co., Ltd. (Shanghai, China). All other reagents and solvents were of analytical grade. The bacterial strains, plasmids and primers used in this study are listed in Table 1. *E. coli* was grown at 37 °C in Luria-Bertani (LB) medium and kanamycin was added at a concentration of 50 µg mL^{-1}, if necessary.

2.2 *Cloning and expression of l-lactate dehydrogenase*

All DNA manipulations were carried out according to standard protocols and instructions. The *l-ldh* gene was amplified from Lactobacillus plantarum subsp. plantarum ATCC 14917, introduced into the pET-28(+) and transformed into *E. coli* BL21(DE3). The total genomic

Table 1. Stains and plasmids in this study.

Strain, plasmid or primer	Relevant characteristics	Source
L. plantarum ATCC 14917	Source of *l-ldh*	CGMCC
E.coli JM109	Cloning host	Novagen
E.coli BL21(DE3)	Expression host	Novagen
pMD19-T (Simple)	Cloning vector	TaKaRa
pET-28a(+)	Expression vector	Novagen
P1(5′→3′)	GCGCC*CATATG*TCAAGCATGCCAAATC	This study
P2(5′→3′)	GCCGCC*CTCGAG*CGTTATTTATTTTCTAATTCAG	This study

The italic character indicates the introduction of restriction sites.

DNA of Lactobacillus plantarum subsp. plantarum ATCC 14917 was extracted with a bacterial genomic DNA extraction kit (Sangon, China) and then used as the template for the amplification of *l-ldh* gene. Two oligonucleotide primers P1 and P2 for *l-ldh* amplification designed based on its nucleotide sequence (GenBank No. ACGZ02000027.1) are listed in Table 1. The amplified DNA fragment was purified and ligated into the pMD19-T by T4 DNA ligase to construct the recombinant plasmid pMD19-T-*ldh*L according to the manufacturer's instructions. The recombinant plasmids were subsequently transferred to competent *E. coli* JM109 cells. DNA sequencing was performed by Sangon Biotech Co. The pMD19-T-*ldh*L was double digested with NdeI and XhoI and then inserted into the expression vector pET-28a. The resulting plasmid, pET28a-*ldh*L, was transformed into *E. coli* BL21(DE3), and positive clones were screened and confirmed by colony PCR and restriction enzyme digestion.

2.3 *Protein expression and SDS-PAGE analysis*

For protein expression, *E. coli* BL21(DE3) harboring recombinant plasmid was incubated in LB media (containing 50 μg mL^{-1} kanamycin) at 37°C with shaking. When the culture reached an optical density of 0.6 at 600 nm, 1 mmol L^{-1} isopropyl-b-D-thiogalactopyranoside (IPTG) was added to induce gene expression. After induction at 25 °C for 5 h, cells were harvested by centrifugation at 12,000 × g for 5 min and then diluted to an OD$_{600}$ of 1 by 100 mmol L^{-1} potassium phosphate buffer (pH 7.0). 1 mL of the resultant suspension was disrupted for 10 min at 30 °C by B-PER (Germany, Thermo). Cell debris was removed by centrifugation at 12,000 × g for 10 min. The obtained supernatant was used for crude enzyme activity assays and Sodium Dodecyl Sulfate-Polyacrylamide Gel Electrophoresis (SDS-PAGE). Protein concentration was determined by SDS-PAGE analysis performed according to standard procedures (Bradford 1976).

2.4 *Preparation of E.coli BL21(DE3)/pET28a-ldhL whole cells*

For whole cell preparation, overnight precultures of strain *E.coli* BL21(DE3)/pET28a-*ldh*L were inoculated into fresh LB liquid medium (50 μg mL^{-1}) with 1% inoculum and the cultures were incubated at 37 °C and 200 rpm. When the culture reached an OD$_{600}$ of 1.2, IPTG was added to induce *l-ldh* gene expression. After induction at 25 °C for 5 h, the culture was centrifuged (10,000 g for 10 min, 4 °C) and the cell pellets washed twice with 100 mmol L^{-1} potassium phosphate buffer (pH 7.0) and resuspended in the same buffer to form cell suspensions of 15 g dry cell wt L^{-1}.

2.5 *Asymmetric reduction of PPA to (S)-PLA by using whole-cell system*

Fed-batch bioconversion was performed with intermittent PPA feeding to increase PLA yield and avoid substrate inhibition. The bioconversion process was performed in 250-mL flasks containing 50 mL of the reaction mixtures. The initial reaction mixtures (50 mL) containing 100 mmol L^{-1} potassium phosphate buffer (pH 7.0), glucose (20 g/L), PPA (50 mmol L^{-1}) and cells (10.2 g DCW L^{-1}) were incubated at 37 °C and 200 rpm for 2.5 h. PPA dissolved in PBS in advance and glucose powders were supplemented to maintain the initial concentration by a pulse-feeding strategy. And then the samples were centrifuged at 10,000 g for 10 min, 4 °C. The concentrations of PPA and PLA in the resulting supernatants were quantitatively analyzed by High-Performance Liquid Chromatography (HPLC). The time course of PLA production from PPA was shown in Figure 2. All experiments were performed in triplicate.

2.6 *Analytical procedures*

PPA, and PLA concentrations were measured by HPLC (Shimadzu, LC-10 AT) equipped with a BioRad Aminex HPX-87H column, a detector (SPD-20 AV, Shimadzu) at 210 nm. The mobile phase was 5 mmol L^{-1} H$_2$SO$_4$ at a flow rate of 0.6 mL min^{-1} at 30 °C.

PLA enantiomeric excess was measured by HPLC (Shimadzu, LC-10 AT) equipped with a CHIRALCEL OJ-RH column (OJRHCD-SJ012), a detector (SPD-20 AV, Shimadzu) at 210 nm. The mobile phase was water/methanol/acetonitrile solvent mixture (900:50:50 v/v) containing 0.15% trifluoroacetic acid at a flow rate of 0.3 mL min^{-1} at 30 °C.

3 RESULTS AND DISCUSSION

3.1 *Cloning and expression of l-ldh and SDS-PAGE analysis*

The sequence analysis indicated that the *l-dh* (963 bp) encoded 320 amino acids. The cloned *l-ldh* showed 100% similarity with the nucleotide sequence of *L. plantarum* ATCC 14917 *l-ldh* (GenBank No.ACGZ02000027), and 100% similarity with its amino acid sequence. Figure 1 shows the SDS-PAGE of the cell-free extract of *E. coli* BL21(DE3)/pET28a-*ldh*L. The recombinant L-LDH product was observed at 41 kDa, which suggested the recombinant enzyme was overexpressed successfully.

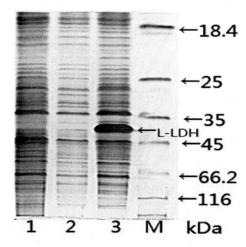

Figure 1. SDS-PAGE analysis of recombinant protein expression of *E. coli* BL21(DE3)/pET28a-*ldh*L. Lane M: protein molecular weight marker; Lane 1: soluble proteins of *E. coli* BL21(DE3); Lane 2: whole cell extract of *E. coli* BL21(DE3)/pET-28a; Lane 3: soluble proteins of *E. coli* BL21(DE3)/pET28a-*ldh*L.

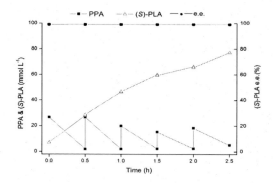

Figure 2. Time course of (*S*)-PLA production in fed-batch bioreduction with intermittent PPA.

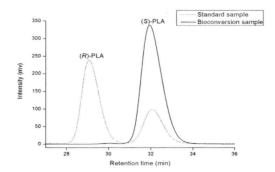

Figure 3. Chiral HPLC analysis of PLA produced by *E.coli* BL21(DE3)/pET28a-*ldh*L.

3.2 *Asymmetric bioreduction of PPA to (S)-PLA catalyzed by recombinant E. coli*

The time course of the bioreduction of PPA to (S)-PLA in an aqueous system catalyzed by *E. coli* BL21(DE3)/pET28a-*ldh*L is shown in Figure 2. After 2.5 h of reaction, 77.26 mmol L^{-1} of (S)-PLA with 99.08% *e.e.* was achieved at 149.25 mmol L^{-1} of PPA.

4 CONCLUSION

So far, few reports has been shown about the production of (S)-PLA. The L-lactate dehydrogenase was successfully instructed and overexpressed in *E.coli*, which exhibited high enantioselectivity and high efficiency in the reduction of PPA to (S)-PLA. Although the (S)-PLA biosynthesis system developed in the present study was proven convenient and highly efficient, further studies on the optimization of the reaction process including cofactor regeneration of NADH, substrate feed batch, aqueous/organic biphasic system, and pH regulation were carried out to further enhance the reaction efficiency and substrate conversion rate.

REFERENCES

Bradford, M.M. 1976. A rapid and sensitive method for the quantitation of microgram quantities of protein utilizing the principle of protein-dye binding. *Analytical Biochemistry* 72(1):248–254.
Dieuleveux, V. 1998. Antimicrobial spectrum and target site of D-3-phenyllactic acid. *International Journal of Food Microbiology* 40(3):177–183.
Hashimoto, Y. 1996. Conversion of a cyanhydrin compound into S-(-)-3-phenyllactic acid by enantioselective hydrolytic activity of *Pseudomonas sp*. BC-18. *Biosci Biotechnol Biochem* 60(8):1279–1283.
Kawaguchi, H. 2014. Simultaneous saccharification and fermentation of kraft pulp by recombinant *Escherichia coli* for phenyllactic acid production. *Biochemical Engineering Journal* 88:188–194.
Lavermicocca, P. 2000. Purification and characterization of novel antifungal compounds from the sourdough *Lactobacillus plantarum* strain 21B. *Applied and Environmental Microbiology* 66(9):4084–4090.
Lavermicocca, P. 2003. Antifungal activity of phenyllactic acid against molds isolated from bakery products. *Applied and Environmental Microbiology* 69(1):634–640.
Li, X. 2008. Purification and partial characterization of *Lactobacillus* species SK007 Lactate Dehydrogenase (LDH) catalyzing Phenylpyruvic acid (PPA) conversion into Phenyllactic Acid (PLA). *J Agric Food Chem* 56(7):2392–2399.
Mu, W. 2009. 3-Phenyllactic acid production by substrate feeding and pH-control in fed-batch fermentation of *Lactobacillus sp*. SK007. *Bioresource Technology* 100(21):5226–5229.
Mu, W. 2012. Recent research on 3-phenyllactic acid, a broad-spectrum antimicrobial compound. *Appl Microbiol Biotechnol* 95(5):1155–1163.
Ohhira, I. 2004. Identification of 3-phenyllactic acid as a possible antibacterial substance produced by *Enterococcus faecalis* TH10. *Biocontrol Science* 9(3):77–81.

Prasuna, M.L. 2012. l-Phenylalanine catabolism and l-phenyllactic acid production by a phototrophic bacterium, *Rubrivivax benzoatilyticus* JA2. *Microbiological Research* 167(9):526–531.

Prema, P. 2010. Production and characterization of an antifungal compound (3-phenyllactic acid) produced by *Lactobacillus plantarum* strain. *Food and Bioprocess Technology* 3(3):379–386.

Ström, K. 2002. *Lactobacillus plantarum* MiLAB 393 produces the antifungal cyclic dipeptides cyclo (L-Phe-L-Pro) and cyclo (L-Phe-trans-4-OH-L-Pro) and 3-phenyllactic acid. *Applied and Environmental Microbiology* 68(9):4322–4327.

Valerio, F. 2004. Production of phenyllactic acid by lactic acid bacteria: an approach to the selection of strains contributing to food quality and preservation. *Fems Microbiology Letters* 233(2):289–295.

Wang, J. 2009. Effects of phenyllactic acid on growth performance, nutrient digestibility, microbial shedding, and blood profile in pigs. *Journal of Animal Science* 87(10):3235–3243.

Zheng, Z. 2011. Efficient conversion of phenylpyruvic acid to phenyllactic acid by using whole cells of *Bacillus coagulans* SDM. *PloS one* 6(4):e19030.

Immunohistochemistry applications in breast carcinoma: The Hypoxia Inducible Factor 1α, Estrogen Receptor, Progesterone Receptor and HER2 expression before and after neoadjuvant chemotherapy

Zuofeng Zhang, Yunai Liang, Chunhua Zhang, Gangping Wang & Shicai Hou
Affiliated Rizhao People's Hospital of Jining Medical University, Rizhao, China

Guangye Pan
Weifang Medical College, Weifang, China

ABSTRACT: Biomarkers such as Hypoxia Inducible Factor 1α (HIF-1α), Estrogen Receptor (ER), Progesterone Receptor (PR) and HER2 profile mammary carcinoma tissue and Malignant Tumor Specific Growth Factor (TSGF), Cancer Antigen 153 (CA153) in serum plays a significant role in patient management and treatment and can be used as prognostic factors. Aim of this study was to study the express of ER, PR, HER2 and HIF-1α in tissue and serum CA153 and TSGF levels in judging the treatment effect of mammary carcinoma. We analyzed the express of above biomarkers in 52 cases mammary carcinoma, and serum CA153 and TSGF levels in 43 cases mammary carcinoma before and after neoadjuvant chemotherapy. Our results demonstrated that before and after therapy, the expression of PR and HIF-1α had decreased significantly ($P < 0.05$). The chemosensitivity of HIF-1α positive expression and TSGF and CA153 positive group were worse than those of HIF-1α nagetive expression and TSGF and CA153 nagetive group ($P < 0.05$). Profiles for ER and HER2 were not significantly different before and after neoadjuvant chemotherapy.

1 INTRODUCTION

Mammary cancer is one of the common malignancies in women and its morbidity is on the rise year by year (Lakhani et al. 2012). In recent years in China, the incidence of mammary carcinoma is relatively high, and the peak incidence is in advance, a large number of patients died of mammary carcinoma complications or serious organ metastasis each year (Wang et al. 2014a). Mammary carcinoma, like most other forms of malignancy, is mainly a multifactorial disease, occurring as a result of the combined effects of environmental and heritable factors (Kabalar et al. 2012 & Wu et al. 2013). Many studies have found that certain biological indicators by detecting abnormal expression of molecules, such as Estrogen Receptor (ER), Progesterone Receptor (PR), proto-oncogene HER2/neu, etc, can guide clinical diagnosis and treatment activities prognosis (Witteveen et al. 2015). The ER, PR and HER2/neu profile of breast carcinoma plays a significant role in patient management and treatment. Because of the increasing utilization of neoadjuvant chemotherapy or hormone therapy, surgically-resected carcinomas often show marked treatment effect. There are several other molecular biomarkers, such as Hypoxia Inducible Factor 1α (HIF-1α), cyclooxygenase-2 and vascular endothelial growth factor and have been confirmed to participate in the evolution of mammary carcinoma (Diers et al. 2013, Kim et al. 2014, Wang et al. 2014a & Witteveen 2015). Over-expression of HIF-1α is a frequent feature of malignant disease and it is commonly associated with poor prognosis and resistance to conventional chemotherapy. Serum biomarkers such as Cancer Antigen 153 (CA153) and malignant Tumor Specific Growth Factor (TSGF) are among the prognostic factors and can be used for follow-up. In recent years, neoadjuvant chemotherapy has additionally become an option for patients with operable tumors who desire breast conservation therapy. Neoadjuvant chemotherapy with agents such as doxorubicin and cyclophosphamide had historically been offered to patients with locally advanced disease with a goal of reducing tumor size to enable surgical resection. Several randomized clinical trials have demonstrated no statistically significant difference in disease-free survival or overall survival in the patients receiving preoperative therapy as opposed to those receiving postoperative chemotherapy (Kinsellal et al. 2012). We composed a panel of potential biomarkers of ER, PR, HER2 and HIF-1α in tissue and CA153 and TSGF levels of serum for mammary carcinoma, in order to assess the subsequent effects.

2 MATERIALS AND METHODS

2.1 Patients

Fifty-two female patients diagnosed mammary carcinoma treated with neoadjuvant therapy after core biopsy diagnosis were included and undergo excision surgery at Rizhao people's Hospital from 2010 to 2014. Permission was obtained from the Local Ethical Committee and all patients signed informed consents to the research. The patients had not been treated with hormone endocrine therapy, anti-neoplastic chemotherapy or radiotherapy during the last six months before needle core biopsy diagnosis. The age of mammary carcinoma patients ranged from 34 to 66 years, mean 51.3 years. Neoadjuvant chemotherapeutic regimens for treated patients included combination anthracyclines, taxanes, and alkylating agents. The express of ER, PR, HER2 and HIF-1αfor both the pretreatment core biopsies and post-treatment mastectomy in tissue specimens and CA153, TSGF, CA125 and CEA levels of serum were detected. The biomarkers of CA153 and TSGF were detected in 43 cases mammary carcinoma with electroc-hemiluminescence method in the clinical laboratory in Rizhao People's Hospital. The cut off values of CA153 and TSGF in serum are 25.00 U/ml and 70.00 U/ml.

2.2 Pathologic study

Immunoreactions were processed using the *Ultra Sensitive*™ *S-P* Kit (Maixin-Bio, China) according to the manufacturer's instructions, and signals were visualized using the DAB substrate, which stains the target protein yellow. The negative controls were used. The primary antibody was replaced with PBS, containing 0.1% bovine serum albumin at the same concentration as the primary antibody. The positive controls were tissues known to express the antigen being studied. ER, PR and HIF-1α immunoreactivity expression the percentage of cancer cells showed nuclear reactivity. For ER, PR and HIF-1α expression the percentage of cancer cells showing a nuclear reactivity was recorded after inspection of all optical fields at 200 and the mean value was used to score each case. In this study, ER, PR and HIF-1α were scored using <10% of tumor staining as the negative cutoff. In addition, cell membrane reactivity for HER2/neu the oncoprotein was evaluated following a similar approach and the mean value was used to score each case. Tumors HER2/neu expressing strong, circumferential membranous staining in more than 10% of invasive carcinoma cells were considered as positive (Lakhani et al. 2012). The pathological diagnosis was verified by histological methods independently by two pathologists, and pathological categorization was determined according to the current WHO standard (Lakhani et al. 2012). The tumor grade was determined according to the modified Bloom-Richardson score (Lakhani et al. 2012).

2.3 Statistical analysis

SPSS ver 17.0 statistical software (SPSS Inc.: Chicago, IL, USA) was used to detect the data. Measurement data between groups was compared with t test, while enumeration data with χ^2 test. The P value was considered to be significant if less than 0.05.

3 RESULTS

3.1 *Comparison of immunohistochemistry express in Pre- and Post-Therapy*

The immunohistochemical *Ultra Sensitive™ S-P* method was used to detect differences in tumor tissue ER, PR, HER2 and HIF-1α expression and the situation before and after chemotherapy in 52 cases of mammary cancer to neoadjuvant chemotherapy. Our results shown of the 52 carcinomas studied, 37 cases (71.75%) were positive for ER by immunohistochemistry both pre- and post—neoadjuvant chemotherapy treatment ($P>0.05$). The studies showed positivity for PR 36(69.23%) cases pre-neoadjuvant therapy and 25(48.10%) positivity post-treatment ($P < 0.05$). For HER2, 16(30.77%) were positive pre-treatment, and 11(21.15%) were positive post-treatment ($P=$). For 52 patients, HIF-1 was positive in 24(46.15%) pre-therapy and in 11(21.15%) post-treatment ($P < 0.05$). The study demonstrated that before and after chemotherapy, the expression of PR and HIF-1α had decreased significantly ($P < 0.05$).

3.2 *The HIF-1α express changed in positive and negative groups in Pre- and Post-Therapy*

We also examined the differences in tumor tissue HIF-1α expression in positive and negative groups and the situation before and after chemotherapy in 52 cases of mammary cancer to neoadjuvant chemotherapy shown in Table 2. Before and after chemotherapy, the expression of HIF-1α has changed significantly from 46.15% down to 21.15%, $P<0.05$. According to the International Union against Cancer (UICC) TNM classification of solid malignant tumors standard (7th ed) and American Joint Committee on Cancer (AJCC) Cancer Staging Manual (7th ed) to evaluate curative effect of therapeutic effect (Edge et al. 2009 & Sobin et al. 2009). We considered Complete Remission (CR) and Partial Response (PR) for the total effective rate. Before neoadjuvant chemotherapy, the chemosensitivity of HIF-1α positive expression (14/24, 58.33%) was worse than the chemosensitivity of HIF-1α nagetive expression (24/28, 85.71%) ($P < 0.05$). HIF-1α can be used as a predictor of mammary cancer chemosensitivity to help develop individualized chemotherapy.

3.3 *Comparison of biomarker levels to HIF-1α express in positive and negative groups*

The TSGF and CA153 serum levels in HIF-1α positive and negative groups in 43 cases of mammary cancer patients were shown in Figure 1.

Table 1. Comparison of immunoreactions before and after therapy (n = 52).

Biomarker	Number of positive before therapy (%)	Number of positive after therapy (%)	Significant
ER	37 (71.75)	37 (71.75)	No
PR	36 (69.23)	25 (48.10)	Yes
HER2	16 (30.77)	11 (21.15)	No
HIF-1α	24 (46.15)	11 (21.15)	Yes

Table 2. Comparison of HIF-1α express before and after therapy (n = 52).

Therapy	HIF-1α Negative	HIF-1α Positive	Significant
Before	28 (53.85%)	24 (46.15%)	Yes
After	41 (78.85%)	11 (21.15%)	

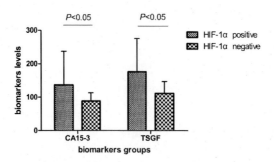

Figure 1. Comparison of biomarker levels in HIF-1α positive and negative groups.

The serum TSGF and CA153 levels in HIF-1α positive patients (176.34±99.63 U/ml, 136.27±101.25 U/ml) were significantly higher than those in negative patients (111.43±35.52 U/ml, 88.48±24.76 U/ml) ($t = 3.15$, $t = 2.43$, respectively, $P < 0.05$).

4 DISCUSSION

Mammary carcinoma is one of the commonly-encountered malignant tumors, and it is the second leading cancer causing death in women (Siegel R. 2012). In recent years, the incidence of mammary carcinoma is significantly on the rise all over the world, hence, how to treat mammary carcinoma, effectively assess the therapeutic effect, correctively evaluate the prognosis and find the postoperative recurrence of patients with mammary carcinoma have been paid attention by more and more scholars all over the world (Deng et al. 2013 & Sun et al. 2013). The ER, PR and proto-oncogene HER2/neu profile of a female mammary carcinoma plays a significant role as a predictive marker in patient management and treatment. Our results shown of the 52 carcinomas studied, 37 cases (71.75%) were positive for ER by immunohistochemistry both before and after neoadjuvant treatment. Immunohistochemistry studies for PR in these 52 patients showed 36 cases (69.23%) positivity for PR before neoadjuvant therapy and 25 (48.10%) positivity after neoadjuvant therapy. For 52 patients with HER2, 16 (30.77%) were positive before neoadjuvant therapy, and 11 (21.15%) were positive after neoadjuvant therapy. For HER2, no significant difference in expression was found in this study in 52 cases breast carcinomas between before and afte neoadjuvant treatment specimens. Because of the increasing utilization of neoadjuvant chemotherapy or hormone therapy, surgically-resected carcinomas often show marked treatment effect. In our study, we examined multiple biological parameters and their relationship with neoadjuvant chemotherapy in mammary carcinoma, and profiles for ER and HER2 were not significantly different in mammary carcinoma before and after neoadjuvant chemotherapy. In this study, statistically significant changes in hormone receptor PR status of primary breast carcinomas following neoadjuvant chemotherapy have been observed. Further investigation is warranted to assess reproducibility of technique and investigate clinical implications of significant loss of PR status in treated patients. Neoadjuvant chemotherapy with agents such as doxorubicin and cyclophosphamide had historically been offered to patients with locally advanced disease with a goal of reducing tumor size to enable surgical resection. Our results shown the TSGF and CA153 serum levels in HIF-1α positive patients (176.34±99.63 U/ml, 136.27±101.25 U/ml) were significantly higher than those in negative patients (111.43±35.52 U/ml, 88.48±24.76 U/ml) ($t = 3.15$, $t = 2.43$, respectively, $P < 0.05$). The serum biomarker levels of CA153 and TSGF in mammary carcinoma patients abnormal express were significantly in before and after neoadjuvant treatment ($P < 0.05$). Serum has long been considered as a rich source for biomarkers and a large number of serum cancer biomarkers. The few biomarkers that are tested for early detection of mammary carcinoma, show limited diagnostic sensitivity and specificity in stand-alone assays (Pitteri et al. 2010

& Zhu et al. 2011). In summary, neoadjuvant chemotherapy in the treatment of primary breast carcinomas has been increasingly used to reduce tumor size prior to surgical breast conservation therapy. Determination of steroid ER and PR hormone receptor and HER2 status in these patients both before and after neoadjuvant treatment has been central in directing patient management and potential response to hormonal and biological agents. This study demonstrates no significant difference in ER or HER2 expression in before and after neoadjuvant treatment mammary carcinoma ($P < 0.05$). However, there was a statistically significant decrease in PR expression by immunohistochemistry following neoadjuvant chemotherapy, the cause and clinical significance of which require further investigation.

In this study, we analyze the expressions of HIF-1α in mammary carcinoma, and detect differences in HIF-1α expression situation pre- and post-therapy of mammary carcinoma to neoadjuvant chemotherapy. In our study, pre- and post—neoadjuvant treatment, the expression of HIF-1α has changed significantly from 46.15% down to 21.15%, $P < 0.05$. Before and after neoadjuvant chemotherapy, the expressions of HIF-1α are positive correlation with chemosensitivity. Before neoadjuvant chemotherapy, the chemosensitivity of HIF-1α positive expression was worse than the chemosensitivity of HIF-1α nagetive expression ($P < 0.05$). Our results implicated the importance of HIF-1α can be used as a predictor of mammary cancer chemosensitivity to help develop individualized chemotherapy. HIF-1α is a novel prognostic marker in determining the aggressive phenotype of mammary cancer and is emerging as a potential target for breast cancer treatment. HIF-1 consists of two subunits, namely HIF-1α and HIF-1β. HIF-1α is a transcription factor forming a heterodimer with a constitutively expressed HIF-1β subunit. These heterodimers regulate target genes by binding to a consensus sequence called hypoxia responsive element (Sueoka et al. 2013). HIF-1α plays an essential role in the regulation of various genes associated with low oxygen consumption. Our previous study indicated that elevated expression of HIF-1α are associated with tumor progression, invasion and metastasis (Wang et al. 2013 & 2014a, b). HIF-1α may be as the predictors for prognosis. In this study, we also detected TSGF and CA153 serum levels in mammary cancer patients HIF-1α positive groups and negative groups. The serum biomarker levels of TSGF and CA153 in HIF-1α positive patients were significantly higher than those in negative groups (t = 3.15, t = 2.43, respectively, $P < 0.05$). HIF-1α, TSGF and CA153 can be used as predictors of mammary cancer chemosensitivity to help develop individualized chemotherapy. Future studies should therefore preferentially select a broader target set of potential biomarkers.

5 CONCLUSION

In this study, we analyze the express of ER, PR, HER2 and HIF-1α in tissue and serum CA153 and TSGF levels in judging the neoadjuvant chemotherapy effect of mammary carcinoma. Our results demonstrated that before and after neoadjuvant chemotherapy, the expression of PR and HIF-1α had decreased significantly. The chemosensitivity of HIF-1α positive expression and TSGF and CA153 positive group are worse than those of HIF-1α nagetive expression and TSGF and CA153 nagetive group. Profiles for ER and HER2 were not significantly different before and after neoadjuvant chemotherapy. These results suggest that combined detection of the indicators above can judge the prognosis better, which is of great importance to guide the treatment for mammary cancer. HIF-1α, TSGF and CA153 can be used as predictors of mammary cancer chemosensitivity to help develop individualized chemotherapy. Future studies should therefore preferentially select a broader target set of potential biomarkers.

ACKNOWLEDGMENTS

The authors acknowledge financial support from the Medicine and Health Care Science and Technology Development Plan Projects Foundation of Shandong Province (No.

2014WS0282, 2014WSA11003), Application Technology Research and Development Project Foundation in Rizhao City (No. 2014SZSH02) as well as the Scientific Research Projects of Jining Medical College (No. JY2013KJ051).

REFERENCES

Deng, Q.Q., Huang, X.E., Ye, L.H, Lu, Y.Y., Liang, Y., & Xiang, J. 2013. Phase II trial of Loubo® (Lobaplatin) and pemetrexed for patients with metastatic breast cancer not responding to anthracycline or taxanes. *Asian Pac J Cancer Prev* 14:413–417.

Diers, A.R., Vayalil, P.K., Oliva, C.R. Griguer, C.E., Darley-Usmar, V., Hurst, D.R., Welch, D.R. & Landar, A. 2013. Mitochondrial Bioenergetics of Metastatic Breast Cancer Cells in Response to Dynamic Changes in Oxygen Tension: Effects of HIF-1α. *PLoS ONE* 8(6): e68348.

Edge, S.B., Byrd, D,R, Compton, C.C. et al. 2009. *American Joint Committee on Cancer (AJCC) Cancer Staging Manual 7th ed* (ed.). New York, Springer.

Hugh, T.J., Dillon, S.A., Taylor, B.A. Pignatelli, M., Poston, G.J. & Kinsella, A.R. 1999. Cadherin-catenin Expression in primary colorectal cancer: a survival analysis. *Br J Cancer* 80:1046–1051.

Kabalar, M.E., Karaman, A., Aylu, B. Ozmen, S.A. & Erdem, I. 2012. Genetic alterations in benign, preneoplastic and malignant breast lesions. *Indian J Pathol Microbiol* 55:319–325.

Kim, M.J., Kim, H.S., Lee, S.H. Yang, Y., Lee, M.S., & Lim, J.S. 2014. NDRG2 controls COX-2/PGE2-mediated breast cancer cell migration and invasion. *Mol Cells* 37(10):759–765.

Kinsella1, M.D., Nassar, A., Siddiqui, M.T., & Cohen, C. 2012. Estrogen Receptor (ER), Progesterone Receptor (PR), and HER2 expression pre- and post—neoadjuvant chemotherapy in primary breast carcinoma: a single institutional experience. *Int J Clin Exp Pathol* 5(6):530–536.

Lakhani, S.R., Ellis, I.O., Schnitt, S.J. et al. 2012. WHO classification of tumours of the breast[M]. Lyon, France: IARC Press:1–75.

Pitteri, S.J., Amon, L.M., Busald, B.T., Zhang, Y., Johnson, M.M., Chin, A., et al. 2010. Detection of elevated plasma levels of epidermal growth factor receptor before breast cancer diagnosis among hormone therapy users. *Cancer Res* 70:8598–8606.

Siegel, R, Naishadham, D. & Jemal, A. 2012. Cancer statistics. *CA Cancer J Clin* 62:10–29.

Sun, M.Q., Meng, A.F., Huang, X.E., et al. 2013. Comparison of psychological infuence on breast cancer patients between breast-conserving surgery and modifed radical mastectomy. *Asian Pac J Cancer Prev* 14:149–152.

Sobin, L.H., Gospodarowicz, M.K., Wittekind, C.h. 2009. *International Union against Cancer (UICC): TNM classification of malignant tumors 7th ed* (ed.). Wiley-Blackwell, Oxford.

Wang, G.P., Zhang, H., Zhang, Z.F., Liang, Y.A., Chen, Y. & Mei L. 2013. Clinical and pathological characteristics of intraductal proliferative lesions and coexist with invasive ductal carcinomas. *Chinses-German J Clin Oncol* 12:574–580.

Wang, G., Qin, Y., Zhang, J. Zhao, J., Liang, Y., Zhang, Z., Qin, M. & Sun, Y. 2014a. Nipple Discharge of CA15–3, CA125, CEA and TSGF as a New Biomarker Panel for Breast Cancer. *International Journal of Molecular Sciences* 15(6):9546–9565.

Wang, G.P., Qin, M.H., Liang, Y.A., Zhang, H., Zhang, Z.F. & Xu, F.L. 2014b. The significance of biomarkers in nipple discharge and serum in diagnosis of breast cance. In: Liquan Xie, Dianjian Huang, eds. *Advanced Engineering and Technology*: 665–671. CRC Press.

Witteveen, A., Kwast, A.B., Sonke, G.S. Ijzerman, M.J. & Siesling, S. 2015. Survival after locoregional recurrence or second primary breast cancer: impact of the disease-free interval. *PLoS One* 10 (4):e0120832.

Wu, D., Zhang, R, Zhao, R., Chen, G., Cai, Y. & Jin, J. 2013. A Novel Function of Novobiocin: Disrupting the Interaction of HIF 1a and p300/CBP through Direct Binding to the HIF1a C-Terminal Activation Domain. *PLoS ONE* 8(5): e62014.

Zhang, S.J., Hu, Y., Qian, H.L. Jiao, S.C., Liu, Z.F., Tao, H.T. & Han, L. 2013. Expression and Significance of ER, PR, VEGF, CA15–3, CA125 and CEA in Judging the Prognosis of Breast Cancer. *Asian Pac J Cancer Prev* 14:3937–3940.

Zhu, C.S., Pinsky, P.F., Cramer, D.W., Ransohoff, D.F., Hartge, P., Pfeiffer, R.M., et al. 2011. A framework for evaluating biomarkers for early detection: Validation of biomarker panels for ovarian cancer. *Cancer Prev. Res* 4:375–383.

Effect of ^{60}Co-γ irradiation on quality and physiology of blueberry storage

C. Wang
Department of Food Science, Shenyang Agricultural University, Shenyang, Liaoning, P.R. China
Academy of Agricultural Sciences, Shenyang, Liaoning, P.R. China

X.T. Li
Academy of Agricultural Sciences, Shenyang, Liaoning, P.R. China

X.J. Meng
Department of Food Science, Shenyang Agricultural University, Shenyang, Liaoning, P.R. China

ABSTRACT: Blueberry fruits are treated by ^{60}Co-γ irradiation with the doses of 0.5, 1.0, 1.5, 2.0, 2.5, 3.0 kGy, and then stored at 5°C for 35 days. The sensory and physiological indexes were measured every 7 days and the results showed that the irradiation treatment has good effects on reducing the decay rate and improving the fruit firmness at refrigerated storage. In the whole storage, blueberry fruit irradiated by 2.5 kGy was the best which decay rate and fruit firmness were 14.53% and 0.29 N respectively. In a word, appropriate ^{60}Co-γ radiation can be used as an efficient way to improve the storage quality of blueberries.

1 INTRODUCTION

Regarded as the king of the world berries (Liu, C. 2003), blueberry fruit is loved deeply by the people for containing a variety of nutrients and biological active substances and having stronger antioxidant capacities and various health care functions. At present, a lot of researches on preservation technology of blueberry were carried out by domestic and foreign scholars (Maria, C.N. et al. 2004, Meng X.J. et al. 2011, Park, H.T. 2002 & Ji, S.J. et al. 2014). However, these technologies still had no ideal preservation effect, especially in the later period of storage appeared a series of problems.

Some researches show that the ^{60}Co-γ irradiation technology of high penetration power can maintain or improve the storage quality and prolong the storage period of fruits and vegetables (Prakash, A. et al. 2002, Wani, A.M. et al. 2007 & 2008, Zu, Z.B. & Li, W.G. 2006, Costa, L. et al. 2006 & Ye, H. et al. 2000). Thus, it is regarded as a preservation technology with the characteristics of high efficiency, energy saving, safety and environmental protection. This experiment aimed at the determination of irradiation dose tolerated by blueberry fruit, then the definition storage effect and physiological mechanism, so it provided the theoretical and practical basis for the application of ^{60}Co-γ irradiation technology in blueberry preservation.

2 MATERIALS AND METHODS

2.1 Fruit material and treatment

"Lan Feng" blueberry hand-picked from Shenyang Shisheng Blueberry-Picking Base was tested as material. The fruits were screened for uniform maturity, size, and color, and without plant diseases, insect pests and mechanical damage. All blueberries were put into 12-air-holes

PET crisper designed to the specification of 105 (mm) × 100 (mm) × 45 (mm) and the thickness is 0.4 (mm). The weight of each crisper blueberry was about 125 g. Then every 12 crispers of blueberry were packaged in cardboard boxes without cover.

2.2 Irradiation processing

Samples were irradiated with a 300 kCi ^{60}Co source at the irradiation doses of 0, 0.5, 1.0, 1.5, 2.0, 2.5, 3.0 kGy for 72 minutes. All samples were rotated from up to down and from front to back during half irradiation processing. Each irradiation dose was repeated 3 times. After irradiation, all blueberry samples were firstly precooled at 10°C for 12 hours then stored at 5 ± 0.5°C. From every irradiation processing, 3 crispers of blueberry were withdrawn every 7 days for further research.

2.3 Measurement of fruit decay rate

Decay rate (%) is equal to the ratio of rotten fruits number and total fruits number.

2.4 Measurement of fruit firmness

According to the method of Zhou, Q. et al. (2014), results were expressed in mg/kg h. Results were expressed in Newton (N).

2.5 Measurement of respiratory intensity

According to the method of Cao, J.K. et al. (2007), results were expressed in mg/kg h.

2.6 Measurement of cell membrane electrolyte leakage

The cell membrane electrolyte leakage was measured by conductivity meter method, which was equal to the ratio of the conductivity value and the total conductivity value.

2.7 Measurement of MDA content

Refer to the method of Martinez-Solano, J. et al. (2005).

2.8 Statistical analysis

Date were subjected to Analysis of Variance (ANOVN) using the software SPSS (SPSS Inc., Chicago, IL) 17.0 for windows. Means were separated using Duncan's multiple range tests. Differences were considered significant at $P < 0.05$.

3 RESULTS

3.1 Effect of irradiation on fruit decay rate

During the cold storage, the changes on the decay rate of blueberry fruit were showed as the Figure 1. In the earlier storage, the decay rates were all at a lower level and had no significant differences. Then with the extension of storage time, CK and 0.5 kGy irradiation treatments were rising rapidly and were obviously higher than the other treatments. On the 28th day of storage, 3.0 kGy suddenly increased to 20.63% closing to CK. In the whole storage, 2.0 and 2.5 kGy have been always kept the lowest level, and 1.0 and 1.5 kGy were the second. On the 35th day of storage, 2.5 kGy was 14.53%, which was 27.21% lower than that of CK. So the irradiation treatment can effectively reduce the rising of the decay rate of blueberry fruit in cold storage.

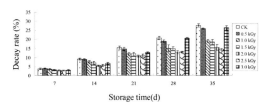

Figure 1. Effect of decay rates of blueberries irradiated by different doses during cold storage.

3.2 *Effect of irradiation on fruit firmness*

As shown in the Figure 2, on the day of irradiation treatment, the firmness of blueberry fruit had no significant changes. With the extension of storage time, CK has been always presenting downward trend, while the irradiation treatments showed a trend of rising first and falling later. Among them, 2.5 kGy was the biggest (0.39 N) and increased 38.90% than initial storage. This phenomenon was probably due to the reason that the self protection of blueberry fruit was inspired by irradiation stress then led to the composition and structure in fruit cell wall repaired. At the end of storage, 0.5 and 3.0 kGy both fell and closed to CK, but the highest (0.29 N) firmness was still 2.5 kGy. Accordingly irradiation treatment can be a good way to improve the firmness of blueberry fruit at refrigerated.

3.3 *Effect of irradiation on respiration intensity*

As showed in the Figure 3, after irradiation treatment the respiratory intensity were all higher than that of CK, and gradually enhanced with the increase of irradiation doses. During the whole storage, CK rose rapidly at first, and began to decrease slowly after reaching the peak on the 21st day of storage. While irradiation treatment declined slowly firstly in early time and rose rapidly again after 7 days. This declination in early time maybe due to low temperature can well alleviate the negative effect of exuberant respiration produced by irradiation treatment and this rose in later time was due to that the physical damages produced by irradiation needing energy to repair. In the middle and later periods of the storage, irradiated blueberry fruit have been always keeping lower than CK and the lowest was 2.5 kGy.

3.4 *Effect of irradiation on cell membrane electrolyte leakage*

As showed in the Figure 4, after irradiation the relative conductivity of irradiated blueberry fruit were little higher than CK. This suggested that the irradiation treatment had a degree of damage on the cell membrane structure of blueberry fruit. During the cold storage, CK was rising continuously, 3.0 kGy changed quite gently, and the other treatments declined at first and rose slowly after reaching the trough. This difference may be caused by that the self protection had a temporary repair effect on the cell membrane. On the 35th day of storage, 2.5 kGy was the lowest. Therefore, appropriate irradiation treatment could reduce the cell membrane damage of the blueberry fruit during cold storage.

3.5 *Effect of irradiation on MDA content*

As showed in the Figure 5, after irradiation the MDA content of CK was closed to 0.5 kGy but was obviously lower than that of the other irradiation treatments. In the whole storage, CK showed a continuous rising trend and was always higher than the irradiation treatment groups. However, the irradiation treatments declined slightly first then rose again later. The declination maybe due to that the falling of active oxygen ion and the decrease of damage in the membrane, while the rose slowly maybe related to that the membrane lipid peroxidation was in a dominant position after the self protection changed gradually weak. By the end of

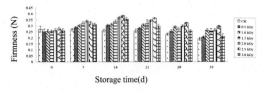

Figure 2. Effect of different irradiation doses on firmness of blueberry fruit during cold storage.

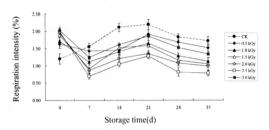

Figure 3. Effect of different irradiation doses on respiration of blueberry fruit during cold storage.

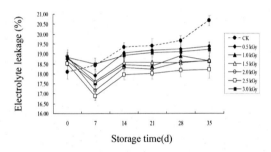

Figure 4. Effect of different irradiation doses on electrolyte leakage of blueberry fruit during cold storage.

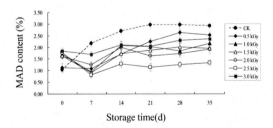

Figure 5. Effect of different irradiation doses on MDA content of blueberry fruit during cold storage.

the cold storage, the content in blueberry fruit irradiated by 2.5 kGy was the lowest. These changes correspond with that of cell membrane electrolyte leakage.

4 DISCUSSION

Studies demonstrated that irradiation treatment not only kills microorganism, but also delays the mature and aging processing of fruit and vegetable by restraining the metabolism.

Usually the decay rate of blueberry was 30–40% stored at 5°C for 35 days, while in this study the decay rates of 0.5–3.0 kGy irradiations were 14.53–26.40% and the lowest one was 2.5 kGy. This is similar to the researches of Zhao, X.N. et al. (1999) and Liu, C. (2003) on strawberry irradiated by ^{60}Co-γ. This experiment about firmness showed that the irradiation treatment hadn't direct influence on the firmness of blueberry fruit, but can improve the fruit firmness during the cold storage and the best dose was 2.5 kGy. But Fu, J.J. & Feng, F.Q. (2003) considered that irradiation treatment had no effect on the firmness of Kiwi fruit sorted at room temperature and Liu, C. (2003) founded that fruit appeared soft collapse phenomenon when the irradiation dose exceeded 5 kGy in the study of strawberry preservation effect. At the same time, the experiments, including respiration intensity, MDA content and cell membrane permeability, showed that irradiation treatment can increase of the respiration intensity in blueberry fruit and led to the enhancement of cell membrane permeability and MDA content while can relieve this kind of damage during the cold temperature. The same conclusion as the decay rate and firmness was drawn that the effect of 2.5 kGy irradiation was the best and the irradiation treatments of 1.0–2.0 kGy were the second.

5 CONCLUSION

By the researches on quality and physiology of blueberry fruit storage at cold temperature, the result showed that the ^{60}Co-γ irradiation treatment has good effects on reduce the decay rate and improve the fruit firmness at refrigerated storage. During the storage time, the storage effect of 2.5 kGy irradiation was the best which decay rate and fruit firmness were 14.53% and 0.29 N respectively, 1.0–2.0 kGy irradiation were the second, and 0.5 and 3.0 kGy were the worst. In a word, appropriate ^{60}Co-γ radiation can be used as an efficient way to improve the storage quality of blueberries.

REFERENCES

Cao, J.k., Jiang, W.B. & Zhao, Y.M. 2007. Experiment guidance of postharvest physiology and biochemistry of fruits and vegetables, *China Light Industry Press*: 122–124.

Chen, Y.T., Guo, D.Q., Yang, Z.Q., et al. 2009. A general review on the effect of ^{60}Co rays irradiation on vitamin in fruits and vegetables, *Journal of Nuclear Agricultural Sciences* 23(2): 302–307.

Costa, L., Vocente, A.R., Civello, P.M., et al. 2006. UV-C treatment delays postharvest senescence in broccoli florets. Postharvest Biology and Technology 39: 204–210.

Fu, J.J. & Feng, F.Q. 2003. Effect of irradiation on the preservation of KiVi fruit, *Acta Agricultae Nucleatae Sinica* 17(5): 367–369.

Hussain, P.R., Meena, R.S., Dar, M.A., et al. 2008. Studies on enhancing the keeping-quality of peach (prunus persica bausch) Cv. Elberta by gamma-irradiation, *Radiation Physics and Chemistry* 77(4): 473–481.

Ji, S.J., Zhou, Q., Ma, C. & Chang, S.C. 2014. Effect of 1-MCP treatment on qualiy changes of blueberry at room temperature, *Food Science* 35(2): 322–327.

Liu, C., 2003. Studies on the strawberry preservation by irradiation and changes of physiological quality, *Journal of Anhui Agricultrual Sciences* 31(5): 744–745.

Maria C.N., Nunes P.E. & Jeffrey K.B. 2004. Quality curves for highbush bluberries as a function of the storage temperature, *Small Fruits Review* 3(3–4): 423–438.

Martinez-Solano, J., Sanchez-Bel, P., Olmos, E., Hellin, E. & Romojaro, F. 2005. Electron beam ionization induced oxidative enzaymatic activities in pepper (Capsicum annuum L.,), associated with ultrstructure celluar damages, *Agric Food Chem* 53: 8593–8599.

Meng, X.J., Jiang, A.L., Hu, W.Z., Tian, M.X. & Liu, C.H. 2011. Effect of plastic box modified atmosphere storage on the physiological and biochemical changes of postharvest blueberry fruits, *Science and Technology of Food Industry* 32(9): 379–383.

Park, H.J. 2002. Fruit and vegetable processing: Improving quality. Woodhead Publishing Limited Cambridge.

Prakash, A., Manley, J., Decosta, S., et al. 2002. The effects of gamma irradiation on the microbiological, physical and sensory qualities of diced tomatoes, *Radiation Physics and Chenmistry* 63: 387–390.

Wani, A.M., Hussain, P.R., Dar, M.A., et al. 2007. Shelf life extension of pear cv. William by gamma irradiation, *Journal of Food Science and Technology* 44(2): 138–142.

Wani, A.M., Hussain, P.R., Meena, R.S., et al. 2008. Effect of gamma-irradiation and refrigerated storage on the improvement of quality and shelf life of pear (Pyrus communis L, Cv. Bartlett/William), *Radiation Physics and Chemistry* 77(8): 983–989.

Ye, H., Chen, J.X., Yu, R.C., Chen, Q.L., Liu, W. 2000 The effect of irradiation on stored straw mushroom and the physiological mechanism. Acta Agricultae Nucleatae Sinica 14(1): 24–28.

Zhao, X.N., Wang, C.B., & Hu, S.X. 2011. Application of Irradiation Technology in fruits and vegetables preservation, *Heilongjiang Agricultural Sciences* 8: 151–153.

Zhao, Y.F., Xie, Z.C., Yu, Y.J. & Xu, Z.C. 1999. Study on the strawberry preservation by irradiation. *Food Science* 2: 54–56.

Zhu, L. & Ling, J.G. 2011. Development of preservation technology of blueberry in the world, *Food and Fermentation Indestries* 31(11): 173–176.

Zu, Z.B. & Li, W.G., 2006. Study on the preservation effect of radiation technology on strawberry, *Food Science and Technology* 5: 114–116.

Determination of eight metal elements in polygonatum tea by microwave digestion-AAS

S.L. Chen, C. Li, A.H. Chen & Y. Shao
Jiangsu Key Laboratory of Food Resource Development and Quality Safe, Xuzhou Institute of Technology, Xuzhou, China

J.W. Li, S.B. Liu & X.W. Zheng
Bozhou Qiancao Pharmaceutical Co. Ltd., Bozhou, China

ABSTRACT: A new application of high resolution continuum source atomic absorption spectrometry has been developed for the determination of Ca, Fe, Zn, Mg, Mn, Cd, Pb and Cu in polygonatum tea. The samples were digested by microwave. Under the selected determination conditions, the correlation coefficients better than 0.995 and the recoveries between 96.5 and 104.7% were obtained for these eight metal elements. The characteristic concentrations were 0.092 mg/L (Ca), 0.17 mg/L (Fe), 0.033 mg/L (Zn), 0.008 mg/L (Mg), 0.15 mg/L (Mn), 0.021 mg/L (Cd), 0.58 mg/L (Pb) and 0.092 mg/L (Cu), respectively. The contents in polygonatum tea were 2.091 ± 0.057 g/kg (Ca), 0.481 ± 0.009 g/kg (Fe), 14.11 ± 0.32 mg/kg (Zn), 1.287 ± 0.025 g/kg (Mg), 46.08 ± 0.98 mg/kg (Mn), 7.852 ± 0.287 μg/kg (Cd), 9.558 ± 0.296 μg/kg (Pb) and 1.256 ± 0.028 mg/kg (Cu), respectively. Therefore, the proposed method was accurate and stable with a high practical value. It provided scientific basis for determination of metal elements in food.

1 INTRODUCTION

Several sample preparation techniques are used for the determination of metal elements in food, such as inductively wet digestion, dry digestion, incomplete digestion and microwave digestion. Wet digestion and dry digestion are two traditional ways, but both of them have obvious disadvantages, for instance, some dense mass of strong acids are used in wet digestion, while dry digestion is a time-consuming process (Afridi, et al, 2007; Gao, 2007). The objective of incomplete digestion is uniform and transparent digestive solutions, which do not need of complete destruction and colorless liquors. As a consequence of that, the sample-process time just need less than 20 min (Liu, et al, 2004). Nevertheless, incomplete digestion also requires a good deal of strong acids and the microemulsification stability after that is an important parameter for accurate analysis. Closed-vessel microwave-assisted digestion needs only 5 mL of HNO_3 and 4 mL H_2O_2 in this study, and its main advantages are the high relative speed, good reproducibility, low blank, low possibility of contamination and the minimum loss of volatile elements (Ren, et al, 2011).

The objective of this study was to develop a simple and robust method for the fast sequential multi-element determination of Ca, Fe, Zn, Mg and Mn (Gao, 2007) in polygonatum tea by High Resolution Continuum Source Flame Atomic Absorption Spectrometry (HR-CS FAAS) (Nunes, et al, 2011, Baysal, et al, 2011 and Ozbek, et al, 2012) and determination of Cd (Li, et al, 2012), Pb (Liu, et al, 2009) and Cu (Gao, et al, 2007) in polygonatum tea by High Resolution Continuum Source Graphite Furnace Atomic Absorption Spectrometry (HR-CS GFAAS) (Borges, et al, 2011, Lepri, et al, 2010 and Ren, et al, 2011) using new features.

2 EXPERIMENTAL

2.1 *Instrumentation*

An Analytik Jena ContrAA 700 High Resolution Continuum Source Atomic Absorption Spectrometer (Analytik Jena, Berlin, Germany) had been used for all measurements in this work. This spectrometer consists of a high-intensity xenon short-arc lamp, a high-resolution Double Echelle Monochromator (DEMON) and a Charge Coupled Device (CCD) array detector (Resano, et al, 2009). All absorption lines in the range from 185 nm to 900 nm are provided by the high-intensity xenon short-arc lamp as the radiation source. The highest resolution of about 2 pm is carried out by DEMON, including a pre-dispersing prism monochromator and a high-resolution echelle grating monochromator. 200 pixels of the linear CCD array detector are used for monitoring all spectral informations on both sides of center wavelength. The flame type was C_2H_2-air and the burner type was 100 mm in the process of determination. The fixed air flow was 470 L/h for determination. The optimized determination conditions of HR-CS FAAS were shown in Table 1. In addition, this instrument is also equipped with an automated sampling accessory (MPE) (Liu, et al, 2014). High purity argon was used as the carrier gas in the process of determination. The sampling volume of 20 μL and the chemical modifier volume of 5 μL were used for determination. The optimal determination parameters of HR-CS GFAAS and graphite furnace temperature programs were shown in Table 2 and Table 3, respectively.

2.2 *Reagents and standards*

The reagents were of guaranteed reagent. Ultrapure water with a resistivity of 18.2 MΩ·cm was obtained from a Milli-Q system (Millipore, Billerica, USA). The stock solutions of Ca,

Table 1. Optimal determination conditions of HR-CS FAAS.

Element	Wavelength (nm)	Spectr.range	Fuel flow (L/h)	Burner height (mm)
Ca	422.6728	200	70	6
Fe	248.3270	200	80	5
Zn	213.8570	200	90	6
Mg	285.2125	200	55	6
Mn	279.4817	200	100	6

Table 2. Optimal determination parameters of HR-CS GFAAS.

Element	Wavelength (nm)	Chemical modifier
Cd	228.8018	$NH_4H_2PO_4$ (10 g/L)
Pb	283.3060	$NH_4H_2PO_4$ (10 g/L)
Cu	324.7540	$Mg(NO_3)_2$ (0.5 g/L)

Table 3. Optimized graphite furnace temperature programs.

Element	Pyrolysis temperature (°C)	Atomization temperature (°C)	Ramp (°C/s)
Cd	600	1200	1400
Pb	800	1500	1500
Cu	1100	2000	1500

Table 4. Microwave digestion program.

Stages	Pressure (Mpa)	Hold (min.)	Power (W)
1	0.2	60	500
2	0.5	60	1000
3	1.0	120	1000
4	1.5	120	1000
5	2.0	60	1000

Fe, Zn, Mg, Mn, Cd, Pb and Cu were purchased from National Chemical Reagent Company (Beijing, China). Calibration solutions of Ca were prepared in the ultrapure water with 0.5% (v/v) HNO_3, 1% (m/m) KCl and 0.5% (m/m) $La(NO_3)_3$ by serial dilution of the stock solutions with 100 mg/L, that of Fe, Zn, Mg and Mn were prepared in the ultrapure water with 0.5% (v/v) HNO_3 and 0.1% (m/m) KCl by serial dilution of the stock solutions with 100 mg/L respectively, that of Cd, Pb and Cu were prepared in the ultrapure water with 0.5% (v/v) HNO_3 by serial dilution of the stock solutions with 100 mg/L respectively. All glasswares were previously soaked overnight in dilute HNO_3 (5% v/v) for cleaning and were rinsed with abundant ultrapure water prior to avoid contamination.

2.3 Microwave digestion

The polygonatum tea was crushed to powder. Approximately a 0.5 g of the crushed sample was preprocessed with 5 mL of HNO_3 and 2 mL H_2O_2 in the PTFE jar, which was the process of heating to 120 °C for 30 min in an air-ventilated oven. The compound (sample and acids) was digested in the intelligent microwave digestion system (Xin-tuo, Shanghai, China) after adding again 2 mL H_2O_2. A five stage program (Table 4) with a maximum pressure of 2.0 Mpa was chosen for achieving complete digestion of the crushed sample within the shorter time. The digestive liquor was diluted to 25 mL with the ultrapure water with 0.5% (v/v) HNO_3 when its volume was less than 3.0 mL. Soon after that, it was hand shaken resulting in a visually homogeneous system. A blank digest was carried out in the same way. Three independent above treatments of the sample were performed for obtaining the average of repetitive determinations of eight metal elements.

3 RESULTS AND DISCUSSION

The analytical characteristic data of HR-CS AAS were shown in Table 5. The calibration function used to determine of Ca, Fe, Zn, Mg and Mn with HR-CS FAAS and Cd, Pb and Cu with HR-CS GFAAS were nonlinear built-up by measuring the absorbance of the calibration solutions in the selected determination conditions. The correlation coefficients were 0.9992 (Ca), 0.9996 (Fe), 0.9989 (Zn), 0.9958 (Mg), 0.9989 (Mn), 0.9976 (Cd), 0.9991 (Pb) and 0.9989 (Cu), respectively. The characteristic concentrations were 0.092 mg/L (Ca), 0.17 mg/L (Fe), 0.033 mg/L (Zn), 0.008 mg/L (Mg), 0.15 mg/L (Mn), 0.021 mg/L (Cd), 0.58 mg/L (Pb) and 0.092 mg/L (Cu), respectively. As can be seen, the proposed method represented one of more sensitive methodologies for determination of these eight metal elements in polygonatum tea.

The proposed means, using microwave digestion-HR-CS AAS, had been applied to determination of Ca, Fe, Zn, Mg, Mn, Cd, Pb and Cu in polygonatum tea and was verified through spike recovery tests (Li, et al, 2012). The results, obtained as the average of three replicates of each element, were shown in Table 6. The contents in polygonatum tea were 2.091 ± 0.057 g/kg (Ca), 0.481 ± 0.009 g/kg (Fe), 14.11 ± 0.32 mg/kg (Zn), 1.287 ± 0.025 g/kg (Mg), 46.08 ± 0.98 mg/kg (Mn), 7.852 ± 0.287 µg/kg (Cd), 9.558 ± 0.296 µg/kg (Pb) and 1.256 ± 0.028 mg/kg (Cu), respectively. The recoveries were 102.5 ± 1.8% (Ca), 104.7 ± 2.1% (Fe), 98.5 ± 1.5% (Zn), 101.6 ± 0.9% (Mg), 97.6 ± 2.2% (Mn), 98.6 ± 3.1% (Cd), 96.5 ± 2.7% (Pb)

Table 5. Analytical characteristic data of HR-CS AAS.

Element	Calibration function (C in mg/L)	Correlation coefficient (R^2)	Characteristic concentration (mg/L)
Ca	$A = (0.0029893 + 0.0472545 \times c)/(1 + 0.0206515 \times c)$	0.9992	0.092
Fe	$A = (-0.0013854 + 0.0256013 \times c)/(1 + 0.0194315 \times c)$	0.9996	0.17
Zn	$A = (-0.0018482 + 0.1311885 \times c)/(1 + 0.1288429 \times c)$	0.9989	0.033
Mg	$A = (0.0410249 + 0.7875324 \times c)/(1 + 0.2081587 \times c)$	0.9958	0.008
Mn	$A = (0.0003392 + 0.0285501 \times c)/(1 + 0.0148045 \times c)$	0.9989	0.15
Cd	$A = (0.0267089 + 0.2082857 \times c)/(1 + 0.1436398 \times c)$	0.9976	0.021
Pb	$A = (0.0072548 + 0.0072305 \times c)/(1 + 0.0039597 \times c)$	0.9991	0.58
Cu	$A = (0.0401687 + 0.0394598 \times c)/(1 + 0.0137857 \times c)$	0.9986	0.092

Table 6. Determination of eight metal elements in polygonatum tea and recoveries (Avg. ± SD of three trials).

Element	Content (µg/kg)	Spiked (µg/kg)	Recoveries (%)
Ca	$(2.091 \pm 0.057) \times 10^6$	2×10^6	102.5 ± 1.8
Fe	$(0.481 \pm 0.009) \times 10^6$	5×10^5	104.7 ± 2.1
Zn	$(1.411 \pm 0.032) \times 10^4$	1×10^4	98.5 ± 1.5
Mg	$(1.287 \pm 0.025) \times 10^6$	1×10^6	101.6 ± 0.9
Mn	$(4.608 \pm 0.098) \times 10^4$	5×10^4	97.6 ± 2.2
Cd	7.852 ± 0.287	10	98.6 ± 3.1
Pb	9.558 ± 0.296	10	96.5 ± 2.7
Cu	$(1.256 \pm 0.028) \times 10^3$	1×10^3	97.6 ± 3.5

and 97.6 ± 3.5% (Cu), respectively. Therefore, the proposed method represented one of the more accurate methodologies for the determination of these eight metal elements.

4 CONCLUSIONS

The developed means provided an exact and sensitive procedure for the determination of Ca, Fe, Zn, Mg and Mn in polygonatum tea by HR-CS FAAS and Cd, Pb and Cu in polygonatum tea by HR-CS GFAAS after microwave digestion. Under the selected determination conditions, the correlation coefficients better than 0.995 and the recoveries between 96.5 and 104.7% were obtained for these eight metal elements. The characteristic concentrations were 0.092 mg/L (Ca), 0.17 mg/L (Fe), 0.033 mg/L (Zn), 0.008 mg/L (Mg), 0.15 mg/L (Mn), 0.021 mg/L (Cd), 0.58 mg/L (Pb) and 0.092 mg/L (Cu), respectively. The contents in polygonatum tea were 2.091 ± 0.057 g/kg (Ca), 0.481 ± 0.009 g/kg (Fe), 14.11 ± 0.32 mg/kg (Zn), 1.287 ± 0.025 g/kg (Mg), 46.08 ± 0.98 mg/kg (Mn), 7.852 ± 0.287 µg/kg (Cd), 9.558 ± 0.296 µg/kg (Pb) and 1.256 ± 0.028 mg/kg (Cu), respectively. Low detection limits, good correlation coefficients and precisions, and high recoveries showed that the proposed method is accurate, reliable and stable.

ACKNOWLEDGEMENTS

This work was financially supported by Anhui science and technology project (1301C063011), the University Scientific Research Projects of Xuzhou Institute of Technology

(No. XKY2013325 and XKY2014214) and Spark Plans of China (No. 2013GA690417, 2013GA690418, 2014GA690103 and 2014GA690105).

REFERENCES

Afridi H.I., Kazi T.G., Arain M.B., et al. 2007. Determination of cadmium and lead in biological samples by three ultrasonic-based samples treatment procedures followed by electrothermal atomic absorption spectrometry. *Journal of AOAC International* 90(2): 470–478.

Baysal A., Akman S. 2011. A practical method for the determination of sulphur in coal samples by high-resolution continuum source flame atomic absorption spectrometry. *Talanta* 85(5): 2662–2665.

Borges A.R., Becker E.M., Lequeux C., et al. 2011. Method development for the determination of cadmium in fertilizer samples using high-resolution continuum source graphite furnace atomic absorption spectrometry and slurry sampling. *Spectrochimica Acta Part B: Atomic Spectroscopy* 66(7): 529–535.

Gao S.Y. 2007. The Determination of Element Content in Burdock by Flame Atomic Absorption Spectrophotometry. *Journal of Xuzhou Institute of Technology* 6(22): 32–35.

Gao S.Y., Cai S.J., Qu D.H. 2007. Determination of Trace Elements (Cu, Se, Cd, Pb) in Garlic by Method of Furnace Atomic Absorption. *Journal of Xuzhou Institute of Technology* 10(22): 61–64.

Lepri F.G., Borges D.L.G., Araujo R.G.O., et al. 2010. Determination of heavy metals in activated charcoals and carbon black for Lyocell fiber production using direct solid sampling high-resolution continuum source graphite furnace atomic absorption and inductively coupled plasma optical emission spectrometry. *Talanta* 81(3): 980–987.

Li Y., Chen S.L., Wang S.L., et al. 2012. Study on Determination Methods of Cadmium Content in Alcoholic Drink. *Journal of Xuzhou Institute of Technology (Natural Sciences Edition)* 4(27): 16–19.

Liu H., Chen S.L., Li C., et al. 2014. Sequence Determination of Cd and Pb in Honey by Incomplete Digestion-High Resolution Continuum Source Graphite Furnace Atomic Absorption Spectrometry. *Applied Mechanics and Materials* 511: 22–27.

Liu H., Wu Q.J. 2009. Determination of Pb in Food Sample by Nano-titanium Dioxide Preconcentration and Separation-furnace Atomic Absorption Spectroscopy. *Journal of Xuzhou Institute of Technology (Natural Sciences Edition)* 1(24): 38–41.

Liu L., Yu M. 2004. Determination of calcium and magnesium in gelatin by noncomplete digestion-flame atomic absorption spectrometry. *Metallurgical Analysis* 5: 015.

Nunes L.S., Barbosa J.T.P., Fernandes A.P., et al. 2011. Multi-element determination of Cu, Fe, Ni and Zn content in vegetable oils samples by high-resolution continuum source atomic absorption spectrometry and microemulsion sample preparation. *Food chemistry* 127(2): 780–783.

Ozbek N., Akman S. 2012. Method development for the determination of fluorine in toothpaste via molecular absorption of aluminum mono fluoride using a high-resolution continuum source nitrous oxide/acetylene flame atomic absorption spectrophotometer. *Talanta* 94(30): 246–250.

Ren T., Zhao L.J., Zhong R.G. 2011. Determination of Aluminum in Wheat Flour Food by Microwave Digestion-High Resolution Continuous Source Graphite Furnace Atomic Absorption Spectrometry. *Spectroscopy and Spectral Analysis* 31(12): 3388–3391.

Resano M., Briceño J., Belarra M.A. 2009. Direct determination of Hg in polymers by solid sampling-graphite furnace atomic absorption spectrometry: a comparison of the performance of line source and continuum source instrumentation. *Spectrochimica Acta Part B: Atomic Spectroscopy* 64(6): 520–529.

Characterization of proteins in Soy Sauce Residue and its hydrolysis by enzymes

Jinlan Zhang
China Meat Research Center, Beijing, China

Jinyang Zhao, Yan Zhao, Chunhui Yang & Wenping Wang
Beijing Academy of Food Sciences, Beijing, China

ABSTRACT: To reveal the biochemical characteristics, proteins in Soy Sauce Residues (SSR) was prepared by centrifugation in combination with lyophilization. The proteins in SSR were identified by SDS-PAGE and MALDI-TOF-MS. Results showed that acidic polypeptides (A1a, A2, A1b, A4) and basic polypeptides (B2, B1a, B1b, B3, B4) of glycinin from soybean proteins were the predominant proteins in SSR. In order to make better use of the residual protein in SSR, the optimal conditions of the cellulase and alcalase were analyzed. The optimal conditions for two-step enzymatic hydrolysis of soy sauce residue were as follows: pre-treatment by 0.01% Celluclast 1.5 L at pH 5.0, 55°C for 1 h, together with treatment by 0.05% Alacase 2.4 L at pH 8.0, 65°C for 2 h. The hydrolysis yield of protein (Soluble nitrogen index) could reach 62.45%. The analysis of the crude protein dissolution rate indicated that enzyme combinations could hydrolyze the proteins of SSR effectively.

1 INTRODUCTION

Low-salt solid-state fermentation processing technology for soy sauce is a unique technology conducted by chinese food technologists (Yang, X. & Cao, L. 2012). Products by this method reach more than 4.4 million tons in the market annually. Soy sauce residues, the main by-product in the soy sauce fermentation process, are obtained after extraction (Yang, B. &Yang, H.S. 2012). About 200–400 thousands of tons soy sauce lees were obtained each year, which account for 5%–10% of the annual production of raw soy sauce (Furukawa, T. & Kokubo, K. 2008). Preliminary results of our work showed that the protein content accounts for approximate 20% of soy sauce residues (Gong, X. & Cheng, Y.Q. 2013; Zhang, Y.F. & Wang, L.J. 2009). It indicated that the unused raw protein mainly remained in the soy sauce residues.

With the outbreak of the international financial crisis, the rising prices of defatted soybean has become a great burden for soy sauce producers. To reduce the cost, how to improve the protein utilization rate in soy sauce production has become the urgent needs for every enterprise (Zhang, H.Z. & Jiang, Y.J. 2008). Therefore, the objective of this work is to identify the proteins and peptides in the soy sauce residues, by SDS-PAGE and MALDI-TOF/TOF MS, and to hydrolyze the proteins by alcalase and cellulase. It will provide a basis for the construction of some effective enzymes to improve the soy sauce production efficiency and the protein utilization.

2 MATERIALS AND METHODS

2.1 *Materials and chemicals*

The soy sauce residues were kindly provided by Beijing Longfeiye Condiment Co., Ltd. (Beijing, China). The protein content was determined to be 21.36 ± 0.52 (w/w).

Alcalase 2.4 L with a nominal activity of 2.4 AU/g, and Celluclast 1.5 L with a nominal activity of 700 EGU/g were obtained from Novozymes (Beijing, China).

Plant Hydrophobic Protein Extraction Kit were purchased from Sigma Co. (St. Louis, MO, USA).

Other chemicals used in this work are obtained from Sinopharm Chemical Reagent Co., Ltd. (Shanghai, China).

2.2 Sample preparation

SSR produced by Low-Salt Solid-State Fermentation (LSSF) was collected, and each type of SSR contained 3 parallel samples. Before analysis, all samples were lyophilized and stored at −20°C.

2.3 Total protein extraction of SSR

SSR of 100 mg were grinded in liquid nitrogen to a fine powder. The samples were transferred to a V-bottom freezing vial stored at −20°C. 10 mL of the prepared methanol solution was added. Allow the mixture to incubate for 5 minutes at −20°C with periodic vortexing. Centrifuge the suspension at 16 000 g for 5 minutes at 4°C to pellet proteins. Discard the supernatant, and add 10 ml of pre-chilled acetone at −20°C. Briefly vortex the sample (15–30 s) and place at −20°C. Centrifuge the suspension at 16 000 g for 5 minutes at 4°C to pellet proteins. Discard the supernatant. Afer drying the pellet in air for 5–10 minutes, weigh the pellet to determine the mass. Add 4 ul of Regent Type 4 Working Solution of Plant Hydrophobic Protein Extraction Kit to each mg of pellet. Allow the solution to incubate for 15 minutes with intermittent vortexing or continuous gentle mixing at ambient temperature. Centrifuge the suspension at 16 000 g for 30 minutes. Remove the supernatant (total protein sample) by pipette and place in a clean, labeled tube. The total protein sample is now ready for SDS-PAGE.

2.4 SDS-PAGE analysis of SSR

Proteins from the SSR were dissolved in aqueous solutions containing 8 M urea, 60 mM DTT, 50 mM Tris, and 1% SDS, and the protein contents were adjusted to approximately 1 mg/mL, respectively. Protein analysis was performed using SDS-PAGE (Gao, X.L. & Sun, P.F. 2013; Gao, X.L. & Zhao, S.W. 2013). The concentrations of the stacking gel and the separating gel were 4 and 10%, respectively.

2.5 MALDI-TOF/TOF MS analysis

Each spot (a, b) in the SDS-PAGE gel (Fig. 1) was cut out and subjected to in-gel trypsin digestion. The peptides was re-dissolved in 0.8 ul 50% acetonitrile, 0.1% trifluoroacetic acid containing 10 mg/mL a-Cyano-4-hydroxycinnamic acid. And analyzed on an UltraflexTM MALDI-TOF/TOF MS (Bruker-Daltonics, Bremen, Germany). As described in Gao, X.L. & Sun, P.F. 2013; Gao, X.L. & Zhao, S.W. 2013, the sequences encoding predicted proteins of unknown function were subjected to BLAST search in NCBI.

2.6 Hydrolysis of the protein in SSR

Ten grammes of soy sauce lees were mixed with 50 mL of deionized water, and the solution was regulated by 1 mol/L NaOH or HCl to the required pH. At the optimal reaction temperature and pH of different enzymes, the solution was placed in water bath with constant temperature oscillator for a certain period of time, and then placed in 85°C thermostat water bath for enzyme inactivation. After 10 min, the solution was centrifuged at 3800 r/min for 10 min, the supernatant was separated and protein extraction rate was determined according to 2.7.

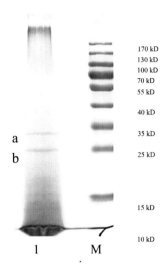

Lane M: standard markers; Lanes 1: SSR.

Figure 1. The representive electrophoresis map of proteins in SSR.

In the first test, the SSR was treated only by Alcalase 2.4 L, and the hydrolysis condition was optimized. While in the second optimization test, the SSR was pre-treated by Celluclast 1.5 L, followed by Alcalase 2.4 L treatment at the optimal conditions.

2.7 *Determination of the extraction rate of SSR protein*

The soy sauce residue was hydrolyzed. The sample was centrifuged at 3800 r/min for 15 min and then the pellet was collected and dried for three times, to determine the protein content. The extraction rate of SSR protein was calculated using the following equation:

$$\text{Protein extraction rate (\%)} = (m_0 - m) \times 100/m_0$$

where m_0 is the total protein content in SSR, and m is the total protein content in SSR after hydrolysis.

The nitrogen content was determined by Shimadzu 4100 Series Total Nitrogen Analyser (Shimadzu, Kyoto, Japan), and expressed as mg/ml. The protein content was calculated as $6.25 \times$ nitrogen content.

2.8 *Statistical analysis*

All determinations were conducted in triplicate, unless specified otherwise. The results were subjected to one-way ANOVA if necessary. Duncan's new multiple range test was performed to determine the significant difference between samples within 95% CI using SPSS 15.0 software (SPSS Inc., Chicago, Illinois, USA).

3 RESULTS AND DISCUSSION

3.1 *SDS-PAGE and MALDI-TOF/TOF MS analysis of proteins in SSR*

The SDS-PAGE profiles of SSR are shown in Figure 1. Totally, two predominant blue zone (band a and b) with approximately molecular weight ranging from 15 to 25 kDa and 25–35 kDa were observed in SSR. To clarify the precise origin and detailed information of the proteins in SSR, MALDI-TOF /TOF MS analysis was conducted. As shown in Figure 1, a total of 2 spots selected from SDS-PAGE gel of SSR (spots a–b) were subjected to MALDI-TOF/TOF MS analysis. Results showed that Spot a were successfully identified as acidic polypeptides (A1a, A2, A1b, A4) of glycinin from soybean proteins, while Spot b were identified as basic polypeptides (B2, B1a, B1b, B3, B4) of glycinin from soybean proteins.

Glycinin is found for storage purpose only (Catsimpoolas, N. 1969). The quaternary structure of glycinin consists of six monomers, where each monomer is composed of an acidic Chain (A) and a basic Chain (B) joined by disulfide linkage (Staswick, P.E. & Hermodson, M.A. 1981). Six different acidic polypeptides (A1a, A1b, A2, A3, A4, A5, and A6) and five basic (B1a, B1b, B2, B3, and B4) polypeptides were found to be homologous in their sequences (Staswick, P.E. & Hermodson, M.A. 1981). Nowadays, most soy sauce is produced by moistened soybeans processed at high pressure and high temperature. Glycinin dissociates on heating up to 100°C in five minutes small soluble aggregates are the only left over and the whole component disappears. With further heating, soluble aggregates precipitates out (Wolf, W.T.T. 1969). During the fermentation of soy sauce, large amounts of free amino acids are liberated from proteins by the hydrolytic action of protease of *Aspergillus oryzae* (Nakadai T. 1985). The glycinin exhibited a typical pH-solubility profile, of which the solubility was minimal at pH 5.0–6.0. The pH of soy sauce was just about 5.0. If the protease activity was not enough to hydrolyze all of the denatured glycinin, the protein might easily remain in the SSR.

The pI of A2 and A4 centered at about pH 5.0 and that of A3 even at pH 4.0. The solubility of A1a and B was much less than that of other acidic polypeptides or glycinin, which may be attributed to the presence of more insoluble aggregates, especially B polypeptide. The insoluble aggregates of B polypeptide, due to hydrophobic interactions, have been observed in many previous literatures (German, B. & Damodaran, S. 1982; Mo, X.Q. & Zhong, Z.K. 2006). The result coincided with the fact of the SSR in high-salt constant temperature fermentation that acidic polypeptide A1a of soy glycinin G1 and basic polypeptide B3 of soy glycinin G4 comprised of the major part of SSR proteins. The much lower solubility of the acidic polypeptide and the basic polypeptide of soy glycinin may be attributed to the presence of more hydrophobic amino acids and structure (Momma, T. & Negoro, T. 1985; Yuan, D.B. &Yang, X.Q. 2009).

3.2 *Optimization of the hydrolysis conditions of SSR by Alcalase 2.4 L*

In order to obtain the best enzymatic hydrolysis conditions by Alcalase 2.4 L FG according to the crude protein dissolution rate, an orthogonal design test ($L_9(3)^4$) was carried out and the results and statistical analysis were shown in Table 1. The K and R values in Table 1 were calculated following the procedures described by Ling J.Y. & Zhang G.Y. 2007. The R value showed that enzyme dosage had the most significant effect on yield, and the order of importance that influenced yield was found to be as follows: incubation temperature > enzyme dosage > pH > incubation time. The K value indicated that the optimal enzymatic hydrolysis conditions were pH 8.0, temperature 65°C, 2 h of incubation and enzyme dosage 0.05%, and a 10.03% higher of the hydrolysis yield (Soluble nitrogen index) was obtained under the optimal conditions.

Degradation of protein macromolecules into small molecular peptides with higher water solubility was supposed to be responsible for the high extraction efficiency (Ortiz, S.E.M. & Wagner, J.R. 2002). Alcalase from *Bacillus licheniformis* (Vernaza, M.G. & Dia, V.P. 2012), was found to be effective to generate soy protein hydrolysates. Temperature, pH, and Enzyme/Substrate (E/S) ratio are among the important parameters to be controlled in the production of protein hydrolysates by microbial proteases (Surówka, K. & Żmudziński, D. 2004).

3.3 *Optimization of the hydrolysis conditions of SSR by cellulase (Celluclast 1.5 L)*

After the hydrolysis of SSR by Alcalase 2.4 L, an orthogonal design test ($L_9(3)^4$) was used to optimize the following hydrolysis conditions by Cellulase (Celluclast 1.5 L), and the results were shown in Table 2 together with statistical analysis.

Table 1. Statistical analysis of the results from orthogonal experiment by Alcalase 2.4 L.

No.	A pH	B Incubation temperature (°C)	C Incubation time (h)	D Alcalase 2.4 L dosage (%)	Yield (soluble nitrogen index) (%)
1	1 (7.5)	1 (45)	1 (0.5)	1 (0.01%)	16.24
2	1	2 (55)	2 (1.0)	2 (0.05%)	25.83
3	1	3 (65)	3 (2.0)	3 (0.10%)	24.65
4	2 (8.0)	1	2	3	18.31
5	2	2	3	1	33.87
6	2	3	1	2	30.68
7	3 (8.5)	1	3	2	19.15
8	3	2	1	3	20.54
9	3	3	2	1	31.04
K1	22.24	17.90	22.49	27.05	
K2	27.62	26.75	25.06	25.22	
K3	23.58	28.79	25.89	21.17	
R	5.38	10.89	3.40	5.88	

Table 2. Statistical analysis of the results from orthogonal experiment by cellulase.

No.	A pH	B Incubation temperature (°C)	C Incubation time (h)	D Celluclast 1.5 L dosage (%)	Soluble nitrogen index (%)
1	1 (5.0)	1 (45)	1 (0.5)	1 (0.005)	48.62
2	1	2 (55)	2 (1.0)	2 (0.01)	60.08
3	1	3 (65)	3 (2.0)	3 (0.015)	41.35
4	2 (5.5)	1	2	3	45.36
5	2	2	3	1	55.25
6	2	3	1	2	40.24
7	3 (6.0)	1	3	2	43.89
8	3	2	1	3	50.74
9	3	3	2	1	38.97
K1	50.02	45.96	46.53	47.61	
K2	46.95	55.36	48.14	48.07	
K3	44.53	40.19	46.83	45.82	
R	5.49	15.17	1.61	2.25	

Based on the results in Table 2, the order of importance that influenced yield by Cellulase was as follows: incubation temperature > pH > Cellulase dosage > incubation time. According to the K value, the optimal condition for the Cellulase is: pH 5.0, incubation temperature 55°C, 1 h of incubation and 0.01% Cellulase (Celluclast 1.5 L). The confirmatory test result demonstrated that, the hydrolysis yield (Soluble nitrogen index) reached 62.45%, which was 22.17% higher than that before optimization.

Furthermore, breaking the linkages between proteins and other components of plant tissues might be another mechanism for the increased dissolution of proteins. J.Q.& Ma, H.L. 2010 have also indicated that ultrasound treatment during proteolysis can facilitate the hydrolysis of wheat germ protein by Alcalase. After the treatment by neutral protease and cellulase, the crude protein dissolution rate in the soy sauce residue reached 9.88% and 12.77%, respectively (Fu, L. & Hou, Z.X. 2012). Chen, M. & Wu, H. 2011 have optimize alkali protease hydrolysis parameters of soy sauce, and the results showed the protein hydrolysis degree improved 55.59% compared with the control. The enzymatic hydrolysis conditions of Alcalase 2.4 L and Novozymes Flavorase were optimized (Shen, H. & Jin, Z.G. 2010), and the Degree of Hydrolysis (DH) was up to 7.67% and 5.47%, with individual enzymatic hydrolysis.

4 CONCLUSIONS

In this work, the proteins in SSR were characterized and their formation mechanism was discussed. The application of enzymes could significantly improve the extraction efficiency of the proteins in SSR. Firstly, it is confirmed that acidic polypeptides (A1a, A2, A1b, A4) and basic polypeptides (B2, B1a, B1b, B3, B4) of glycinin from soybean proteins were the predominant proteins in SSR. Secondly, it is proved Alcalase 2.4 L combined with Celluclast 1.5 L could extract 62.45% of the proteins in SSR.

Further work to decrease the remaining protein in SSR after soy sauce fermentation using specific microbial enzyme(s) is underway. The construction of enzyme system which could effectively degrade the proteins in soy sauce lees will be further studied. Our study provided detailed information on the proteins in soy sauce residues, which will help the further studies on the utilization of soy sauce residues and the development of traditional Chinese fermentated food.

ACKNOWLEDGMENTS

This work is supported by the National Natural Science Foundation of China (Grant No. 31171738) and National High-Tech R&D Program of China (863 Program) (Grant No. 2013AA102105).

REFERENCES

Catsimpoolas, N. 1969. A note on the proposal of an immunochemical system of reference and nomenclature for the major soybean globulins. *Cereal Chemistry* 46: 369–372.
Chen, M. &Wu, H. 2011. Optimization of conditions for soy sauce slag enzymolysis using alkali protease by response surface methodology. *China Condiment* 12(36): 38–43.
Fu, L. & Hou, Z.X. 2012. Utilization rate analysis and distribution of soluble protein in soy sauce residue. *Guangdong Agricultural Sciences* 39(4): 78–80.
Furukawa, T. & Kokubo, K. 2008. Modeling of the permeate flux decline during MF and UF cross-flow filtration of soy sauce lees. *Journal of Membrane Science* 322: 491–502.
Gao, X.L. & Sun, P.F. 2013. Characterization and Formation Mechanism of Proteins in the Secondary Precipitate of Soy Sauce. *European Food Research and Technology* 237: 647–654.
Gao, X.L. & Zhao, S.W. 2013. Isolation, Identification and Amino Acid Composition of Proteins in Soy Sauce Residue. *Modern Food Science and Technology* 29(10): 2512–2516.
German, B. & Damodaran, S. 1982. Thermal dissociation and association behavior of soy proteins. *Journal of Agricultural and Food Chemistry* 30: 807–811.
Gong, X. & Cheng, Y.Q. 2013. Research Progress in the Comprehensive Utilization of Soy Sauce Residue. *Science and Technology of Food Industry* 34(5): 384–387.
Han, Y.F. & Cheng, Y.Q. 2015. Protein denaturation during the steaming process of low-salt solid-stated soy sauce. *China condiment* 40(2): 66–71.
Jia, J.Q. & Ma, H.L. 2010. The use of ultrasound for enzymatic preparation of ACE-inhibitory peptides from wheat germ protein. *Food Chemistry* 119: 336–342.
Li, C.Z. & Yoshimoto, M. 2004. A kinetic study on enzymatic hydrolysis of a variety of pulps for its enhancement with continuous ultrasonic irradiation. *Biochemical Engineering Journal* 19: 155–164.
Ling, J.Y.& Zhang, G.Y. 2007. Supercritical fluid extraction of quinolizidine alkaloids from Sophora flavescens Ait. and purification by high-speed counter-current chromatography. *Journal of Chromatography A* 1145: 123–127.
Mo, X.Q. & Zhong, Z.K. 2006. Soybean glycinin subunits: Characterization of physicochemical and adhesion properties. *Journal of Agricultural and Food Chemistry* 54: 7589–7593.
Momma, T. & Negoro, T. 1985. Glycinin A5 A4B3 mRNA: cDNA cloning and nucleotide sequencing of a splitting storage protein subunit of soybean. European Journal of Biochemistry 149: 491–496.
Nakadai T. 1985. Role of each enzyme produced by soy sauce-koji. *Journal of Japanese Soy Sauce Research Institution* 11(2): 67–79.
Ortiz, S.E.M. & Wagner, J.R. 2002. Hydrolysates of native and modified soy protein isolates: Structural characteristics, solubility and foaming properties. *Food Research International* 35: 511–518.

Shen, H. & Jin, Z.G. 2010. Optimization of the auxiliary ultrasound step by step enzymatic hydrolysis of two enzymes by response surface methodology. *China Brewing* 3(216): 16–22.

Staswick, P.E. & Hermodson, M.A. 1981. Identification of the acidic and basic subunit complexes of glycinin. *Journal of Biological Inorganic Chemistry* 256: 8752–8755.

Surówka, K.& Żmudziński, D. 2004. New protein preparations from soy flour obtained by limited enzymic hydrolysis of extrudates. *Innovative Food Science & Emerging Technologies* 5: 225–234.

Vernaza, M.G. & Dia, V.P. 2012. Antioxidant and anti-inflammatory properties of germinated and hydrolysed Brazilian soybean flours. *Food Chemistry* 134: 2217–2225.

Wolf, W.T.T. 1969. Heat denaturation of soybean 11 S protein. *Cereal Chemistry* 46:14.

Yang, B. & Yang, H.S. 2012. Amino Acid Composition, Molecular Weight Distribution and Antioxidant Activity of Protein Hydrolysates of Soy Sauce Lees. *Food Chemistry* 21(6): 1729–1734.

Yang, X. & Cao, L. 2012. The development status and trend of the soy sauce industry in china. *China condiment* 10(37): 18–20.

Yuan, D.B. & Yang, X.Q. 2009. Physicochemical and functional properties of acidic and basic polypeptides of soy glycinin. *Food Research International* 42: 700–706.

Zhang, H.Z. & Jiang, Y.J. 2008. Application of multi-strain inoculation in koji-making and fermentation in soy sauce production. *China Brewing* 17(9): 1–4.

Zhang, Y.F. & Wang, L.J. 2009. Biochemical changes in low-salt fermentation of solid-state soy sauce. *African Journal of Biotechnology* 8(24): 7028–703.

Study on corrosion behavior of stainless steel 316 in low oxygen concentration supercritical water

J.Q. Yang, S.Z. Wang, Y.H. Li, T. Zhang, L.S. Wang & M. Wang
Key Laboratory of Thermo-Fluid Science and Engineering of MOE, School of Energy and Power Engineering, Xi'an Jiaotong University, Xi'an, Shaanxi, China

ABSTRACT: In this paper, corrosion behavior of stainless steel 316 was studied in low oxygen concentration supercritical water containing 0.1% H_2O_2 using weight measurement, scanning electron microscopy, and X-ray diffractometry. It was found that the mass gain increased linearly with exposure time in 0.1% H_2O_2 supercritical water. At the low oxygen concentration, there were non-continuous oxide powders which mainly contained Fe and O forming on the surface of stainless steel 316. Fe oxide consisted of Fe_2O_3 and Fe_3O_4.

1 INTRODUCTION

1.1 Supercritical water oxidation

Supercritical water is the water whose temperature and pressure are all above the critical point (374.15°C/22.1 MPa). With the temperature increasing, the ionic product, density, and dielectric constant of water drop drastically. However, the solubility of water to organic product increases significantly. As a result, supercritical water is a promising medium for dealing with organic wastes. A wastes treatment system using Supercritical Water Oxidation (SCWO) has a lot of superiorities such as a relative low time, an avoidance of secondary pollution, and an ability of self heating.

1.2 Corrosion problems in SCWO systems

Because of the high temperature and high pressure environment with several aggressive anions in SCWO systems, the metal materials using in SCWO systems, such as heat exchangers and reactors, face a severe corrosion problems. A number of commercialized SCWO plants around the world were forced to shut down in less than 5 years because of the corrosion problems of constructional materials (Marrone, 2013).

There are several forms of corrosion found from the materials using in SCWO systems. The typical forms are pitting, general corrosion, Stress Corrosion Crack (SCC), and intergranular corrosion. General corrosion is the most common types of corrosion whose corrosion rate is liner and could be predicted. Pitting corrosion is often found in sharp and discontinuous places in metal surface, wherever there is a relatively high electrochemical potential, there is always corrosion. As a rule of thumb, an accumulation of pitting can form general corrosion. SCC is the most dangerous types of corrosion which is unpredicted and fatal. The whole SCWO systems would be forced to shut down if SCC occurs.

Both solution parameters and metal parameters can affect corrosion. The most influential parameter is the solubility and dissociation of the aggressive anions and corrosion products (Kritzer, 2004). The solubility and dissociation is affected by the density and ionic product of solution, respectively. As mentioned above, with the temperature increasing, the ionic product and density of water drop drastically. In SCWO systems, stainless steels and nickel-based alloys are widely used as constructional material facing extreme environments. As a result, choosing a kind of corrosion resisting material is important in SCWO systems.

Table 1. Chemical composition of stainless steel 316.

Element	Fe	Cr	Ni	Mn	C	P	S	Si	Mo
Mass %	65	17	12	2	0.08	0.045	0.03	1.00	2.50

Oxidation is one of the most important types of corrosion. Many researchers have done studies on the oxidation of metal in supercritical water. Sun et al (Sun *et al.*, 2009) investigated the oxidation of 316 in the supercritical water containing 2% H_2O_2 and observed a duplex layer oxide structures with Ni-enrichment at oxide/metal interface, which were identified as Fe_3O_4/Fe_2O_3+spinel/Cr_2O_3/Ni-enrichment/316 SS from the outer to inner layer. However, no researches have been done on the oxidation of 316 in low oxygen concentration.

In the paper, oxidation behavior of stainless steel 316 was studied in supercritical water containing 0.1% H_2O_2. The duration time of the corrosion experiment is 24 h, 72 h, and 120 h. After the experiment, the samples were investigated in detail with a Scanning Electron Microscope (SEM) equipped with an Energy-Dispersive X-Ray Spectrometer (EDS) and an X-Ray Diffraction (XRD) analyzer. That is to say, the morphologies and chemical compositions of the samples were examined.

2 SPECIMEN PREPAREMENT AND EXPERIMENT

2.1 *The template file*

The material stainless steel 316 used in the study were cut into small coupons with a dimension of 10 mm*15 mm*4 mm and was polished by abrasive papers from 04# to 07#. Table 1 is the composition of the sample. After the polishing, the samples were soaked in acetone for 1 h to solve the organic matters and then cleaned by ultrasonic cleaner for 10 minutes. Finally, the samples were fixed on a wire and exposed in supercritical water in 510°C/24 MPa with 0.1% and 0.5% H_2O_2 for 24 h, 72 h, and 120 h. The reactor is a batch reactor made of alloy 625, which can stand 650°C and 30 MPa.

After the experiment, the samples were cleaned in the same methods as the former methods. The surface morphologies of oxide films generated after the exposure were examined using SEM, and the chemical compositions of the oxide films were analyzed by EDS and XRD.

3 RESULTS AND DISCUSSION

3.1 *Mass measurement*

From Figure 1, we can conclude that the mass gain of the stainless steel 316 in supercritical water containing 0.1% H_2O_2 increased with the exposure time rising. There was a rapid increase in mass gain from 24 h to 72 h. After the rapid increase, the corrosion rate became liner. As we know, 316 is a steel of the element iron with a small amount of nickel and chromium. At the beginning of oxidation, some sharp places on the surface of metal where the electric potential was relatively low began to dissolve and formed oxide of Fe, Ni, and Cr. Because of the high diffusion velocity of Fe ions, they diffused from the inner space to outer space to react with oxygen firstly. The oxides of Fe which were loose and porous formed on the surface of 316 first. The porous films were not able to prevent oxygen passing through the surface of metal and as a result, the mass gain of stainless steel 316 showed a character of general corrosion.

3.2 *Morphologies of the oxide films*

Figure 1 shows the surface morphologies of the oxide film grown on stainless steel 316 after exposed in supercritical water at 510°C/24 MPa containing 0.1% H_2O_2 for 24 h, 72 h,

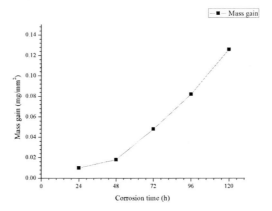

Figure 1. Mass gain as a function of exposure time for stainless steel 316 in supercritical water at 510°C/24 MPa.

(a) 24h   (b)72h   (c)120h

Figure 2. Morphology of sample surface of 316 after exposed in supercritical water at 510°C/24 MPa containing 0.1% H_2O_2 for 24 h, 72 h and 120 h.

and 120 h. We can figure out a significant change on the surface over time. At 24 h, there were only some non-continuous oxide powders generated on the surface and there were no significant changes at 72 h but the size of oxide powders grew a little. However, the size of oxide powders had grown from 0.625 μm at 24 h to 1.563 μm at 120 h. That is to say, stainless steel 316 cannot form a continuous oxide film in supercritical water containing 0.1% H_2O_2 in 120 h. Two major reasons can explain this result. The oxygen concentration was not high enough or the corrosion time was not long enough to form continuous oxide film. Indeed, further studies should be done on the analysis of the structure of the oxide film to figure out the composition of it. We used EDS to scan a piece of oxide powder in Figure 2(b) and the composition of it is listed in Table 2. As we can see, the oxide powder contained mainly O and Fe.

3.3 *XRD measurement*

Figure 3 shows the XRD pattern form stainless steel 316 in supercritical water at 510°C/24 MPa for 120 h. The peaks corresponding to Fe_2O_3 and Fe_3O_4 were found, indicating that the oxide film of 316 was mainly these two oxide. As mentioned above, there were several oxide powders rather than continuous oxide film forming on the surface of the sample. As a rule of thumb, Fe_3O_4 has a porous and loose structure. We can infer that $FeO \cdot Fe_2O_3(Fe_3O_4)$ and Fe_2O_3 were formed because of the low oxygen concentration.

The growth mechanism of oxide films of stainless steel 316 is similar to that in high-temperature water, namely solid-state growth mechanism (Gao et al., 2007). At the beginning, Fe_3O_4 was formed and then, Fe_2O_3 became to nucleate on the surface of Fe_3O_4. As the corrosion time went on, there were more and more Fe_2O_3 forming.

Table 2. EDS results of a piece of oxide powder (corresponding to A zone shown in Fig. 2(b)).

Element	O	Al	Si	Ca	Cr	Fe	Ni	Mo
Mass %	38.7	0.58	1.12	0.27	7.4	46.6	3.38	1.85
Atom %	68.25	0.6	1.23	0.19	4.02	23.55	1.62	0.54

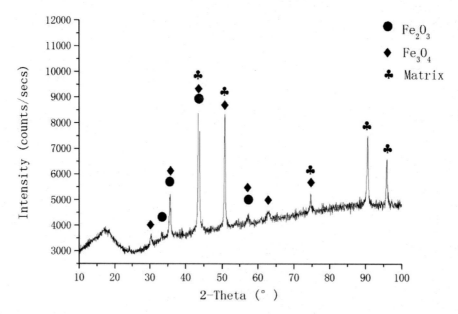

Figure 3. XRD pattern from stainless steel 316 in supercritical water at 510°C/24 MPa for 120 h.

4 CONCLUSIONS

The corrosion behavior of stainless steel 316 in supercritical water containing 0.1% H_2O_2 at 510°C/24 MPa was investigated by using mass gain measurement, SEM equipped with EDS, XTD. The mass gain increased with the exposure time rising and showed a character of general corrosion. At the low oxygen concentration, there were non-continuous oxide powders which mainly contained Fe and O forming on the surface of stainless steel 316. Fe oxide consisted of Fe_2O_3 and Fe_3O_4.

REFERENCES

Gao, X., Wu, X., Zhang, Z., Guan, H. & Han, E. (2007), "Characterization of oxide films grown on 316 L stainless steel exposed to H_2O_2-containing supercritical water", Journal of SuperCritical Fluids, Vol. 42 No. 1, pp. 157–163.

Kritzer, P. (2004), "Corrosion in high-temperature and supercritical water and aqueous solutions: a review", Journal of Supercritical Fluids, Vol. 29 No. 1–2, pp. 1–29.

Marrone, P.A. (2013), "Supercritical water oxidation-Current status of full-scale commercial activity for waste destruction", Journal of Supercritical Fluids, Vol. 79 No. SI, pp. 283–288.

Sun, M., Wu, X., Zhang, Z. & Han, E. (2009), "Oxidation of 316 stainless steel in super critical water", Corrosion Science, Vol. 51 No. 5, pp. 1069–1072.

300 MW steam turbine transformation technology using high-temperature circulating water for heating

Xuetong Wang & Xuedong Wang
Shandong Branch of Huadian Electric Power Research Institute, Jinan, P.R. China

Yue Han
State Grid Shandong Electric Power Research Institute, Jinan, P.R. China

ABSTRACT: This paper introduces the technology and content of the first subcritical 300 MW steam turbine transformation using high-temperature circulating water for direct heating, and illustrates the similarities and differences with the 135 MW–150 MW stage units. On the basis of transformation of 135 MW–150 MW stage units, technical problems that commonality transformation of dual low-pressure cylinder, feeding water pump turbine transformation, and condensate polishing treatment system transformation have been solved on 300 MW turbine unit to heating directly using high-temperature circulating water. After transformation, the power load is 230.4 MW when the unit back-pressure is 54.9kPa; heat capacity is 460.2 MW when the unit inlet steam flow is 1025t/h; unit average heat consumption rate is 3725.65 kJ/kWh, thermal efficiency of cycle is above 96%, and coal consumption for power generation is 139 g/kWh. At least 64,600 tons of standard coal is saved every heating season, and area of heat supply increases 4 million square meters. Small energy intensive boilers are replaced, and the annual saving for standard coal consumption is 28,300 tons. The comprehensive saving for standard coal is 92,900 tons.

1 INTRODUCTION

By increasing back-pressure of steam turbine in operation, the condenser exhaust temperature gets higher, and circulating water outlet temperature is enhanced. Heating system can utilize circulating water which is heated in condenser and joins into heat supply network, thus meeting users' demand for heating. The cooled circulating water is then back into condenser and re-heated again. High back-pressure circulating water heating recovers the heat that is originally wasted from cooling tower into the air, saves the steam for heating, and enhances the thermal efficiency of steam turbine unit (Kao Fang 2010, Zheng Jie 2006 & Wang Xiaohong et al. 2009).

In recent years in the power grid of Shandong area, seven sets of 135, 150, and 300 MW grade units have completed dual back-pressure dual rotor switching transformation on Low-Pressure (LP) cylinder, which operates in High-Pressure (HP) during heating period and normal pressure during non-heating period. After transformation, high-temperature circulating water for direct heating is used during heating period, meeting the demand for large heat load and improving urban residents' heating quality under rapidly developing urbanization. The dual back-pressure dual rotor switching transformation on low-pressure cylinder of 300 MW steam turbine solves complex issues, e.g. dual low-pressure inner cylinder, the elevation up of original cylinder bearing, steam-driving water-feeding pump transformation, and condensate polishing treatment system transformation. This project makes new technology breakthrough in the domestic 300 MW grade unit circulating water heating transformation technology, opens a new route for large-scale units heating transformation, and has great values for popularization.

2 TECHNICAL MANUAL FOR STEAM TURBINE CIRCULATING WATER HEATING BEFORE AND AFTER TRANSFORMATION

2.1 Technical manual for steam turbine heating before and after transformation

For a certain power plant, a 300 MW steam turbine uses dual flow low-pressure cylinder, with the flow stage number of 7. The low-pressure cylinder flow path is transformed, eliminating the last two stages of baffles and rotor blades, in order to realize high back-pressure circulating water heating. Technical manual before and after transformation is shown in Table 1.

2.2 Design system of steam turbine circulating water for direct heating

In heating period, the unit works in high back-pressure condition. The original cooling tower and circulating pump aborts, and circulating water system switches to the circulating water circuit of hot water pipe network set up by circulating pump in heating network, forming a new "excess heat-heating network circulating water" heat exchange system. After the switch completes, circulating water flow decreases to 9700t/h, condensing steam back-pressure goes up from 4.9 to 54 kPa, and LP cylinder exhaust steam temperature rises from 32.5 to 83.3°C. After heating through condenser, heating network circulating water temperature rises from 53 to 80°C. Next, after pressurizing through heat circulation pump, the heating network feeding water is sent into the initial station heater to complete the second heating, which will generate high-temperature hot water and is sent to heat users. The initial station heater uses steam extraction from the IP and LP cylinder link tube of its own unit or adjacent unit. High-temperature hot water returns to the condenser after cooling, which constitutes a complete circulating water cycle. When the unit works in normal back-pressure working condition, heating network circulating pump and network heater abort; the original circulating water system recovers, and condenser back-pressure returns to 4.9 kPa. Circulating water heating system of 300 MW sea water cooling unit is shown in Figure 1.

Table 1. Technical manual before and after transformation for 300 MW unit.

Name	Before transformation	After transformation
Type	C300-16.7/0.79/538/538	CB300-13.37/0.79/0.054/538/538
Steam turbine pattern	Subcritical, single shaft, single re-heat, dual-cylinder dual steam exhaustion, extraction condensing type, HP-IP cylinder combined	Subcritical, single re-heat, single shaft, dual-cylinder dual steam exhaustion, single adjusting extraction, back-pressure turbine
Rated power	300.167 MW	238.928 MW
Rated main steam parameters	16.7MPa/538°C	16.7MPa/538°C
Rated re-heat steam temperature	538°C	538°C
Rated inlet steam flow	915.679t/h	970.004t/h
Maximum power	329.2 MW	261.781 MW
Steam extraction flow for heating	300t/h	300t/h
Steam extraction pressure for heating	0.79 MPa	0.79 MPa
Rated cooling water temperature	20°C	53°C
Maximum cooling water temperature	33°C	80°C
Rated back-pressure	4.9kPa	54kPa
Rated feeding water temperature	275.1°C	276.3°C
Regenerative stage number	3 HP heater + 4 LP heater + 1 deaerator	3 HP heater + 2 LP heater + 1 deaerator
Heat consumption	7952.8kJ/kW.h	3658.8 kJ/kW.h
Length of last stage blade	905 mm	311.5 mm
Condenser circulating water flow	Rated 37000t/h	Rated 9700t/h

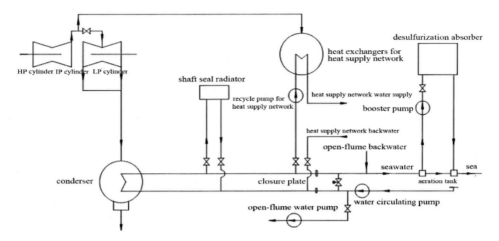

Figure 1. Sketch map of circulating water heating system of 300 MW sea water cooling unit.

3 TECHNICAL FEATURES AND CONTENT OF 300MW UNIT CIRCULATING WATER HEATING TRANSFORMATION

The key technical characteristics of 135 MW grade unit circulating water direct heating transformation are: (1) Using dual low-pressure rotor switching technology; (2) Using the new type of high strength integrally shrouded blade; (3) Condenser is enhanced to meet the safety requirements of both high back-pressure heating and low back-pressure pure condensing working conditions (Wang Xuedong et al. 2012 & 2013).

On the basis of high back-pressure circulating water heating transformation of 135–150 MW stage steam turbine, 300 MW steam turbine transformation is developed. There are research and development needs for new low-pressure rotor under high back-pressure working condition, and for transformation of condenser. In addition, more specific 300 MW unit technical problems should be stressed, including commonality transformation of dual low-pressure inner cylinder, reconstruction of low-pressure cylinder bearings, expanding the running scope of water-feeding pump turbine under variable working condition, and using high-temperature resin in condensate polishing treatment system.

3.1 Improving the reliability of bearings seated on LP cylinder

LP rotor of 300 MW unit applies bearings seated on LP cylinder. The elevation of bearings changes with the exhaust steam temperature. When back-pressure increases, the exhaust temperature goes up to 83°C, even 100°C under variable working condition, thus elevating the center line of LP cylinder bearings, and increasing bush temperature and vibration. Based on the LP rotor structure after transformation, a shafting computational model is established to calculate bearing loads, elevation, rotor deflection, etc. before and after transformation. The above problem will be solved by adjusting bearing elevation and the flow space of shafting and LP cylinder, as well as adding two stages of water spray cooling devices in LP cylinder.

3.2 Commonality transformation and optimization of dual LP inner cylinder

The original LP cylinder uses dual inner cylinder, with the last two stages installing embedded baffles. This cylinder type has difficulty in assembling and disassembling, and has no method to interchange the rotor and baffles. Through optimization design and transformation of LP inner cylinder, the LP rotor of high back-pressure cylinder can use the same cylinder as the original LP rotor does.

The optimization design of LP cylinder flow path is to replace the original LP dual inner cylinder for the new designed integrated inner cylinder structure, and to retain the original

LP outer cylinder. The first 2 × 5 stages of baffles of LP cylinder are installed on the new designed LP static blades carrier ring; the last two stages of baffles are replaced for a detachable structure of which the upper and bottom parts can be respectively disassembled. For heating condition, only change the LP carrier ring, remove the last two stages of baffles and replace with guide baffle plate which can protect the clapboard groove, and further use the new designed high back-pressure rotor for heating. For non-heating condition, the guide baffle plate is removed, and the last two stages of baffles as well as the first five stages of LP carrier ring are installed; the new designed LP inner cylinder works with the original LP rotor. Figure 2 is the optimized schematic diagram of the LP cylinder flow path.

3.3 *Research and development of high back-pressure low-pressure rotor*

The original 2 × 7 stages of low-pressure rotor under pure condensing working condition are transformed into 2 × 5 stages of low-pressure rotor under high back-pressure working condition. The new designed high back-pressure low-pressure rotor remains consistent with the old rotor in total length, axial size, axle diameter, etc. By adjusting the baffle seal diameter of rotor wheel, the new designed rotor and the old rotor have the same deflection characteristics, to ensure that the bearing load distribution and the rotor's rotation characteristics basically keep constant, and to guarantee the stability of shafting system.

3.4 *Transformation of feeding water pump turbine*

The exhaust steam out of feeding water pump turbine goes to host condenser. With the turbine working under high back-pressure condition, exhaust steam temperature and pressure of feeding water pump turbine also increase. The original designed turbine back-pressure is 4.5 kPa–12 kPa; during heating period the condenser back-pressure can be as high as 54 kPa pressure, causing insufficiency of the turbine output. Meanwhile with the host transformation, feeding water pump turbine flow path should be transformed overall. Schemes can be implemented such as research and development of new rotor, transformation of steam chamber nozzle ring, and processing of new guide blade carrier ring. During high back-pressure operation, steam source of feeding water pump turbine is switched to exhaust steam of HP cylinder, and the rotor's range of application is expanded for varied working conditions. Thus the same rotor of feeding water pump turbine can work well in both pure condensing mode for non-heating period and high back-pressure mode for heating period.

3.5 *Enhancement and transformation of condenser*

Condenser works under high and low back-pressure modes, which results in large pressure and temperature changes. After unit transformation, the condenser's water chamber pressure

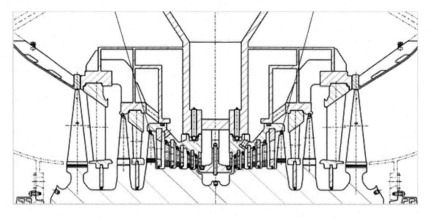

Figure 2. Sketch map of the low pressure cylinder flow part of 300 MW unit.

increases from the original designed 0.4MPa to 0.5–0.6MPa. LP cylinder exhaust steam temperature causes unevenness of condenser tube bundle and shell expansion, and leakage of expansion mouth occurs. The condenser is transformed overall, condenser tube bundles are replaced, arc plate structure is adopted for water chamber, support is added in the internal part, thickness of the tube plate is thickened, and the designed water chamber pressure reaches 1.0MPa. Slip joint is installed after the condenser water chamber, for the tube and water chamber to freely expand when condenser temperature increases.

3.6 *Adoption of high-temperature resin in condensate polishing treatment system*

During circulating water heating period, the temperature of condensing water can be as high as 80°C, and the original condensate polishing treatment system cannot run in such high temperature. To meet the needs of high back-pressure heating condition, domestic intermediate-pressure and high-temperature resin is investigated, and three butyl rubber lining mixed beds are added which can resist 100–120°C high temperature.

4 PERFORMANCE INDEX OF 300MW UNIT AFTER HIGH BACK-PRESSURE TRANSFORMATION

4.1 *Economic index of 300MW unit at high back-pressure operation condition*

Performance test of circulating water heating has been conducted after the first 300 MW unit high back-pressure transformation, in order to check the heating capacity and economic index under heating working condition. Test results are shown in Table 2.

After transformation of 300 MW unit high back-pressure circulating water heating, under 4VWO condition, HP cylinder efficiency is 81.335%, IP cylinder efficiency is 88.594%, LP cylinder efficiency is 93.292%, and the revised heat consumption rate is 3713.098 kJ/kWh. For the biggest pure condensing working condition of 230 MW unit, HP cylinder efficiency is 81.346%, IP cylinder efficiency is 87.232%, LP cylinder efficiency is 93.538%, and revised heat consumption rate is 3738.2 kJ/kWh. For the corresponding condition, the designed HP cylinder efficiency is 82.385%, IP cylinder efficiency is 91.15%, LP cylinder efficiency is 94.131%. As for high back-pressure transformation unit, the LP cylinder efficiency after transforming is slightly lower than the design value.

As can be seen from Table 2, for 4VWO and 230 MW conditions, heat consumption rate is from 3736.193 to 3762.792 kJ/kWh, the revised heat consumption rate is from 3713.098 to 3738.195 kJ/kWh. The average heat consumption rate is 3725.65 kJ/kWh, and unit thermal efficiency is from 96.3 to 97.0%, without too much fluctuation. This is mainly because the unit is under high back-pressure condition, and high-temperature circulating water undertakes great external heat load. The unit cold end parameters are influenced by heat load. For steam extraction heating condition, the power generation rate decreases steam is extracted from the IP and LP cylinder link tube without doing work in LP cylinder. Therefore, both of the unit test heat consumption rate and the revised heat consumption rate are increasing.

4.2 *Test results analysis*

4.2.1 *Unit charged load capacity*
The power rate of unit at high back-pressure circulating water heating condition is influenced greatly by cold end parameters and heat load. For pure condensing test condition, when the heat network circulating water flow reaches 11476t/h, the unit back-pressure is 54.95 kPa; the maximum unit power can be as high as 230.4 MW, reaching the designed power 229.7 MW.

4.2.2 *Economic index at condition of steam extraction for heating*
At steam extraction condition, the designed maximum steam extraction is 270t/h, and maximum steam extraction for test is 246.2t/h, under which condition the unit power output is

Table 2. Test results for 300MW steam turbine circulating water heating transformation.

Parameters	Unit	4VWO condition	230 MW condition	250t/h steam extraction condition
Generator power rate	kW	214482.7	230432.3	223467.7
Main steam temperature	°C	532.404	530.251	531.665
Main steam pressure	MPa	16.1844	16.2252	16.4744
Main steam flow	kg/h	805502.8	873705.4	1005057.5
Re-heat steam temperature	°C	530.74	532.249	531.156
Re-heat steam pressure	MPa	2.7642	3.011	3.3752
Re-heat steam flow	kg/h	625584.1	683614.2	789287.7
HP cylinder exhaust steam temperature	°C	310.007	317.252	326.476
HP cylinder exhaust steam pressure	MPa	3.1139	3.3897	3.8041
HP cylinder exhaust steam flow	kg/h	673764.7	729756.4	830060.1
Condenser back-pressure	kPa	51.274	54.949	40.61
Feeding water temperature	°C	269.376	274.285	280.732
Feeding water pressure	MPa	18.196	18.5276	19.2426
Feeding water flow	kg/h	764291.9	826832.8	945103.5
Heating network circulating water flow	t/h	11220.940	11476.601	11275.04
Circulating water outlet temperature	°C	79.773	81.297	74.037
Circulating water inlet temperature	°C	51.754	51.855	54.147
Steam extraction flow for heating	kg/h	0	0	246178.1
Steam extraction pressure for heating	MPa	/	/	0.8715
Steam extraction temperature for heating	°C	/	/	348.367
HP cylinder efficiency	%	81.335	81.346	83.213
IP cylinder efficiency	%	88.594	87.232	70.648
LP cylinder UEEP efficiency	%	93.292	93.538	90.870
Heat consumption rate	kJ/kW·h	3736.193	3762.792	4070.249
Steam consumption rate	kg/kW·h	3.756	3.792	4.498
Power rate after the 2nd class correction	kW	220300.5	238350.9	217725
Heat consumption rate after the 2nd class correction	kJ/kW·h	3713.098	3738.195	4056.902
Unit thermal efficiency	%	96.954	96.303	88.738

223.47 MW. For this working condition, the unit inlet steam flow is 1005.06t/h, power output is 223.47 MW, condenser back-pressure is 40.61 kPa, and LP cylinder efficiency is 90.870%. Under corresponding working conditions, the unit designed inlet steam flow is 970t/h, power output is 238.93 MW, steam extraction for heating is 190t/h, condenser back-pressure is 54 kPa, and LP cylinder efficiency is 93.906%.

Under heating steam extraction condition, the extracted steam and high-temperature circulating water both supply heat to users. As the steam extraction increases, the LP cylinder steam flow decreases, condenser back-pressure decreases, and the heat supplied by high-temperature circulating water goes down. In the same circumstance of the same main steam flow, the unit power output decreases in order to ensure the economic benefit of the unit and the whole power plant. Optimization measures should be taken on the running modes of high-temperature circulating water heating unit and other heating units. Priority should be given to high-temperature circulating water heating unit, and then steam extraction of an increasing pressure for heating can be used.

4.2.3 *Economic index and social benefit of unit after transformation*
After transformation of 300 MW unit circulating water heating, the average heat consumption rate is 3725.65 kJ/kWh, thermal efficiency of cycle is above 96%, coal consumption is 139 g/kWh, and at least 64,600 tons standard coal is saved every heating period.

After transformation of 300 MW unit circulating water heating, cold source loss is recovered, and heat source capacity is enlarged. When the unit inlet steam flow is 1025t/h, heating capacity is 460.2 MW; area of heat supply increases 4 million square meters, which increases 1.23 million GJ heat every year for urban area. Considering the heating coal consumption difference between small boiler and large heating unit, the annual saving for standard coal consumption is 28,300 tons. The comprehensive saving for standard coal is 92,900 tons, the reduction of SO_2 emission is 2933 tons, and the reduction of NO_X emission is 883 tons.

4.3 *Controlling index at high back-pressure operation condition*

For extraction and condensing unit using high-temperature circulating water heating, external heat load has large impacts on the circulating water flow and the condenser inlet and outlet water temperature, which are significant to the unit running index. Therefore, external heat load also impacts the running mode and running index of the unit.

4.3.1 *Limitation of LP cylinder exhaust steam pressure*

When LP cylinder exhaust steam pressure gets too high and steam flow is low, the LP cylinder exhaust steam volume flow decreases, and blast loss occurs in the last stage of blades, which may cause too high temperature of exhaust steam and turbine dynamic and static friction accident. Turbine back-pressure should be strictly controlled within the limited range according to the controlling curves for LP cylinder exhaust steam pressure (shown in Fig. 3), provided by the manufacturer.

4.3.2 *Limitation of IP cylinder exhaust steam pressure*

Heating steam extraction is attached from the IP and LP cylinder link tube, which controls the heat supply temperature of high-temperature circulating water. As there is strength demand for the last stage of blades in IP cylinder, pressure difference between stages should not exceed the limit. To ensure the safety of IP cylinder blades, different IP cylinder exhaust pressure should be controlled under different main steam flow: steam extraction pressure of link tube between IP and LP cylinder should not be lower than 0.79MPa when the main steam flow is higher than 850t/h; steam extraction pressure should be no lower than the IP cylinder steam extraction pressure under pure condensing working condition when the main steam flow is lower than 850t/h.

4.3.3 *Limitation of IP and LP cylinder exhaust steam temperature*

After unit high back-pressure circulating water heating transformation, the IP cylinder exhaust steam temperature is controlled to be lower than 378°C, and the exhaust steam temperature is lower than 85°C for LP cylinder.

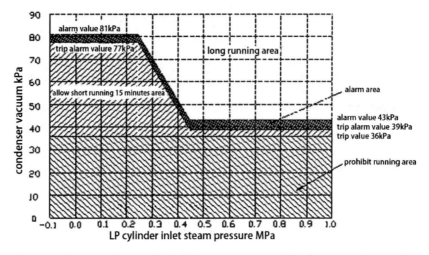

Figure 3. Controlling curves for LP cylinder inlet steam pressure and back-pressure after transformation.

The unit works in high back-pressure condition, and uses high-temperature circulating water for heating, which is influenced by external heat load. At the beginning and end of heating period, there is low heat load, and the unit has difficulty in running and adjusting. On this occasion, steam extraction from IP and LP cylinder link tube and high-temperature circulating water should be sufficiently adjusted to coordinate the unit running mode. Moreover, the IP and LP cylinder exhaust steam temperature can be guaranteed not to exceed the above limitations.

5 CONCLUSIONS

The first subcritical 300 MW extraction condensing steam turbine adopts "dual back-pressure dual rotor switching technology" for high back-pressure transformation, realizing high-temperature circulating water for direct heating. This paper introduces the technology and content of 300 MW high back-pressure transformation, and illustrates the similarities and differences with the 135 MW-150 MW stage units which adopt high back-pressure transformation. Similarities are: (1) applying dual rotor switching technology; (2) high back-pressure transformation for LP cylinder flow part; (3) research and development of LP rotor under high back-pressure condition; (4) enhancement and transformation of condenser. Differences are: (1) commonality transformation of 300 MW unit dual LP inner cylinder. The original LP cylinder uses dual inner cylinder, with the last two stages installing embedded baffles, where there is difficulty in assembling and disassembling. Through optimization design and transformation of LP inner cylinder, the LP rotor of high back-pressure cylinder can use the same cylinder as the original LP cylinder rotor does; (2) 300 MW unit adopts steam driving feeding water pump, and the feeding water pump turbine works under high back-pressure condition with a decreased power output. New rotor and corresponding parts are researched for water pump steam turbine, and the rotor's range of application is expanded for variable working conditions, solving the challenge of insufficient power output of water pump steam turbine under high back-pressure condition; (3) Domestic IP high-temperature resin is investigated, and three butyl rubber lining mixed beds are added which can resist 100–120°C high temperature, thus meeting the requirement of condensate polishing treatment system working under high back-pressure heating condition.

After the first 300 MW unit high-temperature circulating water heating transformation, condenser works under high back-pressure during heating period using high-temperature circulating water for direct heating, and works under low back-pressure during non-heating period, thus satisfying demand for high back-pressure heating in winter and ensuring economical running for the whole year. After transformation, the power load is 230.4 MW when the unit back-pressure is 54.9 kPa; heat capacity is 460.2 MW when the unit inlet steam flow is 1025t/h; unit heat consumption rate is 3725.65kJ/kWh, thermal efficiency of cycle is above 96%, and coal consumption for power generation is 139g/kWh. At least 64,600 tons of standard coal is saved every heating season. Small energy intensive boilers are replaced, and the annual saved standard coal consumption is 28,300 tons. The comprehensive saving for standard coal is 92,900 tons.

REFERENCES

Kao Fang. Analysis of circulating water heat supply reformation in an unit on low vacuum of small condensing steam turbine [J]. Shandong Electric Power, 2010, 3:46–48.

Wang Xiaohong, Sun Chao. Economic analysis on heat supply at low vacuum operation of condenser [J]. Huadian Technology, 2009, vol 31(1):37–39.

Wang Xuedong, Wang Dehua, Zheng Wei, et al. experimental investigation and analysis on reconstruction of high back pressure circulating water heat supply of 150MW unit [J], Turbine Technology, 2012, vol 54(5):397–400.

Wang Xuedong, Yao Fei, Zheng Wei, et al. technical analysis of turbine with two modes of high BP reconstruction for heat supply [J]. Power System Engineering, 2013, vol 29(2):47–50.

Zheng Jie. application of circulating water heat supply technology in an unit on low vaccum running [J]. Energy Conservation Technology, Jul. 2006, No. 4:380–382.

Optimization of fracturing penetration ratio and fracturing time in Fang 48 fault block

P. Guo & Z.W. Zhang
State Key Laboratory of Oil and Gas Reservoir Geology and Exploitation, Southwest Petroleum University, Sichuan, China

L.J. Huo
CNPC Economic and Technology Research Institute, Beijing, China

B. Jiang
CNOOC Research Institute, Beijing, China

Y.Z. Lei
PetroChina Daqing Oil Field Co., Daqing, China

ABSTRACT: Considering that fractures will result in early breakthrough of the injected CO_2, fracturing wasn't implemented in the target block at first. However, after putting into production, due to the overlong well spacing, the oil production was extremely low. Therefore, it is necessary to carry out a study on fracturing optimization. We have selected a typical well group with producers fractured in the main layers and simulated five schemes with different fracturing penetration ratios. By comparing and appraising relevant parameters, the optimal range was obtained. It would enable the producers and the injectors to communicate in a shorter time and the effective displacement system could be built earlier. When selecting optimal fracturing time, the "Fracturing and starting production at the same time" scheme was found to be able to cause the highest ultimate oil recovery, but it also led to the lowest gas replace ratio. Therefore, fracturing time can be further optimized by taking economic factors into consideration.

1 INTRODUCTION

With the deepening of exploration and exploitation of the oil fields in China, "Three-Low" reservoirs (low porosity, low permeability and low abundance) have become the major concern, which are hard to develop and are generally developed by water flooding or depletion at present. Some low permeability reservoirs are unable to be developed or even unable to be touched because water injection is hard to be implemented in this kind of reservoirs. Some other low permeability reservoirs are developed by depletion and their recovery factors are extraordinary low because of the absence of energy supplementation. Therefore, how to develop low permeability reservoirs economically and effectively has become the main concern of oil fields development.

Currently, a lot of researches on low permeability reservoirs fracturing are mainly concerned about the whole reservoir and by systematically considering the effectiveness of different combinations of fracturing parameters, injection-production parameters and well pattern arrangements, and the optimum scheme for integral reservoir fracturing can be obtained (Elbel 1986; Feng 2010; Holditch 1978; Sui & Zhang 2007; Shi et al. 2010; Wang & Feng 2005; Wang et al. 2010; Zhou et al. 2006). Moreover, some researches are carried out on the combined optimization of fracturing fractures and well pattern in order to select the most suitable well pattern while optimizing fracturing parameters (Han et al. 2010; Meehan 1988; Zhang & Cui 2008). Fracturing water injectors to improve water injection rate is an effective method to enhance water flooding effectiveness in low permeability reservoirs, therefore, some optimization studies are conducted on water well fracturing to enlarge the sweep volume of water flooding and to quickly increase single well production and overall production (Chen et al. 2009; Qu et al. 2009). After advanced water injection, some researches have been done to optimize the fracturing technology in order to improve the development situation and stabilize the productivity of oil wells (Yang 2010).

In general, most of the studies related to low permeability reservoirs fracturing are targeted at water flooding reservoirs, there is rarely research done on fracturing gas flooding low permeability reservoirs. This paper presents Fang 48 low permeability reservoir whose gas injection effectiveness was not obvious and a well group was selected whose well spacing is long enough to construct geological model to carry out the optimization study, with the producers being fractured in the main layers and its fracturing penetration ratio and fracturing time optimized so as to obtain the most reasonable fracturing parameters to provide reference for the development of this low permeability reservoir.

2 GEOLOGICAL AND DEVELOPMENT SITUATION OVERVIEW OF THE PILOT AREA

The peripheral oilfields of Daqing belong to the "Three-Low" reservoirs and the untouched reserve is nearly 8.25×10^8 t, among which 3.8×10^8 t is in Fuyu reservoir that takes up 46.06 percent of the total untouched reserves. In the part of Fuyu reservoir which had been developed, the effectiveness of water flooding was not ideal due to the reservoir's poor water absorbing capacity and high water injection pressure, therefore, the strategy was to develop it by CO_2 flooding.

Fang 48 fault block CO_2 flooding pilot area of Songfangtun Oilfield in the peripheral part of Daqing located in the southeast of Songfangtun Oilfield, belongs to Sanzhao sag to the east of Daqing placanticline region. The pilot area lies in the saddle area between two nose-like structures named Songfagntun and Mofantun and is a horst which is sealed by three faults in the north, east and west directions, and there is few faults in the centre. The oil bearing-area of Fang 48 pilot area is 2.49 km^2, and the oil reserve is 105×10^4 t.

The oil bearing formations of the pilot area are mainly developed from F1 Formation, partly from F2 Formation, and much less from F3 Formation. The pilot targeted formation is F1 Formation. The thickness of F1 Formation is about 100 m, and F1 Formation is divided into 7 layers and 19 sedimentation units. The average porosity and permeability of F1 Formation are 12.3% and 1.26×10^{-3} μm^2 respectively, and the main Layer F17 has superior physical properties compared with the other layers, with its average porosity 12.2% and the average permeability 1.34×10^{-3} μm^2.

The average density of the surface oil of Fuyu Formation is 0.869 t/m^3 and oil viscosity is 36.1 mPa·s. Cl$^-$ content of the formation water is 3067.6 mg/L, water salinity is 7158.0 mg/L and the water type is NaHCO$_3$.

In early 2002, gas injection test was implemented in Fang 48 pilot area on a small scale and favorable results were gained. Therefore, test in enlarged pilot area was carried out in 2007. Thirty producers and fifteen injectors were arranged in the enlarged area and the five-spot-pattern with the sizes of 400×250, 300×150 and 300×100 m were adopted. In the preliminary scheme of the development test, ten main Layers, (FI31, FI33, FI34, FI41, FI51, FI52, FI61, FI62, FI71 and FI72) were chosen to be developed simultaneously.

In November 2007, advanced gas injection was carried out successively in injectors, and in April 2009, oil wells were put into production in succession. However, to everyone's disappointment, there was no oil response resulted from gas injection and the average daily oil production was only 0.2 t. The reason for the ineffectiveness of gas injection may be that the well spacing of the test area was initially designed on the basis of fracturing well pattern, but during the implementation process, considering that fractures would result in early breakthrough of the injected CO_2, fracturing was not implemented to all the well groups. After putting into production, because of the overlong well spacing, there was nearly no oil production.

Table 1. Reserve parameters.

Layer	18th Layer (FI7$_1$)	19th Layer (FI7$_2$)	Cumulative
STOOIP of the model, m^3	192992.41	149926.19	342918.6

To establish an effective displacement system and to develop Fang 48 low permeability reservoir effectively, it is necessary to conduct fracturing optimization research and analysis.

3 BASIC INFORMATION OF THE SIMULATED WELL GROUP TEXT AND INDENTING

In this study, on the basis of the overlong well spacing wells in Fang 48 low permeability reservoir, we have selected Fang 184-128 "1 injector and 4 producers" well group with the size of "400 m × 100 m" and built its geological model with the oil wells being fractured in the main layers, namely Layer FI71 and Layer FI72 (in correspondence to the 18th and 19th layers of the model), and then the development situations of the well group has been simulated.

Firstly, reserves of the 18 and 19th layers of the well group model were calculated and the results are given in Table 1.

4 FLUID PHASE BEHAVIOR FITTING

We have selected PVTi Module of the Eclipse software to fit the experimental data of the formation oil PVT test. The fitting mainly involves Heavy Fractions Characterization of formation fluid, Component Lumping, Saturation Pressure Fitting, Experimental Data Fitting of the experiments including Single Flash, Differential Liberation, Constant Composition Expansion and CO_2 Injection Expansion, etc. Finally, PVT parameters reflecting the actual property changes of the formation fluid were obtained.

4.1 Composition of formation fluid

The original composition of the formation fluid of Fang 48 area is presented in Table 2, it is indicated that the fluid is the black oil with high heavy component content. The P-T phase-diagram given in Figure 1 also reveals that the fluid is the heavy black oil.

4.2 Division of pseudo-components

In order to increase the computing speed of numerical simulation, parameters of EOS in the compositional model should be optimized, thus enhancing the prediction precision of crude oil properties. Based on the principle of "Property Similarity of Components" and considering the circumstance of CO_2 injection, the 11 components of the well stream are extended and lumped into 7 pseudo components as follows: CO_2, N_2–^1C, ^2C–^6C, ^7C–^9C, ^{10}C–^{17}C, ^{18}C–^{22}C and ^{23+}C respectively.

4.3 Experimental data fitting of formation oil PVT test

The fitting results of the formation oil PVT test are given in Table 3, the precision of the fitting were relatively high and it indicated the basic requirements of gas injection numerical simulation had been met. Favorable results were obtained from the fitting of Constant Composition Expansion, favorable results about Multistage Degassing and Gas Injection Expansion experiments were also obtained.

Table 2. Original composition of the formation fluid of Fang 48 area.

Original composition	CO_2	N_2	^1C	^2C	^3C	IC_4	NC_4	IC_5	NC_5	^6C	^{7+}C
Moore composition, %	0	0.36	13.99	1.6	0.48	1.6	4.51	3.3	1.27	1.66	71.23
Weight composition, wt %	0	0.05	1.18	0.25	0.11	0.49	1.37	1.25	0.48	0.75	94.07

*Relative density of $^{7+}C = 0.8750$, molecular weight of $^{7+}C = 252.1$.

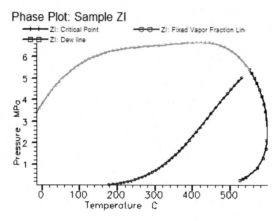

Figure 1. P-T phase-diagram of formation oil.

Table 3. Comparison of saturation pressure and single flash data.

Item	Unit	Experiment	Fitting	Error (%)
Oil viscosity at saturation pressure	MPa·s	5.0	4.5784	8.43
Density of surface degassing oil	g/cm³	0.865	0.836	3.35
Single flash GOR	m³/m³	18	17.64	2
Saturation pressure	MPa	5.200	5.207	0.14

5 SCHEME DESIGN AND RESULTS ANALYSIS

5.1 Scheme design

We built the geological model of Fang 184-128 well group and fractured the producers in the main layers FI71 and FI72 (in correspondence with the 18th and 19th layers of the model), and then five schemes with different penetration ratios (the ratio of half fracture length and well spacing) were designed and simulated. The penetration ratios selected are 0.2, 0.4, 0.6, 0.8 and 1 respectively. The main objective of this study was to analyze qualitatively the impact of different fracture lengths and fracturing times on the development effectiveness. As to the treatment of fractures in the simulation, we have adopted the conductivity coefficients adjusting method, with the fracture transmissibility set as 20 times of the matrix transmissibility. Schematic diagram of the fractured well group is shown in Figure 2.

During the simulation process, the control mode implemented to the injectors and producers is Fixed Bottom-hole Pressure Control. The maximum bottom-hole pressure for the injection well is set at 40 MPa, and for the producers the value is about 5 MPa. Producers shall be shut when its gas oil ratio reaches 3000 m³/m³. Because the well group pattern is the non-uniform five-spot-pattern, F182-126 and F182-128 are a little far away from the central

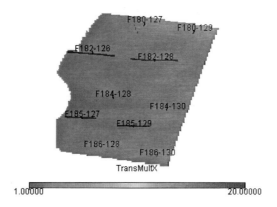

Figure 2. Schematic diagram of the fractured well group.

injection well F184-128, gas injection is controlled by switching the surrounding injectors and adjusting their bottom-hole pressures, consequently, the producers can be flooded uniformly by the injected gas to a certain degree, and the sweep area can be correspondingly enlarged. At the initial stage of the simulation, F184-128 were producing at a fixed bottom-hole pressure of 30 MPa, and the pressure for the other 5 injection wells was 40 MPa. After gas breakthrough had occurred in the producers F185-127 and F185-129, we shut the injectors F186-130, F186-128, F180-127 and F180-129 and increased the injection pressure of F184-128 to make its bottom-hole pressure rise to 40 MPa, and this pressure level had been maintained till the end of the simulation.

5.2 *Results analysis*

The results of this study indicate that, by oil well fracturing and adjusting the horizontal distribution of the injected gas, the sweep area of CO_2 can be enlarged effectively. At the end of the simulation, the horizontal distributions of CO_2 within the confines of F184-128 well group of the non-fracturing scheme and the scheme with the penetration ratio of 0.6 are shown in Figures 3–4 (We have only selected a scheme with an intermediate penetration ratio value to be a representative for the comparison with the non-fracturing scheme).

While observing the two figures, it is obvious that the CO_2 sweep area has been enlarged significantly after the producers have been fractured. Nevertheless, because of the low permeability of the reservoir, the injected gas failed to communicate with the production wells effectively at the early stage, therefore, oil increasing effectiveness of the fractured production wells were not remarkable and the oil replace ratio was very low and displayed a declined tendency at the early stage of gas injection. As the gas injection volume was increasing, producers became effective gradually and the oil replacement ratio displayed a rising tendency. However, after the breakthrough of the injected gas, gas replacement ratio declined again for the reason that a large amount of gas had been produced with oil. The variation tendency of the oil replacement ratio is shown in Figure 5.

After 20 years of simulation, the curves of cumulative gas injection, cumulative oil production and oil replace ratio varied, and variations are given in Figures 5–7. In Figure 6, it is evident that the cumulative oil production volume of those fracturing schemes is apparently bigger than that of the non-fracturing schemes. Furthermore, it can be found that the cumulative gas injection and oil production volume of the schemes with the penetration ratio of 0.4 and 0.6 are apparently bigger than the schemes with larger or smaller penetration ratio values, and although the oil replacement ratio of these two schemes are smaller than that of the other schemes, the difference is very small.

Through careful comparison and investigation, it was found that the extra incremental oil is mainly from F185-129. By further investigating the performance of F185-129, it was

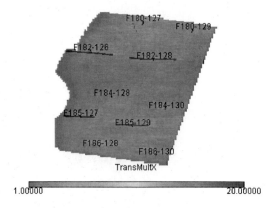

Figure 3. CO_2 distribution of the non-fracturing scheme for F184-128 well group.

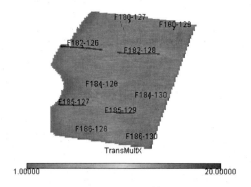

Figure 4. CO_2 distribution of the scheme with the penetration ratio of 0.6 for F184-128 well group.

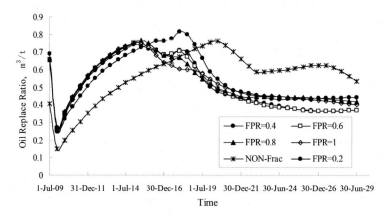

Figure 5. Oil replacement ratio vs. time.
*Non-fracturing scheme and schemes with different Fracturing Penetration Ratios (FPR).

discovered that, when its gas oil ratio was very high before reaching the shut-in limitation of 3000 m³/m³, the production lasted for 7 years and the oil production volume remained stable during the production process. Through repeated testing and verifying we found that the reason is that, when gas breakthrough occurred in F185-129 and its GOR began to go up, we closed the injectors surrounding it, and although the gas supply had been

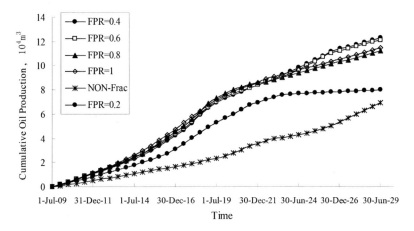

Figure 6. Cumulative oil production vs. time.
*Non-fracturing scheme and schemes with different Fracturing Penetration Ratios (FPR).

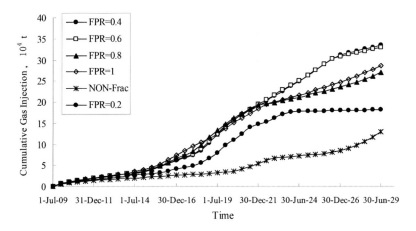

Figure 7. Cumulative gas injection vs. time.
*Non-fracturing scheme and schemes with different Fracturing Penetration Ratios (FPR).

terminated, there was still a large volume of CO_2 accumulated and the higher pressure level maintained. Hence, CO_2 began to be produced with oil as a result of the low permeability of the reservoir. And then, without extra CO_2 supplementation, the GOR of well F185-129 was maintained at the same level for a long time and the oil production remained stable.

If the fractures are too short (smaller penetration ratio), the injected CO_2 will not be able to sweep a large area, therefore the CO_2 cannot communicate with the producers effectively, and the oil increment will be smaller correspondingly. On the other hand, if the fractures are too long (larger penetration ratio), then the gas will break through prematurely and the GOR will increase rapidly, this would bring about negative impact on the oil flow and finally leading to premature well shut-in.

On the basis of the results above, we designed another scheme, namely "Shut injectors after gas breakthrough in all the producers" and obtained its simulating results. By comparing its results with continuous gas injection, it was found that stopping gas injection after gas breakthrough will reduce the cumulative gas injection volume to a large degree and at the same time the oil production will only decreased slightly. The final oil production volume of this scheme is just a little lower than that of continuous gas injection, and its oil replacement ratio stops decreasing and becomes steady gradually (Figs 8–10). So it can be

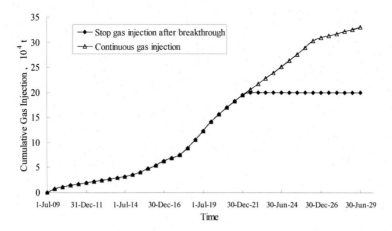

Figure 8. Cumulative gas injection vs. time.
*Stop injection after gas breakthrough and continuous gas injection schemes.

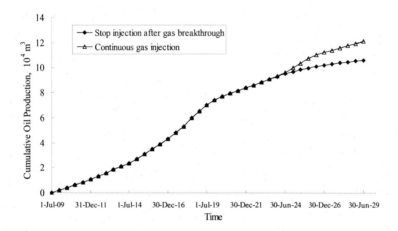

Figure 9. Cumulative oil production vs. time.
*Stop injection after gas breakthrough and continuous gas injection schemes.

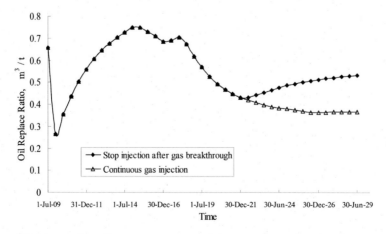

Figure 10. Oil replace ratio vs. time.
*Stop injection after gas breakthrough and continuous gas injection schemes.

concluded that although gas injection is stopped, there is still a large amount of oil that can be produced.

In conclusion, for the oil well fracturing optimization of low permeability reservoir in Fang 48, the optimum range of fracturing penetration ratio is 0.4–0.6. Penetration ratios within this range would enable the producers and injectors to communicate in a shorter time, thus effective displacement system can be built earlier. Shutting injectors after CO_2 breakthrough in all the producers thus allowing the injected CO_2 to diffuse sufficiently, oil production can not only be maintained stable, and the gas oil ratio can also be kept from getting too high to render the production wells to be shut prematurely, which will be detrimental to the ultimate recovery.

6 OPTIMIZATION OF FRACTURING TIME

The above analysis and research reveals that the optimum range of fracturing penetration ratio is 0.4–0.6. Consequently, while optimizing fracturing time, the penetration ratio chosen is 0.6. Seven schemes have been designed to implement fracturing when the cumulative gas injection volume reaches 0 ("Fracturing while producing"), 0.05, 0.1, 0.15, 0.2, 0.3 and 0.45 HCPV respectively. We have predicted the development effectiveness of these schemes by numerical simulation and thereby have gained the best time for fracturing. The final amount of gas injection volume, oil production volume and oil replace ratio of different schemes are given in Figure 11.

As seen in Figure 11, the "Fracturing while producing" scheme (0 HCPV) has gained the highest ultimate oil recovery, because under that circumstance, the effective displacement system can be built earlier. However, its gas replacement ratio is the lowest—the utilization of CO_2 is the lowest. In actual implementation process, we can combine cumulative gas injection and cumulative oil production data with economic factors together to evaluate the economic benefit, therefore the fracturing time can be further optimized and the maximum economic benefits can be obtained.

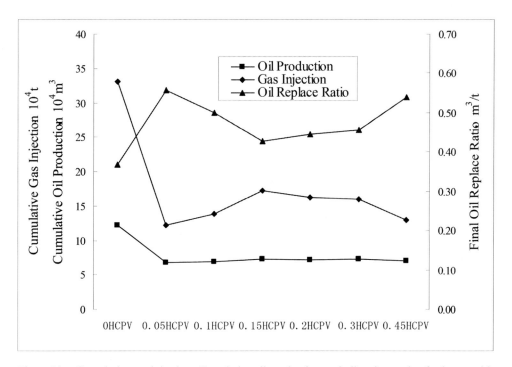

Figure 11. Cumulative gas injection, Cumulative oil production and oil replace ratio of schemes with different fracturing time.

7 CONCLUSION

The well spacing of low permeability fault block CO_2 flooding pilot area in Fang 48 was initially designed on the basis of fractured well pattern, during the implementation process, considering that fractures will result in early breakthrough of the injected CO_2, fracturing was not implemented. After putting into production, because of the overlong well spacing, the oil production was extremely low, so it is necessary to conduct fracturing optimization analysis.

In this study, Fang 184-128 "1 injector and 4 producers" well group was selected as the research subject, its geological model were built and all the oil wells in the main layers were fractured, then 5 schemes were designed and simulated and the optimum range of fracturing penetration ratio is 0.4–0.6. Penetration ratios within this range would enable the production and injection wells to communicate in a short time, thus effective displacement system can be built earlier.

Shutting injection wells after CO_2 breakthrough in all the production wells thus allowing the injected CO_2 to diffuse sufficiently, the oil production can not only be maintained stable, the gas oil ratio can also be kept from getting too high to render the production wells to be shut prematurely, which will be detrimental to the ultimate recovery.

While optimizing fracturing time, it is found that the "Fracturing while producing" scheme has obtained the highest oil recovery, because under that circumstance the displacement system can be built even earlier. But the utilization of CO_2 of that scheme is the lowest and fracturing time can be further optimized by taking economic factors into consideration when actual implementation is carried out.

REFERENCES

Chen, M.F. et al. 2009. Simulation study on parameter optimization of water well fracturing in low permeability sandstone reservoir. *Oil Drilling & Production Technology* 29(1): 54–57.

Elbel, J.L. 1986. Designing hydraulic fractures for efficient reserve recovery. *Spe Unconventional Gas Technology Symposium*.

Feng, X.W. et al. 2010. Optimization of overall fracturing fracture parameters in well block B304. *Journal of Oil and Gas Technology* 32(4): 356–358.

Holditch, S.A. et al. 1978. The optimization of well spacing and fracture length in low permeability gas reservoirs. *Spe Annual Fall Technical Conference & Exhibition*.

Han, X.L. et al. 2010. Optimization of integral fracturing parameters for five-spot pattern with horizontal fractures. *Special Oil and Gas Reservoirs* 17(1): 111–113.

Meehan, D.N. 1988. Effects of reservoir heterogeneity and fracture azimuth on optimization of fracture length and well spacing. *International Meeting on Petroleum Engineering*.

Qu, Z.Q. et al. 2009. Numerical simulation of fracturing in water injection well for south zhuang 74 low permeability reservoir in Shengli Oilfield. *Special Oil and Gas Reservoir* 16(6): 70–73.

Sui, W.B. & Zhang, S.C. 2007. Optimization design of integral fracturing parameters for low-permeability highly faulted reservoirs. *Petroleum Exploitation and Development* 34(1): 98–103.

Shi, Q.X. et al. 2010. Research on integral fracturing of low permeability reservoirs—take Tai 1 Block as an example. *Journal of Yangtze University (Nat Sci Edit)* 7(2): 62–65.

Wang, Q.Y. & Feng L. 2005. Optimized design application of whole fracturing and practice of E&P for the thin oil layers in low-permeability oilfields. *China Petroleum Exploration* 6: 56–60.

Wang, Y.W. et al. 2010. Research on the mathematical mode of integral fracturing in low permeability reservoir. *Drilling and Production Technology* 33(3): 64–71.

Yang, Y. 2010. Optimizing the fracturing treatment of pingbei low grade blocks after advanced water injection. *Journal of Jianghan Petroleum University of Staff and Workers* 23(2): 46–48.

Zhou, J.J. et al. 2006. Study on integral fracture technology of low permeability and thin interbed reservoir in Dagang Oilfield. *Fault-Block Oil & Gas Field* 13(6): 37–39.

Zhang, X.D. & Cui, R.R. 2008. Hydraulic fractures in low permeability reservoirs and optimization of well pattern assemblage. *Journal of Oil and Gas Technology (J.JPI)* 30(4): 124–128.

Supercritical Water Oxidation of acrylic acid wastewater and sludge

L.S. Wang, S.Z. Wang, Y.H. Li, T. Zhang, J.Q. Yang & M. Wang
Key Laboratory of Thermo-Fluid Science and Engineering of Ministry of Education, School of Energy and Power Engineering, Xi'an Jiaotong University, Xi'an, Shaanxi, China

ABSTRACT: SCWO experiments of acrylic acid wastewater and sewage sludge by an intermittent flow SCWO experimental system has been studied. The conditions of the experiment included reaction temperature (T) 793 K and 873 K, oxidation coefficient (α) of 1.05 and 4, a constant pressure (P) of 25 MPa, and a residence time (t) of 1200s. The experimental results showed that the temperature (T) and the oxidation coefficient (α) had great influences on the oxidation reaction and the Chemical Oxygen Demand (COD) could be improved when we increased the temperature (T) and the oxidation coefficient (α). The removal rate of chroma was perfect when the temperature of the experiment was 792 K, but it was bad when the temperature was 873 K. By speculation, we can get the conclusion that the chroma was caused by the corrosion of the tubular reactor and it was the color of Fe^{2+} and Fe^{3+}.

1 INTRODUCTION

Acrylic acid ($CH_2 = CHCOOH$, AA) is an important chemical raw material, and it is widely used in industry such as paints, chemical fibers, papermaking, adhesives, and cleaning agents. It can generate a large number of biodegradable, toxic, and hazardous waste during the manufacturing and application processes. The compositions of the waste are very complex and there are many organics in the waste such as acetic acid, acrylic acid, methacrylic acid, formaldehyde, and acetaldehyde (Gong et al., 2009). The Chemical Oxygen Demand (COD) is very high reaching more than 100 g/L. Thus, it is obvious that the waste will cause serious pollution to the environment without treatment.

Many published studies have been done concerning conventional treatment methods for the waste. Bednarik and Vondruska (Bednarik and Vondruska, 2003) introduced a procedure for the removal of formaldehyde from this wastewater, which consisted of adding urea into wastewater and removing formed solid-reaction products by filtration or centrifugation. Although the removal efficiency of formaldehyde in the process was about 99.5%, the effluent still required further treatment. Certain bacteria, such as *Bacillus thuringiensis, Ralstonia solanacearum*, and *Acidovorax avenae*, could utilize acrylic acid for growth, whereas the removal efficiency could reach 100% only when the initial acrylic acid concentration was below 690.8 mg/L. (Wang and Lee, 2006) Oliviero presented that a complete oxidation of acrylic acid was obtained by Catalytic Wet Air Oxidation (CWAO) using the Ru/CeO_2 catalyst. In addition, Total Organic Carbon (TOC) removal efficiency was 97.7% using the Mn/Ce catalysts at the residence time of 120 min (Silva et al., 2004). However, this method is difficult to deal with acrylic acid production wastewater harmlessly with a low operation cost because CAWO may need an expensive noble metal catalyst and/or a long residence time.

Supercritical Water Oxidation (SCWO) technology is a kind of environmental friendly, efficient technology which can deal with organic wastewater developed in the 1980s. It has many significant advantages, such as high efficiency, thorough reaction, and no secondary pollution, because of the special properties of supercritical water (SCW, $T \geq 374.15$ °C and $P \geq 22.12$ MPa). It was considered as the most efficient method for the treatment of municipal sludge (Wang et al., 2011), landfill leachate (Onwudili et al., 2013), and some of the industrial wastewaters (Sogut et al., 2011). Several researchers attempted to dispose acrylic acid production wastewater by the Supercritical Water Oxidation (SCWO) technology. For instance, Gong summarized that high efficiency (>99%) could be reached in a Transpiring Wall Reactor (TWR) (Gong et al., 2009).

In this study, we carried out the SCWO experiments of acrylic acid production wastewater and sewage sludge by an intermittent flow SCWO experimental system. We studied the influence of some important operation parameters on the destruction of COD in SCWO reaction. The provided information of the study is expected to be valuable for guiding the design of SCWO reactor to treat the acrylic acid production wastewater and sludge.

2 APPARATUSES AND PROCEDURES

We used 30wt% hydrogen peroxide (H_2O_2) as oxidant in this study. The wastewater and the sludge were collected from an acrylic acid manufacturing plant located in municipality of Tianjin. The characteristics of the wastewater and the sludge were shown in Table 1.

In this work, a treatment of acrylic acid production wastewater was carried out by an intermittent flow SCWO experimental system. The reactor in this study is a tubular reactor made by 304 stainless steel. It was designed for maximum reaction condition of 30 MPa and 600 °C and providing a maximum volume of 100 mL. Figure 1 shows the sand bath fluidized bed experimental device. It mainly consisted of an air compressor and a fluidized bed device. The air compressor was used to provide air to the fluidized bed device and the fluidized bed device was used to provide a constant temperature environment. The flow rate of the air was adjust by a rotameter. The measuring errors of pressure and temperature were within

Table 1. The characteristics of the wastewater and the sludge.

	1#	2#	3#	Sludge
COD	114,000 mg/L	11,600 mg/L	111,000 mg/L	234,000 mg/Kg
pH	1.5	2.5	13.2	–

Figure 1. The fluidized bed device.

±0.1 MPa and ±1 K. In the tests, we turned on the air compressor of the sand bath fluidized bed experimental device at first and adjusted the inlet air speed to 1.5 m³/h. The reactor temperature was heated up to the predicted temperature and adjusted by the temperature controller automatically.

We measured quantitative acrylic acid wastewater, sewage sludge, and hydrogen peroxide accurately and added them to the tubular reactor, and then sealed reactor. We vacuumed reactor by the vacuum pump to reduce the influence of the oxygen in air, and then started timing when we put the tubular reactor into the sand bath fluidized bed experimental device. When time is over, we put out the tubular reactor from the sand bath fluidized bed experimental device and put it into the water to cool it to room temperature. Finally, we scavenged reaction tube to get the reaction products and then analyzed the nature of the reaction products.

COD of liquid products were determined via a multiparameter water analyzer (Model NOVA60) with individual Merck cell. The pH of the samples was analyzed in a Sartorius Professional Meter (PP-15, Germany).

COD removal efficiency η is defined as follows (Xu et al., 2009):

$$\eta(\%) = \left[1 - \frac{[COD]_{pro}}{[COD]_{raw}} \times \left(\frac{Q_1 + Q_2}{Q_2}\right) \times 100\%\right] \quad (1)$$

where $[COD]_{raw}$ were the COD of the raw wastewater or the raw sewage sludge (mg/L) which were shown in Table 1. $[COD]_{pro}$ were the COD of the reaction products. Q_1 and Q_2 represented the volume of wastewater and H_2O_2 solution (mL) correspondingly which were obtained by a graduated cylinder.

3 RESULTS AND DISCUSSION

As shown in Table 2, the SCWO experiments of acrylic acid production wastewater and sewage sludge were conducted at temperature (T) of 793 K and 873 K, fixed pressure (P) of 25 MPa, and constant oxidation coefficient (α) of 1.05 and 4 within the reaction time (t) of 1200s. Each experiment repeated three times and the average result was plotted as the final value.

In the study, the operation pressure was set to be 25 MPa. Many researchers have confirmed that pressure influence on SCWO is small when pressure is above the critical pressure of water (22.12 MPa) (Bermejo and Cocero, 2006); even this influence can be neglected, compared with that of reaction temperature. (Segond et al., 2002) Moreover, it will be also comparably dangerous to operate at a higher pressure for a long time. Gong found that the COD or TOC removal efficiency increased slowly with pressure from 22 to 30 MPa (Gong et al., 2009).

Table 2. The results of SCWO experiments.

Sample	Number	T (K)	P (MPa)	α	$[COD]_{pro}$ (mg/L)	η (%)
1#	①	873	25	4	210	99.3
	②	793	25	1.05	2000	96.9
2#	①	873	25	4	300	96.6
	②	793	25	1.05	1180	89.0
3#	①	873	25	4	160	99.4
	②	793	25	1.05	1450	97.7
Sludge	①	873	25	4	170	99.5
	②	793	25	1.05	2650	97.1

As we can see from the results of COD removal efficiency, temperature and oxidation coefficient had powerful influence on the decomposition of organics. It can be seen that the COD removal efficiency at T = 873 K, α = 4 is higher than the COD removal efficiency at T = 793 K, α = 4. We can also see that all the results of the COD removal efficiency were higher than 96.0%. So we could get the conclusion that the organics had been decomposed without any other advanced treatment or catalyst during SCWO process. Correspondingly, SCWO is verified to be an effective method for the treatment of acrylic acid production wastewater and sewage sludge by our experiments, compared with biological and physical-chemical techniques (Bednarik and Vondruska, 2003).

As we can see from the Figure 2, the color of the wastewater and sewage sludge changed obviously. The removal rate of chroma was perfect when T = 793 K, α = 1.05. But when parameters were T = 873 K, α = 4, the removal rate of chroma of the sample 1# was bad. By speculation, the tubular reactor are made by 304 stainless steel and the acid-stage of the sample 1# is very strong. The 304 stainless steel would be corroded in the environment and the Fe^{2+} and/or Fe^{3+} would be soluble in the liquid phase. So we can get the conclusion that the chroma was caused by the corrosion of the tubular reactor and it was the color of Fe^{2+} and/or Fe^{3+}.

Figure 2. The results of sample by SCWO. (1) is the results of sample 1#, (2) is the results of sample 2#, (3) is the results of sample 3#, (4) is the results of the sludge. The "a" is the raw sample of the wastewater and sludge, "b" is the outcome of the SCWO under the condition of T = 873 K, α = 4, "c" is the outcome of the SCWO under the condition of T = 793 K, α = 1.05.

4   CONCLUSIONS

In this study, we carried out the SCWO experiments of acrylic acid wastewater and sewage sludge by an intermittent flow SCWO experimental system. We can see that the COD is lower than 3000 mg/L when the parameters were T = 793 K, α = 1.05 and it can be discharged after further processing, so we can get the conclusion that SCWO is an efficient technology to deal with acrylic acid production wastewater and sewage sludge. The removal rate of chroma was perfect when the temperature of the experiment was 792 K, but the removal rate of chroma was bad when the temperature of the experiment was 873 K. By speculation, we can get the conclusion that the chroma was caused by the corrosion of the tubular reactor and it was the color of Fe^{2+} and/or Fe^{3+}.

REFERENCES

Bednarik, V. & Vondruska, M. 2003. Removal of formaldehyde from acrylic acid production wastewater. Environmental Engineering Science, 20, 703–707.

Bermejo, M.D. & Cocero, M.J. 2006. Supercritical water oxidation: A technical review. Aiche Journal, 52, 3933–3951.

Gong, W.J., Li, F. & Xi, D.L. 2009. Supercritical Water Oxidation of Acrylic Acid Production Wastewater in Transpiring Wall Reactor. Environmental Engineering Science, 26, 131–136.

Onwudili, J.A., Radhakrishnan, P. & Williams, P.T. 2013. Application of hydrothermal oxidation and alkaline hydrothermal gasification for the treatment of sewage sludge and pharmaceutical wastewaters. Environmental Technology, 34, 529–537.

Segond, N., Matsumura, Y. & Yamamoto, K. 2002. Determination of ammonia oxidation rate in sub- and supercritical water. Industrial & Engineering Chemistry Research, 41, 6020–6027.

Silva, A.M., Marques, R.R. & Quinta-Ferreira, R.M. 2004. Catalysts based in cerium oxide for wet oxidation of acrylic acid in the prevention of environmental risks. Applied Catalysis B-Environmental, 47, 269–279.

Sogut, O.O., Yildirim, E.K. & Akgun, M. 2011. The treatment of wastewaters by supercritical water oxidation. Desalination and Water Treatment, 26, 131–138.

Wang, C.C. & Lee, C.M. 2006. Acrylic acid removal by acrylic acid utilizing bacteria from acrylonitrile-butadiene-styrene resin manufactured wastewater treatment system. Water Science and Technology, 53, 181–186.

Wang, S.Z., Guo, Y., Chen, C.M., Zhang, J., Gong, Y.M. & Wang, Y.Z. 2011. Supercritical water oxidation of landfill leachate. Waste Management, 31, 2027–2035.

Xu, D.H., Wang, S.Z., Hu, X., Chen, C.M., Zhang, Q.M. & Gong, Y.M. 2009. Catalytic gasification of glycine and glycerol in supercritical water. International Journal of Hydrogen Energy, 34, 5357–5364.

Study of relationship of shaft seal steam leakage flow with Variable Steam Temperature Experiment parameters

Xuedong Wang
Shandong Branch of Huadian Electric Power Research Institute, Jinan, P.R. China

Yuzhen Hao & Wei Zheng
Shandong Electric Power Research Institute of State Grid, Jinan, P.R. China

Jianli Qu
SEPCO Electric Power Construction Corporation, Jinan, P.R. China

ABSTRACT: For 330MW subcritical unit and 660MW supercritical unit, the relationship of calculation value of shaft seal steam leakage from HP to IP cylinder with the Variable Steam Temperature Experiment (VSTE) parameters is studied. For 330MW subcritical unit, the calculation value of shaft seal steam leakage from HP to IP cylinder occupies from 0.49% to 2.42% of main steam flow respectively when the VSTE is carried out under various temperature differences of Main Steam (MS) and Reheat Steam (RS), as well the practical IP thermal efficiency is from 89.89% to 90.41%. For 660MW supercritical unit, the calculation value of shaft seal steam leakage from HP to IP cylinder occupies from 0.81% to 1.53% of main steam flow respectively when the VSTE is carried out under various temperature differences of MS and RS, as well the practical IP thermal efficiency is from 88.79% to 89.26%. According to the results above, practical IP thermal efficiency decreases when shaft seal steam leakage increases. For 330MW subcritical unit, practical IP thermal efficiency and shaft seal steam leakage change greatly in different condition combinations because the VSTE temperature difference is small. For both of the two capacity units, the deviation of the shaft seal steam leakage is larger, which is obtained by combination of reducing MS temperature condition with rated parameter condition. By calculating flow capacity of shaft seal clearance and comparing design value with experiment result, it shows that the shaft seal steam leakage and practical IP cylinder efficiency can be obtained more reasonably only when the MS and RS temperature difference is above 15°C in VSTE experiment.

1 INTRODUCTION

At present, with the introduction of foreign design and manufacturing technology, domestic design and manufacturing level of steam turbine are greatly enhanced, and internal efficiency of turbine also reaches a higher level. However, steam turbine of various kinds of capacity generally does not achieve design value of internal efficiency after unit operation, which brings about reducing of economic efficiency. There are many influence factors to internal efficiency of steam turbine, of which the most important ones are steam leakage of blade seal, diaphragm seal and shaft seal. For the same seal clearance, the seal leakage flow increases leading to reduction of steam turbine thermal efficiency when the unit operates in higher parameters.

For large capacity steam turbine, the configuration of HP and IP integrated cylinder and reverse of those flow part can greatly balance thrust of rotor, but the integrated cylinder arrangement also makes some steam of HP cylinder leak into the first stage of IP cylinder through shaft seal between HP and IP cylinder. In spite of more shaft seal numbers between

HP and IP cylinders, the shaft seal steam leakage is usually big on account of differential pressure between the leakage point and mixing point, as well as the big shaft diameter of this part. The shaft seal leakage design value of a new unit is commonly 10–20 t/h, which is about 2% of reheat steam flow (Zhong Ping et al. 2006). The quantity of shaft seal steam leakage directly affects steam turbine heat consumption rate and cylinder efficiency. At present, there is no direct measuring method because the location of shaft seal leakage is special. We only indirectly measure the quantity of shaft seal steam leakage by VSTE method. As the adjustment of unit operation parameters is difficult in the experiment, the difference of shaft seal leakage calculation value is larger, and some even reaches more than 10% of the reheat steam flow, exceeding the design value and reasonable range. Through experiment, the article has studied the influence of VSTE parameters on calculation value of shaft seal steam leakage of 330MW subcritical unit and 660MW supercritical unit.

2 MEASUREMENT METHOD OF VSTE

2.1 Basic principle

Since the shaft seal leakage between HP and IP cylinder happens in the interior of HP and IP cylinder, it cannot be measured directly through conventional methods. Now, variable steam temperature method is commonly applied for indirect measurement, which is carried out by changing MS and RS temperature, as well as separately calculating actual efficiency change of IP cylinder.

If such parameters are approximate as open wide of HP and IP main valve as well as adjusting valve, MS flow, RS flow and other related steam parameters, it can be confirmed that the shaft seal leakage flow between HP and IP cylinder and actual IP cylinder efficiency are constant, and the influence of shaft seal leakage on nominal IP cylinder efficiency changes as the variation of difference between specific enthalpy of leaked steam and RS specific enthalpy. By changing RS temperature and MS temperature, the temperature of mixing point before first stage of IP cylinder changes, so does the exhaust steam temperature of IP cylinder. Therefore, the nominal IP cylinder efficiency will correspondingly change (Wang Xuedong et al. 1998, Guo Yongjie et al. 1991 & Zhang Juan et al. 2004) on the base of parameters before IP main steam valve and parameters of IP cylinder exhaust steam point. According to the relationship of shaft seal leakage with nominal IP cylinder efficiency, the shaft seal leakage flow between HP and IP cylinders and actual IP efficiency can be calculated utilizing heat and mass balance method of shaft seal leakage point.

2.2 Method of experiment

On two same load conditions (such as valve point condition), we separately reduce MS temperature and RS temperature, keeping other conditions and parameters unchanged. And then we measure some parameters by thermal experiment, such as parameters after adjusting stage, reheat steam and IP cylinder exhaust steam and so on. Assuming that the ratio (N) of the shaft seal leakage quantity from HP to IP cylinder accounting for MS flow is a certain value, the actual efficiency of IP cylinder (ηi) is obtained by calculating mixing parameter of the first stage of IP cylinder. When N takes different values, we can get different values of ηi, thus we can respectively obtain N-ηi curves for two variable steam temperature conditions, as shown in Figure 1. There must be a node between two curves because the actual IP efficiency is equal on both conditions. The N and ηi of this node respectively stand for the shaft seal steam leakage flow and actual IP cylinder efficiency.

2.3 Key points of experiment

The precondition of this experiment method is that, besides MS temperature and RS temperature, the other parameters are equal in two load conditions, so it is believed that the shaft

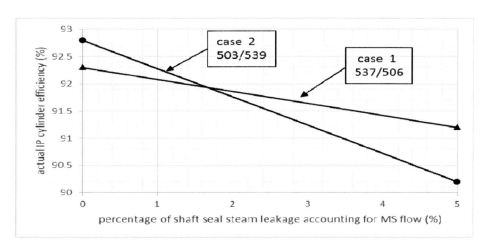

Figure 1. Relation of shaft seal steam leakage with IP cylinder efficiency on VSTE. (Note: Data in the figure is MS temperature and RS temperature, °C).

seal leakage quantity between HP and IP cylinders and the actual IP cylinder efficiency keep the same. Therefore, the two conditions should strictly meet experiment requirements, and operating condition and corresponding parameters should be stable. At the same time, difference of MS and RS temperatures should be as large as possible and generally maintain 20–30°C in order to demonstrate a clear node of two curves.

As the unit operating adjustment is difficult, especially increasing RS temperature and reducing MS temperature being difficult on most of the units, the difference of MS and RS temperatures changes a lot during VSTE. Most experiments cannot reduce MS temperature and increase RS temperature at the same time; only maintaining rated MS temperature is feasible, leading to large difference of calculation value of shaft seal leakage. In order to obtain influence of experiment parameters on shaft seal leakage calculation value, we separately carried out experiments on 330MW subcritical and 660MW supercritical units, respectively keeping MS temperature 30 °C, 20 °C, 15 °C, 0 °C, –5 °C, and –15 °C higher than RS temperature. Then we calculated the nominal IP cylinder efficiency of these conditions and shaft seal steam leakage flow, as well as actual IP cylinder efficiency in different combination of these conditions.

3 VARIABLE STEAM TEMPERATURE EXPERIMENT AND CALCULATION

3.1 Experiment conditions of variable steam temperature

For 330MW subcritical and 660MW supercritical units, we carried out VSTE on three valve point condition, respectively keeping the difference of MS temperature and RS temperature on 30 °C, 20 °C, 15 °C, 0 °C, –5 °C, and –15 °C (Different unit sets up different experiment conditions). Then we combined the above conditions to calculate the relationship of actual IP cylinder efficiency with shaft seal steam leakage between HP cylinder and IP cylinder, and to analyze the relationship of VSTE parameters with shaft seal steam leakage between HP cylinder and IP cylinder.

3.2 Calculation of VSTE data

3.2.1 RS pressure after mixing

When calculating the IP efficiency after mixing RS and shaft seal steam leakage, we adopted RS pressure in front of IP main steam valve, namely, assuming that the steam pressure after mixing point is still equal to RS pressure in front of IP main steam valve.

3.2.2 Specific enthalpy of steam leakage

The process is an isenthalpic and depressurizing process, in which the shaft seal steam leakage between HP and IP cylinders leaks into the first stage of IP cylinder through middle shaft seal clearance behind adjusting stage. That is, the specific enthalpy of steam leakage into IP cylinder is equal to steam enthalpy behind adjusting stage.

3.2.3 Specific enthalpy of mixing steam

G_{leak} stands for shaft seal steam leakage between HP and IP cylinders, i_{leak} stands for enthalpy of shaft seal steam, G_{rh} stands for RS flow, i_{ich} stands for exhaust steam enthalpy of IP cylinder, G_{ms} stands for MS flow, N stands for percentage of G_{leak} accounting for G_{ms}, and i_{mix} stands for enthalpy after mixing shaft seal steam leakage and RS steam. So the specific calculation formulas are:

$$N = G_{leak} \times 100/G_{ms} \qquad (1)$$

$$i_{mix} = [G_{rh} \times i_{ich} + i_{leak} \times N \times G_{ms}]/(G_{rh} + N \times G_{ms}) \qquad (2)$$

Formula (2) shows that i_{mix} is related to N, i_{leak} and i_{ich}.

3.2.4 Actual IP cylinder efficiency

According to IP cylinder efficiency calculating formula, IP cylinder efficiency after steam mixing is basically a linear relationship with N. So we can obtain one linear curve from every experiment condition through giving different shaft seal steam leakage. Respectively assuming that N is from 0 to 12, we can calculate actual IP cylinder efficiency corresponding to the assumed N. Assuming that actual IP cylinder efficiency basically remains unchanged on VSTE conditions, we obtain a relation curve according to N and actual IP cylinder efficiency. As shown in Figure 1, the node of two curves is the actual shaft seal steam leakage percentage between HP and IP cylinders and actual IP cylinder efficiency.

4 EXPERIMENT RESULTS OF VARIABLE STEAM TEMPERATURE CONDITIONS

For 330MW subcritical and 660MW supercritical units, we carried out VSTE on three valve point condition, reducing MS temperature and RS temperature to keep them on various differences. Other parameters remain unaltered.

4.1 VSTE result of 330MW subcritical unit

The subcritical 330MW steam turbine is a single shaft, double cylinders, double steam exhaustion, single reheat, extraction and condensing steam turbine. The experiment data are shown in Table 1. The VSTE is carried out in conditions that the difference of MS and RS temperatures is 18 °C, 0 °C, and −5 °C.

We combine calculation results of different conditions above and then obtain the relationship of shaft seal steam leakage with actual efficiency of IP cylinder, as shown in Table 2 and Figure 2.

4.2 VSTE result of 660MW supercritical unit

The supercritical 660MW steam turbine is a single shaft, three cylinders, four steam exhaustion, single reheat and condensing steam turbine. The experiment data are shown in Table 3. The VSTE is carried out in conditions that the difference of MS and RS temperature is 30°C, 15 °C, 0 °C and −15 °C.

We combine calculation results of different condition above and then obtain the relationship of shaft seal steam leakage with actual efficiency of IP cylinder, as shown in Table 4 and Figure 3.

Table 1. Calculating data of VSTE of 330MW subcritical unit.

Parameter	Unit	Condition 1 Reducing MS Temp.	Condition 2 Reducing RS Temp.	Condition 3 Rated parameter
Generator power	MW	316.9519	317.7140	320.5687
MS pressure	MPa	16.4055	16.4380	16.7054
MS temp.	°C	510.61	533.19	536.39
Adjusting stage pressure	MPa	12.1307	12.1476	12.1276
Adjusting stage temp.	°C	472.61	495.76	499.00
Exhaust steam pressure of HP	MPa	3.8745	3.8614	3.8484
Exhaust steam temp. of HP	°C	312.37	330.85	332.46
RS pressure	MPa	3.5126	3.4985	3.4840
RS temp.	°C	515.74	515.48	535.68
Exhaust steam pressure of IP	MPa	0.5395	0.5374	0.5384
Exhaust steam temp. of IP	°C	266.03	266.21	281.86
Nominal IP cylinder efficiency	%	90.64	90.51	90.53
Difference of MS and RS Temp.	°C	−5.13	17.71	0.71
IP cylinder efficiency (0% steam leakage)	%	90.64	90.51	90.53
IP cylinder efficiency (2% steam leakage)	%	90.02	90.08	90.00
IP cylinder efficiency (4% steam leakage)	%	89.40	89.65	89.47

Table 2. VSTE result of different condition combinations of 330MW subcritical unit.

Item	Unit	Combination of Condition 1 and Condition 2	Combination of Condition 1 and Condition 3	Combination of Condition 2 and Condition 3
Ratio of shaft seal steam leakage between HP and IP cylinder to MS flow	%	1.41	2.42	0.49
Actual efficiency of IP cylinder	%	90.21	89.89	90.41

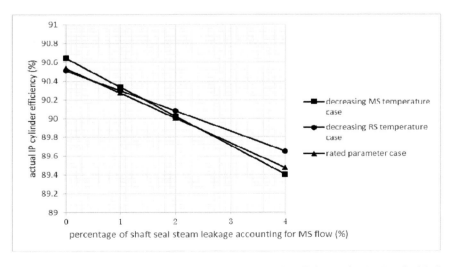

Figure 2. Relationship of shaft seal steam leakage with IP cylinder efficiency of 330MW subcritical unit.

Table 3. Calculating data of VSTE of 660MW supercritical unit.

Parameter	Unit	Condition 1 Reducing RS Temp. 1	Condition 2 Reducing RS Temp. 2	Condition 3 Reducing MS Temp.	Condition 4 Rated parameter
Generator power	MW	647.902	652.049	638.312	649.394
MS temp.	°C	560.85	560.81	528.59	560.80
MS pressure	MPa	24.3412	24.5001	24.4109	24.4899
MS flow	t/h	1895.941	1903.4787	1956.1904	1887.9375
Adjusting stage pressure	MPa	17.4301	17.5406	17.4837	17.5513
Adjusting stage temp.	°C	511.09	511.70	481.44	512.37
RS temp.	°C	530.80	547.67	543.65	560.26
RS pressure	MPa	4.1562	4.1369	4.0812	4.0532
Exhaust steam temp. of HP	°C	318.81	317.16	289.37	314.59
Exhaust steam pressure of HP	MPa	4.4825	4.4559	4.3974	4.371
Exhaust steam temp. of IP	°C	343.93	357.90	353.83	368.35
Exhaust steam pressure of IP	MPa	1.1135	1.107	1.0888	1.0842
Nominal IP cylinder efficiency	%	89.59	89.648	89.862	89.644
Difference of MS and RS Temp.	°C	30.05	13.14	-15.06	0.54
IP cylinder efficiency (0% steam leakage)	%	89.59	89.648	89.862	89.644
IP cylinder efficiency (2% steam leakage)	%	89.17	89.15	89.16	89.08
IP cylinder efficiency (4% steam leakage)	%	88.75	88.65	88.45	88.52

Table 4. VSTE result of different condition combinations of 660MW supercritical unit.

Item	Unit	Combination of Condition 1 and Condition 3	Combination of Condition 2 and Condition 3	Combination of Condition 1 and Condition 4	Combination of Condition 2 and Condition 4	Combination of Condition 3 and Condition 4
Ratio of shaft seal steam leakage between HP and IP cylinder to MS flow	%	0.98	1.08	0.81	0.90	1.53
Actual efficiency of IP cylinder	%	89.18	89.11	89.26	89.21	88.79

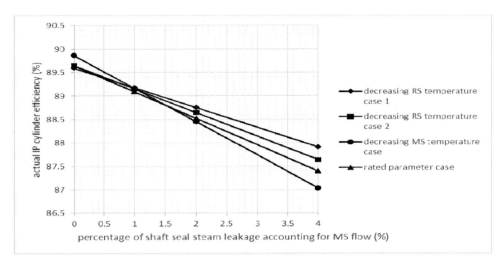

Figure 3. Relationship of shaft seal steam leakage with IP cylinder efficiency of 660MW supercritical unit.

5 ANALYSIS OF VSTE RESULTS

5.1 *Influence of VSTE parameters on calculated value of shaft seal steam leakage*

Analyzing the VSTE results of the above 330MW subcritical and 660MW supercritical units, the shaft seal leakage between HP and IP cylinders and actual IP cylinder efficiency calculated by VSTE method are influenced greatly by VSTE parameters as for the steam turbine with reverse arrangement of HP and IP cylinders. 330MW subcritical unit carried out the VSTE in conditions of different MS and RS temperatures. The calculated shaft seal leakage between HP and IP cylinders is from 0.49% to 2.42% of MS flow, and the actual IP cylinder efficiency is from 89.89% to 90.41% with different MS and RS temperature differences. While for 660MW supercritical unit, the shaft seal leakage between HP and IP cylinders is from 0.81% to 1.53% of MS flow, and the actual IP efficiency is from 88.79% to 89.26%. The experiment results show that, with the shaft seal leakage increasing, the actual IP cylinder efficiency reduces.

For VSTE results of 330MW subcritical unit,, the actual IP cylinder efficiency and shaft seal leakage between HP cylinder and IP cylinder change more largely due to small temperature difference in different condition combinations. Separately to combine reducing RS temperature condition and reducing MS temperature condition with rated parameter condition, the experiment result deviation and fluctuation are larger. When combining the reducing MS temperature condition with rated parameter condition, the shaft seal steam leakage obtained is larger. While for the front four combinations of 660MW supercritical unit, the actual IP cylinder efficiency and shaft seal steam leakage don't change too much; however, for the last combination of reducing MS temperature condition with rated parameter condition, both the actual IP cylinder efficiency and shaft seal steam leakage have obvious deviations.

Mentioned in reference (John A. Booth et al.), in order to get the most accurate results, the gradient difference of the curves of two experiment conditions need to be large enough. If two curves tend to be parallel, slight change of one curve would cause large deviation of the node of two curves, for which reason we need to ensure that temperature difference of MS and RS on two experiment conditions should be as large as possible. From Table 1 to Table 4, the experiment results show that when the temperature difference of two conditions for combination is less than 15 °C, especially for the combination of reducing MS temperature condition with rated parameter condition, the shaft seal leakage quantity is bigger, actual IP cylinder efficiency is lower and deviation of experiment results is larger. Therefore, when VSTE is carried out, we cannot only reduce MS temperature and combine the condition with the rated parameter condition.

5.2 *Theoretical calculation value and design value of shaft seal steam leakage*

The shaft seal steam leakage between HP and IP cylinder can be calculated through the following empirical formula.

When $\frac{p_1}{p_0} \geq \frac{0.85}{\sqrt{z+1.5}}$, that is, the leakage steam velocity is lower than sound speed, the formula is as follow:

$$G_{leak} = 0.360 \mu_l A_l \sqrt{\frac{p_0^2 - p_1^2}{z_l p_0 v_0}} \text{ t/h} \quad (3)$$

When $\frac{p_1}{p_0} < \frac{0.85}{\sqrt{z+1.5}}$, that is, the leakage steam velocity is higher than sound speed, the formula is as follow:

$$G_{leak} = 0.360 \mu_l A_l \sqrt{\frac{p_0}{(z_l + 1.5) v_0}} \text{ t/h} \quad (4)$$

Of which, A_l with unit of cm² stands for shaft seal clearance area between rotor and stator blades, and p_0 and p_1 (MPa) respectively stands for steam pressure before shaft seal and behind shaft seal, that is, high pressure side and low pressure side of the shaft seal. z_l stands for number of shaft seal teeth, v_0 stands for steam specific volume before shaft seal with unit of m³/kg. And u_l stands for coefficient of shaft seal steam leakage flow, related to configuration of shaft seal teeth. From formulas (3) and (4), it is shown that the shaft seal leakage quantity is proportional to the shaft seal clearance area. If turbine shaft diameter and shaft seal clearance are known, meanwhile assuming that the steam leakage parameters change little, we can approximately calculate the shaft seal steam leakage change only by changing the shaft seal clearance.

We utilize the formula above to calculate the shaft seal steam leakage between HP and IP cylinder of 330MW unit, for which HP cylinder and IP cylinder are integrated. When the shaft seal clearance is 0.5 mm, the shaft seal leakage quantity is 6961 kg/h (Hao Yuzhen et al. 2013), accounting for 0.77% of main steam flow. When steam turbine is in overhaul or shaft seal is reconstructed, normally the shaft seal clearance is adjusted to 0.4~0.5 mm. Because steam turbine vibrates after startup, inducing friction of rotor and stator blades, the shaft seal clearance will increase. And when shaft seal clearance reaches 1.0 mm, the shaft seal steam leakage is 13922 kg/h accounting for 1.54% of main steam flow.

In THA condition, the shaft seal steam leakage flow between HP cylinder and IP cylinder is designed as 0.9% (330MW subcritical unit) and 0.986% (660MW supercritical unit) of MS flow. Comparing both values with the data and results in Table 1 to Table 4, we obtain that experiment temperature difference of MS and RS should reach 15°C and −15°C when the VSTE is carried out. That is, when absolute temperature difference of MS and RS exceeds 15°C, the experimental calculation results of the shaft seal leakage flow between HP and IP cylinders and the actual IP cylinder efficiency are close to design value, leading to a relatively accurate calculation result. If VSTE cannot satisfy precondition above, the experimental calculation value of the shaft seal leakage is relatively larger. Thus it is less accurate than design value to calculate unit economic index.

6 CONCLUSION

This article aims at 330MW subcritical unit and 660MW supercritical unit to study the relationship of calculation value of shaft seal steam leakage from HP to IP cylinder with variable steam temperature experiment parameters. We carried out VSTE on three valve point condition, respectively keeping difference of MS temperature and RS temperature on 30 °C, 20 °C, 15 °C, 0 °C, −5 °C, and −15 °C (Different unit sets up different experiment conditions). For 330MW subcritical unit, the calculation value of shaft seal steam leakage from HP to IP Cylinder occupies from 0.49% to 2.42% of main steam flow respectively when the VSTE is carried out under various temperature differences of MS and RS, as well the practical IP thermal efficiency is from 89.89% to 90.41%. For 660MW supercritical unit, the calculation value of shaft seal steam leakage from HP to IP cylinder occupies from 0.81% to 1.53% of main steam flow respectively when the VSTE is carried out under various temperature differences of MS and RS, as well the practical IP thermal efficiency is from 88.79% to 89.26%. According to results above, practical IP thermal efficiency decreases when the shaft seal steam leakage increases.

The VSTE should be carried out under valve fully open condition, as well the parameter stability, temperature difference of MS and RS, and accuracy of RS temperature and IP exhaust temperature all have great influence on the VSTE results. For VSTE results of 330MW subcritical unit, the actual IP cylinder efficiency and shaft seal leakage between HP and IP cylinder change more largely due to small temperature difference in different condition combinations. Separately to combine reducing RS temperature condition and reducing MS temperature condition with rated parameter condition, the experiment result deviation and fluctuation are larger. When combining the reducing MS temperature condition with rated parameter condition, the shaft seal steam leakage flow obtained is larger. While for the front four combinations of 660MW supercritical unit, the actual IP cylinder efficiency and shaft seal steam leakage don't change largely; however, for the last combination of reducing MS temperature condition with rated parameter condition, both the actual IP cylinder efficiency and shaft seal steam leakage have obvious deviations.

By calculating flow capacity of shaft seal clearance and further comparing design value with experimental results of 330MW and 660MW units, we obtain that experiment temperature difference of MS and RS should reach 15°C and −15°C when the VSTE is carried out. That is, when the absolute temperature difference of MS and RS exceeds 15°C, the experimental calculation results of the shaft seal leakage between HP and MP cylinders and the actual IP cylinder efficiency are close to design value, leading to a relatively accurate result. If VSTE can not satisfy precondition above, the experimental calculation value of the shaft seal leakage is relatively larger. Thus it is less accurate than design value to calculate unit economic index.

REFERENCES

Guo Yong-jie, Guo Jian-lin, Li Ming-yuan. On site measurement and analysis of mid-gland leakage of integrated HP and IP cylinders [J]. Power Engineering, 1991, vol 11(2):20–24.
Hao Yu-zhen, Zhen Wei, Ding Jun-qi, et al. Impacts of shaft seal gap on economical efficiency of 300MW unit [J]. Shandong Electric Power, 2013(03):50–54.
John A. Booth and David E. Kautzmann, Estimating the leakage from HP to IP turbine section [R]. General Electric Company.
Wang Xue-dong, Zhang Hu. Measurement and calculation of HP front labyrinth leakage and IP efficiency accounts of 125 MW units [J]. Shandong Electric Power, 1998(03):53–56.
Zhang Juan, Bao Jin-song. Application of the steam temperature variation method for measuring gland-steam leakage between HP and IP cylinders of 300 MW sets [J]. Power Generation Equipment, 2004(02):79–81.
Zhong Ping, Shi Yan-zhou, Wang Zhu-cheng. Study on steam leakage test of intermediate shaft packing for HP&IP combined casing of large-scale steam turbine [J]. Thermal Power Generation, 2006(01):44–47.

Analysis on biomass briquette status quo and development problems

Peng-lai Zuo, Bin-jie Han, Tao Yue, Nan Yang, Chen-long Wang, Xiao-xi Zhang, Yong-hua Ding & Shu-fang Qi
Beijing Municipal Institute of Labour Protection, Beijing, China

ABSTRACT: Burning biomass briquette produces less PM, SO_2 and NOx, and achieves zero CO_2 emission, which is very significant to energy conservation and environmental protection. This paper describes the combustion characteristics and pollutants emissions characteristics of biomass briquette, relevant laws and policies; analyzes the current situation of biomass briquette materials, status quo of furnace development and the development prospects, analyzes the major problems of biomass briquette development and puts forward relevant proposals.

1 INTRODUCTION

Oil, gas, coal and other fossil fuel are the main energy resources for our human being, but these resources are non-renewable energy confronted with the issue of depletion. The utilization of fossil fuels could produce a large amount of smog, CO_2, SO_2, NOx and other harmful gases, which results in acid rain and greenhouse gases. All of these ultimately cause serious environmental issues. Based on the provisions of Copenhagen, the amount of CO_2 emission per unit of GDP should decrease 40%~45% by 2020 (Liu et al. 2011). Compared to that, biomass briquette is an environmental friendly fuel which consists of the wastes from agriculture and forestry, such as straw, rice husk, wood chips, twigs, etc. These raw materials are molded into rod, block or granular fuel by specialized equipments under the specific conditions (Jiang et al. 2010). The renewable energy is an indispensable energy source in the energy structure of European Union and the biomass accounts for nearly 70% of the renewable energy (Huo et al. 2013). Biomass briquette has low ash and low sulfur content and it can generate low dust, SO_2 and NOx emissions. It also has the potential to achieve zero CO_2 emission (Zuo et al. 2014). Therefore, to develop the biomass fuel is of great significance for the conservation of energy, the reduction of emissions and the protection of environment.

2 PROPERTIES OF BIOMASS BRIQUETTE FUEL

2.1 *Combustion characteristics*

The volatile components of biomass are considerably higher than those of coal while the ash, carbon content and calorific value are much lower than those of coal. These fuel characteristics decided its combustion characteristics. The study of the composition and the combustion characteristics of biomass fuel will help further scientifically and reasonably develop and utilize the biomass resources. The results from the related research show that the biomass fuel has significant differences from fossil fuels. Due to the specific properties of biomass fuel, the combustion mechanism, reaction rate and composition of combustion products in the combustion process of the biomass fuel are significantly different from those of fossil fuels, resulting in the different combustion characteristics, compared with fossil fuels. The combustion process of biomass briquette is mainly divided into two individual stages: precipitation of volatiles, combustion of biomass and the residual coke, burnt out. The characteristics of combustion processes are listed as follows.

1. In the pretreatment stage of raw materials for the production of biomass briquette, raw materials have already lost most of the water. The water content of biomass briquette is only 12–15% and the volatile content of that is as high as 60–70%. Once the biomass briquette is delivered into the combustion equipment, it can release a large amount of volatile promptly since the time of preheating and drying is significantly short, leading to combustion of biomass briquette immediately. The combustion of biomass briquette can be treated as the combustion of volatiles in essence. If the combustion condition can be controlled well, the biomass combustion will generate less carbon black particles.
2. Biomass briquette only contains about 20% fixed carbon and the coke porosity of biomass briquette is large, so the coke burning time is shorter, and easy to burn out.
3. The content of alkali metals in biomass, such as potassium (K), sodium (Na), are higher than that in coal and the ash melting point is lower, which results in the formation of slag and coke in the process of combustion of biomass briquette. This can adversely affect the combustion process itself and the safe and stable operation of the combustion equipment. Biomass also contains a large amount of Chlorine (Cl) which could end up in the biomass combustion products. These products have the synergetic effect with the alkali metals in the flue gas, which causes corrosion of the metal exposed in high temperature flue gas. This is one of disadvantages for the utilization of biomass briquette.

2.2 Emission characteristics of pollutants

The pollutant emissions are determined by the composition of biomass fuel. The pollutants from biomass combustion mainly include Sulfur dioxide (SO_2), Nitrogen Oxides (NOx), Particulate Matter (PM), acid gases (HCl) and heavy metals, etc. These pollutants are all from the elements of biomass and related to the biomass characteristics, so biomass briquette has some combustion emissions characteristics as follows.

1. Fly ash is less. The fly ashes of timber and crops straw generally are lower than 4% and 8% respectively and the fly ashes of rice straw and rice husk ash are higher than the former, but generally do not exceed the ash content of high-quality coal. This can simplify the ash removal equipment in the combustion device. The volatiles are the main burning ingredients and account for more than 70%, so the biomass only generates a small amount of soot.
2. Biomass utilization usually does not affect the carbon cycle in the nature. Even if the biomass does not be burnt, it will release CO_2 in the process of natural digestion. As a result, the CO_2 emission from biomass is no more than the absorbed CO_2 in the growth process of biomass, so as to achieve zero emission of CO_2.
3. The content of nitrogen and sulfur in biomass is lower than that in the fossil fuels such as coal. Under ideal conditions NOx and SO_2 emissions from biomass combustion are only 1/5 and 1/10 of those from coal combustion respectively, and pollutant concentrations are much lower than those specified in the coal-fired boiler emissions standards as the reference, which benefits the reduction in emissions of pollutants.
4. Unlike coal, biomass contains less heavy metal, so the heavy metal pollutants are rare in the flue gas.

3 THE DEVELOPMENT STATUS OF BIOMASS BRIQUETTE

3.1 The status quo of the raw material of BMF

A China is an agricultural country and most of the areas are rich in the food crops. The annual output of straw is up to more than 600 million tonnes. The main varieties and their quantities are shown in Table 1. The annual output of straw and availability in different regions are shown in Table 2.

Table 1. The output of various crops and straw in China.

Name	The annual output of crops/billion tonnes	The annual output of straw/billion tonnes	Equivalent to coal/billion tonnes
Rice	1.8523	1.1540	0.4951
Wheat	1.0221	1.3962	0.6981
Corn	1.3961	2.2398	1.1849
Coarse cereals	0.1669	0.1669	0.0835
Bean	0.1788	0.2681	0.1456
Potato	0.3262	0.1631	0.0793
Oil crops	0.2250	0.4501	0.2381
Cotton	0.0477	0.1430	0.0777
Sugarcane	0.6542	0.0654	0.0288
Total	–	6.46	3.03

Table 2. The annual output of straw and the distribution of availability in different regions.

District	The annual output/surplus of straw	District	The annual output/surplus of straw	District	The annual output/surplus of straw
Beijing*	429/251.8	Shanghai*	179/105.1	Tianjin*	302/177.3
Guangdong*	1405/824.7	Shandong	7191/1510	Fujian	6643/1395
Henan	5650/1186.5	Jilin	3371/707.9	Guangxi	1525/320.3
Sichan	4464/937.4	Hebei	4413/926.7	Jiangsu	3549/745.3
Hubei	3202/672.4	Anhui	2924/614	Hunan	2074/435.5
Inner-Mongolia	1698/356.6	Heilongjiang	3824/803	Yunnan	1486/312.1
Xinjiang	1463/307.2	Shanxi	1338/281	Shanxi	1334/280.1
Jiangxi	1304/273.8	Zhejiang	1134/238.1	Guizhou	1123/235.8
Gansu	754/158.3	Liaoning	550/115.5	Hainan	176/37
Ningxia	268/0	Qinghai	205/0	Xizang	50/0

*Multiply 58.7% when calculating the amount of straw residual, multiply 21% when calculating the amount of straw residual in other cities.

It can be analyzed from the above data that the yield of straw is huge in China, which is equivalent to about 300 million tonnes of coal. Part of the straw is used as livestock feed, fertilizer, industrial raw materials and traditional biomass fuel. The rest which reaches 116 million tonnes has not yet been used. Especially in the harvest season, a lot of straws are burnt in the open areas, resulting in the environmental pollution and waste of energy. The reasons behind this include adequate promotion, the blocked channel of collecting straw, and the lack of policies to encourage use of straw.

3.2 *The development status of the boiler*

Biomass boilers refer to those which can burn biomass fuel. The biomass boilers have a number of types according to the different methods of combustion, and they can be divided into fluidized bed boiler and layer combustion furnace, etc.

The biomass fluidized bed boiler has been studied since the late 1980s in China. To improve the combustion efficiency of boiler, the sand is used as the bed material of boiler to form a stable material layer with the dense-phase zone to provide a sufficient preheating and drying heat source for biomass fuel. Using the rotating updrafts formed by strong rotary tangential secondary air in the dilute phase zone, the high temperature flue gas can mix biomass materials particles to further enhance adequate burning of combustible gas and solid particles.

Wuxi boiler factory and the relative companies jointly designed and developed a fluidized bed boiler with the capacity of burning 35 t/h rice husk. The main features of that fluidized bed boiler are list as follows. Firstly, the pneumatic conveying is adopted to deliver the rice husk to avoid the interruption of feeding due to the blockage. Secondly, the soot blower is installed in the tail of the boiler to reduce the deposit of dust on the heating surface. Thirdly, the required flow regime is maintained by adjusting the primary and secondary airflow rate and recirculation of the flue gas in the furnace to expand the scope of application of boiler fuel.

From the perspective of the development status of direct combustion technology of biomass, the fluidized bed boiler has good adaptability to biomass fuels and the adjusting range of load is larger. Since the disturbance of the granular substances in bed are violent, heat transfer and mass transfer conditions are very superior. This can help sufficient mixing of high temperature flue gas air and fuel to provide excellent fire ignition conditions for the biomass fuels with the high moisture content and low heating value. At the same time, due to the long residence time of fuel in bed, the biomass fuel can be burnt fully to improve the efficiency of biomass boilers. In addition, the fluidized bed boiler can maintain the stable burning of biomass at about 850°C so that slag will not easily form after fuel burning and it the generation of harmful gas such as NOx, SOx will be reduced. Therefore the fluidized bed boiler has significant economic and environmental benefits.

4 THE UTILIZATION PROSPECTS OF BIOMASS BRIQUETTE

Compared with natural gas, diesel and other fuel, biomass briquette has low cost. Using 6 t/h steam boiler of biomass briquette designed by Dison new energy group in Guangzhou as an example, its combustion cost is compared with other fuels in Table 3.

Table 3 shows the burning cost of the biomass briquette are lower than that of heavy oil, natural gas and diesel oil by 28%, 40% and 28% respectively.

Currently, with the increasing of standout of the energy security issue and the reinforcement of environmental constraints, the energy conservation and emissions reduction confront with the grim situation. Compared to the conventional energy, biomass, as a new type of clean energy, has abundant resource and its supply is stable. It has carbon neutral, low contents of nitrogen and sulfur, and can release less air pollutions. Biomass briquette can replace coal, natural gas or heavy oil quietly in the fields of heating, gas supply and power generation. Biomass briquette can provide power effectively and significantly reduce pollution to achieve zero emission of CO_2, which conforms to the concept of sustainable development of the current society.

The amount of unemployed biomass has reached to 116 million tonnes. The central and local government departments issue a series of encouragement and preferential policies to implement the renewable energy law and long-term development plan, which will promote the development of industry of biomass briquette fuel and provide a good prospect.

Table 3. The comparison of cost of biomass briquette and other fuels.

Items	Biomass briquette	Heavy oil	Natural gas	Diesel
Calorific value (kcal/kg, kcal/m³)	4100	9600	8500	10200
Sulfur content (%)	0.03~0.06	1~3	0.01~0.02	0.06~0.08
Cost (yuan/kg, yuan/m³)	1.15	4	4.25	6
Boiler efficiency (%)	86	90	90	90
The fuel consumption per tonnes of steam (kg, m³)	172	69	78	65
The fuel cost per tonnes of steam (yuan)	198	276	331	390
Cost comparison base on biomass briquette (%)	–	28	40	49

5 THE MAIN PROBLEM

Compared with the biomass resource in China, although the biomass energy such as biomass briquette fuel has hugely developed in recent years, the development of biomass energy also has some issues due to the scattered biomass resources, the challenge of conversion technology of processing, the unestablished development environment of market.

1. The development train of thought is not clear enough. All relevant departments and all sectors of society attach great importance to the development of biomass energy industry and take a series of measures and action and obtain a positive progress. However most of achievements focus on the development of biomass briquette technology. The studies about the product, market, environmental impact, economic indicators, and related policy research, especially the quantitative research are rare. In some regions the overall consideration and scientific planning for the industry development of biomass energy are deficiency.
2. The lack of accurate investigation and evaluation. The sustainable supply of biomass resources is the basis for the scale development of biomass energy. The utilization potential of biomass energy is great in China, but the type of resource and quantity, availability, resource potential and distribution, etc. still need to investigate and evaluate systematically.
3. The great collection difficulties of raw material. The raw material of biomass of agroforestry is dispersed and seasonal, so the collection of raw material mainly rely on manual work and small machinery currently and the transportation mainly rely on general transport. The lack of complete professional collection, transportation, storage and supply system for the raw materials results in the low efficiency for the above chains. It is difficult to satisfy the requirement of large-scale utilization of the biomass briquette fuel.
4. The biomass fuel technology still has some limitations. Currently the serious abrasion of mould for curing molding technology of biomass, high energy consumption and the deficient of combustion equipment hinder the industrialization of this technology. In addition, the low level of comprehensive utilization and conversion efficiency need to be enhanced. The technology level of manufacture of the combustion equipment of the biomass briquette and other aspects need to be improved. For example: since the heating surface of boiler is easy to wear, the particle size of the charging are strict in the fluidized bed; the speed of volatilization of biomass fuel is fast in the layer combustion boiler, so a large amount of air need to add during the burning. If the fuel and air cannot be mixed in time, it will cause insufficient air supply so that it is difficult to guarantee the biomass fuel to burn adequately. Thus it affects the combustion efficiency of boiler.
5. The lack of operating mechanism of market and the low degree of industrialization. The experience of construction management of biomass energy project is deficient. The standard related to product, equipment, engineering construction and project operation are defective. The construction of the test certification system is lag. The supervision of market and technical are deficiency. Currently, the utilization projects of biomass energy including the established projects and under construction projects in Beijing are invested by government basically. The market incentive mechanism in the form of tax and price subsidies is in the process of study and development. Although a lot of government investment effectively promotes the rapid development of biomass energy, it causes the unclear responsibility in the process of operation management and operation maintenance issue such as sustainability.
6. The faultiness of support policies from government. A series of regulations and supporting measures for the management are enacted, but the legal system is not perfect and the reasonable and effective incentive policies are lack in the aspects of finance, financial system and the open market.

6 SUGGESTIONS

1. According to the emission characteristics of biomass briquette, the management department should enact the independent emission standards of pollutions from biomass briquette fuel boiler to make it distinguish from the emissions standards of coal-fired boiler.
2. To guarantee the discharging of pollutants from the biomass briquette fuel boiler reach the standard, the national standard of product quality of biomass briquette fuel should be enacted by the related departments.
3. For straw burning, law enforcement and penalties should be strengthened. At the same time the relevant incentive economic policies should be formulated to guide farmers to return the straw to farm or process the molding fuel in order to reduce the emission of atmospheric pollutants caused by scattered burning.
4. The issues of high content of moisture and tar in the flue gas of biomass briquette fuel boiler, which does not favor the operation of the bag dust collector, should be solved by strengthening the technical research. To strengthen the design research of the supporting boiler of biomass briquette fuel and to adopt the reasonable technology of air distribution mode, the emissions of nitrogen oxides can be reduced.
5. The concept that biomass briquette fuel is clean energy should be verified so that the funds and policy supports should be provided for promotion and application.

ACKNOWLEDGEMENTS

This work was financially supported by "Study on PM and SO_2 control technology route of industrial boilers (2015A008)" from Chinese Academy for Environmental Planning and "Study on technology route for biomass briquette fired boilers meeting the emission standard of air pollutants for boiler".

REFERENCES

Huo Li-li, Yao Zong-lu, Tian Yi-shui, et al. Development and Reference of Biomass Solid Fuel in Poland (in Chinese) [J]. Renewable Energy Resources, 2013, 31(12):130–136.

Jiang Jian-chun, Biomass Briquette Development History and Future (in Chinese) [J]. Beijing: 2010.

Liu Guo-hua. The Prospect of the Technology and Utilization of Biomass Briquette Fuel (in Chinese) [J]. Applied Energy Technology, 2011(1):44–47.

Peng-lai Zuo, Bin-jie Han, Tao Yue, et al. Tests of Air Pollutants Emissions from Biofuels-fired Boilers and Analysis on Abatement Potential (in Chinese) [J]. Annual Conference Hosted by Chinese Society for Environmental Sciences, 2014:5418–5422.

Adsorption of uranium from aqueous solution by bamboo charcoal

Xueying Xiong, Diyun Chen, Jiawei Zhao & Zhe Li
Institute of Environment Science and Engineering, Guangzhou University, Guangzhou, China
Guangdong Provincial Key Laboratory of Radionuclides Pollution Control and Resources, Guangzhou, China

ABSTRACT: The adsorption behaviors of U (VI) ion on bamboo charcoal were investigated with various chemical methods. Parameters studied include the effects of pH, particle size, initial ion concentration, contact time and temperature by batch method. The results showed that bamboo charcoal could remove U (VI) ions effectively from aqueous solution. The loading of U (VI) ions was strongly dependent on pH and the optimal adsorption pH value is 3. In the batch system, the bamboo charcoal exhibited the highest U (VI) ion uptake rate as 98%, at an initial pH value of 3. The results revealed that bamboo charcoal was a good choice as a bio-sorbent for the recovery of uranium from aqueous solution.

1 INTRODUCTION

Radioactive metals due to accidental release from nuclear power plants could cause groundwater contamination. The accidental releases of the radioactive contaminated gas or water can easily permeate to groundwater system in nature. In addition to the accidental release, dismantlement of medical instruments containing radioisotopes, radioactive waste produced from nuclear power plants and nuclear fission products routinely released by nuclear testing could also cause groundwater contamination. The amount of the radioactive contamination over self-purification capacity of nature will be continuously widespread to the drinking water source through underground water system. The amount of damage depends on the amount of radiation to which you are exposed, which is related to the amount of activity in the radioactive material and the length of time that you are exposed. Most of the information regarding health effects from exposure to radiation comes from exposures for only short time periods. At the same time, it also has strong radiation toxicity and chemical toxicity. It is a hardly degradable substance by biological or chemical method. And enrichment of uranium in the environment can be accumulated. Therefore, further study of the effect of uranium to the environment and ecology is very important.

Compared with most other types of adsorption materials, carbon material has excellent ability to resist radiation and thermal stability, and has a certain advantage in dealing with radioactive waste liquid. At the same time, it is easy to improve the selectivity and adsorption capacity of various ions through a simple physical or chemical processing to the surface of carbon materials, Therefore, carbon materials have a wide application in water pollution control. Bamboo charcoal is the new product developed in recent years, mainly composed of carbon, hydrogen, oxygen and other elements, with a hexagonal, hard, close and porous molecular structure. Compared with charcoal, bamboo charcoal has special micro structure and larger specific surface area, thus has strong adsorption ability. It is a kind of good water treatment agent in wastewater treatment. With good carbon microbes can effectively improve the efficiency of wastewater treatment. In the industrial production, bamboo charcoal can be used for the adsorption of pigment dyeing industrial wastewater of Ni^{2+}, Cr^{2+}, Pb^{2+}, Cu^{2+}, such as heavy metal ions, organic wastewater of phenol, aniline and so on. About the treatment of wastewater containing uranium with bamboo charcoal is less. Bamboo charcoal has the advantages of economical efficiency and is widely applicable. Therefore, it is essential to carry out the research of adsorption of low concentration of uranium wastewater with bamboo charcoal.

Our bamboo charcoal was applied by Guizhou Sino Quarry and Mine Industries Corp. Ltd. The particle size was less than 75 μm. 0.1179 g U_3O_8 was dissolved in 100 ml concentrated nitric acid of pH = 2 resulting in a total uranium concentration for the experiment of 1 mg/ml.

2 MATERIALS AND METHODS

2.1 Apparatus

The U (VI) ion concentration was determined with Shanghai precision scientific instrument 721G UV-VIS spectrophotometer at 578 nm. PHSJ-4A pH meter was used for measuring pH of solutions. The samples were shaken in the SHA-B temperature constant shaking machine. The water used in the present work was purified using Mol-research analysis-type ultrapure water machine. The bamboo charcoal was weighted by BT 423S precision electronic balance.

2.2 Adsorption experiments

Batch adsorption experiments were conducted at room temperature (25°C) in 250 mL vials. The vials were capped quickly (to minimize CO_2 exchange) and were shaken for essential time. An aliquot of the supernatant was withdrawn and immediately filtered. Experiments were run in a certain range of pH, initial U (VI) ion concentrations, time. A amount of treated bamboo charcoal was weighted and added into a conical flask with stopper, then, a desired volume of required amount of standard solution of uranium were added. The flask was shaken in a shaker at room temperature. All of the experiments were completed in duplicates. The adsorption capacity (Q) and adsorption rate (E) of U (VI) ion on bamboo charcoal were calculated with the following formula:

$$E = \frac{c_o - c_e}{c_e} \times 100\% \qquad (1)$$

$$Q = \frac{(c_o - c_e)V}{m} \qquad (2)$$

where c_o (mg/ml) = initial concentrations; c_e (mg/ml) = equilibrium concentrations; V (ml) = total volume of solution; and m (g) = the mass of bamboo charcoal.

3 RESULTS AND DISCUSSION

3.1 Effect of contact time on adsorption

About 0.5 g of bamboo charcoal with size less than 150 μm were used in this experiments. The U (VI) solution in our batch experiments were maintained at 5 mg/L with a pH of 3. And the pH of the solution was adjusted using NaOH or HNO_3. The influence of contact time for uranium adsorption is shown in Figure 1 from 5 min to 60, the adsorption rate of uranium in the solution increased sharply by contact with bamboo charcoal. At the 60 min, the adsorption rate reached a maximum rate of 97.64%. After 60 min, the adsorption rate get down slightly, but the change is not big. Presumably, the adsorption of bamboo charcoal to uranium is initially happened on the surface of particle, so the adsorption rate is higher, when adsorption reached a certain rate, the uranium ions started internal diffusion from the surface of the bamboo charcoal bamboo charcoal gap into the internal pores. The transfer of uranium ions in the bamboo charcoal is slow, so the growth rate of the overall adsorption

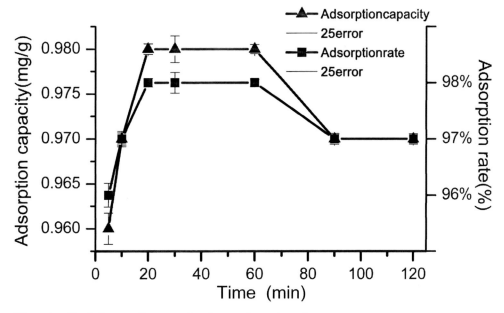

Figure 1. The influence of contact time for uranium adsorption.

effect is reduced. But it can reach adsorption equilibrium as time goes by. Finally, we chose 60 min as a optimal time.

3.2 Effect of solution pH on adsorption

The effect of pH (ranging from 1 to 6) on U (VI) sorption onto bamboo charcoal are presented in Figure 2. As in alkaline environment, uranium ions is very easy to precipitate, Consequently the pH was designed from1 to 6. For bamboo charcoal, the extent of U (VI) removal was high in the pH ranging from 1 to 3 and it decreased sharply at pH below 4. Maximum U (VI) adsorption occurred under the pH of 3. The pH-dependent sorption effect is due to the ionization of both the adsorbate and the adsorbent causing repulsion at the surface and decreasing the net U (VI) adsorption. Due to the strong acid condition, a solution of H^+ may compete with uranium adsorption, as a result the adsorption rate decreased. And H^+ also can make bamboo charcoal protons. With the electrostatic repulsion of uranyl ion solution, the research is not suitable under the environment of strong acid condition.

3.3 Effect of quantity of bamboo charcoal on adsorption

The effect of quantity of bamboo charcoal (0.5, 1, 2, 3, 4, 5 g) on U (VI) sorption are presented in Figure 3. As it is show below the influence of quantity of bamboo charcoal for U (VI) removal is decreased sharply with the quantity rise from 1 g to 2 g. Obviously, the quantity of bamboo should be set less, further study need to be done in this part.

3.4 Other effects on adsorption

Except for the effects above, it also do some other research before experiment, which help to decided which effect should be done much more carefully. The data haven't been shown here, it can be found that the particular size of bamboo charcoal, the temperature of environment do not have much impact on the study. Finally, we choose to show the data above to show our result. Within the 100 mesh, bamboo charcoal particle size is smaller, the better the adsorption efficiency, but the growth rate is very slow. So we don't show the figure out.

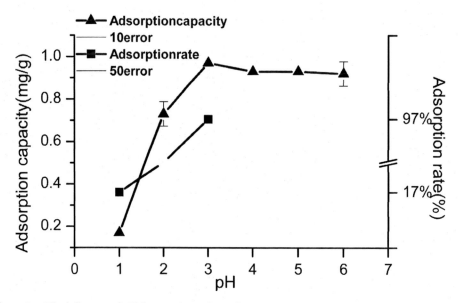

Figure 2. The influence of pH for uranium adsorption.

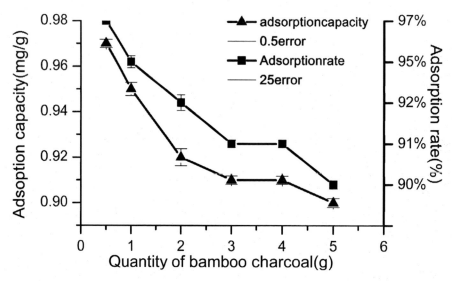

Figure 3. The influence of quantity of bamboo charcoal for uranium adsorption.

Initial concentration of uranium in our study is very low, it will weaken the bamboo charcoal adsorption effect. With the increase of initial concentration of uranium, uranium adsorption increased significantly, while the increase of the adsorption rate is also very slow, which means that higher concentration is not equal high efficiency. When the initial concentration of uranium is confirmed, improve the quantity of bamboo charcoal can promote the adsorption, but large quantity of bamboo charcoal will slow down the adsorption rate.

3.5 *Conclusions*

In this study, batch studies for the adsorption of U (VI) ions from aqueous solutions had been carried out using modified bamboo charcoal as adsorbent. The bamboo charcoal showed highest percentage of adsorption. The adsorption was influenced by the contact time and pH of uranium solution. The optional time for adsorption is 60 min and the optimal pH value is 3. What's more, because of the limit time, the individual error may be in the experiments, especially the low concentration of uranium has certain limitation in the data. But the experiment still has certain reliability. In conclusion, bamboo charcoal can be efficiently used for the removal of U (VI) ions from aqueous solutions.

REFERENCES

ASTDR (Agency for Toxic Substances and Disease Registry). 2004. Toxicological profile for cobalt. US Department of Health and Human Services, Public Health Service.

Cheng, J.F. 2011. Bamboo charcoal removal effect of Pb^{2+} in water research. Journal of energy and environment.

Chen, Q. 2010. Study on the adsorption of lanthanum (III) from aqueous solution by bamboo charcoal, Journal of Rare Earths: 125.

Cheung, W.H. & Lau1 S.S.Y. & Leung S.Y. 2012. Characteristics of Chemical Modified Activated Carbons from Bamboo Scaffolding, Chinese Journal of Chemical Engineering: 515–523.

Cai, X. & Liu, Y. & Zhang, Y. 2012. Nuclear accidents occurring radioactive wastewater treatment research progress. Journal of Chemistry: 483–488.

Chen, Y.M. & Chen, J.F. 2011. Bamboo charcoal adsorption performance and application research progress. Chinese Journal of Ningxia Engineering: 316–319.

Liu, J. & Chen, D.Y. & Zhang J. 2012. The study of the mechanism and adsorption properties of uranium on attapulgite. Journal of Environmental Science: 2889–2894.

Li, Q.L. & Chen, Q.S. & Li, X.Y. 2008. Bamboo charcoal adsorption performance of Cr (VI) study. Journal of Fujian normal university (natural science edition): 50–53.

Li, X.Y & Zhang, W.M. & Liu, Y.B. 2011. Bamboo charcoal on the adsorption of Cu^{2+} in wastewater Chemical Industry Environmental Protection: 26–29.

Li, X.L. & Song. Q. 2011. Carbon materials for the adsorption of uranium. Progress in Chemistry: 1446–1453.

Miao, J.T. 2011. Summary of radioactive waste water treatment technology. Journal of Information Science and Technology: 480.

Tan, H.C. & Wang, D.S. 2004. Processing and utilization of bamboo charcoal. Chinese Journal of Yunnan Forestry: 21~22.

Tan, Z.Q. Qiu J.R., Zeng H.C., Liu H., Xiang J. 2011. Removal of elemental mercury by bamboo charcoal impregnated with H_2O_2, Fuel: 1471–1475.

Wang, Z.L. 2008. Bamboo charcoal adsorption experiments of heavy metal ions Ni^{2+}. Journal of Southwest Water Supply and Drainage: 19–22.

Wei, G.W. & Xu, L.C. 2007. Low concentration of uranium wastewater treatment technology and its research progress. Journal of Uranium Mining and Metallurgy: 90–95.

Zhan, A.L. 2007. Sulfate-reducing bacteria governance acid method of groundwater pollution in uranium research. Hengyang: South China University of Municipal Engineering: 2.

Resources, Environment and Engineering II – Xie (Ed.)
© 2016 Taylor & Francis Group, London, ISBN 978-1-138-02894-4

Research progress of phytoremediation for heavy metal wastewater

M.L. Ji, J. Yan & H.X. Li
College of Environmental Science and Engineering, Guilin University of Technology, Guilin, Guangxi, China

ABSTRACT: Remediation of heavy metal wastewater is a big project. Phytoremediation has been paid an increasing attention for its special features such as high efficiency, low cost, no secondary pollution. This paper briefly introduces the pollution status of heavy metal wastewater, and expounded the concept and content of phytoremediation. Meanwhile, the mechanism, studies and application status of phytoremediation were discussed. And the development prospects of this field were given.

1 INTRODUCTION

Water is an important part of the human ecological environment, which has a high sensitivity to the environment. With the rapid development of industry and agriculture, more and more pollutants were discharged into water environment in the process of human activities, which leading to a serious environmental problem. Heavy metals can easily enter food chain in the process of bio-accumulation, posing a serious threat to the health of animals and human beings. Heavy metal pollution in water has become one of the most serious environmental problems in the world today. How to scientifically and effectively solve the pollution of heavy metals in water have become research hot spots of governments and environmental researchers.

2 SOURCES OF HEAVY METAL WASTEWATER

Heavy metal wastewater mainly comes from the mine pit drainage, waste-rock yard rain flooding, concentrator tailings drainage, non-ferrous metal smelting plant dust removal drainage, non-ferrous metal processing plant acidic washing water, electroplate factory plated parts washings, steel plant pickling drainage, and electrolysis, pesticide, medicine, tobacco, paint, pigment and other industries. For example, mercury-containing industry wastewater mainly discharged from the chemical industry, metallurgy, machinery and other industrial wastewater, the wastewater discharged from the chlor-alkali industry, plastics industry, electronic industry, the amalgamated gold production are the main sources of mercury-containing wastewater (Huang, et al. 2010); Chromium and its compounds are widely used in industry, a series of industries such as metallurgy, chemical industry, mineral engineering, electroplating, chromium, pigment, pharmaceutical, light and textile industry, the production of chromium salt and chromium compounds will produce large amounts of chromium-containing wastewater (Zhou, 2010).

The types, contents and speciation of heavy metals in wastewater vary from different types of production process, presenting great differences.

3 HARMFULNESS OF HEAVY METAL WASTEWATER

The environmental pollution of heavy metals mainly refer to the significant biological toxicity of arsenic, mercury, cadmium, lead, chromium, copper, cobalt, nickel, tin, vanadium and so on. Though some heavy metals are essential for human beings, it will be a threat to human health if there is a lack of one or several of them. It was evidenced that copper deficiency could lead to collagen and protein synthesis disorders and hyperuricemia, resulting in cardiovascular abnormalities and congenital defects or acquired artery elastic tissue structure and function abnormalities (Huang & Huang, 2010). However, some heavy metals (such as mercury, lead, cadmium and chromium, etc.) can present significant toxicity at a low level of intake. Evidences were given that chromium in water could accumulate in fish bones, presenting the toxicity of Cr^{+3} was higher than that of Cr^{+6}. The concentration at 3.0 mg/L in freshwater which had a lethal effect on fish; the concentration at 0.01 mg/L could make some aquatic organisms to death and inhibit the self-purification of water (Chen, 2012).

Heavy metals cannot be biodegraded into harmless substances. Discharged into water, the toxics will last for periods and be hard to be degraded, which will not only harm to the animals and plants, but also participate in food chain cycle, and eventually accumulate in organism to produce genetic, reproductive and other toxicity. This will greatly influence people's health and sustainable development.

4 TREATMENT TECHNOLOGIES OF HEAVY METAL WASTEWATER

At present, the treatment technologies of heavy metal wastewater mainly include three types in the world (Bai, 2013).

The first type is to remove the heavy metal ions from wastewater by chemical reaction. These methods can be used solely or in a combination process according to water quality and quantity. The second is to absorb, concentrate and separate the heavy metals from wastewater under the chemical form maintain unchanged. The third is to remove the heavy metal ions from wastewater by flocculation, absorption and accumulation of microorganisms or plants.

Traditional physical and chemical treatments of wastewater containing heavy metal ions are not ideal because of the complication of the operation process and secondary pollution, especially when heavy metal ions are at a lower concentration, it will be difficult to put into practice due to the relatively high operation and raw material costs. While the biotechnology (biosorption, phytoremediation, etc.) removes heavy metals from wastewater with the help of flocculation, absorption and accumulation of microorganisms or plants has been paid an increasing attention as the traits of rich in raw materials, low cost and other advantages.

5 PHYTOREMEDIATION

Phytoremediation is a method utilizing absorption, precipitation and accumulation of higher plants to reduce heavy metals content in contaminated soil and surface water for the purpose of pollution control and environmental restoration. The use of plants for treatment of heavy metals, mainly compose of three parts. One is using hyperaccumulators to absorb, precipitate and accumulate toxic metals from wastewater; the second is using hyperaccumulators to reduce the activity of toxic metals, which can reduce the opportunities that heavy metals leach into groundwater or spread through air carrier; and the third is using hyperaccumulators to extract heavy metals from soil or water, and accumulate, transport them to the roots and above-ground parts of plants that can be harvested. The plants that can be used in phytoremediation are herbaceous plants, woody plants and so on.

5.1 *Hyperaccumulator*

Hyperaccumulators are also called hyperaccumulating plants or super accumulation plants. These plants can excessively absorb and accumulate heavy metals. The special plant was first discovered by Italian botanist Cesalpino in 1583 which was growing on "black rock" in Tuscany, and this was the earliest reports on hyperaccumulators (Yan, 2008). In 1977, Brooks proposed the concept of hyperaccumulators (Brooks, 1977). The definition of

Hyperaccumulators referred to the plants that could excessively accumulate heavy metals from soil and transfer to the above-ground parts (Mei, 2013). The currently reference standards for hyperaccumulators was proposed by Baker and Brooks. The plants were not hyperaccumulators unless their leaves or above-ground parts (dry weight) Cd was up to 100 μg/g, Co, Cu, Ni, Pb up to 1000 μg/g, and Mn, Zn over 10000 μg/g (Baker, 1989).

Subsequently, researches on hyperaccumulators increased gradually. Phytoremediation proposed as a method for the treatment of contaminated soil. Practical experimental researches and engineering application had shown the great commercial prospects of phytoremediation. The foreign countries entered in this field earlier, given that the world's first hyperaccumulator was first discovered by Minguizzi and Vergnan. They found a plant (*Alyssum bertolonii*) in soils containing Ni up to 10 mg/g (dry weight) in the leaves (Minguizzi, 1948). While hyperaccumulators reported in China were less. To date it has been reported that *Pteris vittata* (Chen, 2002) and *P. cretica* (Wei, 2002) in China that have a strong enrichment ability of arsenic, and *Leersia hexandra* (Zhang, 2006) that of chromium.

Hyperaccumulators usually have the following characteristics (Li, 2007): (1) high level of tolerance for high concentration of metals; (2) high level of accumulation for heavy metals; (3) growing faster; (4) high level of biomass; (5) a positive response to agronomic regulation; and (6) developed root system.

5.2 *Physiological and molecular mechanisms of hyperaccumulators to absorb heavy metals*

5.2.1 *Tolerance and detoxification mechanisms to heavy metals*

Tolerance means a plant has some specific physiological mechanisms, ensuring that the plant can survive under high stress of heavy metals without being damaged. At the same time, the heavy metals are highly concentrated in the plant. Tolerance of hyperaccumulators relates to the regionalization of heavy metals in plant cells, namely the distribution of heavy metals are in cell wall and vacuole, by which the toxicity of heavy metals are reduced. Hyperaccumulators can produce some new proteins to form the resistance mechanism against the heavy metal stress, while these proteins play a complexation role in plant. (Chen, 2008).

5.2.2 *Activation and absorption mechanisms to heavy metals*

Hyperaccumulators can change the rhizosphere, effectively activate non-dynamic of heavy metals in soil, improving its bioavailability. McGrath et al. (McGrath, 2001) have found that the content of NH_4NO_3 distorted state of Zn in rhizosphere and non-rhizosphere soil were both significantly decreased after *T. caerulescens* and non-hyperaccumulator *Thlaspiochroleuhatcum* were cultivated, while the decrease was more significant in rhizosphere than that in the counterpart, and the reduction of dynamic was only 10% of the total absorption of *T. caerulescens*.

5.2.3 *Transport and accumulation mechanisms to heavy metals*

Hyperaccumulators have a strong ability to transport heavy metals. It was found that the content of heavy metals in vivo is higher than that of normal plants. The content of heavy metals in the above-ground part is higher than that in the root. This is related to the transport and accumulation mechanisms of heavy metals in plants. Metal ions transfer from the root to the above-ground part is mainly controlled by 2 processes: (1) transfer xylem parenchyma to catheter; and (2) transport in the catheter, which is affected by root pressure and transpiration stream (Chen, 2012).

5.3 *The influence factors of hyperaccumulators absorb heavy metals*

5.3.1 *Physical and chemical factors*

Plants absorb heavy metals will be affected by some physical and chemical factors such as temperature, pH, medium conditions and so on. Lu Yuanyuan et al. (Lu, 2013) studied the root system of the hyperaccumulator *Leersia hexandra swanz* on the absorption characteristics of Cr (VI) at a low temperature of 2°C through pot experiments. The results showed that, the low temperature (2°C) significantly inhibited the absorption of Cr (VI) by the root of *Leersia hexanda swartz*. After 48 h, compared with the contrast, the concentration of chromium in the root of *Leersia hexanda swartz* was reduced by 69.8%, which indicated that the absorption of Cr (VI) by the root of *Leersia hexanda swartz* was an active process that required energy.

5.3.2 *Antagonistic and synergistic effects of metal ions*

The absorption of heavy metals by majority of hyperaccumulators is selective. While heavy metals pollution are mostly complex pollution, so it is necessary to consider the interactions of metal ions about the capacity of plants to absorb heavy metals. We can take advantage of this trend to control reaction conditions artificially, to achieve the better purification effect.

5.3.3 *The speciation of heavy metals*

The speciation of heavy metals can affect the absorption of heavy metal ions by hyperaccumulators. The content changes of heavy metals in soil solution depend on adsorption-desorption, dissolution-precipitation and balance of oxidation-reduction of heavy metals. Heavy metals, which are in the states of adsorption and precipitation, can exchange each other under certain conditions. In general, the reduction of pH can make heavy metals in the state of adsorption release into the soil solution, thus increasing the absorption of heavy metals by plants. Changing the forms of heavy metals has a great influence on transferring heavy metal ions from the roots of plant to the shoots. (Dai, 2007).

6 RESEARCH STATUS OF PHYTOREMEDIATION

Significant advantage of phytoremediation is that it can be implemented in situ in the project, and thus reducing the disturbance to soil properties and the impact on the surrounding ecological environment. Hence, it deserves the name of "green remediation method". Cultivating plants can purify and beautify the environment. Removing heavy metal pollutants from soil, while it offers an opportunity to recycle precious resources from plant residues over accumulated with heavy metals to have direct economic benefits.

At present, the researches of phytoremediation mainly focus on some aspects.

The one is searching, screening and breeding of hyperaccumulators. The majority of hyperaccumulators used in phytoremediation present defects such as short, slow growth, and low biomass. Therefore, searching and developing of hyperaccumulators with high biomass and high enrichment ability is the primary task of phytoremediation (Wang, 2013). The second is the application of molecular biology and gene engineering. The core work is the application of transgenic technology, by which genes presenting metal tolerance and accumulation capability of natural hyperaccumulators are transplanted to the plants of high biomass and fast growth rate to provide transgenic plants practical value, thus overcoming the disadvantages of natural hyperaccumulators and improve the practicability of phytoremediation. The third is plant-microbial remediation technology. The action principle of combination of plants and microbial is that the root system provides the necessary living sites for microorganisms when the plant grows; at the same time, the vigorous growth of microorganisms enhance the degradation of pollutants to make the plants have a better growth space. Plant-microbial system promotes the rapid degradation and transformation of pollutants (Han, 2012). The last but not the least, safe disposal of hyperaccumulators (Ren, 2013). At present, removing the apoptosis of plants from the system mainly adopts the method of manual harvesting. The plants which are used to treat heavy metals contaminated wastewater are not allowed to be exposed to environment casually, and the extraction of heavy metals from plant tissues is also a huge project. These are the problems urgently to be solved in phytoremediation.

7 CONCLUSION

At present, the majority of the world's industrialized countries are facing serious problems of heavy metals pollution, it is urgent to clean up the excessive heavy metals in environment.

Phytoremediation is attracting more and more attention because of its green, economic, convenient and potent. However, it is a very new research field until now, and not yet matured enough in practice. Heavy metals stress to plants in various aspects, while the resistance mechanism of hyperaccumulators is very complex. Although some progress has been made on the physiological mechanism of the tolerance of plants to heavy metals, there are still some problems. Therefore, it will take time for phytoremediation to be progressed from the laboratory to industrial application. How to improve the tolerance of plants to heavy metals, increase or decrease the absorption of heavy metals, and to be applied in phytoremediation and agricultural production in heavy metals contaminated area still need to be considered. With the rapid development of social economy and biological technology as well as the improvement of environmental quality, phytoremediation technology can be more widely applied in treatment of heavy metals wastewater.

REFERENCES

Bai, Y.B. et al. 2013. Research progress of heavy metal wastewater treatment technology. *Pollution Control Technology* 26(3):36–40.
Brooks, R.R. et al. 1977. Detection of nickeliferous rocks by analysis of herbarium specimens of indicator plants. *Journal of Geochemical Exploration* 7:49–57.
Baker, A.J.M. & Brooks, R.R. 1989. Terrestrial high plants which hyperaccumulate metallic elements-a review of their distribution, ecology and phytochemistry. *Biorecovery* 1(2):81–126.
Chen, L. 2012. Overview of waste water with chrome treatment technology. *Ferro-Alloys* (2):41–42.
Chen, Y.P. 2008. Research trends on heavy metals hyperaccumulators. *Environmental Science and Management* 33(3):21–22.
Chen, T.B. et al. 2002. Arsenic hyperaccumulator *Pteris vittata* and its characteristics of arsenic enrichment. *Chinese science bulletin* 47(3):207–210.
Chen, F.Y. et al. 2012. Research advances on screening of hyperaccumulator and tolerant plant species of Pb-Zn. *Journal of Central South University of Forestry & Technology* 32(12):92–96.
Dai, Y. et al. 2007. Mechanism and influencing factors of phytoremediation of heavy metals contamination by hyperaccumulators. *Henan Agricultural Sciences* (4):11–12.
Huang, M.R. et al. 2010. Study on treatment of wastewater containing mercury. *Chemical Engineering Design* 20(2):33–35.
Huang, Z.M. & Huang, X. 2010. Trace elements and human health. *Studies of Trace Elements and Health* 27(27):58–62.
Han, Y.H. et al. 2012. The treatment of plant-microbial remediation technology of eutrophication of water. *Technology of Water Treatment* 38(3):4–5.
Li, K.T. 2007. Research and utilization of hyperaccumulators. *Bulletin of Biology* 42(9):5–6.
Lu, Y.Y. et al. 2013. Mechanism of Cr (VI) uptake by hyperaccumulator *Leersia hexandra Swartz*. *Journal of Agro-Environment Science* 32(11):2140–2144.
Mei, J. et al. 2013. The research progress of Cd-hyperaccumulator in contaminated soil remediation. *Energy and Energy Conservation* 2:80–82.
Minguizzi, C. & Vergnan, O. 1948. Conteruto di nichel nell ceneri di Alyssum bertolonni. *Desv Atti Soc Tosc Sci Nat Mem* 55:49–74.
McGrath, S.P. et al. 2001. Plant and rhizosphere processes involved in phytoremediation of metal contaminated soils. *Plant and Soil* 232:207–214.
Ren, X.J. et al. 2013. Research progress of phytoremediation technology applied in sewage ecological system. *Journal of Henan Institute of Science and Technology* 41(3):68–69.
Wei, Z.Y. et al. 2002. Cretan brake (*Pteris cretica L.*): an arsenic-accumulating plant. *Journal of ecology* 22(5):777–778.
Wang, Q.H. & Que, X.E. 2013. Phytoremediation—a green approach to environmental clean-up. *Chinese Journal of Eco-Agriculture* 21(2):261–266.
Yan, Y. et al. 2008. Advances in the mechanisms of heavy metal tolerance and accumulation in hyperaccumulators. *Guihaia* 28(4):505–510.
Zhou, Q.L. et al. 2010. Current situation and prospect on the processing craft for Cr (VI)-containing wastewater treatments. *Jiangxi Energy* (2):29–33.
Zhang, X.H. et al. 2006. *Leersia hexandra Swartz*: a newly discovered hygrophyte with chromium hyperaccumulator properties. *Journal of ecology* 26(3):950–953.

Ecosystem health assessment of Dongyang River basin

Tao Wu, Yong Zhang, Guo-jun Jiang, Xue-feng Xie & Hua-jing Bian
College of Geography and Environmental Science, Zhejiang Normal University, Jinhua, China

ABSTRACT: This article takes Dongyang River basin as the research object to carry out watershed ecosystem health assessment. According to Watershed Ecosystem Health Assessment Technical Guidelines of Ministry of Environmental Protection, including water habitat structure, aquatic organisms, water ecological pressure, land ecological pattern, land ecological function, and land ecological pressure, a total of 6 categories, 17 evaluation indexes are established. The evaluation results show that: the ecosystem health of Dongyang River basin can be classified as excellent, good, and general, accounting for 6%, 60%, and 34% of the total area, respectively. The main limiting factor of aquatic ecosystems was water quality index, while land ecological limiting factors were mainly forest coverage rate, the proportion of construction land, important habitat retention, and water conservation functions.

1 INTRODUCTION

Ecosystem health is an ecosystem that has stability and sustainability, and in terms of time can maintain its organizational structure, self-regulation, and resilience of the threat. Ecosystem health can be defined through three characteristics which are ecosystem vitality, organizational structure, and resilience. Basin, as a distinct geographical unit, plays an important role in the regulation of runoff, water conservation, water purification, and maintenance of regional ecological balance.

In this paper, Ministry of Environmental Protection (MEP) issues a "Technical Guidelines for Assessing ecological health Basin (Trial)" as guidance and constructs evaluation model, taking Zhejiang Province, Dongyang River basin as an example to conduct basin ecosystem health assessment. The evaluation results provide essential and supportive information for the ecological red line in Zhejiang Province, the components of ecological compensation system, and the development of the "Refers to the sewage, flood, water logging, water supply and water saving".

2 THE STUDY AREA

Dongyang River rises in the valley between Longcongwujian and Yanwujian of Mount Dapan in Pan'an county and crosses Dongyang and Yiwu city, finally ending in Jinhua City. The length of this river is 165.5 km, and Dongyang River drainage area is 3378.5 km^2. This area has subtropical monsoon climate with an annual average temperature of 17.2 °C and an annual precipitation of 1419.9 mm (Data from the Statistical Yearbook of Dongyang County, Figure 1 comes from XIE Xuefeng, 2014).

3 DATA SOURCES AND METHODS

3.1 *Data sources*

The evaluation data set is divided into digital image data set, monitoring data set, and statistical data set.

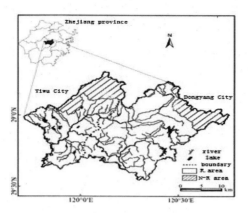

Figure 1. Sketch of study area.

Digital image data set includes: ETM+, SPOT-5, DEM, land use database, soil type vector database; monitor data set include: water quality section monitoring data, rainfall data, river runoff data, point source, and surface source emissions data. Statistical data set includes: 2012 Dongyang and Yiwu City Statistical Yearbook, water conservancy census data of Dongyang, and Yiwu City and so on.

3.2 Research methods

3.2.1 Evaluation unit
Basin evaluation unit takes minimum natural units within the basin as the basis, merging the smallest natural unit according to the natural conditions consistency of watersheds and similar characteristics. Meanwhile, we consider spatial overlay relationship and combine the basin natural geographical unit, administrative divisions unit, and basin environmental management unit. Taking the natural geographical unit as primary one, we combine it with the administrative unit, referring to the "focus on water pollution prevention plan (2011–2015)" and controlling unit's division; evaluation unit size substantially is consistent with the township zone.

3.2.2 Evaluation system
Considering the evaluation of aquatic organisms, taking the macrobenthos and benthic diatom diversity index as one of the assessment indicators, it will be sensitive to acidification and eutrophication of rivers.

Water ecological health evaluation index includes three categories: habitat structure, aquatic organisms, and ecological pressure, a total of eight targets and terrestrial ecological health assessment indicators include three categories, which are the ecological structure, ecological function, and ecological stress, a total of nine indicators. The weigh to re-evaluation of the decisions of the evaluation of the contribution of watershed ecosystem health condition, has a direct impact on the accuracy of the evaluation.

3.2.3 Index standardization
As there are many evaluation types and different units, they found it difficult to compare directly. Therefore, according to various indicators of the impact on the ecological health of the watershed size and correlation of the evaluation, if it is given to standardized scores, scores during 0–1. And referring to MEP "Watershed Health Assessment Guide", the measured value of the evaluation will be divided into five levels.

3.2.4 Evaluation model
Watershed Health Index, (*WHI*) *a*s a measure of a regional ecosystem health of each model, can be based on evaluation data to determine the health of ecosystems, and its value is between 0 and 100.

WHI model for each object layer: $WHI_i = \sum_i^n L_i * W_i$ (1)

In formula, WHI_i is ecological health level of each object layer; L_i is the normalized value of each index, W_i is the weight of each evaluation.

WHI model for watershed: $WHI = \sum_i^2 EHI_i * L_i$ (2)

In formula, WHI is ecological safety level of basin; WHI_i is ecological health level of each object layer; L_i is weight of object layer.

4 RESULTS AND ANALYSIS

According to the level of river basin, ecosystem health assessment values refer to MEP "watershed health assessment guidelines". Watershed ecosystem health will be divided into excellent (80–100), good (60–80), general (40 to 60), not bad (20 to 40), and bad (0–20) five levels which make the spatial distribution of thematic maps with using GIS spatial analysis module.

4.1 Water ecosystem health evaluation

By the index of aquatic habitat structure, aquatic organisms, and aquatic ecosystems pressure calculation analysis, ecological health of river basin waters include excellent evaluation unit area of about 22% of the total evaluation unit, a good evaluation units of the total area of 42%, typically 36% of the evaluation units of the total evaluation unit.

In many regions of human activity, such as Hengdian town and other evaluation units, water quality degradation leads to indicators of low aquatic organisms low, aquatic ecosystems in good health. Flowing main stream through the city, due to the large amount discharge of sewage and industrial waste, results in poor water quality conditions, which possess seriously affected or even not existing GradeVand IVclass and aquatic biological populations. Take Wu Ning county evaluation unit as an example (Fig. 2, D-season stands for dry season).

4.2 Land ecosystem health evaluation

Overall Dongyang River basin land ecosystem health situation is somewhat bad, no excellent evaluation unit, the area of good account for 67%, the moderate area account for 31%, not bad area accounted for 2%. On the distribution of space, the mountains better than the basin, towns than urban areas, Dongyang than Yiwu.

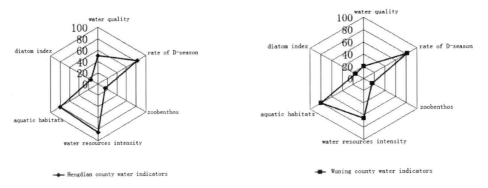

Figure 2. Aquatic ecological indicators radar chart in some evaluation unit in Dongyang River basin.

From Figure 3 we can find that health rating of outstanding units, such as Dongyang county is high, especially forest cover, land for construction, an important habitat retention and Water Conservation index are higher than the other evaluation units. Rating of good health evaluation unit, the main limiting factor is an important habitat retention and water conservation index, Cantabile Town and other evaluation units. In healthy level of general evaluation unit, the high proportion of construction land, forest coverage and Water Conservation index is lower than its main limiting factor. (Fig. 3, S-pollution, P-pollution, LF, CHPA and WC stand for surface pollution, point pollution, landscape fragmentation, critical habitats protected area and water conservation respectively).

4.3 Comprehensive assessment of the ecological health of the Dongyang River

As Dongyang River basin ecosystem health assessment shows, excellent evaluation unit accounts for only 6%, good area accounts for 60%, and general area accounts for 34%. Watershed ecosystem health distribution exhibits the tendency to deteriorate step by step from the upstream to downstream, Dongyang range gradually transits from excellent to good and general, Yiwu territory transits from good to general; south river watershed upstream of Hengdian and Nanma region' aquatic ecosystems is damaged severely. The evaluation grade is general, lower than each surrounding evaluation unit.

With the development of urbanization further acclerating, the proportion of construction land in the region will likely expand, in the industrial and agricultural output further increased pressure, pressure of reduction load index is enormous, Dongyang River basin water quality objectives can't be fundamentally improved in short time.

4.4 Analysis of ecological environment problems in Dongyang River basin

4.4.1 Water quality of main stream is not bad

Main stream of Dongyang River basin directly crosses major urban areas and towns of Dongyang and Yiwu city. This area has a large population density and intensive industrial activities which brings a huge amount of industrial and domestic sewage. Secondly, ever though there is large-scale precipitation in Dongyang River basin, all size of reservoirs that come from tributaries in the upstream area cut off the river runoff, the ecological base flow in the river is difficult to protect resulting in the Dongyang River and Yiwu River during the dry season have few ecological runoff and the quantity of sewage discharged flowing into rivers is runoff in its entirety. In this case, even if the emission of agencies all along the river meets emission standards. The concentration of sewage pollutants is still much higher than the environmental requirement quality of surface water.

4.4.2 High proportion of constructive land, low efficiency of water conservation

Constructive lands directly relate to the basin's ecological functions and ecological pressure. The transfer matrix analysis of land cover types from 2000 to 2010 shows that total area of traffic, residence and industry, mine land has increased 8748 hectares. The high rate of

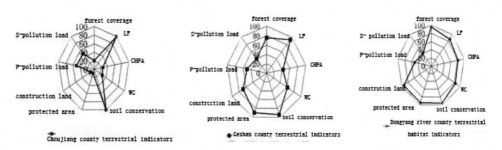

Figure 3. Terrestrial ecological indicators radar chart in some evaluation unit in Dongyang River basin.

expansion of construction land put great ecological pressure on forests, grass, farmlands, and water ecosystems, which decreased all kinds of habitat areas. Meanwhile, the decrease of habitats increased the degree of habitat fragmentation and reduced the stability of the ecosystem. Land hardening changed the region's hydro-thermal conditions and reduced the time of peak concentration, which easily lead to floods and ecological disasters.

4.4.3 *Large emissions of point and surface sources*

The main reason of Dongyang River basin's poor water quality lies in pollutants emission largely. The point source COD emission of Dongyang River basin in 2012 is 24,558.75 tons, the total ammonia emission is 3802.17 tons, the total amount of non-point source discharge of COD is 4009.52 tons, the total ammonia emissions is 582.06 tons, and industrial waste water emission COD shows an upward trend. Pollution source of Dongyang River basin is mainly point source, within the scope of Dongyang City, the main pollutant come from urban life and rural life source. Pharmaceutical manufacturing, textile, chemical materials, chemical products manufacturing, and communications equipment manufacturing industry are the main polluting industries. Pollution of Yiwu City mainly comes from the urban life source, such as textiles and paper products industry. With population growth and industrial production boosting, pollution load emission targets of Dongyang River basin will face serious challenges.

5 CONCLUSION

1. The status of eco-heath in the waters of Dongyang River basin is in common and above level. The excellent evaluation units cover 38% of the total area while common ones cover 48%.
2. The overall conditions of basin and land area in Dongyang River are like this: mountainous areas are superior to basins, villages and towns are superior to urban areas, and Dongyang city is superior to Yiwu city.
3. The primary causes of the health level difference in Dongyang River basin are the interference of human activities, which results in such a situation: the health level in hilly areas is high while that in plain areas is low; the health level in upstream areas is high while that in middle and downstream areas featured with dense population is lower.

ACKNOWLEDGMENTS

This study was supported by social development projects in public welfare research for Science and Technology Department of Zhejiang Province under Grant No. 2013C33029, by the Xinmiao talent project for Zhejiang province under Grant No. 2014R404058.

REFERENCES

Deshon J.E., 1995, Development and Application of the Invertebrate Community Index (ICI) // Davis WS, Simon TP, Biological Assessment and Criteria: Tools for Water Resource Planning and Decision Making. Boca Raton, FL: Lewis Publishers, 217–243.
Karr J.R., 1991, Biological Integrity: A Long-Neglected Aspect of Water Resource Management [J], *Ecological Applications*, 1:66–84.
Karr J.R., Assessment of Biotic Integrity Using Fish Communities [J], *Fisheries*, 1981, 6:21–27.
Kleynhans CJ, 1996, A Qualitative Procedure for the Assessment of the Habitat Integrity Status of the Luvuvhu River [J], *Aquatic Ecosystem Health*, 5:41–54.
Liu Yong, Guo Huaicheng, Lake—Basin Ecosystem Management [M], *Beijing: Science Press*, 2008, 106–132.
Ladson A.R., White L.J., Doolan J.A., 1999, Development and Testing of an Index of Steam Condition for Waterway Management in Australia [J], *Fresh Water Biology*, 41:453–468.

Ohio E.P.A., 1990, The Use of Biocriteria in the Ohio EPA Surface Water Monitoring and Assessment Program Columbus, *OH: Ohio Environmental Protection Agency, Ecological Assessment Section, Division of Water Quality Planning and Assessments.*

Rapport D.J. 1989, What Constitute Ecosystem Health [J], *Perspectives in Biology and Medicine*, 33:120–132.

Rapport D.J., Costanza R., McMichael A.J., 1998, Assessing Ecosystem Health, Trends in Ecology & Evolution, 13(10):397.

Wang Wenjie, Zhang Zhe, Wang Wei, 2012, Framework and Method System of Watershed Ecosystem Health Assessment: Framework and Indicator System [J], *Journal of Environmental Engineering Technology.* 2(4):271–277.

Xie Xuefeng, Wu Tao, Xiao Cui, Ecological Security Assessment of the Dongyang River Watershed Using PSR Modeling [J], Resources Science, 2014, 36(8):1702–1711.

Treatment of sludge and wastewater mixture by Supercritical Water Oxidation

T. Zhang, S.Z. Wang, Z.Q. Zhang, J.Q. Yang & M. Wang
Key Laboratory of Thermo-Fluid Science and Engineering of MOE, School of Energy and Power Engineering, Xi'an Jiaotong University, Xi'an, Shaanxi, China

ABSTRACT: Mixture of sludge and wastewater treated by Supercritical Water Oxidation technique (SCWO) was studied in intermittent equipment at 440~460 °C, 25 MPa, reaction residence time 1~20 mins. Experimental results showed that SCWO is a high efficiency organic waste treatment and disposal technique. Removal rate of COD was obviously increased as temperature, residence time and oxidation coefficient extend. At the condition of 600 °C, 25 MPa and three times the oxidation coefficient, removal rate of COD reached 99.5%.

1 INTRODUCTION

Supercritical water oxidation (Bermejo and Cocero, 2006, Guo et al., 2010) is a new hydrothermal oxidation technology which is developed in recent years. Water above its critical point (T_c = 374 °C, P_c = 22.1 MPa) shows many unique properties which are different from the normal water, such as small dielectric constant, weak hydrogen bonds, low fluid viscosity and high diffusion coefficient, etc. In this condition water becomes a complete non-polar solvent. Supercritical water is an excellent medium for the rapid destruction of organic wastes by oxidation, because inorganic salt is almost insoluble in supercritical water but the supercritical water has a strong ability to dissolve non-polar molecule such as O_2, CO_2 or organic matters, which means that water can form homogeneous with O_2, CO_2 to eliminate the resistance of heat and mass transfer between the phase interface. The oxidation reaction of organic wastes in the supercritical water has been reported in many studies at home and abroad (Guo et al., 2010, Veriansyah and Kim, 2007) (Vadillo et al., 2013). The reaction is rapid and complete and the final product is inorganic small molecule compounds such as CO_2, H_2O, N_2 and salts, etc, which has no secondary pollution. As pointed in the "energy and environment", which is one of the six major areas listed by the national key technology, SCWO is the most promising treatment technology.

2 EXPERIMENTAL

2.1 *Apparatus and materials*

The experimental system and process in this work refer to previous work of our team.

The treatment object was the mixture of the sludge and wastewater provided by Tianjin De Xincheng environmental protection technology co., LTD. Table 1 shows the composition of the mixture.

2.2 *Analytical methods*

The pollutants concentrations of the mixture and liquid effluents were characterized by analyzing the Chemical Oxygen Demand (COD) and ammonia nitrogen (NH3-N). COD was analyzed by potassium dichromate method GB 11914-11914; NH3-N and the chloride ion

Table 1. Composition of the mixture of the sludge and wastewater.

Component	COD (mg/L)	NH4-N (mg/L)	Cl⁻ (mg/L)	Total nitrogen (mg/L)	Moisture content (%)	Total salt (mg/l)
Mixture	47140	933.39	4600	2000	0.93	61238

content were determined by silver nitrate titration GB 11896-89; total salt was detected by gravimetric method.

3 RESULT AND DISCUSSION

3.1 *Effect of temperature*

The conversion of the COD X_{cod} is calculated by the following formula (Du et al., 2013):

$$X_{COD}(\%) = \frac{[COD]_0 - [COD]_t}{[COD]_0} \times 100 \qquad (1)$$

where $[COD]_0$ is the initial COD concentration of the wastewater (mg/l), and $[COD]_t$ is the residual COD concentration of the liquid effluent after the reaction (mg/l).

The effect of temperature on the destruction of pollutants for the mixture is illustrated in Table 2, Table 3 and Figure 1.

As shown in Figure 2, the temperature has a significant effect on the degradation rate of the organic matter. The conversion of the COD increases with the temperature. The X_{cod} is 97.2% when the temperature is 440 °C, while the X_{cod} reaches 99.6% when the temperature is 600 °C, which means the COD is almost completely destructed. But the concentration of NH3-N increases with the temperature, it may due to the stable property of the ammonia nitrogen and the transformation of the nitrogen material to ammonia nitrogen. These results indicated that COC can be easily destructed at moderate temperature, whereas NH_3-N is an inert compound in the process of SCWO.

The oxidation of organic pollutants is an irreversible process, the reaction rate constant k increases with temperature, so the reaction rate will speed up and the final degradation rate of the organic will also increase; on the other hand, the increase of the temperature will lead to the smaller density of supercritical water, so that the concentration of the reactants is reduced leading to slower reaction rate. It can be inferred from the results that the influence of the former is greater than the latter. The increase of the temperature leads to improving the requirements of the equipment and COD removal rate increases flatten out after the temperature reaches a certain degree, so the temperature should not be too high.

3.2 *Effect of oxidation coefficient*

The oxidant in this work is H_2O_2 and the oxidation coefficient is defined as:

$$n = \frac{[H_2O_2]_0}{[H_2O_2]_t} \qquad (2)$$

where $[H_2O_2]_0$ is the concentration of hydrogen peroxide fed into the reactor (mg/l), and $[H_2O_2]_t$ is the stoichiometric requirement concentration of H_2O_2 to obtain complete oxidation of the feed based on $[COD]_0$ of wastewater (mg/l). The experimental results are shown in Table 4 and 5.

As shown in Figure 2, the concentration of COD reduces continuously and the concentration of NH3-N rises slightly with the increase of the oxidation coefficient. The addition of the oxidation promotes the transformation of total nitrogen. At the condition of 600 °C, 25 MPa, the conversion of the COD reaches 99.56% when the oxidation coefficient is 1.

Table 2. Effect of the temperature on the destruction of the COD.

	Temperature/ (°C)	Pressure/ (MPa)	Oxidation coefficient	Residence time/min	Initial COD/ (mg·L^{-1})	COD after treatment/(mg·L^{-1})	Conversion of COD/%
1#	440	25	1.1	10	32061	900	97.2
2#	520	25	1.1	10	32061	315	99.0
3#	600	25	1.1	10	32061	140	99.6

Table 3. Effect of the temperature on the destruction of the NH$_3$-N.

	Temperature/ (°C)	Pressure/ (MPa)	Oxidation coefficient	Residence time	Initial NH3-N/ (mg·L^{-1})	NH$_3$-N after treatment/(mg·L^{-1})	Conversion of NH3-N/%
1#	440	25	1.1	10	635	1475	−132.3
2#	520	25	1.1	10	635	1932	−204.3
3#	600	25	1.1	10	635	1080	−70.1

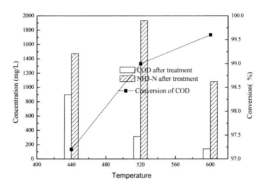

Figure 1. Effect of temperature (°C) on the destruction of pollutants for the mixture.

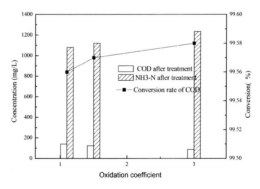

Figure 2. Effect of oxidation coefficient on the destruction of pollutants for the mixture.

The quantity of oxidation is not the bigger the better, because SCWO belongs to free radical reaction, increase the concentration of H$_2$O$_2$ at the beginning of the reaction generates more free radicals leading to the reaction rate faster but the rate flattens out gradually. Some studies at home and abroad have found that the degradation rate of the organic waste doesn't increase monotonously with the increase of oxidation coefficient, there is a optimal value and the influence of oxidation coefficient on the X$_{cod}$ becomes very small. So it is not reasonable to increase the oxygen content constantly in order to get higher X$_{cod}$.

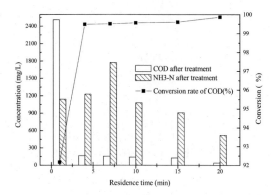

Figure 3. Effect of residence time on the destruction of pollutants for the mixture.

Table 4. Effect of the temperature on the destruction of the COD.

	Temperature/ (°C)	Pressure/ (MPa)	Oxidation coefficient	Residence (time/min)	Initial COD/ (mg·L^{-1})	COD after treatment/(mg·L^{-1})	Conversion of COD/%
1#	600	25	1.1	10	32061	140	99.56
2#	600	25	1.5	10	28721	123	99.57
3#	600	25	3	10	20651	86	99.58

Table 5. Effect of the temperature on the destruction of the NH$_3$-N.

	Temperature/ (°C)	Pressure/ (MPa)	Oxidation coefficient	Residence time	Initial NH3-N/ (mg·L^{-1})	NH3-N after treatment/(mg·L^{-1})	Conversion of NH3-N/%
1#	600	25	1.1	10	635	1080	−70.1
2#	600	25	1.5	10	569	1121	−174.1
3#	600	25	3	10	409	1235	−171.9

Table 6. Effect of the residence time on the destruction of the COD.

	Temperature/ (°C)	Pressure/ (MPa)	Oxidation coefficient	Residence (time/min)	Initial COD/ (mg·L^{-1})	COD after treatment/(mg·L^{-1})	Conversion of COD/%
1#	600	25	1.1	1	32061	2510	92.17
2#	600	25	1.1	4	32061	165	99.48
3#	600	25	1.1	7	32061	155	99.51
4#	600	25	1.1	10	32061	140	99.56
5#	600	25	1.1	15	32061	128	99.60
6#	600	25	1.1	20	32061	41	99.87

3.3 *Effect of residence time*

The experimental results are shown in Table 6 and 7.

We can seen from the Figure 2, with the extension of the reaction time the concentration of NH3-N increases at first then decreases. The main reason is that the total nitrogen transforms to NH3-N firstly then NH3-N begins to degrade with the further extension of residence time.

The reaction time has a significant influence on the removal rate of organic pollutants. The X_{cod} increases with the residence time and the X_{cod} becomes flat after the reaction time

Table 7. Effect of the residence time on the destruction of the NH$_3$-N.

	Temperature/ (°C)	Pressure/ (MPa)	Oxidation coefficient	Residence time	Initial NH$_3$-N/ (mg·L^{-1})	NH$_3$-N after treatment/(mg·L^{-1})	Conversion of NH$_3$-N/%
1#	600	25	1.1	1	635	1142	−80.0
2#	600	25	1.1	4	635	1229	−93.6
3#	600	25	1.1	7	635	1775	−179.6
4#	600	25	1.1	10	635	1080	−70.1
5#	600	25	1.1	15	635	909	−43.1
6#	600	25	1.1	20	635	521	18.0

reaching 10 min. Predictably, the X_{cod} might increase with the residence time, but the experimental can be seen that the influence of reaction time on the X_{cod} is very small when the reaction time exceeds a certain value.

3.4 Effect of pressure

Although there are a great number of studies about oxidation kinetics in supercritical water, the influence on pressure in the oxidation rate is still not clear. Some works indicated that oxidation efficiency increases when density, that is pressure, is elevated (Koo et al., 1997, Thornton and Savage, 1992a, Thornton and Savage, 1992b, Gopalan and Savage, 1995) (Thornton et al., 1992); others concluded that high pressure is detrimental for the oxidation rate80 and there are authors who indicated that pressure has no effect on the SCWO rate. (Oshima et al., 1998) All these experiments were performed in the temperature range from 370 to 480 °C and pressure range from 18.7 to 28.2 MPa. Li et al. (Krajnc and Levec, 1996) indicated that pressure influence in SCWO was very small compared to that of temperature and it could be neglected.

4 CONCLUSIONS

The temperature has a significant effect on the degradation rate of the organic matter. The X_{cod} is 97.2% when the temperature is 440 °C, while the X_{cod} reaches 99.6% when the temperature is 600 °C, which means the COD is almost completely destructed; the concentration of COD reduces continuously with the increase of the oxidation coefficient. At the condition of 600 °C, 25 MPa, the conversion of the COD reaches 99.56% when the oxidation coefficient is 1; the reaction time has a significant influence on the removal rate of organic pollutants. The X_{cod} increases significantly with the residence time, the X_{cod} reaches 99.48% when the reaction time reaches 4 minutes; the influence on pressure in the oxidation rate is still not clear.

Experimental results showed that SCWO is a high efficiency organic waste treatment and disposal technique.

REFERENCES

Bermejo, M.D. & Cocero, M.J. 2006. Supercritical water oxidation: A technical review. *Aiche Journal*, 52, 3933–3951.

Du, X., Zhang, R., Gan, Z.X. & Bi, J.C. 2013. Treatment of high strength coking wastewater by supercritical water oxidation. *Fuel*, 104, 77–82.

Gopalan, S. & Savage, P.E. 1995. A Reaction Network Model for Phenol Oxidation in Supercritical Water. *Aiche Journal*, 41, 1864–1873.

Guo, Y., Wang, S.Z., Xu, D.H., Gong, Y.M., MA, H.H. & Tang, X.Y. 2010. Review of catalytic supercritical water gasification for hydrogen production from biomass. *Renewable & Sustainable Energy Reviews*, 14, 334–343.

Koo, M., Lee, W.K. & Lee, C.H. 1997. New reactor system for supercritical water oxidation and its application on phenol destruction. *Chemical Engineering Science,* 52, 1201–1214.

Krajnc, M. & Levec, J. 1996. On the kinetics of phenol oxidation in supercritical water. *Aiche Journal,* 42, 1977–1984.

Oshima, Y., Hori, K., Toda, M., Chommanad, T. & Koda, S. 1998. Phenol oxidation kinetics in supercritical water. *Journal of Supercritical Fluids,* 13, 241–246.

Thornton, T.D., Ladue, D.E. & Savage, P.E. 1992. Phenol Oxidation in Supercritical Water—Formation of Dibenzofuran, Dibenzo-P-Dioxin, and Related-Compounds—Comment. *Environmental Science & Technology,* 26, 1850–1850.

Thornton, T.D. & Savage, P.E. 1992a. Kinetics of Phenol Oxidation in Supercritical Water. *Aiche Journal,* 38, 321–327.

Thornton, T.D. & Savage, P.E. 1992b. Phenol Oxidation Pathways in Supercritical Water. *Industrial & Engineering Chemistry Research,* 31, 2451–2456.

Vadillo, V., Sanchez-Oneto, J., Portela, J.R. & DE LA Ossa, E.J.M. 2013. Problems in Supercritical Water Oxidation Process and Proposed Solutions. *Industrial & Engineering Chemistry Research,* 52, 7617–7629.

Veriansyah, B. & Kim, J.D. 2007. Supercritical water oxidation for the destruction of toxic organic wastewaters: A review. *Journal of Environmental Sciences-China,* 19, 513–522.

Author index

Ai, C.M. 235
Apeltauer, T. 15
Arai, K. 21

Belyaev, L. 77
Bian, H.-j. 493
Bian, Y.H. 261
Budik, O. 15

Cai, X.Q. 357
Cancellara, D. 1
Chen, A.H. 423
Chen, D.Y. 481
Chen, Q. 141
Chen, R. 363
Chen, S.L. 383, 423
Chen, Y.-h. 337
Chen, Y.J. 255
Chen, Y.P. 209
Chung, W. 203
Coufalik, P. 83

Dasek, O. 83
Daskova, J. 27
Diao, Y.F. 69
Ding, Y.-h. 475
Du, H.K. 157
Duan, W.-w. 369
Duan, Z.-h. 369

Fang, Y.-F. 105
Feng, L. 267
Fu, S.F. 357

Gao, H.X. 47
Gao, P. 273
Gong, C. 53
Guo, P. 449

Han, B.-j. 475
Han, M. 157
Han, Y. 441
Hao, Y.Z. 465
He, H. 191
Hong, G.J. 91

Hou, N. 41
Hou, S.C. 411
Hou, Y. 285
Hou, Z.M. 59
Hou, Z.-X. 53
Hu, S.Q. 285
Hu, X. 191
Huang, G.-H. 53
Huang, H. 279
Huang, J. 351
Huang, Z.Q. 405
Huo, L.J. 449
Hyzl, P. 83

Ji, M.L. 487
Jia, B. 41
Jian, J. 123
Jiang, B. 449
Jiang, B.W. 243
Jiang, G.-j. 493
Jiang, L. 123, 185
Jiang, S.Y. 383
Jiang, W. 191
Jiang, Y.S. 397
Jin, P. 243
Jung, W. 203

Krcmova, I. 83
Kudrna, J. 27, 83

Lee, H. 203
Lei, Y.Z. 449
Lei, Z. 255
Li, B.X. 35
Li, C. 35
Li, C. 423
Li, H.X. 487
Li, J.W. 423
Li, L. 317
Li, L.L. 279
Li, N.J. 149
Li, S.Q. 389
Li, W. 375
Li, X.T. 417
Li, Y. 285

Li, Y. 383
Li, Y.H. 437, 459
Li, Y.S. 59, 69
Li, Z. 481
Li, Z.Z. 363
Liang, X. 91
Liang, Y.N. 411
Liu, C.J. 389
Liu, G.X. 91
Liu, P. 243
Liu, P.J. 343
Liu, S.B. 423
Liu, T. 41
Liu, W.Q. 323
Liu, X. 41
Liu, X. 397
Liu, Y. 129
Liu, Y. 305
Liu, Y. 383
Liu, Y.L. 59
Lu, H. 197

Matuszkova, R. 15
Meng, X.J. 389
Meng, X.J. 417
Min, R. 111
Mo, Z.P. 363
Morimoto, E. 21
Morozov, V. 9, 77, 229
Mu, J.J. 389

Nekulova, P. 27
Ni, Y.L. 117
Novikova, E. 229

Pan, G.Y. 411
Pan, Y. 209
Pang, Q.X. 331
Park, C. 203
Pasquino, V. 1
Peng, X. 267

Qi, B. 405
Qi, S.-f. 475
Qin, X.R. 279

Qin, X.S. 111
Qiu, L. 169
Qu, J.L. 465

Radimsky, M. 15
Ricciardi, E. 1
Rong, X.-l. 223

Sarkar, S. 135
Shan, J.X. 267
Shang, F.-f. 369
Shao, Y. 423
Shen, Y. 117
Song, C.-f. 223
Song, S.H. 267
Song, Y. 203
Sperka, P. 83
Stehlik, D. 83
Sun, J.H. 185
Sun, L.C. 331
Sun, N. 323
Sun, Q.H. 285

Tang, L. 323
Tang, Z. 249
Tang, Z.Y. 243
Tian, Q.B. 163

Volkova, I. 77, 229

Wang, C. 417
Wang, C.-l. 475
Wang, G.P. 411
Wang, H.J. 235
Wang, J.D. 235
Wang, L. 405
Wang, L.S. 437, 459
Wang, M. 197
Wang, M. 437, 459, 499
Wang, S. 297
Wang, S.Z. 437, 459, 499
Wang, T. 375
Wang, W.P. 429
Wang, X. 91
Wang, X.D. 465
Wang, X.T. 441
Wang, X.D. 441

Wang, X.G. 123, 185
Wang, Y. 405
Wang, Z.G. 217
Wei, G.-b. 177
Wei, X.P. 123
Wen, X.-c. 177
Wu, A.X. 235
Wu, H.Y. 357
Wu, T. 493
Wu, W. 249
Wu, X.F. 273

Xiao, H. 311
Xiao, Y.F. 69
Xie, L.Q. 91
Xie, L.Q. 99
Xie, X.-f. 493
Xin, G. 389
Xiong, C. 197
Xiong, X.Y. 481
Xu, Y. 209
Xu, Y. 337
Xue, S. 169

Yan, G.X. 323
Yan, J. 487
Yan, X. 311
Yan, Y.-z. 177
Yang, C.H. 429
Yang, G.-x. 369
Yang, J.Q. 437, 459, 499
Yang, N. 475
Yang, Y.M. 291
Ye, Z.J. 291
Yin, K.L. 47
Yin, Z.C. 149
Yu, F. 279
Yu, J.F. 117
Yu, J.J. 111
Yu, L.-j. 177
Yu, M. 305
Yu, Q. 369
Yu, Q.H. 397
Yue, T. 475

Zhai, W. 123
Zhang, B. 389

Zhang, C. 41
Zhang, C.H. 411
Zhang, D. 35
Zhang, D. 217
Zhang, H.J. 123, 185
Zhang, H.W. 217
Zhang, J.J. 185
Zhang, J.L. 429
Zhang, N. 331
Zhang, P.P. 91
Zhang, R. 217
Zhang, R.Q. 157
Zhang, T. 437, 459, 499
Zhang, W. 141
Zhang, X. 249
Zhang, X.-x. 475
Zhang, X.-y. 337
Zhang, Y. 493
Zhang, Y.L. 59, 69
Zhang, Z.F. 411
Zhang, Z.Q. 499
Zhang, Z.W. 449
Zhao, G. 297
Zhao, J.W. 481
Zhao, J.Y. 429
Zhao, M.J. 243
Zhao, P. 223
Zhao, Y. 429
Zhdanov, A. 9, 77, 229
Zhen, W. 465
Zheng, X.W. 423
Zheng, Y. 53, 169
Zhong, H. 297
Zhou, C.J. 297
Zhu, W. 35
Zhu, Y. 405
Zhu, Y.H. 99
Zhu, Y.H. 273
Zhu, Y.X. 383
Zhuang, K.Z. 357
Zou, L. 197
Zuo, P.-l. 475
Zuo, Z.Y. 279